Digital Systems Using Verilog

International Edition

Charles H. Roth, Jr.
The University of Texas at Austin

Lizy Kurian John
The University of Texas at Austin

Byeong Kil Lee
The University of Texas at San Antonio

Australia • Brazil • Japan • Korea • Mexico • Singapore • Spain • United Kingdom • United States

**Digital Systems Design Using Verilog
International Edition**
Charles H. Roth, Jr., Lizy Kurian John,
and Byeong Kil Lee

Publisher, Global Engineering:
    Timothy L. Anderson

Development Editor: Eavan Cully

Media Assistant: Ashley Kaupert

Team Assistant: Sam Roth

Marketing Manager: Kristin Stine

Director, Content and Media Production:
    Sharon L. Smith

Senior Content Project Manager:
    Kim Kusnerak

Production Service/Compositor:
    MPS Limited

Copyeditor: Harlan James

Proofreader: Lori Martinsek

Indexer: Shelly Gerger-Knechtl

Senior Art Director: Michelle Kunkler

Internal Designer: Carmela Pereira

Cover Designer: Tin Box Studio, Inc.

Cover Image: © Raimundas/Shutterstock;
    Digital VIsion

Intellectual Property
    Analyst: Christine Myaskovsky
    Project Manager: Sarah Shainwald

Text and Image Permissions Researcher:
    Kristiina Paul

Manufacturing Planner: Doug Wilke

© 2016 Cengage Learning

WCN: 01-100-101

CENGAGE and CENGAGE LEARNING are registered trademarks of Cengage Learning, Inc., within the United States and certain other jurisdictions.

ALL RIGHTS RESERVED. No part of this work covered by the copyright herein may be reproduced, transmitted, stored, or used in any form or by any means graphic, electronic, or mechanical, including but not limited to photocopying, recording, scanning, digitizing, taping, Web distribution, information networks, or information storage and retrieval systems, except as permitted under Section 107 or 108 of the 1976 United States Copyright Act, without the prior written permission of the publisher.

> For permission to use material from this text or product,
> submit all requests online at **www.cengage.com/permissions**.
> Further permissions questions can be emailed to
> **permissionrequest@cengage.com**.

Library of Congress Control Number: 2014947845

ISBN: 978-1-305-12074-7

Cengage Learning International Offices

Asia
www.cengageasia.com
tel: (65) 6410 1200

Brazil
www.cengage.com.br
tel: (55) 11 3665 9900

Latin America
www.cengage.com.mx
tel: (52) 55 1500 6000

Australia/New Zealand
www.cengage.com.au
tel: (61) 3 9685 4111

India
www.cengage.co.in
tel: (91) 11 4364 1111

UK/Europe/Middle East/Africa
www.cengage.co.uk
tel: (44) 0 1264 332 424

**Represented in Canada by
Nelson Education, Ltd.**
tel: (416) 752 9100 / (800) 668 0671
www.nelson.com

Cengage Learning is a leading provider of customized learning solutions with office locations around the globe, including Singapore, the United Kingdom, Australia, Mexico, Brazil, and Japan. Locate your local office at:
www.cengage.com/global

For product information: **www.cengage.com/international**
Visit your local office: **www.cengage.com/global**
Visit our corporate website: **www. cengage.com**

Printed in the United States of America
Print Number: 01   Print Year: 2014

Contents

Preface vii

Chapter 1 Review of Logic Design Fundamentals 1

1.1 Combinational Logic 1
1.2 Boolean Algebra and Algebraic Simplification 3
1.3 Karnaugh Maps 7
1.4 Designing with NAND and NOR Gates 11
1.5 Hazards in Combinational Circuits 13
1.6 Flip-Flops and Latches 15
1.7 Mealy Sequential Circuit Design 17
1.8 Design of a Moore Sequential Circuit 25
1.9 Equivalent States and Reduction of State Tables 28
1.10 Sequential Circuit Timing 30
1.11 Tristate Logic and Busses 47
Problems 48

Chapter 2 Introduction to Verilog® 58

2.1 Computer-Aided Design 59
2.2 Hardware Description Languages 62
2.3 Verilog Description of Combinational Circuits 64
2.4 Verilog Modules 68
2.5 Verilog Assignments 73
2.6 Procedural Assignments 74
2.7 Modeling Flip-Flops Using Always Block 78
2.8 Always Blocks Using Event Control Statements 82
2.9 Delays in Verilog 84
2.10 Compilation, Simulation, and Synthesis of Verilog Code 87
2.11 Verilog Data Types and Operators 93
2.12 Simple Synthesis Examples 98
2.13 Verilog Models for Multiplexers 102
2.14 Modeling Registers and Counters Using Verilog Always Statements 104

2.15 Behavioral and Structural Verilog   112
2.16 Constants   124
2.17 Arrays   125
2.18 Loops in Verilog   127
2.19 Testing a Verilog Model   129
2.20 A Few Things to Remember   133
      Problems   136

Chapter 3 Introduction to Programmable Logic Devices   158

3.1 Brief Overview of Programmable Logic Devices   158
3.2 Simple Programmable Logic Devices (SPLDs)   161
3.3 Complex Programmable Logic Devices (CPLDs)   176
3.4 Field-Programmable Gate Arrays (FPGAs)   180
     Problems   205

Chapter 4 Design Examples   210

4.1 BCD to 7-Segment Display Decoder   211
4.2 A BCD Adder   212
4.3 32-Bit Adders   214
4.4 Traffic Light Controller   220
4.5 State Graphs for Control Circuits   225
4.6 Scoreboard and Controller   226
4.7 Synchronization and Debouncing   230
4.8 A Shift-and-Add Multiplier   232
4.9 Array Multiplier   238
4.10 A Signed Integer/Fraction Multiplier   241
4.11 Keypad Scanner   255
4.12 Binary Dividers   264
     Problems   277

Chapter 5 SM Charts and Microprogramming   288

5.1 State Machine Charts   288
5.2 Derivation of SM Charts   293
5.3 Realization of SM Charts   306
5.4 Implementation of the Dice Game   309
5.5 Microprogramming   314
5.6 Linked State Machines   327
     Problems   329

Chapter 6 Designing with Field Programmable Gate Arrays   341

6.1 Implementing Functions in FPGAs   341
6.2 Implementing Functions Using Shannon's Decomposition   347

- 6.3 Carry Chains in FPGAs   352
- 6.4 Cascade Chains in FPGAs   353
- 6.5 Examples of Logic Blocks in Commercial FPGAs   355
- 6.6 Dedicated Memory in FPGAs   357
- 6.7 Dedicated Multipliers in FPGAs   368
- 6.8 Cost of Programmability   369
- 6.9 FPGAs and One-Hot State Assignment   371
- 6.10 FPGA Capacity: Maximum Gates versus Usable Gates   373
- 6.11 Design Translation (Synthesis)   375
- 6.12 Mapping, Placement, and Routing   385
- Problems   390

Chapter 7 Floating-Point Arithmetic   399

- 7.1 Representation of Floating-Point Numbers   399
- 7.2 Floating-Point Multiplication   406
- 7.3 Floating-Point Addition   417
- 7.4 Other Floating-Point Operations   425
- Problems   426

Chapter 8 Additional Topics in Verilog   431

- 8.1 Verilog Functions   431
- 8.2 Verilog Tasks   435
- 8.3 Multivalued Logic and Signal Resolution   437
- 8.4 Built-in Primitives   439
- 8.5 User-Defined Primitives   442
- 8.6 SRAM Model   445
- 8.7 Model for SRAM Read/Write System   446
- 8.8 Rise and Fall Delays of Gates   450
- 8.9 Named Association   451
- 8.10 Generate Statements   452
- 8.11 System Functions   455
- 8.12 Compiler Directives   457
- 8.13 File I/O Functions   460
- 8.14 Timing Checks   463
- Problems   464

Chapter 9 Design of a RISC Microprocessor   473

- 9.1 The RISC Philosophy   473
- 9.2 The MIPS ISA   476
- 9.3 MIPS Instruction Encoding   482
- 9.4 Implementation of a MIPS Subset   485
- 9.5 Verilog Model   492
- Problems   508

Chapter 10 Hardware Testing and Design for Testability 514

10.1 Testing Combinational Logic 515
10.2 Testing Sequential Logic 518
10.3 Scan Testing 522
10.4 Boundary Scan 525
10.5 Built-In Self-Test 538
Problems 549

Appendix A 554

Appendix B 562

References 564

Index 567

Preface

This textbook is intended for a senior-level course in digital systems design. The book covers both basic principles of digital system design and the use of a hardware description language, Verilog, in the design process. After basic principles are covered, a variety of examples are used to illustrate the principles. Many digital system design examples, ranging in complexity from a simple binary adder to a microprocessor, are included in the text.

Students using this textbook should have completed a course in the fundamentals of logic design, including both combinational and sequential circuits. Although no previous knowledge of Verilog is assumed, students should have programming experience using a modern higher-level language such as C. A course in assembly language programming and basic computer organization is also very helpful, especially for Chapter 9.

Because students typically take their first course in logic design two years before this course, most students need a review of the basics. For this reason, Chapter 1 includes a review of logic design fundamentals. Most students can review this material on their own, so it is unnecessary to devote much lecture time to this chapter. However, a good understanding of timing in sequential circuits and the principles of synchronous design is essential to the digital system design process. A detailed treatment of timing analysis is added in Chapter 1. If your students have a strong background in logic design, you may wish to skip most of chapter 1, however the timing discussion towards the end of the chapter will be beneficial for most students.

Chapter 2 starts with an overview of modern design flow. It also summarizes various technologies for implementation of digital designs. Then, it introduces the basics of Verilog, and this hardware description language is used throughout the rest of the book. Additional features of Verilog are introduced on an as-needed basis, and more advanced features are covered in Chapter 8. From the start, we relate the constructs of Verilog to the corresponding hardware. Some textbooks teach Verilog as a programming language and devote many pages to teaching the language syntax. Instead, our emphasis is on how to use Verilog in the digital design process. The language is very complex, so we do not attempt to cover all its features. We emphasize the basic features that are necessary for digital design and omit some of the less-used features.

Verilog is very useful in teaching top-down design. We can design a system at a high level and express the behavior in Verilog. We can then simulate and debug the designs at this level before proceeding with the detailed logic design. However, no design is complete until it has actually been implemented in hardware and the hardware has been tested. For this reason, we recommend that the course include some lab exercises in which designs are implemented in hardware. We introduce simple programmable logic devices (PLDs) in Chapter 3 so that real hardware can be used early in the course if desired. Chapter 3 starts with an overview of programmable logic devices and presents simple programmable logic devices first, followed by an introduction to complex programmable logic devices (CPLDs) and Field Programmable Gate Arrays (FPGAs). There are many products in the market, and it is good for students to learn about commercial products. However, it is more important for them to understand the basic principles in the construction of these programmable devices. Hence we present the material in a generalized fashion, with references to specific products as examples. The material in this chapter also serves as an introduction to the more detailed treatment of FPGAs in Chapter 6.

Chapter 4 presents a variety of design examples, including both arithmetic and non-arithmetic examples. Simple examples such as a BCD to 7-segment display decoder to more complex examples such as game scoreboards, keypad scanners and binary dividers are presented. The chapter presents common techniques used for computer arithmetic, including carry look-ahead addition, and binary multiplication and division. Use of a state machine for sequencing the operations in a digital system is an important concept presented in this chapter. Synthesizable Verilog code is presented for the various designs. A variety of examples are presented so that instructors can select their favorite designs for teaching.

Use of sequential machine charts (SM charts) as an alternative to state graphs is presented in Chapter 5. We show how to write Verilog code based on SM charts and how to realize hardware to implement the SM charts. Then, the technique of microprogramming is presented. Transformation of SM charts for different types of microprogramming is discussed. Then, we show how the use of linked state machines facilitates the decomposition of complex systems into simpler ones. The design of a dice-game simulator is used to illustrate these techniques. If an instructor skips chapter 5, it does not seriously affect the use of the following chapters. A couple of SM charts have been used in Chapters 8 and 9, however, a simple explanation may be sufficient to make them understandable.

Chapter 6 presents issues related to implementing digital systems in Field Programmable Gate Arrays. A few simple designs are first hand-mapped into FPGA building blocks to illustrate the mapping process. Shannon's expansion for decomposition of functions with several variables into smaller functions is presented. Features of modern FPGAs like carry chains, cascade chains, dedicated memory, dedicated multipliers, etc., are then presented. Instead of describing all features in a selected commercial product, the features are described in a general fashion. Once students understand the fundamental general principles, they will be able to understand and use any commercial product they have to work with. Some commercial products are also discussed so students can learn about the state of the art FPGA chips. This chapter also presents an introduction to the processes and algorithms in the software design flow. Synthesis, mapping, placement, and routing processes are briefly described. Optimizations during synthesis are illustrated.

Basic techniques for floating-point arithmetic are described in Chapter 7. A simple floating-point format with 2's complement numbers is presented and then the IEEE standard floating-point formats are presented. A floating-point multiplier example is presented starting with development of the basic algorithm, then simulating the system using Verilog, and finally synthesizing and implementing the system using an FPGA. Some instructors may prefer to cover Chapter 8 and 9 before teaching Chapter 7. Chapter 7 can be omitted without loss of any continuity.

By the time students reach Chapter 8, they should be thoroughly familiar with the basics of Verilog. At this point we introduce some of the more advanced features of Verilog and illustrate their use. A memory model with tri-state output busses is presented to illustrate the use of tri-state devices.

Chapter 9 presents the design of a microprocessor, starting from the description of the instruction set architecture (ISA). The processor is an early RISC processor, the MIPS R2000. The important instructions in the MIPS ISA are described and a subset is then implemented. The design of the various components of the processor, such as the instruction memory module, data memory module and register file are illustrated module by module. These components are then integrated together and a complete processor design is presented. The model can be tested with a test bench, or can be synthesized and implemented on an FPGA. In order to test the design on an FPGA, one will need to write input-output modules for the design. This example requires understanding of the basics of assembly language programming and computer organization.

The important topics of hardware testing and design for testability are covered in Chapter 10. This chapter introduces the basic techniques for testing combinational and sequential logic. Then scan design and boundary-scan techniques, which facilitate the testing of digital systems, are described. The chapter concludes with a discussion of built-in self-test (BIST). Verilog code for a boundary-scan example and for a BIST example is included. The topics in this chapter play an important role in digital system design, and we recommend that they be included in any course on this subject. Chapter 10 can be covered any time after the completion of Chapter 8.

This book is the result of many years of teaching a senior course in digital systems design at the University of Texas at Austin. Throughout the years, the technology for hardware implementation of digital systems has kept changing, but many of the same design principles are still applicable. In the early years of the course, we handwired modules consisting of discrete transistors to implement our designs. Then integrated circuits were introduced, and we were able to implement our designs using breadboards and TTL logic. Now we are able to use FPGAs and CPLDs to realize very complex designs. We originally used our own hardware description language together with a in-house simulator. When VHDL was adopted as an IEEE standard and became widely used in industry, we switched to VHDL. There is a similar book available with VHDL code examples, which has been in use for several years. The popularity of Verilog in the industry and student demand motivated us to do this Verilog book.

All of the Verilog code in this textbook has been tested using the Modelsim simulator. The Modelsim software is available in a student edition, and we recommend its use in conjunction with this text. All of the Verilog code in this textbook is available on the book web site. The website also provides two software packages, LogicAid and SimuAid, which are useful in teaching digital system design. Instruction manuals and examples of using this software are on the web site.

This textbook is also available online through Cengage Learning's MindTap, a personalized learning program. Students who purchase the MindTap version have access to the book's MindTap Reader and are able to complete homework and assessment material online, through their desktop, laptop, or iPad. If you are using a Learning Management System (such as Blackboard or Moodle) for tracking course content, assignments, and grading, you can seamlessly access the MindTap suite of content and assessments for this course.

In MindTap, instructors can:

- Personalize the Learning Path to match your course syllabus by rearranging content or appending original material to the online content.
- Connect a Learning Management System portal to the online course and Reader
- Customize online assessments and assignments
- Track student progress and comprehension
- Promote student engagement through interactivity and exercises

Additionally, students can listen to the text through ReadSpeaker, take notes, create their own flashcards, highlight content for easy reference, and check their understanding of the material through practice quizzes and homework.

Acknowledgments

We would like to thank the many individuals who have contributed their time and effort to the development of this textbook. Over many years we have received valuable feedback from the students in our digital systems design courses. We would especially like to thank the faculty members who reviewed the previous edition and offered many suggestions for its improvement. These faculty include:

Jerry Trahan
James Peckol
Steven Barrett

We also wish to acknowledge Prof. Nur Touba's comments on various parts of the book. Special thanks go Swati Meherishi, Eavan Cully, and Rose Kernan during various steps of the publication process. It was a pleasure to work with you all. We also take this opportunity to express our gratitude to Shuang Song and other student assistants who helped with the word processing, Verilog code testing and illustrations. Santos Gomez, Eula Tolentino, Frank Weng helped towards the Verilog version and Roger Chen, WIlliam Earle, Manish Kapadia, Matt Morgan, Elizabeth Norris, Arif Mondal and Raman Suri helped towards the VHDL edition.

Charles. H. Roth, Jr.
Lizy K. John
Byeong Kil Lee

Review of Logic Design Fundamentals

This chapter reviews many of the logic design topics normally taught in a first course in logic design. Some of the review examples that follow are referenced in later chapters of this text. For more details on any of the topics discussed in this chapter, the reader should refer to a standard logic design textbook such as Roth and Kinney, *Fundamentals of Logic Design*, 6th ed. (Cengage Learning, 2010). First, we review combinational logic and then sequential logic. Combinational logic has no memory, so the present output depends only on the present input. Sequential logic has memory, so the present output depends not only on the present input but also on the past sequence of inputs. The sections on sequential circuit timing and synchronous design are particularly important, since a good understanding of timing issues is essential to the successful design of digital systems.

1.1 Combinational Logic

Some of the basic gates used in logic circuits are shown in Figure 1-1. Unless otherwise specified, all the variables that we use to represent logic signals will be two-valued, and the two values will be designated 0 and 1. We will normally use positive logic, for which a low voltage corresponds to a logic 0 and a high voltage corresponds to a logic 1. When negative logic is used, a low voltage corresponds to a logic 1 and a high voltage corresponds to a logic 0.

For the AND gate of Figure 1-1, the output $C = 1$ if and only if the input $A = 1$ *and* the input $B = 1$. We will use a raised dot or simply write the variables side by side to indicate the AND operation; thus $C = A$ AND $B = A \cdot B = AB$. For the OR gate, the output $C = A$ OR $B = 1$ if and only if the input $A = 1$ *or* the input $B = 1$ (inclusive OR). We will use + to indicate the OR operation; thus $C = A$ OR $B = A + B$. The NOT gate, or inverter, forms the complement of the input; that is, if $A = 1$, $C = $ NOT $A = 0$, and if $A = 0$, $C = 1$. We will use a prime (') to indicate the complement (NOT) operation, so $C = $ NOT $A = A'$. The exclusive-OR (XOR) gate has an output $C = 1$ if $A = 1$ and $B = 0$ or if $A = 0$ and $B = 1$. The symbol $\oplus$ represents exclusive OR, so we write

$$C = A \text{ XOR } B = AB' + A'B = A \oplus B \tag{1-1}$$

1.1 Combinational Logic

FIGURE 1-1: Basic Gates

AND: $C = AB$

OR: $C = A + B$

NOT: $C = A'$

Exclusive OR: $C = A \oplus B$

The behavior of a combinational logic circuit can be specified by a truth table that gives the circuit outputs for each combination of input values. As an example, consider the full adder of Figure 1-2, which adds two binary digits (X and Y) and a carry (C_{in}) to give a sum (*Sum*) and a carry out (C_{out}). The truth table specifies the adder outputs as a function of the adder inputs. For example, when the inputs are $X = 0$, $Y = 0$, and $C_{in} = 1$, adding the three inputs gives $0 + 0 + 1 = 01$, so the sum is 1 and the carry out is 0. When the inputs are 011, $0 + 1 + 1 = 10$, so *Sum* = 0 and $C_{out} = 1$. When the inputs are $X = Y = C_{in} = 1$, $1 + 1 + 1 = 11$, so *Sum* = 1 and $C_{out} = 1$.

FIGURE 1-2: Full Adder

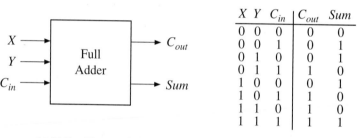

(a) Full adder module   (b) Truth table

We will derive algebraic expressions for *Sum* and C_{out} from the truth table. From the table, *Sum* = 1 when $X = 0$, $Y = 0$, and $C_{in} = 1$. The term $X'Y'C_{in}$ equals 1 only for this combination of inputs. The term $X'YC_{in}' = 1$ only when $X = 0$, $Y = 1$, and $C_{in} = 0$. The term $XY'C_{in}'$ is 1 only for the input combination $X = 1$, $Y = 0$, and $C_{in} = 0$. The term XYC_{in} is 1 only when $X = Y = C_{in} = 1$. Therefore, *Sum* is formed by ORing these four terms together:

$$Sum = X'Y'C_{in} + X'YC_{in}' + XY'C_{in}' + XYC_{in} \tag{1-2}$$

Each of the terms in this Sum of Products (SOP) expression is 1 for exactly one combination of input values. In a similar manner, C_{out} is formed by ORing four terms together:

$$C_{out} = X'YC_{in} + XY'C_{in} + XYC_{in}' + XYC_{in} \tag{1-3}$$

Each term in equations (1-2) and (1-3) is referred to as a *minterm*, and these equations are referred to as *minterm expansions*. These minterm expansions can also be written in *m*-notation or decimal notation as follows:

$$Sum = m_1 + m_2 + m_4 + m_7 = \Sigma m(1, 2, 4, 7)$$
$$C_{out} = m_3 + m_5 + m_6 + m_7 = \Sigma m(3, 5, 6, 7)$$

The decimal numbers designate the rows of the truth table for which the corresponding function is 1. Thus $Sum = 1$ in rows 001, 010, 100, and 111 (rows 1, 2, 4, 7). A logic function can also be represented in terms of the inputs for which the function value is 0. Referring to the truth table for the full adder, $C_{out} = 0$ when $X = Y = C_{in} = 0$. The term $(X + Y + C_{in})$ is 0 only for this combination of inputs. The term $(X + Y + C_{in}')$ is 0 only when $X = Y = 0$ and $C_{in} = 1$. The term $(X + Y' + C_{in})$ is 0 only when $X = C_{in} = 0$ and $Y = 1$. The term $(X' + Y + C_{in})$ is 0 only when $X = 1$ and $Y = C_{in} = 0$. C_{out} is formed by ANDing these four terms together:

$$C_{out} = (X + Y + C_{in})(X + Y + C_{in}')(X + Y' + C_{in})(X' + Y + C_{in}) \quad (1\text{-}4)$$

C_{out} is 0 only for the 000, 001, 010, and 100 rows of the truth table and therefore must be 1 for the remaining four rows. Each of the terms in the Product of Sums (POS) expression in (1-4) is referred to as a *maxterm*, and (1-4) is called a *maxterm expansion*. This *maxterm expansion* can also be written in decimal notation as

$$C_{out} = M_0 \cdot M_1 \cdot M_2 \cdot M_4 = \prod M(0, 1, 2, 4)$$

where the decimal numbers correspond to the truth table rows for which $C_{out} = 0$.

1.2 Boolean Algebra and Algebraic Simplification

The basic mathematics used for logic design is Boolean algebra. Table 1-1 summarizes the laws and theorems of Boolean algebra. They are listed in dual pairs and can easily be verified for two-valued logic by using truth tables. These laws and theorems can be used to simplify logic functions so they can be realized with a reduced number of components.

A very important law in Boolean algebra is DeMorgan's law. DeMorgan's law (1-16, 1-16D) can be used to form the complement of an expression on a step-by-step basis. The generalized form of DeMorgan's law (1-17) can be used to form the complement of a complex expression in one step. Equation (1-17) can be interpreted as follows: To form the complement of a Boolean expression, replace each variable by its complement; also replace 1 with 0, 0 with 1, OR with AND, and AND with OR. Add parentheses as required to assure the proper hierarchy of operations. If AND is performed before OR in F, then parentheses may be required to ensure that OR is performed before AND in F'.

Example

$$F = X + E'K \, (C \, (AB + D') \cdot 1 + WZ' \, (G'H + 0))$$
$$F' = X' \, (E + K' + (C' + (A' + B') \, D + 0) \, (W' + Z + (G + H') \cdot 1))$$

The boldface parentheses in F' were added when an AND operation in F was replaced with an OR. The dual of an expression is the same as its complement, except that the variables are not complemented.

TABLE 1-1: Laws and Theorems of Boolean Algebra

Operations with 0 and 1:

$X + 0 = X$ (1-5) $\quad X \cdot 1 = X$ (1-5D)
$X + 1 = 1$ (1-6) $\quad X \cdot 0 = 0$ (1-6D)

Idempotent laws:

$X + X = X$ (1-7) $\quad X \cdot X = X$ (1-7D)

Involution law:

$(X')' = X$ (1-8)

Laws of complementarity:

$X + X' = 1$ (1-9) $\quad X \cdot X' = 0$ (1-9D)

Commutative laws:

$X + Y = Y + X$ (1-10) $\quad XY = YX$ (1-10D)

Associative laws:

$(X + Y) + Z = X + (Y + Z)$ (1-11) $\quad (XY)Z = X(YZ) = XYZ$ (1-11D)
$\quad\quad\quad\quad\quad\;\; = X + Y + Z$

Distributive laws:

$X(Y + Z) = XY + XZ$ (1-12) $\quad X + YZ = (X + Y)(X + Z)$ (1-12D)

Simplification theorems:

$XY + XY' = X$ (1-13) $\quad (X + Y)(X + Y') = X$ (1-13D)
$X + XY = X$ (1-14) $\quad X(X + Y) \quad\quad\;\, = X$ (1-14D)
$(X + Y')Y = XY$ (1-15) $\quad XY' + Y \quad\quad\;\;\, = X + Y$ (1-15D)

DeMorgan's law:

$(X + Y + Z + \cdots)' = X'Y'Z' \cdots$ (1-16) $\quad (XYZ \cdots)' = X' + Y' + Z' + \cdots$ (1-16D)
$[f(X_1, X_2, \ldots, X_n, 0, 1, +, \cdot)]' = f(X_1', X_2', \ldots, X_n', 1, 0, \cdot, +)$ (1-17)

Duality:

$(X + Y + Z + \cdots)^D = XYZ \cdots$ (1-18) $\quad (XYZ \cdots)^D = X + Y + Z + \cdots$ (1-18D)
$[f(X_1, X_2, \ldots, X_n, 0, 1, +, \cdot)]^D = f(X_1, X_2, \ldots, X_n, 1, 0, \cdot, +)$ (1-19)

Theorem for multiplying out and factoring:

$(X + Y)(X' + Z) = XZ + X'Y$ (1-20) $\quad XY + X'Z = (X + Z)(X' + Y)$ (1-20D)

Consensus theorem:

$XY + YZ + X'Z = XY + X'Z$ (1-21) $\quad (X + Y)(Y + Z)(X' + Z)$ (1-21D)
$\quad\quad\quad\quad\quad\quad\quad\quad\quad\quad\quad\quad\quad\quad\;\, = (X + Y)(X' + Z)$

Four ways of simplifying a logic expression using the theorems in Table 1-1 are as follows:

1. *Combining terms.* Use the theorem $XY + XY' = X$ to combine two terms. For example,
$$ABC'D' + ABCD' = ABD' \quad [X = ABD', Y = C]$$
When combining terms by this theorem, the two terms to be combined should contain exactly the same variables, and exactly one of the variables should appear complemented in one term and not in the other. Since $X + X = X$, a given term may be duplicated and combined with two or more other terms. For example, the expression for C_{out} (Equation 1-3) can be simplified by combining the first and fourth terms, the second and fourth terms, and the third and fourth terms:
$$\begin{aligned} C_{out} &= (X'YC_{in} + XYC_{in}) + (XY'C_{in} + XYC_{in}) + (XYC_{in}' + XYC_{in}) \\ &= YC_{in} + XC_{in} + XY \end{aligned} \quad (1\text{-}22)$$
Note that the fourth term in (1-3) was used three times.

The theorem can still be used, of course, when X and Y are replaced with more complicated expressions. For example,
$$(A + BC)(D + E') + A'(B' + C')(D + E') = D + E'$$
$$[X = D + E', Y = A + BC, Y' = A'(B' + C')]$$

2. *Eliminating terms.* Use the theorem $X + XY = X$ to eliminate redundant terms if possible; then try to apply the consensus theorem ($XY + X'Z + YZ = XY + X'Z$) to eliminate any consensus terms. For example,
$$A'B + A'BC = A'B \quad [X = A'B]$$
$$A'BC' + BCD + A'BD = A'BC' + BCD \quad [X = C, Y = BD, Z = A'B]$$

3. *Eliminating literals.* Use the theorem $X + X'Y = X + Y$ to eliminate redundant literals. Simple factoring may be necessary before the theorem is applied. For example,
$$\begin{aligned} A'B + A'B'C'D' + ABCD' &= A'(B + B'C'D') + ABCD' & \text{(by (1-12))} \\ &= A'(B + C'D') + ABCD' & \text{(by (1-15D))} \\ &= B(A' + ACD') + A'C'D' & \text{(by (1-10))} \\ &= B(A' + CD') + A'C'D' & \text{(by (1-15D))} \\ &= A'B + BCD' + A'C'D' & \text{(by (1-12))} \end{aligned}$$

The expression obtained after applying 1, 2, and 3 will not necessarily have a minimum number of terms or a minimum number of literals. If it does not and no further simplification can be made using 1, 2, and 3, deliberate introduction of redundant terms may be necessary before further simplification can be made.

4. *Adding redundant terms.* Redundant terms can be introduced in several ways, such as adding XX', multiplying by $(X + X')$, adding YZ to $XY + X'Z$ (consensus theorem), or adding XY to X. When possible, the terms added

should be chosen so that they will combine with or eliminate other terms. For example,

$$WX + XY + X'Z' + WY'Z' \quad \text{(Add } WZ' \text{ by the consensus theorem.)}$$
$$= WX + XY + X'Z' + WY'Z' + WZ' \quad \text{(eliminate } WY'Z')$$
$$= WX + XY + X'Z' + WZ' \quad \text{(eliminate } WZ')$$
$$= WX + XY + X'Z'$$

When multiplying out or factoring an expression, in addition to using the ordinary distributive law (1-12), the second distributive law (1-12D) and theorem (1-20) are particularly useful. The following is an example of multiplying out to convert from a product of sums to a sum of products:

$$(A + B + D)(A + B' + C')(A' + B + D')(A' + B + C')$$
$$= (A + (B + D)(B' + C'))(A' + B + C'D') \quad \text{(by (1-12D))}$$
$$= (A + BC' + B'D)(A' + B + C'D') \quad \text{(by (1-20))}$$
$$= A(B + C'D') + A'(BC' + B'D) \quad \text{(by (1-20))}$$
$$= AB + AC'D' + A'BC' + A'B'D \quad \text{(by (1-12))}$$

Note that the second distributive law (1-12D) and theorem (1-20) were applied before the ordinary distributive law. Any Boolean expression can be factored by using the two distributive laws (1-12 and 1-12D) and theorem (1-20). As an example of factoring, read the steps in the preceding example in the reverse order.

The following theorems apply to exclusive-OR:

$$X \oplus 0 = X \tag{1-23}$$
$$X \oplus 1 = X' \tag{1-24}$$
$$X \oplus X = 0 \tag{1-25}$$
$$X \oplus X' = 1 \tag{1-26}$$
$$X \oplus Y = Y \oplus X \quad \text{(commutative law)} \tag{1-27}$$
$$(X \oplus Y) \oplus Z = X \oplus (Y \oplus Z) = X \oplus Y \oplus Z \quad \text{(associative law)} \tag{1-28}$$
$$X(Y \oplus Z) = XY \oplus XZ \quad \text{(distributive law)} \tag{1-29}$$
$$(X \oplus Y)' = X \oplus Y' = X' \oplus Y = XY + X'Y' \tag{1-30}$$

The expression for *Sum* (equation (1-2)) can be rewritten in terms of exclusive-OR by using (1-1) and (1-30):

$$Sum = X'(Y'C_{in} + YC_{in}') + X(Y'C_{in}' + YC_{in})$$
$$= X'(Y \oplus C_{in}) + X(Y \oplus C_{in})' = X \oplus Y \oplus C_{in} \tag{1-31}$$

The simplification rules that you studied in this section are important when a circuit has to be optimized to use a lower number of gates. The existence of equivalent forms also helps when mapping circuits into particular target devices where only certain types of logic (e.g., NAND only or NOR only) are available.

1.3 Karnaugh Maps

Karnaugh maps (K-maps) provide a convenient way to simplify logic functions of three to five variables. Figure 1-3 shows a four-variable Karnaugh map. Each square in the map represents one of the 16 possible minterms of four variables. A 1 in a square indicates that the minterm is present in the function, and a 0 (or blank) indicates that the minterm is absent. An X in a square indicates that we don't care whether the minterm is present or not. Don't cares arise under two conditions: (1) The input combination corresponding to the don't care can never occur, and (2) the input combination can occur, but the circuit output is not specified for this input condition.

The variable values along the edge of the map are ordered so that adjacent squares on the map differ in only one variable. The first and last columns and the top and bottom rows of the map are considered to be adjacent. Two 1s in adjacent squares can be combined by eliminating one variable using $xy + xy' = x$. Figure 1-3 shows a four-variable function with nine minterms and two don't cares. Minterms $A'BC'D$ and $A'BCD$ differ only in the variable C, so they can be combined to form $A'BD$, as indicated by a loop on the map. Four 1s in a symmetrical pattern can be combined to eliminate two variables. The 1s in the four corners of the map can be combined as follows:

$$(A'B'C'D' + AB'C'D') + (A'B'CD' + AB'CD') = B'C'D' + B'CD' = B'D'$$

as indicated by the loop. Similarly, the six 1s and two Xs in the bottom half of the map combine to eliminate three variables and form the term C. The resulting simplified function is

$$F = A'BD + B'D' + C$$

FIGURE 1-3: Four-Variable Karnaugh Maps

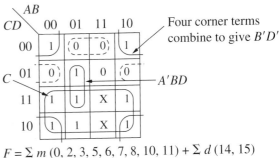

$F = \Sigma\, m\,(0, 2, 3, 5, 6, 7, 8, 10, 11) + \Sigma\, d\,(14, 15)$
$= C + B'D' + A'BD$

(a) Location of minterms   (b) Looping terms

The minimum sum-of-products representation of a function consists of a sum of prime implicants. A group of one, two, four, or eight adjacent 1s on a map represents a prime implicant if it cannot be combined with another group of 1s to eliminate a variable. A prime implicant is essential if it contains a 1 that is not contained in any other prime implicant. When finding a minimum sum of products from a map, essential prime implicants should be looped first, and then a minimum number of prime implicants to cover the remaining 1s should be looped. The Karnaugh map shown in Figure 1-4 has five prime implicants and three essential prime implicants. $A'C'$ is essential because minterm m_1 is not covered by any other prime implicant. Similarly, ACD is essential because of m_{11}, and $A'B'D'$ is essential because of m_2. After the essential prime implicants have been looped, all 1s are covered except m_7. Since m_7 can be covered by either prime implicant $A'BD$ or BCD, F has two minimum forms:

$$F = A'C' + A'B'D' + ACD + A'BD$$
and
$$F = A'C' + A'B'D' + ACD + BCD$$

FIGURE 1-4: Selection of Prime Implicants

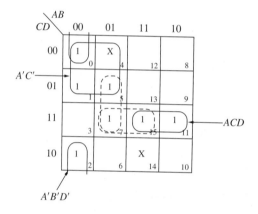

When don't cares (Xs) are present on the map, the don't cares are treated like 1s when forming prime implicants, but the Xs are ignored when finding a minimum set of prime implicants to cover all the 1s. The following procedure can be used to obtain a minimum sum of products from a Karnaugh map:

1. Choose a minterm (a 1) that has not yet been covered.
2. Find all 1s and Xs adjacent to that minterm. (Check the n adjacent squares on an n-variable map.)
3. If a single term covers the minterm and all the adjacent 1s and Xs, then that term is an essential prime implicant, so select that term. (Note that don't cares are treated as 1s in steps 2 and 3 but not in step 1.)
4. Repeat steps 1, 2, and 3 until all essential prime implicants have been chosen.
5. Find a minimum set of prime implicants that cover the remaining 1s on the map. (If there is more than one such set, choose a set with a minimum number of literals.)

To find a minimum product of sums from a Karnaugh map, loop the 0s instead of the 1s. Since the 0s of F are the 1s of F', looping the 0s in the proper way gives the minimum sum of products for F', and the complement is the minimum product of sums for F. For Figure 1-3, we can first loop the essential prime implicants of

$F'(BC'D'$ and $B'C'D$, indicated by dashed loops), and then cover the remaining 0 with AB. Thus the minimum sum for F' is

$$F' = BC'D' + B'C'D + AB$$

from which the minimum product of sums for F is

$$F = (B' + C + D)(B + C + D')(A' + B')$$

Simplification Using Map-Entered Variables

Two four-variable Karnaugh maps can be used to simplify functions with five variables. If functions have more than five variables, *map-entered variables* can be used. Consider a truth table as shown in Table 1-2. There are six input variables and one output variable. Only certain rows of the truth table have been specified. To completely specify the truth table, 64 rows will be required. The input combinations not specified in the truth table result in an output of 0.

TABLE 1-2: Partial Truth Table for a 6-Variable Function

A	B	C	D	E	F	G
0	0	0	0	X	X	1
0	0	0	1	X	X	X
0	0	1	0	X	X	1
0	0	1	1	X	X	1
0	1	0	1	1	X	1
0	1	1	1	1	X	1
1	0	0	1	X	1	1
1	0	1	0	X	X	X
1	0	1	1	X	X	1
1	1	0	1	X	X	X
1	1	1	1	X	X	1

Karnaugh map techniques can be extended to simplify functions such as this using map-entered variables. Since E and F are the input variables with the greatest number of don't cares (X), a Karnaugh map can be formed with A, B, C, D and the remaining two variables can be entered inside the map. Figure 1-5 shows a 4-variable map with variables E and F entered in the squares in the map. When E appears in a square, this means that if $E = 1$, the corresponding minterm is present in the function G, and if $E = 0$, the minterm is absent. Rows 5 and 6 in the truth table result in the E in the box corresponding to minterm 5 and minterm 7. Row 7 results in the F in the box corresponding to minterm 9. Thus, the map represents the 6-variable function

$$G(A, B, C, D, E, F) = m_0 + m_2 + m_3 + Em_5 + Em_7 + Fm_9 \\ + m_{11} + m_{15} (+ \text{ don't care terms})$$

where the minterms are minterms of the variables A, B, C, D. Note that m_9 is present in G only when $F = 1$.

FIGURE 1-5: Simplification Using Map-Entered Variables

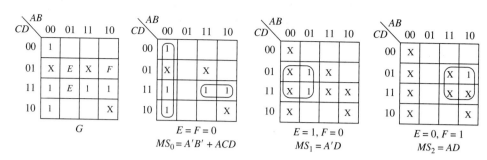

Next we will discuss a general method of simplifying functions using map-entered variables. In general, if a variable P_i is placed in square m_j of a map of function F, this means that $F = 1$ when $P_i = 1$ and the variables are chosen so that $m_j = 1$. Given a map with variables $P_1, P_2, \ldots$ entered into some of the squares, the minimum sum-of-products form of F can be found as follows: Find a sum-of-products expression for F of the form

$$F = MS_0 + P_1 MS_1 + P_2 MS_2 + \cdots \tag{1-32}$$

where

- MS_0 is the minimum sum obtained by setting $P_1 = P_2 = \cdots = 0$.
- MS_1 is the minimum sum obtained by setting $P_1 = 1$, $P_j = 0$ ($j \neq 1$), and replacing all 1s on the map with don't cares.
- MS_2 is the minimum sum obtained by setting $P_2 = 1$, $P_j = 0$ ($j \neq 2$), and replacing all 1s on the map with don't cares.

Corresponding minimum sums can be found in a similar way for any remaining map-entered variables.

The resulting expression for F will always be a correct representation of F. This expression will be a minimum sum provided that the values of the map-entered variables can be assigned independently. On the other hand, the expression will not generally be a minimum sum if the variables are not independent (for example, if $P_1 = P_2'$).

For the example of Figure 1-5, maps for finding MS_0, MS_1, and MS_2 are shown where E corresponds to P_1 and F corresponds to P_2. Note that it is not required to draw a map for $E=1$, $F=1$, because $E=1$ already covers cases with $E=1$, $F=0$ and $E=1$, $F=1$. The resulting expression is a minimum sum of products for G:

$$G = A'B' + ACD + EA'D + FAD$$

After some practice, it should be possible to write the minimum expression directly from the original map without first plotting individual maps for each of the minimum sums.

1.4 Designing with NAND and NOR Gates

In many technologies, implementation of NAND gates or NOR gates is easier than that of AND and OR gates. Figure 1-6 shows the symbols used for NAND and NOR gates. The *bubble* at a gate input or output indicates a complement. Any logic function can be realized using only NAND gates or only NOR gates.

FIGURE 1-6: NAND and NOR Gates

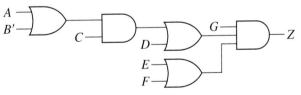

Conversion from circuits of OR and AND gates to circuits of all NOR gates or all NAND gates is straightforward. To design a circuit of NOR gates, start with a product-of-sums representation of the function (circle 0s on the Karnaugh map). Then find a circuit of OR and AND gates that has an AND gate at the output. If an AND gate output does not drive an AND gate input and an OR gate output does not connect to an OR gate input, then conversion is accomplished by replacing all gates with NOR gates and complementing inputs if necessary. Figure 1-7 illustrates the conversion procedure for

$$Z = G(E + F)(A + B' + D)(C + D) = G(E + F)[(A + B')C + D]$$

Conversion to a circuit of NAND gates is similar, except that the starting point should be a sum-of-products form for the function (circle 1s on the map) and the output gate of the AND-OR circuit should be an OR gate.

FIGURE 1-7: Conversion to NOR Gates

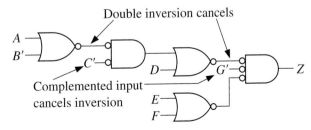

Even if AND and OR gates do not alternate, we can still convert a circuit of AND and OR gates to a NAND or NOR circuit, but it may be necessary to add extra inverters so that each added inversion is canceled by another inversion. The following procedure may be used to convert to a NAND (or NOR) circuit:

1. Convert all AND gates to NAND gates by adding an inversion bubble at the output. Convert OR gates to NAND gates by adding inversion bubbles at the inputs. (To convert to NOR, add inversion bubbles at all OR gate outputs and all AND gate inputs.)
2. Whenever an inverted output drives an inverted input, no further action is needed, since the two inversions cancel.
3. Whenever a non-inverted gate output drives an inverted gate input or vice versa, insert an inverter so that the bubbles will cancel. (Choose an inverter with the bubble at the input or output, as required.)
4. Whenever a variable drives an inverted input, complement the variable (or add an inverter) so the complementation cancels the inversion at the input.

In other words, if we always add bubbles (or inversions) in pairs, the function realized by the circuit will be unchanged. To illustrate the procedure, we will convert Figure 1-8(a) to NANDs. First, we add bubbles to change all gates to NAND gates (Figure 1-8(b)). The highlighted lines indicate four places where we have added only a single inversion. This is corrected in Figure 1-8(c) by adding two inverters and complementing two variables.

FIGURE 1-8: Conversion of AND-OR Circuit to NAND Gates

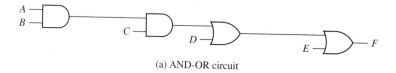

(a) AND-OR circuit

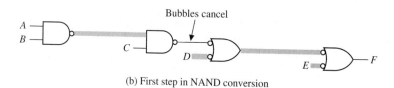

(b) First step in NAND conversion

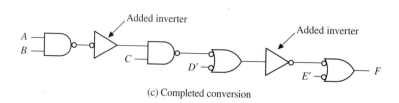

(c) Completed conversion

1.5 Hazards in Combinational Circuits

When the input to a combinational circuit changes, unwanted switching transients may appear in the output. These transients occur when different paths from input to output have different propagation delays. If, in response to an input change and for some combination of propagation delays, a circuit output may momentarily go to 0 when it should remain a constant 1, we say that the circuit has a static 1-*hazard*. Similarly, if the output may momentarily go to 1 when it should remain a 0, we say that the circuit has a static 0-*hazard*. If, when the output is supposed to change from 0 to 1 (or 1 to 0), the output may change three or more times, we say that the circuit has a *dynamic hazard*.

Consider the two simple circuits in Figure 1-9. Figure 1-9(a) shows an inverter and an OR gate implementing the function $A + A'$. Logically, the output of this circuit is expected to be a 1 always, however a delay in the inverter gate can cause static hazards in this circuit. Assume a non-zero delay for the inverter, and that the value of A just changed from 1 to 0. There is a short interval of time until the inverter delay has passed when both inputs of the OR gate are 0 and hence the output of the circuit may momentarily go to 0. Similarly in the circuit in Figure 1-9(b), the expected output is always 0; however, when A changes from 1 to 0, a momentary 1 appears at the output of the inverter because of the delay. This circuit hence has a static 0-hazard. The hazard occurs because both A and A' have the same value for a short duration after A changes.

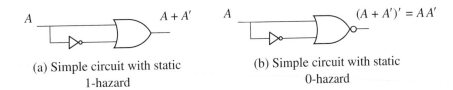

FIGURE 1-9: Simple Circuits Containing Hazards

(a) Simple circuit with static 1-hazard

(b) Simple circuit with static 0-hazard

A static 1-hazard occurs in a sum-of-product implementation when two minterms differing by only one input variable are not covered by the same product term. Figure 1-10(a) illustrates another circuit with a static 1-hazard. If $A = C = 1$, the output should remain a constant 1 when B changes from 1 to 0. However, as shown in Figure 1-10(b), if each gate has a propagation delay of 10 ns, E will go to 0 before D goes to 1, resulting in a momentary 0 (a 1-hazard appearing in the output F). As seen on the Karnaugh map, there is no loop that covers both minterm ABC and $AB'C$. So if $A = C = 1$ and B changes from 1 to 0, BC immediately becomes 0, but until an inverter delay passes, AB' does not become a 1. Both terms can momentarily go to 0, resulting in a glitch in F. If we add a loop corresponding to the term AC to the map and add the corresponding gate to the circuit (Figure 1-10(c)), this eliminates the hazard. The term AC remains 1 while B is changing, so no glitch can appear in the output. In general, non-minimal expressions are required to eliminate static hazards.

FIGURE 1-10: Elimination of 1-Hazard

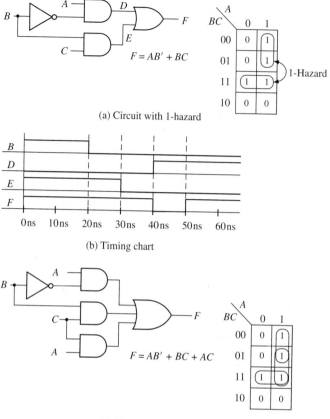

To design a circuit that is free of static and dynamic hazards, the following procedure may be used:

1. Find a sum-of-products expression (F') for the output in which every pair of adjacent 1s is covered by a 1-term. (The sum of all prime implicants will always satisfy this condition.) A two-level AND-OR circuit based on this F' will be free of 1-, 0-, and dynamic hazards.

2. If a different form of circuit is desired, manipulate F' to the desired form by using simple factoring, DeMorgan's law, and so forth. **Treat each x_i and x_i' as independent variables to prevent introduction of hazards.**

Alternatively, you can start with a product-of-sums expression in which every pair of adjacent 0s is covered by a 0-term.

Given a circuit, you can identify the static hazards in it by writing an expression for the output in terms of the inputs exactly as it is implemented in the circuit and manipulating it to a sum-of-products form, treating x_i and x_i' as independent variables. A Karnaugh map can be constructed, and all implicants corresponding to each term can be circled. If any pair of adjacent 1's is not covered by a single term, a static 1-hazard can occur. Similarly, a static 0-hazard can be identified by writing a product-of-sums expression for the circuit.

1.6 Flip-Flops and Latches

Sequential circuits commonly use flip-flops as storage devices. There are several types of flip-flops such as Delay (D flip-flops, J-K flip-flops, Toggle (T) flip-flops, and the like. Figure 1-11 shows a clocked D flip-flop. This flip-flop can change state in response to the rising edge of the clock input. The next state of the flip-flop after the rising edge of the clock is equal to the D input before the rising edge. The *characteristic equation* of the flip-flop is therefore $Q^+ = D$, where Q^+ represents the next state of the Q output after the active edge of the clock and D is the input before the active edge.

FIGURE 1-11: Clocked D Flip-Flop with Rising-Edge Trigger

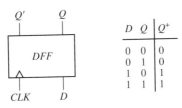

D	Q	Q^+
0	0	0
0	1	0
1	0	1
1	1	1

Figure 1-12 shows a clocked J-K flip-flop and its truth table. Since there is a bubble at the clock input, all state changes occur following the falling edge of the clock input. If $J = K = 0$, no state change occurs. If $J = 1$ and $K = 0$, the flip-flop is set to 1, independent of the present state. If $J = 0$ and $K = 1$, the flip-flop is always reset to 0. If $J = K = 1$, the flip-flop changes state. The characteristic equation, derived from the truth table in Figure 1-12, using a Karnaugh map is

$$Q^+ = JQ' + K'Q. \tag{1-33}$$

FIGURE 1-12: Clocked J-K Flip-Flop

J	K	Q	Q^+
0	0	0	0
0	0	1	1
0	1	0	0
0	1	1	0
1	0	0	1
1	0	1	1
1	1	0	1
1	1	1	0

A clocked T flip-flop (Figure 1-13) changes state following the active edge of the clock if $T = 1$, and no state change occurs if $T = 0$. The characteristic equation for the T flip-flop is

$$Q^+ = QT' + Q'T = Q \oplus T \tag{1-34}$$

A J-K flip-flop is easily converted to a T flip-flop by connecting T to both J and K. Substituting T for J and K in (1-33) yields (1-34).

FIGURE 1-13: Clocked T Flip-Flop

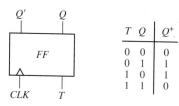

T	Q	Q^+
0	0	0
0	1	1
1	0	1
1	1	0

Two NOR gates can be connected to form an unclocked S-R (set-reset) flip-flop, as shown in Figure 1-14. An unclocked flip-flop of this type is often referred to as an S-R latch. If $S = 1$ and $R = 0$, the Q output becomes 1 and $P = Q'$. If $S = 0$ and $R = 1$, Q becomes 0 and $P = Q'$. If $S = R = 0$, no change of state occurs. If $R = S = 1$, $P = Q = 0$, which is not a proper flip-flop state, since the two outputs should always be complements. If $R = S = 1$ and these inputs are simultaneously changed to 0, oscillation may occur. For this reason, S and R are not allowed to be 1 at the same time. For purposes of deriving the characteristic equation, we assume the $S = R = 1$ never occurs, in which case $Q^+ = S + R'Q$. In this case, Q^+ represents the state after any input changes have propagated to the Q output.

FIGURE 1-14: S-R Latch

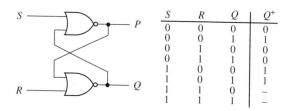

S	R	Q	Q^+
0	0	0	0
0	0	1	1
0	1	0	0
0	1	1	0
1	0	0	1
1	0	1	1
1	1	0	–
1	1	1	–

A gated D latch (Figure 1-15), also called a transparent D latch, behaves as follows: If the gate signal $G = 1$, then the Q output follows the D input ($Q^+ = D$). If $G = 0$, then the latch holds the previous value of Q ($Q^+ = Q$). Essentially, the device will not respond to input changes unless $G = 1$; it simply "latches" the previous input just before G became 0. Some refer to the D latch as a level-sensitive D flip-flop. Essentially, if the gate input G is viewed as a clock, the latch can be considered as a device that operates when the clock level is high and does not respond to the inputs when the clock level is low. The characteristic equation for the D latch is $Q^+ = GD + G'Q$. Figure 1-16 shows an implementation of the D latch using gates. Since the Q^+ equation has a 1-hazard, an extra AND gate has been added to eliminate the hazard.

FIGURE 1-15: Transparent D Latch

G	D	Q	Q^+
0	0	0	0
0	0	1	1
0	1	0	0
0	1	1	1
1	0	0	0
1	0	1	0
1	1	0	1
1	1	1	1

FIGURE 1-16: Implementation of D Latch

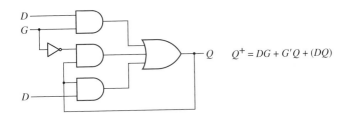

1.7 Mealy Sequential Circuit Design

There are two basic types of sequential circuits: Mealy and Moore. In a Mealy circuit, the outputs depend on both the present state and the present inputs. In a Moore circuit, the outputs depend only on the present state. A general model of a Mealy sequential circuit consists of a combinational circuit, which generates the outputs and the next state, and a state register, which holds the present state (see Figure 1-17). The state register normally consists of D flip-flops. The normal sequence of events is (1) the X inputs change to a new value; (2) after a delay, the corresponding Z outputs and next state appear at the output of the combinational circuit; and (3) the next state is clocked into the state register and the state changes. The new state feeds back into the combinational circuit, and the process is repeated.

FIGURE 1-17: General Model of Mealy Sequential Machine

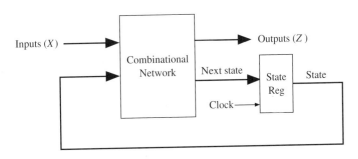

Mealy Machine Design Example 1: Sequence Detector

To illustrate the design of a clocked Mealy sequential circuit, let us design a sequence detector. The circuit has the form indicated in the block diagram in Figure 1-18.

FIGURE 1-18: Block Diagram of a Sequence Detector

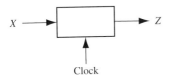

The circuit will examine a string of 0s and 1s applied to the X input and generate an output $Z=1$ only when the input sequence is 101. The input X can change only between clock pulses. The output $Z=1$ coincides with the last 1 in 101. The circuit does not reset when a 1 output occurs. A typical input sequence and the corresponding output sequence are shown here:

$$X = 0\ 0\ 1\ 1\ 0\ 1\ 1\ 0\ 0\ 1\ 0\ 1\ 0\ 1\ 0\ 0$$
$$Z = 0\ 0\ 0\ 0\ 0\ 1\ 0\ 0\ 0\ 0\ 0\ 1\ 0\ 1\ 0\ 0$$

Let us construct a *state graph* for this sequence detector. We will start in a reset state designated S_0. If a 0 input is received, we can stay in state S_0 as the input sequence we are looking for does not start with 0. However, if a 1 is received, the circuit should go to a new state. Let us denote that state as S_1. When in S_1, if we receive a 0, the circuit must change to a new state (S_2) to indicate that the first two inputs of the desired sequence (10) have been received. If a 1 is received in state S_2, the desired input sequence is complete and the output should be a 1. The output will be produced as a Mealy output and will coincide with the last 1 in the detected sequence. Since we are designing a Mealy circuit, we are not going to go to a new state that indicates the sequence 101 has been received. When we receive a 1 in S_2, we cannot go to the start state since the circuit is not supposed to reset with every detected sequence. But the last 1 in a sequence can be the first 1 in another sequence; hence, we can go to state S_1. The partial state graph at this point is indicated in Figure 1-19.

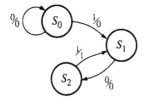

FIGURE 1-19: Partial State Graph of the Sequence Detector

When a 0 is received in state S_2, we have received two 0s in a row and must reset the circuit to state S_0. If a 1 is received when we are in S_1, we can stay in S_1 because the most recent 1 can be the first 1 of a new sequence to be detected. The final state graph is shown in Figure 1-20. State S_0 is the starting state, state S_1 indicates that a sequence ending in 1 has been received, and state S_2 indicates that a sequence ending in 10 has been received. Converting the state graph to a state table yields Table 1-3. In row S_2 of the Table, an output of 1 is indicated for input 1.

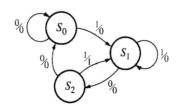

FIGURE 1-20: Mealy State Graph for Sequence Detector

TABLE 1-3: State Table for Sequence Detector

Present State	Next State		Present Output	
	X = 0	X = 1	X = 0	X = 1
S_0	S_0	S_1	0	0
S_1	S_2	S_1	0	0
S_2	S_0	S_1	0	1

Next, *state assignment* is performed, whereby specific flip-flop values are associated with specific states. There are two techniques to perform state assignment: (i) one-hot state assignment and (ii) encoded state assignment. In one-hot state assignment, one flip-flop is used for each state. Hence three flip-flops will be required if this circuit is to be implemented using the one-hot approach. In encoded state assignment, just enough flip-flops to have a unique combination for each state are sufficient. Since we have three states, we need at least two flip-flops to represent all states. We will use encoded state assignment in this design. Let us designate the two flip-flops as A and B. Let the flip-flop states $A = 0$ and $B = 0$ correspond to state S_0; $A = 0$ and $B = 1$ correspond to state S_1, and $A = 1$ and $B = 0$ correspond to state S_2. Now, the transition table of the circuit can be written as in Table 1-4.

TABLE 1-4: Transition Table for Sequence Detector

AB	A^+B^+		Z	
	X = 0	X = 1	X = 0	X = 1
00	00	01	0	0
01	10	01	0	0
10	00	01	0	1

From this table, we can plot the K-maps for the next states and the output Z. The next states are typically represented by A^+ and B^+. The 3 K-maps are shown in Figure 1-21.

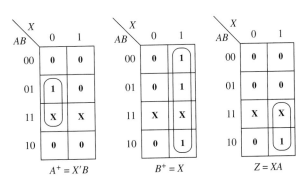

FIGURE 1-21: K-Maps for Next States and Output of Sequence Detector

The next step is deriving the flip-flop inputs to obtain the next states shown in Figure 1-21. If D flip-flops are used, one simply needs to give the expected next state of the flip-flop to the flip-flop input. So, for flip-flop A, $D_A = A^+$ and $D_B = B^+$. The resulting circuit is shown in Figure 1-22.

FIGURE 1-22: Circuit for Mealy Sequence Detector

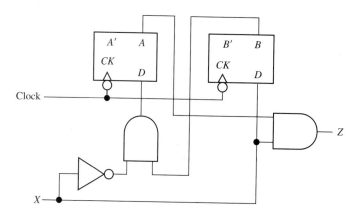

Mealy Machine Design Example 2: BCD to Excess-3 Code Converter

As an example of a more complex Mealy sequential circuit, we will design a serial code converter that converts an 8-4-2-1 binary-coded-decimal (BCD) digit to an excess-3-coded decimal digit. The input (X) will arrive serially with the least significant bit first. The outputs will be generated serially as well. Table 1-5 lists the desired inputs and outputs at times t_0, t_1, t_2, and t_3. After receiving four inputs, the circuit should reset to its initial state, ready to receive another BCD digit.

TABLE 1.5: Code Converter

X Input (BCD)				Z Output (excess -3)			
t_3	t_2	t_1	t_0	t_3	t_2	t_1	t_0
0	0	0	0	0	0	1	1
0	0	0	1	0	1	0	0
0	0	1	0	0	1	0	1
0	0	1	1	0	1	1	0
0	1	0	0	0	1	1	1
0	1	0	1	1	0	0	0
0	1	1	0	1	0	0	1
0	1	1	1	1	0	1	0
1	0	0	0	1	0	1	1
1	0	0	1	1	1	0	0

The excess-3 code is formed by adding 0011 to the BCD digit. For example,

$$\begin{array}{r} 0\ 1\ 0\ 0 \\ +\ 0\ 0\ 1\ 1 \\ \hline 0\ 1\ 1\ 1 \end{array} \qquad \begin{array}{r} 0\ 1\ 0\ 1 \\ +\ 0\ 0\ 1\ 0 \\ \hline 1\ 0\ 0\ 0 \end{array}$$

If all of the BCD bits are available simultaneously, this code converter can be implemented as a combinational circuit with four inputs and four outputs. Here, however, the bits arrive sequentially, one bit at a time. Hence this code converter must be implemented sequentially.

Let us now construct a state graph for the code converter (Figure 1-23(a)). Let us designate the start state as S_0. The first bit arrives, and one needs to add 1 to this bit, as it is the least significant bit (LSB) of 0011, the number to be added to the BCD digit to obtain the excess-3 code.

At t_0, we add 1 to the LSB, so if $X = 0$, $Z = 1$ (no carry), and if $X = 1$, $Z = 0$ (carry = 1). Let us use S_1 to indicate no carry after the first addition, and S_2 to indicate a carry of 1 after the addition to the LSB.

At t_1, we add 1 to the next bit, so if there is no carry from the first addition (state S_1), $X = 0$ gives $Z = 0 + 1 + 0 = 1$ and no carry (state S_3), and $X = 1$ gives $Z = 1 + 1 + 0 = 0$ and a carry (state S_4). If there is a carry from the first addition (state S_2), then $X = 0$ gives $Z = 0 + 1 + 1 = 0$ and a carry (S_4), and $X = 1$ gives $Z = 1 + 1 + 1 = 1$ and a carry (S_4).

At t_2, 0 is added to X, and transitions to S_5 (no carry) and S_6 are determined in a similar manner. At t_3, 0 is again added to X, and the circuit resets to S_0.

FIGURE 1-23: State Graph and Table for Code Converter

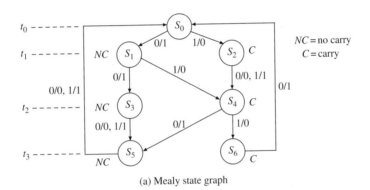

(a) Mealy state graph

PS	NS		Z	
	X=0	X=1	X=0	X=1
S_0	S_1	S_2	1	0
S_1	S_3	S_4	1	0
S_2	S_4	S_4	0	1
S_3	S_5	S_5	0	1
S_4	S_5	S_6	1	0
S_5	S_0	S_0	0	1
S_6	S_0	–	1	–

(b) State table

Figure 1-23(b) gives the corresponding state table. At this point, we should verify that the table has a minimum number of states before proceeding (see Section 1-9). Then state assignment must be performed. Since this state table has seven states, three flip-flops will be required to realize the table in encoded state assignment. In the one-hot approach, one flip-flop is used for each state. Hence seven flip-flops will be required if this circuit is to be implemented using the one-hot approach.

The next step is to make a state assignment that relates the flip-flop states to the states in the table. In the sequence detector example, we simply did a straight binary state assignment. Here we are going to look for an optimal assignment. The best state assignment to use depends on a number of factors. In many cases, we should try to find an assignment that will reduce the amount of required logic. For some types of programmable logic, a straight binary state assignment will work just as well as any other. For programmable gate arrays, a one-hot assignment may be preferred. In recent years, with the abundance of transistors on silicon chips, the emphasis on optimal state assignment has been reduced.

In order to reduce the amount of logic required, we will make a state assignment using the following guidelines (see *Fundamentals of Logic Design*, 7th ed., Cengage Learning, 2014 for details):

I. States that have the same next state (NS) for a given input should be given adjacent assignments (look at the columns of the state table).
II. States that are the next states of the same state should be given adjacent assignments (look at the rows).
III. States that have the same output for a given input should be given adjacent assignments.

Using these guidelines tends to clump 1s together on the Karnaugh maps for the next state and output functions. The guidelines indicate that the following states should be given adjacent assignments:

I. (1, 2), (3, 4), (5, 6)   (in the $X = 1$ column, S_1 and S_2 both have NS S_4; in the $X = 0$ column, S_3 and S_4 have NS S_5, and S_5 and S_6 have NS S_0)

II. (1, 2), (3, 4), (5, 6)   (S_1 and S_2 are NS of S_0; S_3 and S_4 are NS of S_1; and S_5 and S_6 are NS of S_4)

III. (0, 1, 4, 6), (2, 3, 5)

Figure 1-24(a) gives an assignment map, which satisfies the guidelines, and the corresponding transition table (Figure 1-24(b)). Since state 001 is not used, the next state and outputs for this state are don't cares. The next state and output equations are derived from this table in Figure 1-25. Figure 1-26 shows the realization of the code converter using NAND gates and D flip-flops.

FIGURE 1-24: State Assignment for BCD to Excess-3 Code Converter

(a) Assignment map

Q_2Q_3 \ Q_1	0	1
00	S_0	S_1
01		S_2
11	S_5	S_3
10	S_6	S_4

(b) Transition table

$Q_1Q_2Q_3$	$Q_1^+ Q_2^+ Q_3^+$ $X=0$	$Q_1^+ Q_2^+ Q_3^+$ $X=1$	Z $X=0$	Z $X=1$
000	100	101	1	0
100	111	110	1	0
101	110	110	0	1
111	011	011	0	1
110	011	010	1	0
011	000	000	0	1
010	000	xxx	1	x
001	xxx	xxx	x	x

1.7 Mealy Sequential Circuit Design

FIGURE 1-25: Karnaugh Maps for Figure 1-23

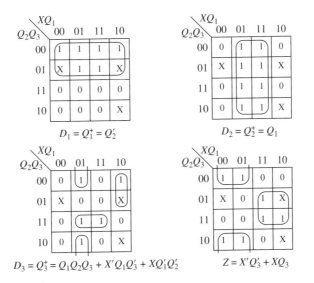

FIGURE 1-26: Realization of Code Converter

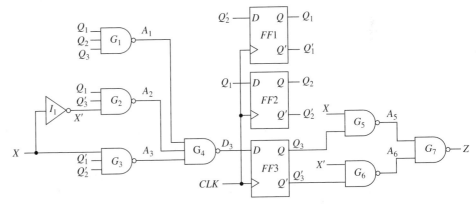

If J-K flip-flops are used instead of D flip-flops, the input equations for the J-K flip-flops can be derived from the next state maps. Given the present state flip-flop (Q) and the desired next state (Q^+), the J and K inputs can be determined from Table 1-6, also known as the excitation table. This table is derived from the truth table in Figure 1-12.

TABLE 1-6 Excitation Table for a J-K Flip-Flop

Q	Q^+	J	K	
0	0	0	X	(No change in Q; J must be 0, K may be 1 to reset Q to 0.)
0	1	1	X	(Change to $Q = 1$; J must be 1 to set or toggle.)
1	0	X	1	(Change to $Q = 0$; K must be 1 to reset or toggle.)
1	1	X	0	(No change in Q; K must be 0, J may be 1 to set Q to 1.)

Figure 1-27 shows the derivation of J-K flip-flops for the state table of Figure 1-23 using the state assignment of Figure 1-24. First, we derive the J-K input equations for flip-flop Q_1 using the Q_1^+ map as the starting point. From the preceding table, whenever Q_1 is 0, $J = Q_1^+$ and $K = X$. Therefore, we can fill in the

$Q_1 = 0$ half of the J_1 map the same as with Q_1^+ and the $Q_1 = 0$ half of the K_1 map as all Xs. When Q_1 is 1, $J_1 = X$ and $K_1 = (Q_1^+)'$. So, we can fill in the $Q_1 = 1$ half of the J_1 map with Xs and the $Q_1 = 1$ half of the K_1 map with the complement of the Q_1^+. Since half of every J and K map consists of don't cares, we can avoid drawing separate J and K maps and read the Js and Ks directly from the Q^+ maps, as illustrated in Figure 1-27(b). This shortcut method is based on the following: If $Q = 0$, then $J = Q^+$, so loop the 1s on the $Q = 0$ half of the map to get J. If $Q = 1$, then $K = (Q^+)'$, so loop the 0s on the $Q = 1$ half of the map to get K. The J and K equations will be independent of Q, since Q is set to a constant value (0 or 1) when reading J and K. To make reading the Js and Ks off the map easier, we cross off the Q values on each map. In effect, using the shortcut method is equivalent to splitting the four-variable Q^+ map into two three-variable maps, one for $Q = 0$ and one for $Q = 1$.

FIGURE 1-27: Derivation of J-K Input Equations

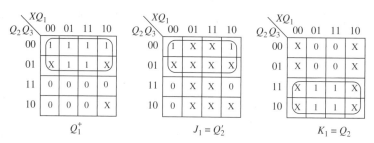

(a) Derivation using separate J–K maps

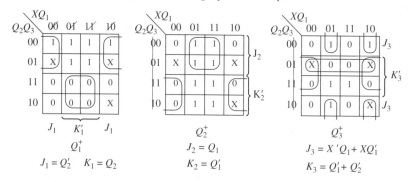

(b) Derivation using the shortcut method

The following summarizes the steps required to design a sequential circuit:

1. Given the design specifications, determine the required relationship between the input and output sequences. Then find a state graph and state table.
2. Reduce the table to a minimum number of states. First eliminate duplicate rows by row matching; then form an implication table and follow the procedure in Section 1.9.
3. If the reduced table has m states ($2^{n-1} < m \leq 2^n$), n flip-flops are required. Assign a unique combination of flip-flop states to correspond to each state in the reduced table. This is the encoded state assignment technique. Alternately, a one-hot assignment with m flip-flops can be used.
4. Form the transition table by substituting the assigned flip-flop states for each state in the reduced state tables. The resulting transition table specifies the

next states of the flip-flops and the output in terms of the present states of the flip-flops and the input.
5. Plot next-state maps and input maps for each flip-flop and derive the flip-flop input equations. Derive the output functions.
6. Realize the flip-flop input equations and the output equations using the available logic gates.
7. Check your design using computer simulation or another method.

Steps 2 through 7 may be carried out using a suitable Computer Aided Design (CAD) program.

1.8 Design of a Moore Sequential Circuit

In a Moore circuit, the outputs depend only on the present state. Moore machines are typically easier to design and debug than Mealy machines, but they often contain more states than equivalent Mealy machines. In Moore machines, there are no outputs that happen during the transition. The outputs are associated entirely to the state.

Moore Machine Design Example 1: Sequence Detector

As an example of a simple Moore circuit, let us design a sequence detector. The circuit will examine a string of 0s and 1s applied to the X input and generate an output $Z=1$ only when the input sequence is 101. The input X can change only between clock pulses. The circuit does not reset when a 1 output occurs.

As in the Mealy machine example, we start in a reset state designated S_0 in Figure 1-28. If a 0 input is received, we can stay in state S_0 as the input sequence we are looking for does not start with 0. However, if a 1 is received, the circuit goes to a new state, S_1. When in S_1, if we receive a 0, the circuit must change to a new state (S_2) to remember that the first two inputs of the desired sequence (10) have been received. If a 1 is received in state S_2, the circuit should go to a new state to indicate that the desired input sequence is complete. Let us designate this new state as S_3. In state S_3, the output must have a value of 1. The outputs in states S_0, S_1, and S_2 must be 0s. The sequence 100 resets the circuit to S_0. A sequence 1010 takes the circuit back to S_2 because another 1 input would cause Z to become 1 again.

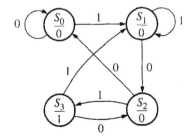

FIGURE 1-28: State Graph of the Moore Sequence Detector

The state table corresponding to the circuit is given by Table 1-7. Note that there is a single column for output because the output is determined by the present state and does not depend on X. Note that this sequence detector requires one more state than the Mealy sequence detector in Table 1-3, which detects the same input sequence.

TABLE 1-7: State Table for Sequence Detector

Present State	Next State		Present Output (Z)
	X = 0	X = 1	
S_0	S_0	S_1	0
S_1	S_2	S_1	0
S_2	S_0	S_3	0
S_3	S_2	S_1	1

Because there are four states, two flip-flops are required to realize the circuit. Using the state assignment $AB = 00$ for S_0, $AB=01$ for S_1, $AB=11$ for S_2, and $AB=10$ for S_3, the following transition table is obtained.

TABLE 1-8: Transition Table for Moore Sequence Detector

AB	A^+B^+		Z
	X = 0	X = 1	
00	00	01	0
01	11	01	0
11	00	10	0
10	11	01	1

The output function $Z = AB'$. Note that Z depends only on the flip-flop states and is independent of X, while for the corresponding Mealy machine, Z was a function of X. (It was equal to AX). The transition table can be used to write the next state maps and inputs to the flip-flops can be derived.

Moore Machine Design Example 2: NRZ to Manchester Code Converter

As another example of designing a more complex Moore sequential machine, we will design a converter for serial data. Binary data is frequently transmitted between computers as a serial stream of bits. Figure 1-29 shows three different coding schemes for serial data. The example shows transmission of the bit sequence 0, 1, 1, 1, 0, 0, 1, 0. With the NRZ (nonreturn-to-zero) code, each bit is transmitted for one bit time without any change. In contrast, for the RZ (return-to-zero) code, a 0 is transmitted as 0 for one full bit time, but a 1 is transmitted as a 1 for the first half of the bit time, and then the signal returns to 0 for the second half. For the Manchester code, a 0 is transmitted as 0 for the first half of the bit time and a 1 for the second half, but a 1 is transmitted as a 1 for the first half and as a 0 for the second half. Thus, the Manchester encoded bit always changes in the middle of the bit time.

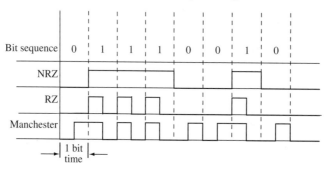

FIGURE 1-29: Coding Schemes for Serial Data Transmission

1.8 Design of a Moore Sequential Circuit

We will design a Moore sequential circuit that converts an NRZ-coded bit stream to a Manchester-coded bit stream (Figure 1-30). To do this, we will use a clock ($CLOCK2$) that is twice the frequency of the basic bit clock. If the NRZ bit is 0, it will be 0 for two $CLOCK2$ periods, and if it is 1, it will be 1 for two $CLOCK2$ periods. Thus, starting in the reset state (S_0), the only two possible input sequences are 00 and 11, and the corresponding output sequences are 01 and 10. When a 0 is received, the circuit goes to S_1 and outputs a 0; when the second 0 is received, it goes to S_2 and outputs a 1. Starting in S_0, if a 1 is received, the circuit goes to S_3 and outputs a 1, and when the second 1 is received, it must go to a state with a 0 output. Going back to S_0 is appropriate since S_0 has a 0 output and the circuit is ready to receive another 00 or 11 sequence. When in S_2, if a 00 sequence is received, the circuit can go to S_1 and then back to S_2. If a 11 sequence is received in S_2, the circuit can go to S_3 and then back to S_0. The corresponding Moore state table has two don't cares, which correspond to input sequences that cannot occur.

FIGURE 1-30: Moore Circuit for NRZ-to-Manchester Conversion

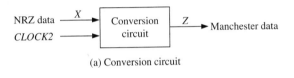

(a) Conversion circuit

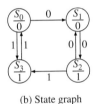

(b) State graph

Present State	Next State X=0	Next State X=1	Present Output (Z)
S_0	S_1	S_3	0
S_1	S_2	—	0
S_2	S_1	S_3	1
S_3	—	S_0	1

(c) State table

Figure 1-31 shows the timing chart for the Moore circuit. Note that the Manchester output is shifted one clock time with respect to the NRZ input. This shift occurs because a Moore circuit cannot respond to an input until the active edge of the clock occurs. This is in contrast to a Mealy circuit, for which the output can change after the input changes and before the next clock.

FIGURE 1-31: Timing for Moore Circuit

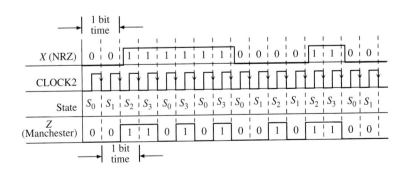

1.9 Equivalent States and Reduction of State Tables

The concept of equivalent states is important for the design and testing of sequential circuits. It helps to reduce the hardware consumed by circuits. Two states in a sequential circuit are said to be *equivalent* if we cannot tell them apart by observing input and output sequences. Consider two sequential circuits, N_1 and N_2 (see Figure 1-32). N_1 and N_2 could be copies of the same circuit. N_1 is started in state s_i, and N_2 is started in state s_j. We apply the same input sequence, $\underline{X}$, to both circuits and observe the output sequences, $\underline{Z}_1$ and $\underline{Z}_2$. (The underscore notation indicates a sequence.) If $\underline{Z}_1$ and $\underline{Z}_2$ are the same, we reset the circuits to states s_i and s_j, apply a different input sequence, and observe $\underline{Z}_1$ and $\underline{Z}_2$. If the output sequences are the same for all possible input sequences, we say the s_i and s_j are equivalent ($s_i \equiv s_j$). Formally, we can define equivalent states as follows: $s_i \equiv s_j$ if and only if, for every input sequence $\underline{X}$, the output sequences $\underline{Z}_1 = \lambda_1(s_i, \underline{X})$ and $\underline{Z}_2 = \lambda_2(s_j, \underline{X})$ are the same. This is not a very practical way to test for state equivalence since, at least in theory, it requires input sequences of infinite length. In practice, if we have a bound on number of states, then we can limit the length of the test sequences.

FIGURE 1-32: Sequential Circuits

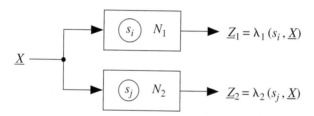

A more practical way to determine state equivalence uses the state equivalence theorem: $s_i \equiv s_j$ if and only if for every single input X, the outputs are the same and the next states are equivalent. When using the definition of equivalence, we must consider all input sequences, but we do not need any information about the internal state of the system. When using the state equivalence theorem, we must look at both the output and the next state, but we need to consider only single inputs rather than input sequences.

The table of Figure 1-33(a) can be reduced by eliminating equivalent states. First, observe that states a and h have the same next states and outputs when $X = 0$ and also when $X = 1$. Therefore, $a \equiv h$ so we can eliminate row h and replace h with a in the table. To determine if any of the remaining states are equivalent, we will use the state equivalence theorem. From the table, since the outputs for states a and b are the same, $a \equiv b$ if and only if $c \equiv d$ and $e \equiv f$. We say that c–d and e–f are implied pairs for a–b. To keep track of the implied pairs, we make an implication chart, as shown in Figure 1-33(b). We place c–d and e–f in the square at the intersection of row a and column b to indicate the implication. Since states d and e have different outputs, we place an × in the d–e square to indicate that $d \not\equiv e$. After completing the implication chart in this way, we make another pass through the chart. The e–g square contains c–e and b–g. Since the c–e square has an ×, $c \equiv e$, which implies $e \not\equiv g$, so we × out the e–g square. Similarly, since $a \equiv g$, we × out the f–g square. On the next pass through the chart, we × out all the squares that contain e–g or f–g as implied pairs (shown on the chart with dashed ×s). In the next pass, no additional squares are ×ed out, so the process terminates. Since all the squares corresponding to nonequivalent states have been ×ed out, the coordinates of the remaining squares indicate equivalent state pairs. From the first column, $a \equiv b$, from the third column, $c \equiv d$, and from the fifth column, $e \equiv f$.

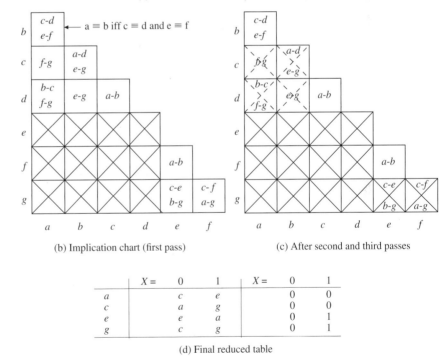

FIGURE 1-33: State Table Reduction

The implication table method of determining state equivalence can be summarized as follows:

1. Construct a chart that contains a square for each pair of states.
2. Compare each pair of rows in the state table. If the outputs associated with states i and j are different, place an × in square i–j to indicate that $i \not\equiv j$. If

the outputs are the same, place the implied pairs in square *i–j*. (If the next states of *i* and *j* are *m* and *n* for some input *x*, then *m–n* is an implied pair.) If the outputs and next states are the same (or if *i–j* implies only itself), place a check ($\checkmark$) in square *i–j* to indicate that $i \equiv j$.
3. Go through the table square by square. If square *i–j* contains the implied pair *m–n*, and square *m–n* contains an ×, then $i \not\equiv j$, and an × should be placed in square *i–j*.
4. If any ×s were added in step 3, repeat step 3 until no more ×s are added.
5. For each square *i–j* that does not contain an ×, $i \equiv j$.

If desired, row matching can be used to partially reduce the state table before constructing the implication table. Although we have illustrated this procedure for a Mealy table, the same procedure applies to a Moore table.

Two sequential circuits are said to be equivalent if every state in the first circuit has an equivalent state in the second circuit, and vice versa.

Optimization techniques such as this are incorporated in CAD tools. The importance of state minimization has slightly diminished in recent years due to the abundance of transistors on chips; however, it is still important to do obvious state minimizations to reduce the circuit's area and power.

1.10 Sequential Circuit Timing

The correct functioning of sequential circuits involves several timing issues. Propagation delays of flip-flops, gates and wires; setup times and hold times of flip-flops; clock synchronization; clock skew; and the like become important issues in designing sequential circuits. In this section, we look at various topics related to sequential circuit timing.

1.10.1 Propagation Delays, Setup, and Hold Times

There is a certain amount of time, albeit small, that elapses from the time the clock changes to the time the *Q* output changes. This time, called *propagation delay* or **clock-to-Q delay** of the flip-flop is indicated in Figure 1-34. The propagation delay can depend on whether the output is changing from high to low or vice versa. In the figure, the propagation delay for a low-to-high change in *Q* is denoted by t_{plh}, and for a high-to-low change it is denoted by t_{phl}.

For an ideal *D* flip-flop, if the *D* input changed at exactly the same time as the active edge of the clock, the flip-flop would operate correctly. However, for a real flip-flop, the *D* input must be stable for a certain amount of time before the active edge of the clock. This interval is called the *setup time* (t_{su}). Furthermore, *D* must be stable for a certain amount of time after the active edge of the clock. This interval is called the *hold time* (t_h). Figure 1-34 illustrates setup and hold times for a *D* flip-flop that changes state on the rising edge of the clock. *D* can change at any time during the shaded region on the diagram, but it must be stable during the time interval t_{su} before the active edge and for t_h after the active edge. If *D* changes at any time during the forbidden interval, it cannot be determined whether the flip-flop will change state. Even worse, the flip-flop may malfunction and output a short pulse or even go into oscillation.

1.10 Sequential Circuit Timing

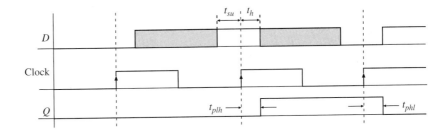

FIGURE 1-34: Setup and Hold Times for D Flip-Flop

Flip-flops typically have a setup time about 3–10x of the propagation delay of an inverter (NOT) gate. The hold times are typically 1–2x of the delay of an inverter. Minimum values for t_{su} and t_h and maximum values for t_{plh} and t_{phl} can be obtained from manufacturers' data sheets or ASIC (Application Specific Integrated Circuit) libraries accompanying design tools.

1.10.2 Timing Conditions for Proper Operation

In a synchronous sequential circuit, state changes occur immediately following the active edge of the clock. The maximum clock frequency for a sequential circuit depends on several factors. The clock period must be long enough so that all flip-flop and register inputs will have time to stabilize before the next active edge of the clock. Propagation delays and setup and hold times create complications in sequential circuit timing.

Static timing analysis (STA) is a method of validating the timing performance of a design by checking all possible paths for timing violations under worst-case conditions. A static analysis path starts at a source flip-flop (or at a primary input) and terminates at a destination flip-flop (or primary output). A static timing path between two flip-flops starts at the input to the source flip-flop and terminates at the input of the destination flip-flop. It does not go through the destination flip-flop. The path terminates when it encounters a clocked device. If a signal goes from register (flip-flop) A to register B and then to register C, the signal contains two paths. The timing paths in a synchronous digital system can be classified into 4 types:

 I. Register to register paths (i.e., flip-flop to flip-flop)
 II. Primary input to register paths (i.e., input to flip-flop)
 III. Register to primary output paths (i.e., flip-flop to output)
 IV. Input to output paths (i.e., no flip-flop)

Question: Identify the static timing paths in the following circuit:

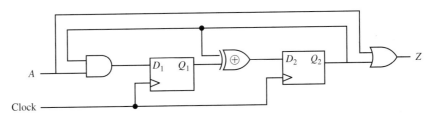

FIGURE 1-35: A Circuit to Illustrate Timing Paths

There are six static timing paths in this circuit:

I. From A to D_1 (primary input to flip-flop)
II. From D_1 to D_2 including the XOR (flip-flop to flip-flop)
III. From D_2 via XOR to D_2 (flip-flop to flip-flop)
IV. From D_2 to D_1 via AND (flip-flop to flip-flop)
V. From D_2 to Z via the OR gate (flip-flop to output)
VI. From A to Z via the OR gate (input to output)

The most complicated paths are the flip-flop to flip-flop paths; the other paths can be treated as special cases of this type of path.

Static timing analysis checks how the data arrives with respect to clock. It detects setup and hold-time violations in the design so that they can be corrected. A **setup time violation** occurs if the data changes just before the clock without providing enough setup time for the flip-flop. A **hold-time violation** occurs if the data changes just after the clock without providing enough hold time for the flip-flop.

Slack is the amount of time still left before a signal will violate a setup or hold-time constraint. Paths must have a positive or zero slack in order to have no violations. Paths that have a zero or very small slack are the speed-limiting paths in the design, because any small changes in clock or gate delays will lead to violations in such circuits. Paths that have a negative slack time have already violated a setup or hold constraint.

Static timing analysis considers the worst possible timing scenarios, but not the logical operation of the circuit. In comparison with circuit simulation, static timing analysis is faster because it doesn't need to simulate multiple test vectors.

Timing Rules for Flip-Flop to Flip-Flop Paths

For a circuit of the general form of Figure 1-36, assume that the maximum propagation delay through the combinational circuit is t_{cmax} and the maximum **clock-to-Q delay** or propagation delay from the time the clock changes to the time the flip-flop output changes is t_{pmax}, where t_{pmax} is the maximum of t_{plh} and t_{phl}. Also assume that the minimum propagation delay through the combinational circuit is t_{cmin} and the minimum clock-to-Q delay or propagation delay from the time the clock changes to the time the flip-flop output changes is t_{pmin}, where t_{pmax} data is launched from flip-flop 1's D (i.e., D_1) to FF_1's Q (i.e., Q_1) at the positive edge of clock at FF_1 (i.e., CK_1). Data is captured at FF_2's D (i.e., D_2) at the positive clock edge at FF_2 (i.e., CLK_2). FF_1 is called the launching flip-flop, and FF_2 is called the capturing flip-flop. There are two rules this circuit has to meet in order to ensure proper operation.

Rule No. 1: Setup time rule for flip-flop to flip-flop path: Clock period should be long enough to satisfy flip-flop setup time.

For proper synchronous operation, the data launched by FF_1 at edge E_1 of clock CK_1 should be captured by FF_2 at edge E_2 of clock CK_2. The clock period should be long enough to allow the first flip-flop's outputs to change and the combinational circuitry to change while still leaving enough time to satisfy the setup time. Once the clock CK_1 arrives, it could take a delay of up to t_{pmax} before FF_1's output changes. Then it could take a delay of up to t_{cmax} before the output of the combinational circuitry changes. Thus the maximum time from the active edge E_1 of the clock CK_1 to the time the change in Q_1 propagates to the second flip-flop's input (i.e., D_2) is $t_{pmax} + t_{cmax}$. In

1.10 Sequential Circuit Timing

FIGURE 1-36: Flip-Flop to Flip-Flop Path via Combinational Logic

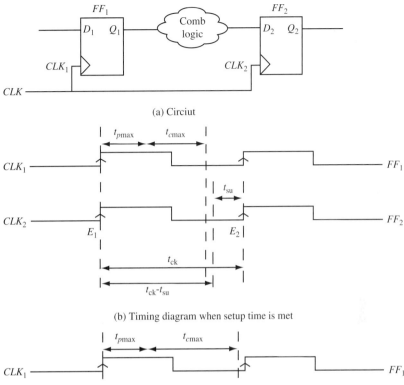

(a) Circiut

(b) Timing diagram when setup time is met

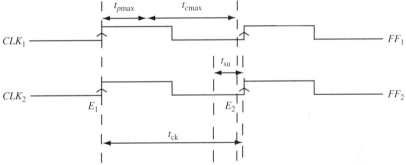

(c) Timing diagram when setup time is violated

order to ensure proper flip-flop operation, the combinational circuit output must be stable at least t_{su} before the end of the clock E_2 reaches FF_2. If the clock period is t_{ck},

$$t_{ck} \geq t_{pmax} + t_{cmax} + t_{su} \qquad (1\text{-}34\text{-}a)$$

Equation (1-34-a) relates the clock frequency of operation of the circuit with setup time of the flip-flops. Therefore, setup time violations can be solved by changing the clock frequency. The difference between t_{ck} and $(t_{pmax} + t_{cmax} + t_{su})$ is referred to as the **setup time margin**. The setup margin has to be zero or positive in order to have a circuit pass timing checks. Figure 1-36 (b) illustrates a situation in which setup time constraint is met, and Figure 1-36 (c) illustrates a situation when setup time constraint is violated. One can check for setup time violations by checking whether

$$t_{ck} - t_{pmax} - t_{cmax} - t_{su} \geq 0 \qquad (1\text{-}34\text{-}b)$$

When a designer creates a design, typically the flip-flops and gates are selected from a vendor's design library. Hence parameters such as t_{pmax} and t_{su} are generally fixed for a designer. Of course the designer can check whether a different design library with more desirable t_{pmax} and t_{su} is available for use, but in general, the strategy during timing analysis is to adjust the clock frequency of the circuit or the overall combinational delay of the logic. Often, the clock frequency specification comes from the customer's requirements or the architecture teams; therefore, the designers often have to "meet" timing by ensuring correct combinational delays.

Rule No. 2 Hold-time rule for flip-flop to flip-flop path: Minimum circuit delays should be long enough to satisfy flip-flop hold time.

For proper synchronous operation, the data launched by flip-flop 1 on edge E_1 of clock CK_1 should not be captured by flip-flop 2 on edge E_1 of clock CK_1. This can be understood by thinking about Rule No. 1. According to Rule No. 1, in Figure 1-37 at edge E_2, FF_2 should capture the data launched by FF_1 on the previous edge (i.e., edge E_1). For this to happen successfully, the old data should remain stable at edge E_2 until FF_2's hold time elapses. When FF_2 is capturing this old data at edge E_2, FF_1 has started to launch new data on edge E_2, which should be captured by FF_2 only at edge E_3. A hold-time violation could occur if the data launched by FF_1 at E_2 is fed through the combinational circuit and causes D_2 to change too soon after the clock edge E_2. The new data being launched by FF_1 takes at least t_{pmin} time to pass through FF_1 and at least t_{cmin} to pass through the combinational circuitry. Hence, the hold time is satisfied if

$$t_{pmin} + t_{cmin} \geq t_h \qquad (1\text{-}35)$$

Figure 1-37 illustrates a situation where hold-time is satisfied. When checking for hold-time violations, the worst case occurs when the timing parameters have their minimum values. Since $t_{pmin} > t_h$ for normal flip-flops, a hold-time violation due to Q changing does not usually occur.

FIGURE 1-37: Timing Diagrams Illustrating Hold Time in Flip-Flop Path

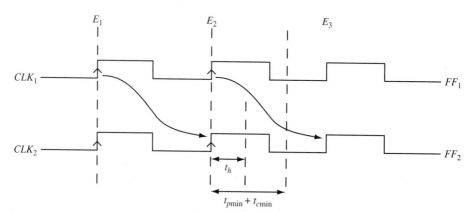

One should note that equation (1-35) does not have the clock frequency in it. Therefore, **if a circuit has a hold-time violation, it cannot be corrected by changing the clock frequency of the circuit.** To correct a hold-time violation, the circuit must be redesigned. In general, to avoid hold-time violations, one needs more combinational delays. Note that this is the opposite of what is desired to meet setup time constraints.

Designing shift registers and counters by chaining together flip-flops is very easy from a functional perspective, however, it is very difficult to meet hold-time constraints, because combinational circuit delay is zero. One way to correct such designs is by inserting buffers between flip-flops as in Figure 1-38.

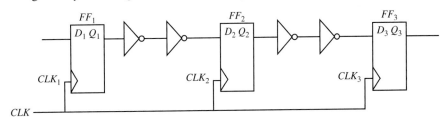

FIGURE 1-38: Shift Register with Buffers for Meeting Hold-Time Constraints

Timing Rules for Input to Flip-Flop Paths

Now let us consider a timing path from primary input to flip-flop as in Figure 1-39. The changes in primary input X should happen such that the value propagates to the flip-flop input satisfying both setup and hold-time constraints. In other words, flip-flop setup time and hold time dictate when primary inputs are allowed to change.

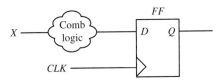

FIGURE 1-39: Input to Flip-Flop Path Timing

Rule No. 3 Setup time rule for input to flip-flop path: External input changes to the circuit should satisfy flip-flop setup time.

A setup time violation could occur if the X input to the circuit changes too close to the active edge of the clock. When the X input to a sequential circuit changes, we must make sure that the input change propagates to the flip-flop inputs such that the setup time is satisfied before the active edge of the clock. If X changes at time t_x before the active edge of the clock (see Figure 1-40), then it could take up to the maximum propagation delay of the combinational circuit before the change in X propagates to the flip-flop input. There should still be a margin of t_{su} left before the edge of the clock. Hence, the setup time is satisfied if

$$t_x \geq t_{cxmax} + t_{su} \tag{1-36}$$

where t_{cxmax} is the maximum propagation delay from X to the flip-flop input.

Rule No. 4 Hold-time rule for input to flip-flop path: External input changes to the circuit should satisfy flip-flop hold times.

In order to satisfy the hold time, we must make sure that X does not change too soon after the clock. If a change in X propagates to the flip-flop input in zero time, X should not change for a duration of t_h after the clock edge. Fortunately, it takes some positive propagation delay for the change in X to reach the flip-flop. If t_{cxmin} is the minimum propagation delay from X to the flip-flop input, changes in X will not reach the flip-flop input until at least a time of t_{cxmin} has elapsed after the clock edge. So, if X changes at time t_y after the active edge of the clock, then the hold time is satisfied if

$$t_y \geq t_h - t_{cxmin} \tag{1-37}$$

If t_y is negative, X can change before the active clock edge and still satisfy the hold time.

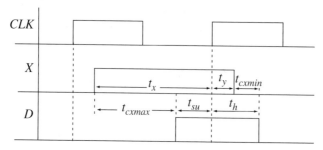

FIGURE 1-40: Setup and Hold Timing for Changes in X

Given a circuit, one can determine the safe frequency of operation and safe regions for input changes using the foregoing principles.

Consider a simple circuit of the form of Figure 1-41(a). The output of a D flip-flop is fed back to its input through an inverter. From a timing perspective, this circuit is equivalent to the circuit in Figure 1-36(a). Assume a clock as indicated by the waveform CLK in Figure 1-41(b). If the current output of the flip-flop is 1, a value of 0 will appear at the flip-flop's D input after the propagation delay of the inverter. Assuming that the next active edge of the clock arrives after the setup time has elapsed, the output of the flip-flop will change to 0. This process will continue yielding the output Q of the flip-flop to be a waveform with twice the period of the clock. Essentially the circuit behaves as a frequency divider.

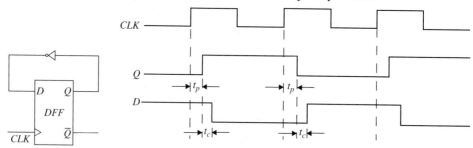

FIGURE 1-41: Simple Frequency Divider

(a) A frequency divider   (b) Frequency divider timing diagram

If we increase the frequency of the clock slightly, the circuit will still work, yielding half of the increased frequency at the output. However, if we increase the frequency to be very high, the output of the inverter may not have enough time to stabilize and meet the setup time requirements. Similarly, if the inverter was very fast and fed the inverted output to the D input extremely quickly, timing problems will occur, because the hold time of the flip-flop may not be met. So one can easily see a variety of ways in which timing problems could arise from propagation delays and setup- and hold-time requirements.

Timing Rules Nos. 1 and 2 can be applied to this circuit, and it can be seen that the maximum clock frequency of this circuit for proper operation can be derived from equation (1-34). If the minimum clock period is denoted by t_{ckmin},

$$t_{ckmin} = t_{pmax} + t_{cmax} + t_{su}$$

Hence maximum clock frequency f_{max} is given by:

$$f_{max} + 1/(t_{pmax} + t_{cmax} + t_{su}) \qquad (1\text{-}38)$$

If the minimum and maximum delays of the inverter are 1 ns and 3 ns, and if t_{pmin} and t_{pmax} are 5 ns and 8 ns, the maximum frequency at which it can be clocked can be derived using equation 1-38. Assume that the setup and hold times of the flip-flop are 4 ns and 2 ns. For proper operation, $t_{ck} \geq t_{pmax} + t_{cmax} + t_{su}$. In this example, t_{pmax} for the flip-flops is 8 ns, t_{cmax} is 3 ns, and t_{su} is 4 ns. Hence,

$$t_{ck} \geq 8 + 3 + 4 = 15 \text{ ns}$$

The maximum clock frequency is then $1/t_{ck} = 66.67$ MHz. One should also make sure that the hold-time requirement is satisfied. Hold-time requirement means that the D input should not change before 2 ns after the clock edge. This will be satisfied if $t_{pmin} + t_{cmin} \geq 2$ ns. In this circuit, t_{pmin} is 5 ns and t_{cmin} is 1 ns. Thus the Q output is guaranteed not to change until 5 ns after the clock edge and at least 1 ns more should elapse before the change can propagate through the inverter. Hence the D input will not change until 6 ns after the clock edge, which automatically satisfies the hold-time requirements. Since there are no external inputs, these are the only timing constraints that we need to satisfy.

Now consider a circuit as shown in Figure 1-42(a). Assume that the delay of the combinational circuit is in the range 2 to 4 ns, the flip-flop propagation delays are in the range 5 to 10 ns, the setup time is 8 ns, and hold time is 3 ns. In order to satisfy the setup time, the clock period has to be greater than $t_{pmax} + t_{cmax} + t_{su}$. So

$$t_{ck} \geq 10 + 4 + 8 = 22 \text{ ns}$$

FIGURE 1-42: Safe Regions for Input Changes

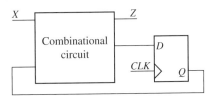

(a) A sequential circuit

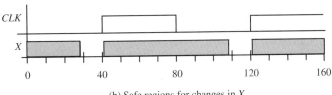

(b) Safe regions for changes in X

The hold-time requirement is satisfied if the output does not change until 3 ns after the clock. Here, the output is not expected to change until $t_{pmin} + t_{cmin}$. Since t_{pmin} is 5 ns and t_{cmin} is 2 ns, the output is not expected to change until 7 ns, which automatically satisfies the hold-time requirement. This circuit has external inputs that allow us to identify safe regions where the input X can change using requirements (iii) and (iv) in the foregoing list. The X input should be stable for a duration of $t_{cxmax} + t_{su}$ (i.e., 4 ns + 8 ns) before the clock edge. Similarly, it should be stable for a duration of $t_h - t_{cxmin}$ (i.e., 3 ns − 0.2 ns) after the clock edge. Thus, the X input should not change 12 ns before the clock edge and 1 ns after the clock edge. Although the hold time is 3 ns, we see that the input X can change 1 ns after the clock edge, because it takes at least another 2 ns (minimum delay of combinational circuit) before the input change can propagate to the D input of the flip-flop. The shaded regions in the waveform for X indicate safe regions where the input signal X may change without causing erroneous operation in the circuit.

In a typical sequential circuit, there are often millions of timing paths that need to be considered in deriving the maximum clock frequency. The maximum frequency must be determined by locating the longest path among all the timing paths in the circuit.

Example

Consider the circuit in Figure 1-43 with the following minimum/maximum delays:

CLK-to-Q for flip-flop A: 7 ns/9 ns

CLK-to-Q for flip-flop B: 8 ns/10 ns

CLK-to-Q for flip-flop C: 9 ns/11 ns

Combinational logic: 3 ns/4 ns

Setup time for flip-flops: 2 ns

Hold time for flip-flops: 1 ns

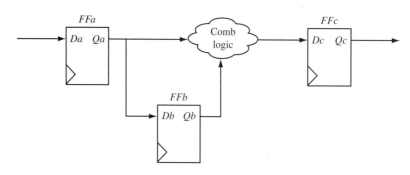

FIGURE 1-43: Circuit with Three Flip-Flops

Compute the delays for all timing paths in this circuit and determine the maximum clock frequency allowed in this circuit.

Answer: Remember that a timing path starts at either a primary input or at the input of a flip-flop. A path terminates at the input of a flip-flop or at a primary output.

Delay for path from flip-flop A to B = $t_{clk\text{-}to\text{-}Q(A)} + t_{su}(B)$ = 9 ns + 2 ns = 11 ns

Delay for path from flip-flop A to C = $t_{clk\text{-}to\text{-}Q(A)} + t_{combo} + t_{su}(C)$ = 9 ns + 4 ns + 2 ns = 15 ns

Delay for path from flip-flop B to C = $t_{clk\text{-}to\text{-}Q(B)} + t_{combo} + t_{su}(C)$ = 10 ns + 4 ns + 2 ns = 16 ns

Delay for path from input to flip-flop A = $t_{su}(A)$ = 2 ns = 2 ns

Delay for path from flip-flop C to output = $t_{clk\text{-}to\text{-}Q(C)}$ = 11 ns

Since the delay for path from B to C is the largest of the path delays, the maximum clock frequency is determined by this delay of 16 ns. The frequency is $1/t_{min} = 1/16\,\text{ns} = 62.5\,\text{MHz}$.

Example

Consider the circuit in Figure 1-35 with the following minimum/maximum delays:

CLK-to-Q for flip-flop 1: 5 ns/8 ns

CLK-to-Q for flip-flop 2: 7 ns/9 ns

XOR Gate: 4 ns/6 ns

AND Gate: 1 ns/3 ns

Setup time for flip-flops: 5 ns

Hold time for flip-flops: 2 ns

(a) What is the minimum clock period that this circuit can be safely clocked at?

Answer: Since XOR gate delay is higher than the AND gate delay, and the second flip-flop's delay is greater than that of the first flip-flop, the path from the second flip-flop to input of the second flip-flop via the XOR is the longest path. This path determines the maximum clock frequency. The maximum frequency is dictated by

$$f_{max} = 1/(t_{flip\text{-}flop\text{-}max} + t_{XORmax} + t_{su})$$

$$= 1/(9 + 6 + 5) = 1/20\,\text{ns} = 50\,\text{MHz}$$

(b) What is the earliest time after the rising clock edge that input A can safely change?

Answer: The earliest time after the rising clock edge that A can safely change can be obtained from equation (1-37)

$$t_y = t_h - t_{ANDmin} = 2\,\text{ns} - 1\,\text{ns} = 1\,\text{ns}$$

(c) What is the latest time before the rising clock edge that input A can safely change?

Answer: The latest time before the rising clock edge that A can safely change can be obtained from equation (1-36)

$$t_x = t_{ANDmax} + t_{su} = 3\,\text{ns} + 5\,\text{ns} = 8\,\text{ns}$$

1.10.3 Glitches In Sequential Circuits

Sequential circuits often have external inputs that are asynchronous. Temporary false values called glitches can appear at the outputs and next states. For example, if the state table of Figure 1-23(b) is implemented in the form of Figure 1-17, the timing waveforms are as shown in Figure 1-44. Propagation delays in the flip-flop have been neglected; hence state changes are shown to coincide with clock edges. In this example, the input sequence is 00101001, and X is assumed to change in the middle of the clock pulse. At any given time, the next state and Z output can be read from the next state table. For example, at time t_a, State = S_5 and $X = 0$, so Next State = S_0 and $Z = 0$. At time t_b following the rising edge of the clock,

State = S_0 and X is still 0, so Next State = S_1 and Z = 1. Then X changes to 1, and at time t_c Next State = S_2 and Z = 0. Note that there is a *glitch* (sometimes called a false output) at t_b. The Z output momentarily has an incorrect value at t_b, because the change in X is not exactly synchronized with the active edge of the clock. The correct output sequence, as indicated on the waveform, is 1 1 1 0 0 0 1 1. Several glitches appear between the correct outputs; however, these are of no consequence if Z is read at the right time. The glitch in the next state at t_b (S_1) also does not cause a problem, because the next state has the correct value at the active edge of the clock.

The timing waveforms derived from the circuit of Figure 1-26 are shown in Figure 1-45. They are similar to the general timing waveforms given in Figure 1-44 except that State has been replaced with the states of the three flip-flops, and a propagation delay of 10 ns has been assumed for each gate and flip-flop.

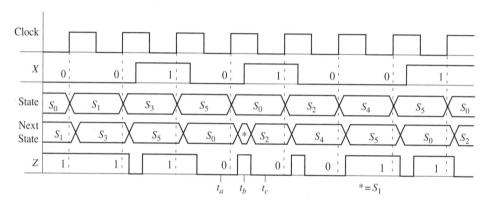

FIGURE 1-44: Timing Diagram for Code Converter

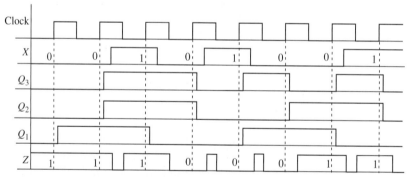

FIGURE 1-45: Timing Diagram for Figure 1-26

1.10.4 Synchronous Design

One of the most commonly used digital design techniques is *synchronous design*. In this type of design, a clock is used to synchronize the operation of all flip-flops, registers, and counters in the system. Synchronous circuits are more reliable than asynchronous circuits. In synchronous circuits, events are expected to occur immediately following the active edge of the clock. Outputs from one part have a full clock cycle to propagate to the next part of the circuit. Synchronous design philosophy makes design and debugging easier as compared with asynchronous techniques. But synchronous designs consume more power than asynchronous designs because of the power consumed in the clock distribution network. Although asynchronous designs can reduce power consumption, it is very difficult to get timing issues under control; hence, despite their high power consumption, designers favor synchronous designs.

Figure 1-46 illustrates a synchronous digital system. Assume that the system is built from several modules or devices. The devices could be flip-flops, registers, counters, adders, multipliers, and so forth. All of the sequential devices are synchronized with respect to the same clock in a synchronous system. A traditional way to view a digital system is to consider it as a control section plus a data section. The various devices shown in Figure 1-46 are part of the data section. The control section is a sequential machine that generates control signals to control the operation of the data section. For example, if the data section contains a shift register, the control section may generate signals that determine when the register is to be loaded (*Ld*) and when it is to be shifted (*Sh*). A common clock synchronizes the operation of the control and data sections. The data section may generate status signals (not shown in this figure) that affect the control sequence. For example, if a data operation produces an arithmetic overflow, then the data section might generate a condition signal *V* to indicate an overflow. The control section is also called *controller* and the data section is often called *architecture* or *data path*.

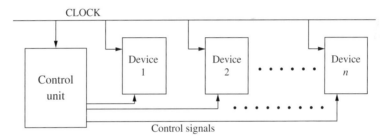

FIGURE 1-46: A Synchronous Digital System

In a synchronous digital system, one desires to see all changes happen immediately at the active edge of the clock, but that might not happen in a practical circuit. Modern integrated circuits (ICs) are fabricated at feature sizes such as or smaller than 0.1 microns. Modern microprocessors are clocked at several gigahertz. In these chips, wire delays are significant as compared with the clock period. Even if two flip-flops are connected to the same clock, the clock edge might arrive at the two flip-flops at different times due to unequal wire delays. If unequal amounts of combinational circuitry (e.g., buffers or inverters) are used in the clock path to different devices, that also could result in unequal delays, making the clock reach different devices at slightly different times. This problem is called **clock skew**. **Clock skew** refers to the absolute time difference in clock signal arrival between two points in the clock network. Clock skew is often caused by delays in the interconnect within the clock distribution network. It can also be caused by the combinational logic used to selectively gate the clock of certain devices.

Timing Rules for Circuits with Skew

When clock skew is present in a circuit, the timing rules Nos. 1 and 2 get appropriately modified. A positive skew means the capturing flip-flop gets the clock delayed with reference to the launching flip-flop. For a circuit with a positive skew as shown in Figure 1-47(a), the timing rules are as follows:

Rule No. 5: $t_{ck} \geq t_{pmax} + t_{cmax} - t_{skew} + t_{su}$ (1-39)

Rule No. 6: $t_{pmin} + t_{cmin} \geq t_h + t_{skew}$ (1-40)

Positive skew is good for setup time, but it is bad for hold time.

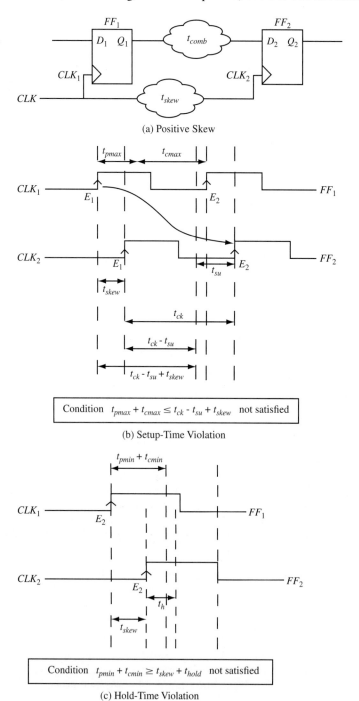

FIGURE 1-47: Illustration of Skew and Timing Violations

Negative skew means that the launching flip-flop gets the clock delayed with reference to the capturing flip-flop. Negative skew is illustrated in Figure 1-48.

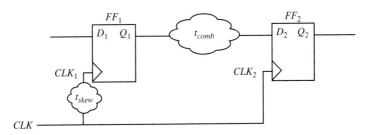

FIGURE 1-48: Negative Skew - CLK1 Is Delayed with Respect to CLK2

For a circuit that has a negative skew, the timing rules are given by the following equations:

$$t_{ck} \geq t_{pmax} + t_{cmax} + t_{skew} + t_{su} \qquad (1\text{-}41)$$

$$t_{pmin} + t_{cmin} \geq t_{h} - t_{skew} \qquad (1\text{-}42)$$

Negative skew is good for hold time, but it is bad for setup time.

Example

Consider the circuit shown in Figure 1-47(a) with the following delays:

CLK-to-Q for Flip-flops: 7 ns/9 ns
Combinational Delay: 4 ns/6 ns
Setup Time for Flip-Flops: 5 ns
Hold Time for Flip-Flops: 2 ns

(a) If skew for the second flip-flop is 3 ns, what is the maximum clock frequency? Compare it with the clock frequency if no skew is present.

Answer: This is a case of positive skew.

$$t_{ck} = t_{pmax} + t_{cmax} - t_{skew} + t_{su}$$
$$= 9\,\text{ns} + 6\,\text{ns} - 3\,\text{ns} + 5\,\text{ns}$$
$$= 17\,\text{ns}$$

The maximum clock frequency when skew is present is 1/17 ns (i.e., 58.82 MHz), whereas without skew the circuit could handle only a maximum frequency of 1/20 ns (i.e., 50 MHz).

(b) What is the biggest skew that the circuit in Figure 1-47(a) can take while meeting the hold-time constraint for this circuit?

Answer:
$$t_{pmin} + t_{cmin} \geq t_h + t_{skew}$$
$$7\,\text{ns} + 4\,\text{ns} \geq 2\,\text{ns} + t_{skew}$$
$$9\,\text{ns} \geq t_{skew}$$

Skew must be less than 9 ns.

(c) If skew for the first flip-flop in Figure 1-48 is 3 ns, what is the maximum clock frequency? Compare it with the clock frequency if no skew is present.

Answer:
$$t_{ck} = t_{pmax} + t_{cmax} + t_{skew} + t_{su}$$
$$= 9\,\text{ns} + 6\,\text{ns} + 3\,\text{ns} + 5\,\text{ns}$$
$$= 23\,\text{ns}$$

The maximum clock frequency when skew is present is 1/23 ns (i.e., 43.47 MHz), whereas without skew the circuit could handle a maximum frequency of 1/20 ns (i.e., 50 MHz).

(d) What is the biggest skew that the circuit in Figure 1-48 can take while meeting the hold-time constraint for this circuit?

Answer:
$$t_{pmin} + t_{cmin} \geq t_h - t_{skew}$$
$$7\,\text{ns} + 4\,\text{ns} + t_{skew} \geq 2\,\text{ns}$$

Since the first flip-flop's clock is delayed by t_{skew} and it takes an additional 11 ns to reach the second flip-flop, there is no possibility this signal change can cause a hold-time violation.

$$t_{skew} \geq -9\,\text{ns}$$

If the skew at flip-flop 1 increases, there will be no hold-time violation, but of course the maximum allowable clock frequency will reduce.

There are also problems that occur due to glitches in control signals. Consider Figure 1-49, which illustrates the operation of a digital system that uses devices that change state on the falling edge of the clock. Several flip-flops may change state in response to this falling edge. The time at which each flip-flop changes state is determined by the propagation delay for that flip-flop. The changes in flip-flop states in the control section will propagate through the combinational circuit that generates the control signals, and some of the control signals may change as a result. The exact times at which the control signals change depend on the propagation delays in the gate circuits that generate the signals as well as the flip-flop delays. Thus, after the falling edge of the clock, there is a period of uncertainty during which control signals may change. Glitches and spikes may occur in the control signals due to hazards. Furthermore, when signals are changing in one part of the circuit, noise may be induced in another part of the circuit. As indicated by the cross-hatching in Figure 1-49, there is a time interval after each falling edge of the clock in which there may be noise in a control signal (CS), and the exact time at which the control signal changes is not known.

If we want a device in the data section to change state on the falling edge of the clock only if the control signal $CS = 1$, we can AND the clock with CS, as shown in Figure 1-50(a). This technique is called clock gating. The transitions will occur in synchronization with the clock CLK except for a small delay in the AND gate. The gated CLK signal is clean because the clock is 0 during the time interval in which the switching transients occur in CS.

FIGURE 1-49: Timing Chart for System with Falling-Edge Devices

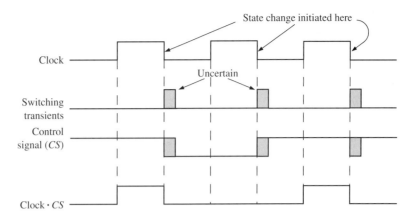

Gating the clock with the control signal, as illustrated in Figure 1-50(a) can solve some synchronization problems. However clock gating can also lead to clock skew and additional timing problems in high-speed circuits. Instead of gating the clock with the control signal, it is more desirable to use devices with clock enable (CE) pins and feed the control signal to the enable pin, as illustrated in Figure 1-50(b). Many registers, counters, and other devices used in synchronous systems have an enable input. When enable = 1, the device changes state in response to the clock, and when enable = 0, no state change occurs. Use of the enable input eliminates the need for a gate on the clock input, and associated timing problems are avoided.

FIGURE 1-50: Techniques Used to Synchronize Control Signals

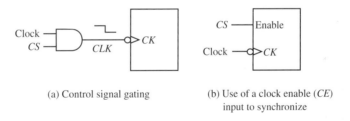

(a) Control signal gating
(b) Use of a clock enable (CE) input to synchronize

We discourage designers from gating clocks or feeding the output of combinational circuits to clock inputs. While clock skew from wire delays is unavoidable to some extent, clock skew due to combinational circuitry in the clock path can easily be avoided. Circuits as in Figure 1-51 should be avoided as much as possible to minimize timing problems.

FIGURE 1-51: Examples of Circuits to Avoid

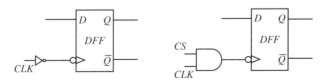

If devices do not have enables and synchronous operation cannot be obtained without clock gating, one should pay attention to gating the clocks correctly. A device with negative edge triggering can be made to function correctly by ANDing the clock signal with the control signal as shown in Figure 1-50(a). In the following paragraphs, we describe issues associated with control signal gating for positive-edge–triggered devices.

Figure 1-52 illustrates the operation of a digital system that uses devices that change state on the rising edge of the clock. In this case, the switching transients that result in noise and uncertainty will occur following the rising edge of the clock. The cross-hatching indicates the time interval in which the control signal CS may be noisy. If we want a device to change state on the rising edge of the clock when $CS = 1$, transition is expected at (a) and (c), but no change is expected at (b) since $CS = 0$ when the clock edge arrives. In order to create a gated control signal, it is tempting to AND the clock with CS, as shown in Figure 1-53(a). The resulting signal, which goes to the CK input of the device, may be noisy and timed incorrectly. In particular, the CLK_1 pulse at (a) will be short and noisy. It may be too short to trigger the device, or it may be noisy and trigger the device more than once. In general, it will be out of synchronization with the clock, because the control signal does not change until after some of the flip-flops in the control circuit have changed state. The rising edge of the pulse at (b) again will be out of synch with the clock, and it may be noisy. But even worse, the device will trigger near point (b) when it should not trigger there at all. Since $CS = 0$ at the time of the rising edge of the clock, triggering should not occur until the next rising edge, when $CS = 1$.

FIGURE 1-52: Timing Chart for System with Rising-Edge Devices

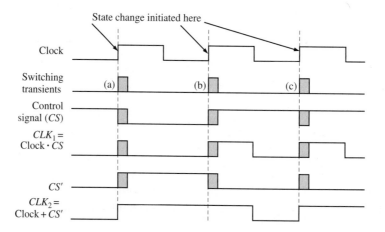

For a rising-edge device, if one changed the AND gate in Figure 1-50 to NAND gate as in Figure 1-53(b), it would be incorrect because the synchronization will happen at the wrong edge. The correct way to gate the control signal will be as in Figure 1-54, which will result in the CK input to the device having a positive edge only when the control signal is positive and the clock is going to have a positive edge. The CK input is then

$$CLK_2 = (CS \cdot clock')' = CS' + clock$$

The last waveform in Figure 1-45 illustrates this gated control signal. Even though this circuit can solve the synchronization problem, we encourage designers to refrain from gating clocks at all, if possible.

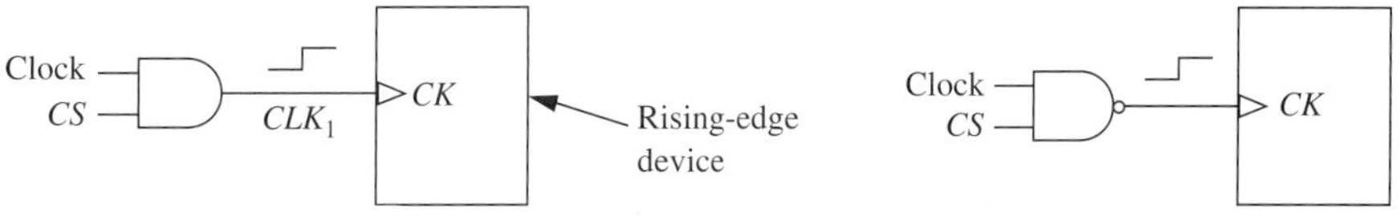

FIGURE 1-53: Incorrect Clock Gating for Rising-Edge Devices

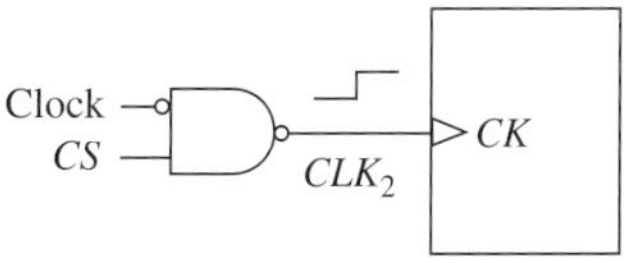

(a) With AND gate

(b) With NAND gate

FIGURE 1-54: Correct Control Signal Gating for Rising-Edge Device

In summary, synchronous design is based on the following principles:

- Method: All clock inputs to flip-flops, registers, counters, and the like are driven directly from the system clock.
- Result: All state changes occur immediately following the active edge of the clock signal.
- Advantage: All switching transients, switching noise, and the like occur between clock pulses and have no effect on system performance.

Asynchronous design is generally more problematical than synchronous design. Since there is no clock to synchronize the state changes, problems may arise when several state variables must change at the same time. A race occurs if the final state depends on the order in which the variables change. Asynchronous design requires special techniques to eliminate problems with races and hazards. On the other hand, synchronous design has several disadvantages: In high-speed circuits where the propagation delay in the wiring is significant, the clock signal must be carefully routed so that it reaches all the clock inputs at essentially the same time (i.e., to minimize clock skew). The maximum clock rate is determined by the worst-case delay of the longest path. Because the system inputs may not be synchronized with the clock, use of synchronizers may be required. Synchronous systems also consume more power than asynchronous systems. The clock distribution circuitry in synchronous chips often consumes a significant fraction of the chip's power.

1.11 Tristate Logic and Busses

Normally, if we connect the outputs of two gates or flip-flops together, the circuit will not operate properly. It can also cause damage to the circuit. Hence, when one needs to connect multiple gate outputs to the same wire or channel, one way to do that is by using tristate buffers. Tristate buffers are gates with a high-impedance state (hi-Z) in addition to high and low logic states. The high-impedance state is equivalent to an open circuit. In digital systems, transferring data back and forth between several system components is often necessary. Tristate busses can be used to facilitate data transfers between registers. When several gates are connected on to a wire, what one expects is that at any one point, one of the gates is going to actually drive the wire and the other gates should behave as if they were not connected to the wire. The high-impedance state achieves this.

Tristate buffers can be inverting or non-inverting. The control input can be active high or active low. Figure 1-55 shows four kinds of tristate buffers. B is the control input used to enable or disable the buffer output. When a buffer is enabled, the output (C) is equal to the input (A) or its complement. However, we can connect two tristate buffer outputs, provided that only one output is enabled at a time.

FIGURE 1-55: Four Kinds of Tristate Buffers

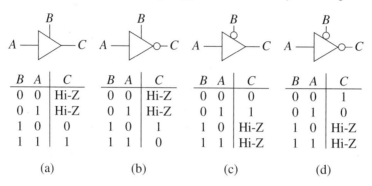

Figure 1-56 shows a system with three registers connected to a tristate bus. Each register is 8 bits wide, and the bus consists of 8 wires connected in parallel. Each tristate buffer symbol in the figure represents 8 buffers operating in parallel with a common enable input. Only one group of buffers is enabled at a time. For example, if $Enb = 1$, the register B output is driven onto the bus. The data on the bus is routed to the inputs of register A, register B, and register C. However, data is loaded into a register only when its load input is 1 and the register is clocked. Thus, if $Enb = Ldc = 1$, the data in register B will be copied into register C when the active edge of the clock occurs. If $Eni = Lda = Ldb = 1$, the input data will be loaded in registers A and B when the registers are clocked.

FIGURE 1-56: Data Transfer Using Tristate Bus

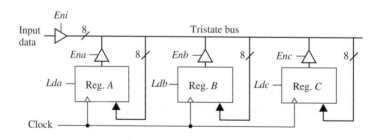

Problems

1.1 Simplify Z using a 4-variable map with map-entered variables. $ABCD$ represents the state of a control circuit. Assume that the circuit can never be in state 0100, 0001, or 1001.

$$Z = BC'DE + ACDF' + ABCD'F' + ABC'D'G + B'CD + ABC'D'H'$$

Problems

1.2 For the following functions, find the minimum sum of products using 4-variable maps with map-entered variables. In **(a)** and **(b)**, m_i represents a minterm of variables $A, B, C,$ and D.

(a) $F(A, B, C, D, E) = \Sigma m(0, 4, 6, 13, 14) + \Sigma d(2, 9) + E(m_1 + m_{12})$
(b) $Z(A, B, C, D, E, F, G) = \Sigma m(2, 5, 6, 9) + \Sigma d(1, 3, 4, 13, 14)$
$\qquad + E(m_{11} + m_{12}) + F(m_{10}) + G(m_0)$
(c) $H = A'B'CDF' + A'CD + A'B'CD'E + BCDF'$
(d) $G = C'E'F + DEF + AD'E'F' + BC'E'F + AD'EF'$

Hint: Which variables should be used for the map sides and which variables should be entered into the map?

1.3 Identify the static 1-hazards in the following circuit. State the condition under which each hazard can occur. Draw a timing diagram (similar to Figure 1-10(b)) that shows the sequence of events when a hazard occurs.

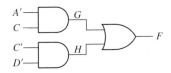

1.4 A full subtracter computes the difference of three inputs X, Y, and B_{in}, where $Diff = X - Y - B_{in}$. When $X < (Y + B_{in})$, the borrow output B_{out} is set. Fill in the truth table for the subtracter and derive the sum-of-products and product-of-sums equations for $Diff$ and B_{out}.

1.5 Write out the truth table for the following equation:

$$F = (A \oplus B) \cdot C + A' \cdot (B' \oplus C)$$

1.6 (a) Find all the static hazards in the following circuit. State the condition under which each hazard can occur.
(b) Redesign the circuit so that it is free of static hazards. Use gates with at most three inputs.

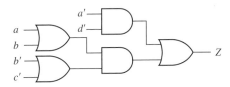

1.7 (a) Show how you can construct a T flip-flop using a J-K flip-flop.
(b) Show how you can construct a J-K flip-flop using a D flip-flop and gates.
(c) Show how you can construct a D flip-flop using a J-K flip-flop and gates.

1.8 Construct a clocked D flip-flop, triggered on the rising edge of CLK, using two transparent D latches and any necessary gates. Complete the following timing diagram, where Q_1 and Q_2 are latch outputs. Verify that the flip-flop output changes to D after the rising edge of the clock.

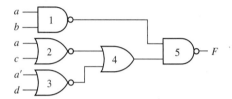

1.9 **(a)** Find all the static hazards in the following circuit. For each hazard, specify the values of the input variables and which variable is changing when the hazard occurs. For one of the hazards, specify the order in which the gate outputs must change.

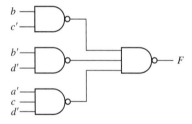

(b) Design a NAND-gate circuit that is free of static hazards to realize the same function.

1.10 Find all of the 1-hazards in the given circuit. Indicate which changes are necessary to eliminate the hazards.

1.11 Derive the state transition table and flip-flop input equations for a modulo-6 counter that counts 000 through 101 and then repeats. Use J-K flip-flops.

1.12 Derive the state transition table and D flip-flop input equations for a counter that counts from 1 to 6 (and back to 1 and continues).

1.13 A D flip-flop has a propagation delay from clock to Q of 15 ns. The setup time of the flip-flop is 10 ns, and the hold time is 2 ns. A clock with a period of 50 ns (low until 25 ns, high from 25 to 50 ns, and so on) is fed to the clock input of the flip-flop. The flip-flop is positive edge triggered. D goes up at 20, down at 40, up at 60, down at 80, and so on. Draw timing diagrams illustrating the clock, D, and Q until 100 ns. If outputs cannot be determined (because of not satisfying setup and hold times), indicate it by placing XX in that region.

1.14 A D flip-flop has a setup time of 5 ns, a hold time of 3 ns, and a propagation delay from the rising edge of the clock to the change in flip-flop output in the range of 6 to 12 ns. An OR gate delay is in the range of 1 to 4 ns.

(a) What is the minimum clock period for proper operation of the following circuit?

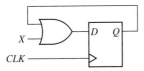

(b) What is the earliest time after the rising clock edge at which X is allowed to change?

1.15 In the following circuit, the XOR gate has a delay in the range of 2 to 16 ns. The D flip-flop has a propagation delay from clock to Q in the range 12 to 24 ns. The setup time is 8 ns, and the hold time is 4 ns.

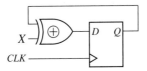

(a) What is the minimum clock period for proper operation of the circuit?
(b) What are the earliest and latest times after the rising clock edge at which X is allowed to change and still have proper synchronous operation? (Assume minimum clock period from (a).)

1.16 Consider the following circuit where the combinational circuit is represented by COMB and clock skew is represented by t_{skew}.

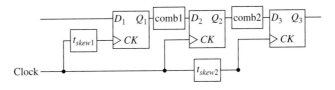

Given the following parameters:

FF setup time = 20 ns
FF hold time = 10 ns
FF propagation delay = 5 to 10 ns
Tcomb 1 = 5 ns to 7 ns
Tcomb 2 = 6 ns to 11 ns

(a) What is the minimum clock period with $t_{skew1} = t_{skew2} = 0$?
(b) Now set Tcomb1 = 1 to 4 ns. Is there a setup time violation for the middle flip-flop? If no, what is the setup time margin?
(c) Now set Tcomb1 = 1 to 4 ns. Is there a hold-time violation for the middle flip-flop? If no, what is the hold-time margin?

(d) What are the minimum values of t_{skew1} and t_{skew2} that will fix the violations?
(e) What is the minimum clock period after violations have been fixed?

1.17 Consider the following circuit where the combinational circuit is represented by COMB and clock skew is represented by tskew.
Given the following parameters:

FF setup time = 10 ns
FF hold time = 2 ns
FF propagation delay = 12 to 20 ns
Tcomb 1 = 5 ns to 7 ns
Tcomb 2 = 6 ns to 11 ns

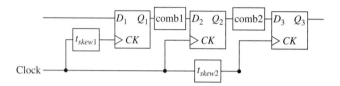

(a) What is the minimum clock period with $t_{skew1} = 0$; $t_{skew2} = 3$?
(b) **Now set Tcomb1 = 1 to 4 ns.** Is there a setup time violation for the middle flip-flop? If no, what is the setup time margin?
(c) **Now set Tcomb1 = 1 to 4 ns.** Is there a hold-time violation for the middle flip-flop? If no, what is the hold-time margin?
(d) What are the minimum values of t_{skew1} and t_{skew2} that will fix the violations?
(e) What is the minimum clock period after violations have been fixed?

1.18 A simple binary counter has only a clock input (Ck_1). The counter increments on the rising edge of Ck_1.

(a) Show the proper connections for a signal *En* and the system clock (*CLK*), so that when *En* = 1, the counter increments on the rising edge of *CLK* and when *En* = 0, the counter does not change state.
(b) Complete the following timing diagram. Explain in terms of your diagram why the switching transients that occur on *En* after the rising edge of *CLK* do not affect the proper operation of the counter.

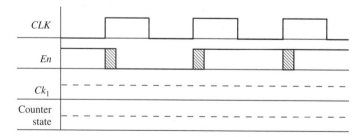

1.19 (a) Do the following two circuits have essentially the same timing?
(b) Draw the timing for Q_a and Q_b given the timing diagram.

(c) If your answer to (a) is no, show what change(s) should be made in the second circuit so that the two circuits have essentially the same timing (do not change the flip-flop).

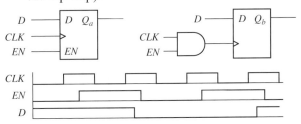

1.20 Reduce the following state table to a minimum number of states.

Present State	Next State		Output	
	$X = 0$	$X = 1$	$X = 0$	$X = 1$
A	B	G	0	1
B	A	D	1	1
C	F	G	0	1
D	H	A	0	0
E	G	C	0	0
F	C	D	1	1
G	G	E	0	0
H	G	D	0	0

1.21 A Mealy sequential circuit is implemented using the circuit shown in Problem 1.22. Assume that if the input X changes, it changes at the same time as the falling edge of the clock.

(a) Complete the timing diagram shown here. Indicate the proper times to read the output (Z). Assume that "delay" is 0 ns and that the propagation delay for the flip-flop and XOR gate has a nominal value of 10 ns. The clock period is 100 ns.

(b) Assume the following delays: XOR gate—10 to 20 ns; flip-flop propagation delay—5 to 10 ns; setup time—5 ns; and hold time—2 ns. Also assume that the "delay" is 0 ns. Determine the maximum clock rate for proper synchronous operation. Consider both the feedback path that includes the flip-flop propagation delay and the path starting when X changes.

(c) Assume a clock period of 100 ns. Also assume the same timing parameters as in **(b)**. What is the maximum value that "delay" can have and still achieve proper synchronous operation? That is, the state sequence must be the same as for no delay.

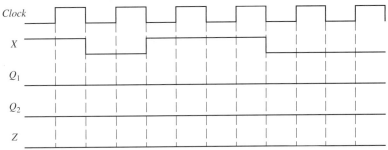

1.22 In the following circuit, the XOR gate has a delay in the range of 2 to 16 ns. The *D* flip-flop has a propagation delay from clock to *Q* in the range 12 to 24 ns. The setup time is 8 ns, and the hold time is 4 ns.

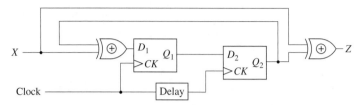

(a) Assume delay = 0 ns and compute the maximum frequency at which this circuit can be safely clocked.

(b) Assume delay = 5 ns and compute the maximum frequency at which this circuit can be safely clocked.

(c) Assume delay = –5 ns (i.e., the first flip gets the clock delayed 5 ns as compared with the second flip-flop) and compute the maximum frequency at which this circuit can be safely clocked.

(d) Assume delay = 0 ns and compute the earliest time and latest times after or before the rising clock edge that *X* is allowed to change and still have proper synchronous operation?

(e) Assume delay = 5 ns and compute the earliest time and latest times after or before the rising clock edge at which *X* is allowed to change and still have proper synchronous operation?

(f) Assume delay = –5 ns and compute the earliest time and latest times after or before the rising clock edge at which *X* is allowed to change and still have proper synchronous operation?

1.23 A Mealy sequential machine has the following state table:

PS	NS		Z	
	X = 0	X = 1	X = 0	X = 1
1	2	3	0	1
2	3	1	1	0
3	2	2	1	0

Complete the following timing diagram. Clearly mark on the diagram the times at which you should read the values of Z. All state changes occur after the rising edge of the clock. Assume the machine is initialized to state 1

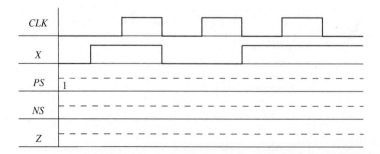

1.24 Referring to Figure 1-56, specify the values of *Eni*, *Ena*, *Enb*, *Enc*, *Lda*, *Ldb*, and *Ldc* so that the data stored in Reg. C will be copied into Reg. A and Reg. B when the circuit is clocked.

1.25 A sequential circuit consists of a PLA and a D flip-flop, as shown in the following diagram.

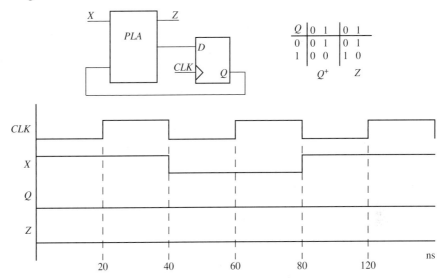

(a) Complete the timing diagram assuming that the propagation delay for the PLA is in the range 5 to 10 ns and the propagation delay from clock to output of the D flip-flop is 5 to 10 ns. Use cross-hatching on your timing diagram to indicate the intervals in which Q and Z can change, taking the range of propagation delays into account.

(b) Assuming that X always changes at the same time as the falling edge of the clock, what is the maximum setup and hold time specification that the flip-flop can have and still maintain proper operation of the circuit?

1.26 A D flip-flop has a propagation delay from clock to Q of 7 ns. The setup time of the flip-flop is 10 ns, and the hold time is 5 ns. A clock with a period of 50 ns (low until 25 ns, high from 25 to 50 ns, and so on) is fed to the clock input of the flip-flop. Assume a 2-level AND-OR circuitry between the external input signals and the flip-flop inputs. Assume gate delays are between 2 and 4 ns. The flip-flop is positive edge triggered.

(a) Assume the D input equals 0 from $t = 0$ until $t = 10$ ns, 1 from 10 until 35, 0 from 35 to 70, and 1 thereafter. Draw timing diagrams illustrating the clock, D, and Q until 100 ns. If outputs cannot be determined (because of not satisfying setup and hold times), indicate it by *XX* during the region.

(b) The D input of the flip-flop should not change between ___ ns before the clock edge and ___ ns after the clock edge.

(c) External inputs should not change between ___ ns before the clock edge and ___ ns after the clock edge.

1.27 Two flip-flops are connected as shown in the following diagram. The delay represents wiring delay between the two clock inputs, which results in clock skew. This can cause possible loss of synchronization. The flip-flop propagation delay from clock to Q is $10\,\text{ns} < t_p < 15\,\text{ns}$, and the setup and hold times for D_1 are always satisfied.

(a) What is the maximum value that the delay can have and still achieve proper synchronous operation? Draw a timing diagram to justify your answer.

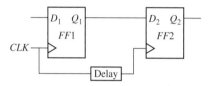

(b) Assuming that the delay is <3 ns, what is the minimum allowable clock period?

1.28 A Moore sequential circuit has one input and one output. The output goes to 1 when the input sequence 111 has occurred, and the output goes to 0 if the input sequence 000 occurs. At all other times, the output holds its value.

For example,
Derive a Moore state graph and table for the circuit.

$X = 0\ 1\ 0\ 1\ 1\ 1\ 0\ 1\ 0\ 0\ 0\ 1\ 1\ 1\ 0\ 0\ 1\ 0\ 0\ 0$
$Z = 0\ 0\ 0\ 0\ 0\ 1\ 1\ 1\ 1\ 1\ 0\ 0\ 0\ 1\ 1\ 1\ 1\ 1\ 1\ 0$

1.29 A sequential circuit has one input (X) and two outputs (D and B). X represents a 4-bit binary number N, which is input least significant bit first. D represents a 4-bit binary number equal to $N - 2$, which is output least significant bit first. At the time the fourth input occurs, $B = 1$ if $N - 2$ is negative; otherwise, $B = 0$. The circuit always resets after the fourth bit of X is received.

(a) Derive a Mealy state graph and table with a minimum number of states (six states).
(b) Try to choose a good state assignment. Realize the circuit using J-K flip-flops and NAND gates. Repeat using NOR gates. (Work this part by hand.)
(c) Check your answer to (b) using the *LogicAid* program. Also use the program to find the NAND solution for two other state assignments.

1.30 A sequential circuit has one input (X) and two outputs (S and V). X represents a 4-bit binary number N, which is input least significant bit first. S represents a 4-bit binary number equal to $N + 2$, which is output least significant bit first. At the time the fourth input occurs, $V = 1$ if $N + 2$ is too large to be represented by 4 bits; otherwise, $V = 0$. The value of S should be the proper value, not a don't care, in both cases. The circuit always resets after the fourth bit of X has been received.

(a) Derive a Mealy state graph and table with a minimum number of states (six states).
(b) Try to choose a good state assignment. Realize the circuit using D flip-flops and NAND gates. Repeat using NOR gates. (Work this part by hand.)

(c) Check your answer to **(b)** using the *LogicAid* program. Also use the program to find the NAND solution for two other state assignments.

1.31 A sequential circuit has one input (X) and two outputs (Z_1 and Z_2). An output $Z_1 = 1$ occurs every time the input sequence 010 is completed provided that the sequence 100 has never occurred. An output $Z_2 = 1$ occurs every time the input sequence 100 is completed. Note that once a $Z_2 = 1$ output has occurred, $Z_1 = 1$ can never occur, but *not* vice versa.

(a) Derive a Mealy state graph and table with a minimum number of states (eight states).
(b) Try to choose a good state assignment. Realize the circuit using J-K flip-flops and NAND gates. Repeat using NOR gates. (Work this part by hand.)
(c) Check your answer to **(b)** using the *LogicAid* program. Also use the program to find the NAND solution for two other state assignments.

1.32 A synchronous sequential circuit has one input and one output. If the input sequence 0101 or 0110 occurs, an output of two successive 1s will occur. The first of these 1s should occur coincident with the last input of the 0101 or 0110 sequence. The circuit should reset when the second 1 output occurs. For example,

input sequence: X = 010011101010 101101 ...
output sequence: Z = 000000000011 000011 ...

(a) Derive a Mealy state graph and table with a minimum number of states (six states).
(b) Try to choose a good state assignment. Realize the circuit using J-K flip-flops and NAND gates. Repeat using NOR gates. (Work this part by hand.)
(c) Check your answer to **(b)** using the *LogicAid* program. Also use the program to find the NAND solution for two other state assignments.

Introduction to Verilog®

As integrated circuit technology has improved to allow more and more components on a chip, digital systems have continued to grow in complexity. While putting a few transistors on an integrated circuit (IC) was a miracle when it happened, technology improvements have advanced the **VLSI** (very large scale integration) field continually. The early integrated circuits belonged to **SSI** (small-scale integration), **MSI** (medium-scale integration), or **LSI** (large-scale integration) categories depending on the density of integration. SSI referred to ICs with 1 to 20 gates, MSI referred to ICs with 20 to 200 gates, and LSI referred to devices with 200 to a few thousand gates. Many popular building blocks such as adders, multiplexers, decoders, registers, and counters are available as MSI standard parts. When the term VLSI was coined, devices with 10,000 gates were called VLSI chips. The boundaries between the different categories are fuzzy today. Many modern microprocessors contain more than 100 million transistors. Compared to what was referred to as VLSI in its initial days, modern integration capability could be described as ULSI (ultra large scale integration). Despite the changes in integration ability and the fuzzy definition, the term VLSI remains popular.

As digital systems have become more complex, detailed design of the systems at the gate and flip-flop level has become very tedious and time-consuming. Two or three decades ago, digital systems were created using hand-drawn schematics, bread-boards, and wires, which were connected to the bread-board. Now, hardware design often involves no-hands-on tasks with bread-boards and wires.

In this chapter, first we present an introduction to computer-aided design. Then, an introduction to hardware description languages is presented. Basic features of Verilog® are presented, and examples are provided to illustrate how digital hardware is described, simulated, and synthesized using Verilog. Advanced features of Verilog are presented later in Chapter 8.

2.1 Computer-Aided Design

Computer-aided design (CAD) tools have advanced significantly during the past decade, and nowadays digital design is performed using a variety of software tools. Prototypes or even final designs can be created without discrete components and interconnection wires.

Figure 2-1 illustrates the steps in modern digital system design. Like any engineering design, the first step in the design flow is formulating the problem, stating the **design requirements,** and arriving at the **design specification**. The next step is to **formulate the design** at a conceptual level, either at a block diagram level or at an algorithmic level.

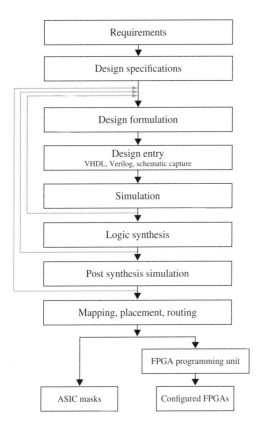

FIGURE 2-1: Design Flow in Modern Digital System Design

Design entry is the next step in the design flow. Previously, this would have been a hand-drawn schematic or blueprint. Now with CAD tools, the design conceptualized in the previous step needs to be entered into the CAD system in an appropriate manner. Designs can be entered in multiple forms. A few years ago, CAD tools used to provide a graphical method to enter designs. This was called **schematic capture**. The schematic editors typically were supplemented with a library of standard digital building blocks such as gates, flip-flops, multiplexers, decoders, counters, registers, and so forth. ORCAD (a company that produced design automation tools) provided a very popular schematic editor. Nowadays, **hardware description languages** (HDLs) are used to enter designs in textual form. Two popular HDLs are VHDL and Verilog.

A hardware description language (HDL) allows a digital system to be designed and debugged at a higher level of abstraction than schematic capture. In schematic capture, a designer inputs a schematic with gates, flip-flops, and standard MSI building blocks. However, with HDLs, the details of the gates and flip-flops do not need to be handled during early phases of design. A design can be entered in what is called a **behavioral description** of the design. In a behavioral HDL description, one specifies only the general working of the design at a flow-chart or algorithmic level without associating to any specific physical parts, components, or implementations. Another method to enter a design in VHDL and Verilog is the **structural description** entry. In structural design, specific components or specific implementations of components are associated with the design. A structural VHDL or Verilog model of a design can be considered as a textual description of a schematic diagram that you would have drawn interconnecting specific gates, flip-flops, and other modules.

Once the design has been entered, it is important to simulate it to confirm that the conceptualized design does function correctly. Initially, one should perform the **simulation** at the high-level behavioral model. This early simulation unveils problems in the initial design. If problems are discovered, the designer goes back and alters the design to meet the requirements.

Once the functionality of the design has been verified through simulation, the next step is **synthesis**. Synthesis means conversion of the higher-level abstract description of the design to actual components at the gate and flip-flop levels. Use of computer-aided design tools to do this conversion, also called synthesis, is standard practice in the industry now. The output of the synthesis tool, consisting of a list of gates and a list of interconnections, specifying how to interconnect them, is often referred to as a **netlist**. Synthesis is analogous to writing software programs in a high-level language such as C and then using a compiler to convert the programs to machine language. Just as a C compiler can generate optimized or unoptimized machine code, a synthesis tool can generate optimized or unoptimized hardware. The synthesis software generates different hardware implementations, depending on algorithms embedded in the software to perform the translation and optimization. A synthesis tool is nothing but a compiler to convert design descriptions to hardware, and it is not unusual to name synthesis packages with phrases such as "design compiler," "silicon compiler," and the like.

The next step in the design flow is **post-synthesis simulation**. The earlier simulation at a higher level of abstraction does not take into account specific implementations of the hardware components that the design is using. If post-synthesis simulation unveils problems, one should go back and modify the design to meet timing requirements. Arriving at a proper design implementation is an iterative process.

Next, a designer moves into specific realizations of the design. A design can be implemented in several different target technologies. It could be a completely custom IC, or it could be implemented in a standard part that is easily available from

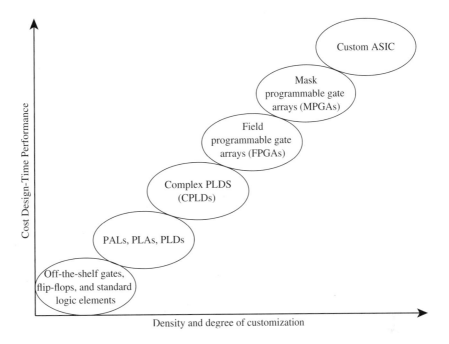

FIGURE 2-2: Spectrum of Design Technologies

a vendor. The target technologies that are commonly available now are illustrated in Figure 2-2.

At the lowest level of sophistication and density is an old-fashioned printed circuit board with off-the-shelf gates, flip-flops, and other standard logic-building blocks. Slightly higher in density are programmable logic arrays (PLAs), programmable array logic (PALs), and simple programmable logic devices (SPLDs). PLDs with higher density and gate count are called complex programmable logic devices (CPLDs). In addition, there are the popular field programmable gate arrays (FPGAs) and mask programmable gate arrays (MPGAs), or simply gate arrays. The highest level of density and performance is a fully custom application-specific integrated circuit (ASIC).

Two most common target technologies currently are FPGAs and ASICs. The initial steps in the design flow are largely the same for either realization. Towards the final stages in the design flow, different operations are performed depending on the target technology. This is indicated in Figure 2-1. The design is **mapped** into specific target technology and **placed** into specific parts in the target ASIC or FPGA. The paths taken by the connections between components are decided during the **routing**. If an ASIC is being designed, the routed design is used to generate a photomask that will be used in the integrated circuit (IC) manufacturing process. If a design is to be implemented in an FPGA, the design is translated to a format specifying what is to be done to various programmable points in the FPGA. In modern FPGAs, programming simply involves writing a sequence of 0s and 1s into the programmable cells in the FPGA, and no specific programming unit other than a personal computer (PC) is required to program an FPGA.

2.2 Hardware Description Languages

Hardware description languages (HDLs) are a popular mode of design entry for digital circuits and systems. There are two popular HDLs—VHDL and Verilog. Before the advent of HDLs, designers used graphical schematics and schematic capture tools to document and simulate digital circuits. A need was felt to create a textual method of documenting circuits and feeding them into simulators in the textual form as opposed to a graphic form. This book uses the Verilog language for illustrating principles of modern digital system design. Those interested in VHDL can find the equivalent material in Roth and John, *Digital Systems Design Using VHDL*, 2nd ed. (Cengage Learning, 2008).

Verilog is a hardware description language used to describe the behavior and/or structure of digital systems. Verilog is a general-purpose hardware description language that can be used to describe and simulate the operation of a wide variety of digital systems, ranging in complexity from a few gates to an interconnection of many complex integrated circuits. While the competing language VHDL was originally developed under funding from the Department of Defense (DoD), Verilog was developed by the industry. It was initially developed as a proprietary language by a company called Gateway Design Automation around 1984.

In 1990, Cadence acquired Gateway Design Automation and became the owner of Verilog. Cadence marketed it as a language and as a simulator, but it remained proprietary. At this time, Synopsis was promoting the concept of top-down design using Verilog. Cadence realized that it needed to make Verilog open in order to prevent the industry from shifting to VHDL and hence opened up the language. An organization called Open Verilog International (OVI) was formed, which helped to create a vendor-independent language specification for Verilog, clarifying many of the confusions around the proprietary specification. This was followed by an effort to create an IEEE standard for Verilog. The first Verilog IEEE Standard was created in 1995, which was revised in 2001 and 2005. Synopsis created synthesis tools for Verilog around 1988.

HDLs can describe a digital system at several different levels—behavioral, data flow, and structural. For example, a binary adder could be described at the behavioral level in terms of its function of adding two binary numbers without giving any implementation details. The same adder could be described at the data flow level by giving the logic equations for the adder. Finally, the adder could be described at the structural level by specifying the gates and the interconnections between the gates that comprise the adder.

HDLs lead naturally to a top-down design methodology, in which the system is first specified at a high level and tested using a simulator. After the system is debugged at this level, the design can gradually be refined, eventually leading to a structural description closely related to the actual hardware implementation. HDLs are designed to be technology independent. If a design is described in HDL and implemented in today's technology, the same HDL description could be used as a starting point for a design in some future technology.

Verilog has its syntactic roots in C whereas VHDL has its syntactic roots in ADA. Some find Verilog easier or less intimidating to learn due to its similarity with C, while many find VHDL to be excellent for supporting design and documentation of large systems. VHDL and Verilog enjoy approximately 50/50 global market share. Both languages can accomplish most requirements for digital design rather easily. But Verilog is touted by many to be slightly more supportive for synthesis and VHDL is touted to be slightly more elegant for simulation of very large systems at a higher level of abstraction. Often design companies continue to use what they are used to; hence, Verilog users continue to use Verilog and VHDL users continue to use VHDL. If one knows one of these languages, it is not difficult to transition to the other. VHDL has more conceptual elegance for the higher-level abstractions, while Verilog has enjoyed its popularity from its syntactic similarity to C.

More recently, there also have been efforts in system design languages such as System C, Handel-C, and System Verilog. System C is created as an extension to C++; hence, some who are more comfortable with general-purpose software find it less intimidating. These languages are primarily targeted at describing large digital systems at a higher level of abstraction. They are primarily used for verification and validation. When different parts of a large system are designed by different teams, one team can use a system-level behavioral description of the block being designed by the other team during the initial design process. Problems that might otherwise become obvious only during system integration may become evident in early stages, reducing the design cycle time for large systems. System-level simulation languages are used during the design of large systems. Efforts to synthesize hardware from high level-languages are steadily progressing, and tools to convert models written in languages such as C++ to hardware are emerging.

Learning a Language

There are several challenges when you learn a new language, whether it be a language for common communication (English, Spanish, French, etc.), a computer language such as C, or a special-purpose language such as Verilog. If it is not your first language, you typically have a tendency to compare it with a language you know. In the case of Verilog, if you already know another hardware description language, it is good to compare it with Verilog, but you should be careful when comparing it with languages such as C. VHDL and Verilog have a very different purpose from languages such as C, and a comparison with C is not a meaningful activity. We will be describing the language assuming it is your first HDL; however, we will assume basic knowledge of computer languages such as C and the basic compilation and execution flow.

When one learns a new language, one needs to study the alphabet of the new language, its vocabulary, grammar, syntax rules, and semantics of language descriptions. The process of learning Verilog is not much different. One needs to learn the alphabet, vocabulary or lexical elements of the language, syntax (grammar and rules), and semantics (meaning of descriptions). The lexical elements of the language include various **identifiers, reserved words**, special symbols, and literals. We have listed these in Appendix B. The syntax or grammar determines what combinations of lexical elements can be combined to make valid Verilog descriptions. These are the rules that govern the use of different Verilog constructs. Then one needs to understand the semantics or meaning of Verilog descriptions. It is here that one understands what descriptions represent combinational hardware versus sequential hardware. And just as fluency in a natural language comes by speaking, reading, and writing the language, mastery of Verilog comes by repeated use of the language to create models for various digital systems.

Since Verilog is a hardware description language, it differs from an ordinary programming language in several ways. Most importantly, Verilog has statements that execute concurrently since they must model real hardware in which the components are all in operation at the same time. It is popularly used for the purposes of describing, documenting, simulating, and automatically generating hardware. Hence, its constructs are tailored for these purposes. We will present the various methods to model different kinds of digital hardware using examples in the following sections.

Common Abbreviations	
VHDL:	VHSIC hardware description language
VHSIC:	very-high-speed integrated circuit
HDL:	hardware description language
CAD:	computer-aided design
EDA:	electronic design automation
LSI:	large-scale integration
MSI:	medium-scale integration
SSI:	small-scale integration
VLSI:	very-large-scale integration
ULSI:	ultra-large-scale integration
ASCII:	American standard code for information interchange
ISO:	International Standards Organization
ASIC:	application-specific integrated circuit
FPGA:	field-programmable gate array
PLA:	programmable logic array
PAL:	programmable array logic
PLD:	programmable logic device
CPLD:	complex programmable logic device
STA:	static timing analysis

2.3 Verilog Description of Combinational Circuits

The biggest difficulty in modeling hardware using a general-purpose computer language is representing concurrently operating hardware. Computer programs that you are normally accustomed to are sequences of instructions with a well-defined order. At any point of time during execution, the program is at a specific point in its flow, and it encounters and executes different parts of the program sequentially. In order to model combinational circuits that have several gates (all of which are always working simultaneously), one needs to be able to "simulate" the execution of several parts of the circuit at the same time.

2.3 Verilog Description of Combinational Circuits

Verilog models combinational circuits by what are called **concurrent statements or continuous assignments**. Concurrent statements (continuous assignments) are statements that are always ready to execute. These are statements that are evaluated any time and every time a signal on the right side of the statement changes.

We will start by describing a simple gate circuit in Verilog. If each gate in the circuit of Figure 2-3 has a 5 ns propagation delay, the circuit can be described by two Verilog statements as shown, where A, B, C, D, and E are signals. A signal in Verilog usually corresponds to a signal in a physical system. The symbol "&&" represents the AND gate and the symbol "||" represents the OR. The #5 indicates a delay symbol of 5 ns. The symbol "=" is the signal assignment operator, which indicates that the value computed on the right side is assigned to the signal on the left side. The **assign** statement is used to assign a value, as shown in Figure 2-3. When the statements in Figure 2-3 are simulated, the first statement will be evaluated any time A or B changes, and the second statement will be evaluated any time C or D changes. Suppose that initially $A = 1$ and $B = C = D = E = 0$. If B changes to 1 at time 0, C will change to 1 at time = 5 ns. Then E will change to 1 at time = 10 ns (assuming timescale is 1 ns).

FIGURE 2-3: A Simple Gate Circuit

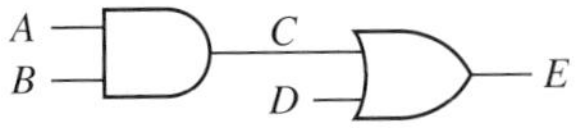

```
assign #5 C = A && B;
assign #5 E = C || D;
```

Verilog signal assignment statements like the ones in the foregoing example are examples of concurrent statements or continuous assignments. The Verilog simulator monitors the right side of each concurrent statement, and any time a signal changes, the expression on the right side is immediately reevaluated. The new value is assigned to the signal on the left side after an appropriate delay. This is exactly the way the hardware works. Any time a gate input changes, the gate output is recomputed by the hardware and the output changes after the gate delay.

Unlike a sequential program, the order of the above concurrent statements is unimportant. If we write

```
assign #5 E = C || D;
assign #5 C = A && B;
```

the simulation results would be exactly the same as before. In general, a signal assignment statement has the form

```
assign [#delay] signal_name = expression;
```

The expression is evaluated when the statement is executed, and the signal on the left side is scheduled to change after delay. The square brackets indicate that #delay is optional; they are not part of the statement. If #delay is omitted, then the signal is updated immediately. Unlike in VHDL, Verilog simulators do not display delta delays for continuous assign statements. The delta delay is an infinitesimally small delay used to maintain/indicate sequentiality between dependent events happening at the same time. Note that the time at which the statement executes and the time at which the signal is updated are not the same if delay is specified.

Even if a Verilog program has no explicit loops, concurrent statements execute repeatedly as if they were in a loop. Figure 2-4 shows an inverter with the output connected back to the input. If the output is 0, then this 0 feeds back to the input and the inverter output changes to 1 after the inverter delay, which is assumed to be 10 ns. Then the 1 feeds back to the input, and the output changes to 0 after the inverter delay. The signal *CLK* will continue to oscillate between 0 and 1 as shown in the waveform. The corresponding concurrent Verilog statement will produce the same result. If *CLK* is initialized to 0 the statement executes and *CLK* changes to 1 after 10 ns. Since *CLK* has changed, the statement executes again, and *CLK* will change back to 0 after another 10 ns. This process will continue indefinitely.

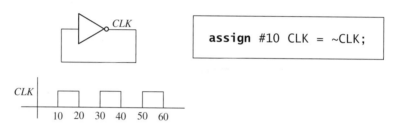

FIGURE 2-4: Inverter with Feedback

The statement in Figure 2-4 generates a clock waveform with a half period of 10 ns. On the other hand, if the concurrent statement

assign CLK = ~CLK;

is used, time will never advance to 1 ns.

In general, **Verilog is case sensitive**; that is, capital and lower-case letters are treated as different by the compiler and by the simulator. Thus, the statements

assign #10 Clk = ~Clk;

and

assign #10 CLK = ~CLK;

would result in two different clocks. Signal names and other **Verilog identifiers** may contain letters, numbers, the underscore character (_), and the dollar sign ($). An identifier must start with a letter or underscore character, and it cannot start with a number or a $ sign. The dollar sign ($) is reserved as the first character for **system tasks**. The following are valid identifiers:

adder
Mux_input
_error_code
Index_bit
vector_sz
_$five
Count
XOR

The following are invalid identifiers:

```
4bitadder
$error_code
```

Every Verilog statement must be terminated with a semicolon. Spaces, tabs, and carriage returns are treated in the same way. This means that a Verilog statement can be continued over several lines, or several statements can be placed on one line. In a line of Verilog code, anything following a double slash (//) is treated as a comment to the end of the line. Comments for more than one line start with "/*" and end with "*/". Words such as **and, or**, and **always** are reserved words (or keywords) that have a special meaning to the Verilog compiler. In this text, we will put all reserved words in boldface type. Verilog reserved words (keywords) are shown in Appendix B.

Figure 2-5 shows three gates that have the signal A as a common input and the corresponding Verilog code. The three concurrent statements execute simultaneously whenever A changes, just as the three gates start processing the signal change at the same time. However, if the gates have different delays, the gate outputs can change at different times. If the gates have delays of 2 ns, 1 ns, and 3 ns, respectively, and A changes at time 5 ns, then the gate outputs D, E, and F can change at times 7 ns, 6 ns, and 8 ns, respectively. The Verilog statements work in the same way. Even though the statements execute simultaneously at 5 ns, the signals D, E, and F are updated at times 7 ns, 6 ns, and 8 ns.

FIGURE 2-5: Three Gates with a Common Input and Different Delays

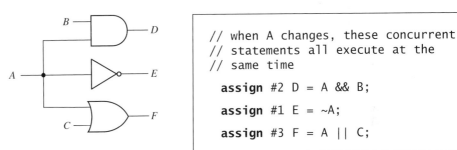

```
// when A changes, these concurrent
// statements all execute at the
// same time
assign #2 D = A && B;
assign #1 E = ~A;
assign #3 F = A || C;
```

In the foregoing examples, every signal is of type **wire** (or *net*), and it generally has a value of 0 or 1 (or 1'b0, 1'b1). In general, the net values in Verilog are represented as *<number of bits>'<base><value>*. The values on nets can be represented as binary, decimal, or hexadecimal indicated by b, d, and h respectively.

In digital design, we often need to perform the same operation on a group of signals. A one-dimensional array of bit signals is referred to as a vector. If a 4-bit vector named B has an index range 0 through 3, then the 4 elements of the **wire** or **reg** data types are designated $B[0]$, $B[1]$, $B[2]$, and $B[3]$. One can declare a multiple bit wire using a statement such as

```
wire B[3:0];
```

The statement B = 4'b1100 assigns 1 to $B[3]$, 1 to $B[2]$, 0 to $B[1]$, and 0 to $B[0]$.

Figure 2-6 shows an array of four AND gates. The inputs are represented by 4-bit vectors *A* and *B*, and the output by 4-bit vector *C*, where the && (logical AND operator) is used. Although we can write four Verilog statements to represent the four gates, it is much more efficient to write a single Verilog statement that performs the & (bitwise AND operator) operation on the vectors *A* and *B*. When applied to vectors, the & operator performs the bitwise AND operation on corresponding pairs of elements.

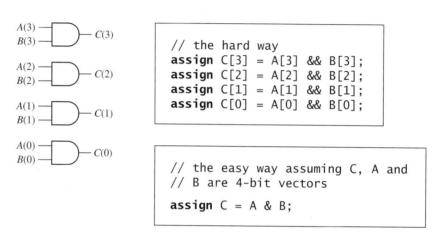

FIGURE 2-6: Array of AND Gates

```
// the hard way
assign C[3] = A[3] && B[3];
assign C[2] = A[2] && B[2];
assign C[1] = A[1] && B[1];
assign C[0] = A[0] && B[0];
```

```
// the easy way assuming C, A and
// B are 4-bit vectors
assign C = A & B;
```

2.4 Verilog Modules

The general structure of a Verilog code is a module description. A module is a basic building block that declares the input and output signals and specifies the internal operation of the module. As an example, consider Figure 2-7. The **module** declaration has the name two_gates and specifies the inputs and outputs. *A*, *B*, and *D* are input signals, and *E* is an output signal. The signal *C* is declared within the module as a **wire** since it is an internal signal. The two concurrent statements that describe the gates are placed and the module ends with **endmodule**. All the input and output signals are listed in the module statement without specifying whether they are input or output.

FIGURE 2-7: Verilog Module with Two Gates

```
module two_gates (A, B, D, E);
output E;
input A, B, D;
wire C;
    assign C = A && B; // concurrent
    assign E = C || D; // statements
endmodule
```

The **module** I/O declaration part can be considered as the black box picture of the module being designed and its external interface; that is, it represents the interconnections from this module to the external world as in Figure 2-8.

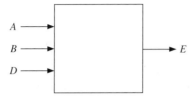

FIGURE 2-8: Black Box View of the 2-Gate Module

Just as in this simple example, when we describe a system in Verilog, we must specify input and output signals and also specify the functionalities of the module that are part of the system (see Figure 2-9). Each module declaration includes a list of interface signals that can be used to connect to other modules or to the outside world. We will use module declarations of the form:

```
module module-name (module interface list);
[list-of-interface-ports]
...
[port-declarations]
...
[functional-specification-of-module]
...
endmodule
```

The items enclosed in square brackets are optional. The `list-of-interface-ports` normally has the following form:

```
type-of-port list-of-interface-signals
{; type-of-port list-of-interface-signals};
```

The curly brackets indicate zero or more repetitions of the enclosed clause. Type-of-port indicates the direction of information; whether information is flowing into the port or out of it. Input port signals are of keyword **input**, output port signals are of keyword **output**, and bidirectional signals are of keyword **inout**. Also, list-of-ports can be combined with the module interface list.

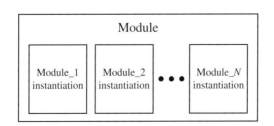

FIGURE 2-9: Verilog Program Structure

In the port-declarations section, we can declare internal signals that are used within the module. The module contains other module instances that describe the operation of the module.

Next, we will write a Verilog module for a full adder. A full adder adds two bit inputs and a carry input to generate a sum bit and a carry output bit. As shown in Figure 2-10, the port declaration specifies that X, Y, and C_{in} are input signals of type bit and that C_{out} and Sum are output signals of type bit.

FIGURE 2-10: Verilog Module for a Full Adder

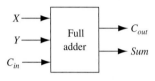

```
module FullAdder(X, Y, Cin, Cout, Sum);
output Cout, Sum;

input X, Y, Cin;

    assign #10 Sum = X ^ Y ^ Cin;
    assign #10 Cout = (X && Y) || (X && Cin) || (X && Cin);
endmodule
```

In this example, the Verilog assignment statements for Sum and C_{out} represent the logic equations for the full adder. The specified equations are

$$\text{Sum} = X \oplus Y \oplus \text{Cin}$$
$$\text{Cout} = XY \oplus Y\text{Cin} + X\text{Cin}$$

Several other architectural descriptions such as a truth table or an interconnection of gates could have been used instead. In the C_{out} equation, parentheses are required around (X && Y) since Verilog does not specify an order of precedence for the logic operators except the NOT operator.

Four-Bit Full Adder

Next, we will show how to use the FullAdder module defined previously as a **module** in a system, which consists of four full adders connected to form a 4-bit binary adder (see Figure 2-11). We first declare the 4-bit adder as another module (see Figure 2-12). Since the inputs and the sum output are four bits wide, we declare them as a 4-bit vector and they are dimensioned [3:0]. (We could have used a range [1:4] instead).

FIGURE 2-11: 4-Bit Binary Adder

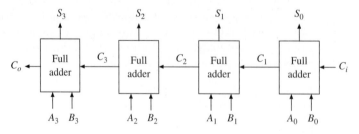

Next, we instantiate the FullAdder module within the module of Adder4 (Figure 2-12).

Following the I/O port declaration, we declare a 3-bit internal carry signal C as a data type **wire**. After that, we create several instances of the FullAdder component. (In CAD jargon, we "instantiate" four copies of the FullAdder.) Each copy of FullAdder has a port map. The port map corresponds one-to-one with the signals in the component port. Thus, *A[0]*, *B[0]*, and *Ci* correspond to the inputs *X*, *Y*, and C_{in}, respectively. *C[1]* and *S[0]* correspond to the C_{out} and *Sum* outputs of the adder for least significant bit. Unconnected ports can be omitted. In case the signals are not connected to the ports by name, the order of the signals in the port map must be the same as the order of the signals in the port of the module declaration. In this example, we use the ports in order, a method called **positional association**. The other method called **named association** is described in Chapter 8. Note that the order of the signals in named association can be in any order as long as the signals in the module are connected to the ports by name.

FIGURE 2-12: Structural Description of a 4-Bit Adder

```
module Adder4 (S, Co, A, B, Ci);
output [3:0] S;
output Co;
input [3:0] A, B;
input Ci;

wire [3:1] C;    // C is an internal signal

// instantiate four copies of the FullAdder

FullAdder FA0 (A[0], B[0], Ci,  C[1], S[0]);
FullAdder FA1 (A[1], B[1], C[1], C[2], S[1]);
FullAdder FA2 (A[2], B[2], C[2], C[3], S[2]);
FullAdder FA3 (A[3], B[3], C[3], Co,   S[3]);

endmodule
```

In preparation for simulation, we can place the modules for the FullAdder and for Adder4 together in one project and compile. Some tools may require them to be in the same file.

All of the simulation examples in this text use the **ModelSim Verilog simulator** from Mentor Graphics. Most other Verilog simulators use similar command files and can produce output in a similar format. The simulator command file is usually called the **do file**. We will use the following simulator commands to test Adder4:

```
add list A B Co C Ci S    // put these signals on the output list
force A 1111              // set the A inputs to 1111
force B 0001              // set the B inputs to 0001
force Ci 1                // set Ci to 1
run 50ns                  // run the simulation for 50ns
force Ci 0
force A 0101
force B 1110
run 50ns
```

We have chosen to run the simulation for 50 ns for each input set, since this is more than enough time for the carry to propagate through all of the full adders. The simulation results for the above command list are:

ns	delta	a	b	co	c	ci	s
0	+0	1111	0001	x	xxx	1	xxxx
10	+0	1111	0001	x	xx1	1	xxx1
20	+0	1111	0001	x	x11	1	xx01
30	+0	1111	0001	x	111	1	x001
40	+0	1111	0001	1	111	1	0001
50	+0	0101	1110	1	111	0	0001
60	+0	0101	1110	1	110	0	0101
70	+0	0101	1110	1	100	0	0111
80	+0	0101	1110	1	100	0	0011

The listing shows how the carry propagates one position every 10 ns. The x values can be avoided by initializing the S, C, and Co values to 0. Under that condition, the simulation progresses as follows

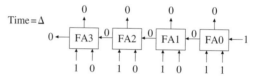

The sum and carry are computed by each FA and appear at the FA outputs 10 ns later:

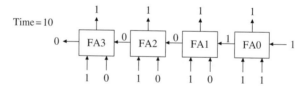

Since the inputs to FA1 have changed, the outputs change 10 ns later:

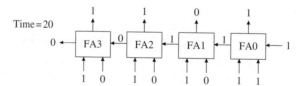

The final simulation results are:

$1111 + 0001 + 1 = 0001$ with a carry of 1 (at time = 40 ns) and
$0101 + 1110 + 0 = 0011$ with a carry of 1 (at time = 80 ns).

The simulation stops at 80 ns since no further changes occur after that time.

Use of "Inout" Mode

Let us consider the example in Figure 2-13. Assume that all variables are 0 @ 0 ns, but A changes to 1 @ 10 ns.

FIGURE 2-13: Verilog Code Illustrating Use of Output as an Input Signal

```
module gates (A, B, C, D, E);
input   A, B, C;
output  D, E;
assign #5 D = A || B;   // statement 1
assign #5 E = C || D;   // statement 2 uses D as an input
endmodule
```

The code in Figure 2-13 will actually compile, simulate, or synthesize in most tools even though *D* is declared only as an output. Statement 2 uses *D* as an input. In VHDL, *D* should be strictly in either **in** or **inout** mode, but Verilog is not that strict in compilation. The **output** mode can also be used as an input in a statement inside the same module, but **inout** has to be used as in Figure 2-14 if *D* has to be used as input and output by other modules.

FIGURE 2-14: Verilog Code Illustrating Use of Mode Inout

```
module gates (A, B, C, D, E);
input   A, B, C;
output  E;
inout   D;
assign #5 D = A || B;   // statement 1
assign #5 E = C || D;   // statement 2
endmodule
```

All signals remain at 0 until time 10 ns. The change in *A* at 10 ns results in statement 1 reevaluating. The value of *D* becomes 1 at time equal to 15 ns. The change in *D* at time 15 ns results in statement 2 reevaluating. Signal *E* changes to 1 at time 20 ns. The description represents TWO gates, each with a delay of 5 ns.

2.5 Verilog Assignments

There are two types of assignment in the Verilog: continuous assignments and procedural assignments. Continuous assignments are used to assign values for combinational logic circuits. The **assign** keyword can be used after the net is separately declared, which is referred to as explicit continuous assignments. Implicit continuous assignments assign the value in declaration without using the **assign** keyword. The following examples show the difference between explicit and implicit continuous assignments.

```
wire C;
assign C = A || B;      // explicit continuous assignment
wire D = E && F;        // implicit continuous assignment
```

Procedural assignments are used to model registers and finite state machines using the **always** keyword. More details on procedural assignment are explained in the following section.

2.6 Procedural Assignments

The concurrent statements from the previous section are useful in modeling combinational logic. Combinational logic constantly reacts to input changes. In contrast, synchronous sequential logic responds to changes dependent on the clock. Many input changes might be ignored since output and state changes occur only at valid conditions of the clock. Modeling sequential logic requires primitives to model selective activity conditional on clock, edge-triggered devices, sequence of operations, and so forth. There are two types of procedural assignments in Verilog. **Initial** blocks execute only once at time zero, whereas **always** blocks loop to execute over and over again. In other words, the initial block execution and always block execution starts at time 0. Always block waits for the event, whereas initial block just executes all the statements without waiting. Verilog also has a procedural assign statement that can be used inside the always block, but we do not use them in this book.

In this unit, we will learn **initial** and **always** statements, which help to model sequential logic. Initial blocks are useful in simulation and verification, but only always blocks are synthesized.

Initial Statements

An initial statement has the following basic form:

```
initial
begin
   sequential-statements
end
```

Always Statements

An always statement has the following basic form:

```
always @(sensitivity-list)
begin
   sequential-statements
end
```

When an always statement is used, the statements between the **begin** and the **end** are executed sequentially rather than concurrently. The expression in parentheses after the word **always** is called a sensitivity list, and the process executes whenever any signal in the sensitivity list changes. The symbol "@" should be used before the sensitivity list. For example, if the always statement has the sensitivity list @(A, B, C), then it executes whenever any one of A, B, or C changes. Whenever one of the signals in the sensitivity list changes, the sequential statements in the always block are executed in sequence one time. In earlier versions of Verilog, "or" is used to specify more than one element in the sensitivity list. In Verilog 2001, comma (,) is also used in the sensitivity list. Starting with Verilog 2001, an always statement can be used with a * in the sensitivity list to cause the always block to execute whenever any signal changes. When a process finishes executing, it goes back to the beginning and waits for a signal on the sensitivity list to change again.

The variables on the left-hand side element of an = or <= in an always block should be defined as **reg** data type. Any other data type including **wire** is illegal.

The assignment operator "=" indicates concurrent execution when used outside an always block. When the statements

```
C = A && B;    // concurrent statements
E = C || D;    // when used outside always block
```

are used outside an always block, the order of the statements does not matter. But when used in an always block, they become sequential statements executed in the order they are written.

Blocking and Non-Blocking Assignments

Sequential statements can be evaluated in two different ways in Verilog—blocking assignments and non-blocking assignments. A **blocking statement** must complete the evaluation of the right-hand side of a statement before the next statements in a sequential block are executed. Operator "=" is used for representing the blocking assignment. The meaning of "blocking" is that a blocking assignment has to complete before the next statement starts execution (i.e., it *blocks* the next assignment in the sequential block from starting evaluation). A **non-blocking statement** allows assignment evaluation without blocking the sequential flow. In other words, several assignments can be evaluated at the same time. Operator "<=" is used for representing the non-blocking assignment.

For instance, consider the situation if they are in an always block, as shown here:

```
always @(A, B, D)
begin
   C = A && B;    // Blocking operator is used
   E = C || D;    // Statements execute sequentially
end
```

The block executes once when any of the signals *A*, *B*, or *D* changes. The first statement updates the value of *C* before the second statement starts execution; hence, the second statement uses the new value of *C* as input. If *C* or *E* changes when the block executes, then the always block will not execute a second time because *C* is not on the sensitivity list. Operator "=" is used for representing what Verilog calls the **blocking assignment**, which *blocks* the next assignment in the sequential block. It should be noticed that the assignment operator "=" has a blocking nature inside the always block but a non-blocking or concurrent nature outside the always block.

The operator "<=" is to evaluate several assignments at the same time without blocking the sequential flow. Consider the following code:

```
always @(A, B, D)
begin
   C <= A && B;    // Statements execute simultaneously because
   E <= C || D;    // non-blocking operator is used
end
```

The block executes once when any of the signals A, B, or D changes. Both statements execute simultaneously with the values of A, B, C, and D at the beginning of the always block. The first statement does not update the value of C before the second statement starts execution; hence, the second statement uses the old value of C as input. If C changes when the block executes, then the always block will not execute a second time because C is not on the sensitivity list. Operator "<=" is used for representing what Verilog calls the **non-blocking assignment** inside an always statement. It should be noticed that the concurrent operations occur with "=" outside the always block, but with "<=" inside the always block. C and E should be defined as **reg** data type since **reg** is the only legal type on the left-hand side element of an = or <= in an always block.

Figure 2-15 shows a comparison of blocking assignments and non-blocking assignments. The **posedge** keyword of Verilog is used for an edge-triggered functionality in the sensitivity list. Signals A and B are used for a blocking statement, and C and D are applied for a non-blocking statement. Assume the initial values of input signals are $A=C=1'b1$ and $B=D=1'b0$. In the case of blocking assignments, both A and B will become $1'b0$. Since the second assignment will not be evaluated until the completion of the first assignment, the newly evaluated A signal will be assigned to B in the second assignment. On the other hand, non-blocking assignments will start evaluating all statements in a sequential block at the same time; the result will be independent of the assignment order. Therefore, the result of non-blocking assignments will be $C=1'b0$ and $D=1'b1$. So the signals swap in the case of C and D, but not in the case of A and B.

Always statements can be used for modeling combinational logic and sequential logic; however, always statements are not necessary for modeling combinational logic. They are, however, required for modeling sequential logic. One should be very careful when using always statements to represent combinational logic. If

FIGURE 2-15: Blocking and Non-Blocking Assignments

```
module sequential_module (A, B, C, D, clk);
input    clk;
output A, B, C, D;
reg      A, B, C, D;

always @(posedge clk)
begin
    A = B;          // blocking statement 1
    B = A;          // blocking statement 2
end

always @(posedge clk)
begin
    C <= D;         // non-blocking statement 1
    D <= C;         // non-blocking statement 2
end

endmodule
```

any of the input signals are accidentally omitted from the sensitivity list, there can be mismatches between synthesis and simulation and a lot of confusion. Hence, the common practice starting with Verilog 2001 of using the **always @*** statement if a combinational circuit is desired, which avoids accidental errors such as these. If the sensitivity list is "*" then the block will get triggered for any input signal changes.

Consider the code in Figure 2-16, where an always statement is used to model two cascaded gates. D and E should be defined as **reg** since **reg** is the only legal type on the left-hand side element of an = or <= in an always block. Also, D should be defined as output if the output of the first gate is desired externally. If inout is used for D, you will have a compile error, since D is of **reg** type. Normally, **input** and **inout** ports can be only net (or wire) data type. The statement order is important here because blocking assignment is used.

FIGURE 2-16: Verilog Code for Combinational Logic with Blocking Assignments in an Always Block

```
module two_gates (A, B, C, D, E);
input   A, B, C;
output  D, E;

reg D, E;

always @(*)
begin
  #5 D = A || B;   // blocking statement 1
  #5 E = C || D;   // blocking statement 2
end

endmodule
```

Let us assume that all variables are 0 @ 0 ns. Then, A changes to 1 @ 10 ns. That causes the module to execute. The statements inside the always statement execute once sequentially. D becomes 1 @ 15 ns, and E becomes 1 @ 20 ns.

Section 2.14 has additional examples illustrating the distinction between blocking and non-blocking operators inside always statements. While sequential logic can be modeled using the blocking operator "=," it is generally advised not to do so. A good coding practice while writing synthesizable code is to use non-blocking assignments (i.e., "<=") in always blocks intended to create sequential logic and the blocking operator "=" in always blocks intended to create combinational logic.

Another rule to remember is not to mix blocking and non-blocking assignments in the same always block. When each always block is written, think whether you want sequential logic or combinational logic and then use blocking assignments if combinational logic is desired.

Sensitivity List

Both combinatorial always blocks and sequential always blocks have a sensitivity list that includes a list of events. An always block will be activated if one of the events occurs. In the combinatorial logic, the sensitivity list includes all signals that are used in the condition statement and all signals on the right-hand side of the assignment.

On the other hand, the sensitivity list in sequential circuit contains three kinds of edge-triggered events: clock, reset, and set signal event. The sensitivity list can be specified using **@(*)** if a combinational circuit is desired, indicating that the block must be triggered for any input signal changes. If sensitivity list is omitted at the always keyword, delays or time-controlled events must be specified inside the always block. More details on this form of always block are presented in Section 2.8.

Wire and Reg

The two Verilog data types that we have used so far are **wire** and **reg** (more on data types is presented in Section 2.11). The **wire** acts as real wires in circuit designs. The **reg** is similar to wires, but can store information just like registers. The declarations for **wire** and **reg** signals should be done inside a module but outside any initial or always block. The initial value of a wire is *z* (high impedance), and the initial value of a reg is *x* (unknown).

The **wires** are either single bit or multiple bits in Verilog. The wires cannot store any information. They can be used only in modeling combinational logic and must be driven by something. The wires are a data type that can be used on the left-hand side of an assign statement but cannot be used on the left-hand side of = or <= in an always @ block.

The data type **reg** is used where the assigned data needs to be stored until the next assignment. If you want to assign your output in sequential code (within an always block), you should declare it as a **reg**. Otherwise, it should be a **wire** by default. One can use **reg** to model both combinational and sequential logic. Data type **reg** is the only legal type on the left-hand side element of an = or <= in an always block or initial block (normally used in test benches). It cannot be used on the left-hand side of an assign statement. Section 2.11 presents more on data types.

The default Verilog HDL data value set is a 4-value system consisting of four basic values:

0 represents a logic zero, or a false condition.

1 represents a logic one, or a true condition.

x represents an unknown logic value.

z represents a high-impedance state (often called the tristated value).

The 4-valued logic is described in more detail in Chapter 8.

For better understanding of sequential statements and operation of always statements, several more examples will be presented. In the following section, we explain how simple flip-flops can be modeled using always statements, and then we explain the basics of the Verilog simulation process.

2.7 Modeling Flip-Flops Using Always Block

A flip-flop can change state either on the rising or on the falling edge of the clock input. This type of behavior is modeled in Verilog by an always block. For a simple D flip-flop with a Q output that changes on the rising edge of *CLK*, the corresponding code is given in Figure 2-17.

FIGURE 2-17: Verilog Code for a Simple D Flip-Flop

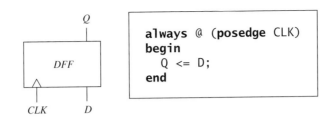

In Figure 2-17, on the rising edge of *CLK*, the always block executes once through and then waits at the start of the block until *CLK* changes again. The **sensitivity list** of the always block tests for a rising edge of the clock, and *Q* is set equal to *D* when a rising edge occurs. The expression posedge (or negedge) is used to accomplish the functionality of an edge-triggered device. If CLK is changed from 0 to 1 it is a rising edge. If CLK changes from 1 to 0, it indicates a falling edge.

If the flip-flop has a delay of 5 ns between the rising edge of the clock and the change in the *Q* output, we would replace the statement Q <= D; with Q <= #5 D; in the foregoing always block.

The statements between **begin** and **end** in an always block operate as sequential statements. In the previous example, Q <= D; is a sequential statement that executes only following the rising edge of *CLK*. In contrast, the concurrent statement assign Q = D; executes whenever *D* changes. If we synthesize the foregoing code, the synthesizer infers that *Q* must be a flip-flop since it changes only on the rising edge of *CLK*. If we synthesize the concurrent statement assign Q = D; the synthesizer will simply connect *D* to *Q* with a wire or a buffer.

In Figure 2-17, note that *D* is not on the sensitivity list because changing *D* will not cause the flip-flop to change state. Figure 2-18 shows a transparent latch and its Verilog representation. Both *G* and *D* are on the sensitivity list since if *G* = 1, a change in *D* causes *Q* to change. If *G* changes to 0, the always block executes, but *Q* does not change. For the sensitivity list, both (*G* or *D*) and (*G*, *D*) are acceptable.

FIGURE 2-18: Verilog Code for a Transparent Latch

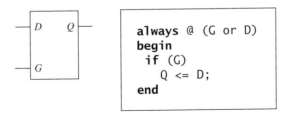

If a flip-flop has an active-low asynchronous clear input (*ClrN*) that resets the flip-flop independently of the clock, then we must modify the code of Figure 2-17 so that it executes when either *CLK* or *ClrN* changes. To do this, we add *ClrN* to the sensitivity list. The Verilog code for a D flip-flop with asynchronous clear is given in Figure 2-19. Since the asynchronous *ClrN* signal overrides *CLK*, *ClrN* is tested first and the flip-flop is cleared if *ClrN* is 0. Otherwise, *CLK* is tested, and *Q* is updated if a rising edge has occurred.

FIGURE 2-19: Verilog Code for a D Flip-Flop with Asynchronous Clear

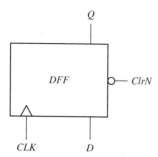

```
always @ (posedge CLK or negedge ClrN)
begin
  if (~ClrN)
    Q <= 0;
  else
    Q <= D;
end
```

In the foregoing examples, we have used two types of sequential statement: signal assignment statements and **if** statements. The basic **if** statement has the form

```
if (condition)
  sequential statements1
else
  sequential statements2
```

The condition is a Boolean expression that evaluates to TRUE or FALSE. If it is TRUE, `sequential statements1` are executed; otherwise, `sequential statements2` are executed.

Verilog **if** statements are sequential statements that can be used within an always block (or an initial block), but they cannot be used as concurrent statements outside of an always block. The most general form of the **if** statement is

```
if (condition)
  sequential statements
  // 0 or more else if clauses may be included
else if (condition)
  sequential statements}
[else sequential statements]
```

The curly brackets indicate that any number of **else if** clauses may be included, and the square brackets indicate that the **else** clause is optional. The example of Figure 2-20 shows how a flow chart can be represented using nested **if**s or the

FIGURE 2-20: Equivalent Representations of a Flow Chart Using Nested Ifs and Else Ifs

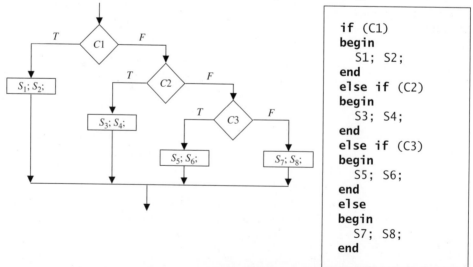

```
if (C1)
begin
  S1; S2;
end
else if (C2)
begin
  S3; S4;
end
else if (C3)
begin
  S5; S6;
end
else
begin
  S7; S8;
end
```

2.7 Modeling Flip-Flops Using Always Block 81

FIGURE 2-21: J-K Flip-Flop

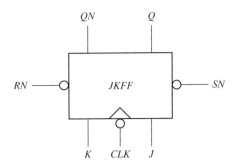

equivalent using **else if**s. In this example, *C1*, *C2*, and *C3* represent conditions that can be true or false, and $S_1, S_2, \ldots, S_8$ represent sequential statements. If more than one statement needs to be in an **if** block, **begin** and **end** should be used.

Next, we will write a Verilog module for a J-K flip-flop (Figure 2-21). This flip-flop has active-low asynchronous preset (*SN*) and clear (*RN*) inputs. State changes related to *J* and *K* occur on the falling edge of the clock. In this chapter, we use a suffix N to indicate an active-low (negative-logic) signal. For simplicity, we will assume that the condition *SN* = *RN* = 0 does not occur.

The Verilog code for the J-K flip-flop is given in Figure 2-22. The input and output signals are listed after the **module** statement. We define a **reg** *Qint* as an internal signal that represents the state of the flip-flop internal to the module. The two concurrent statements, statement4 and statement5, transmit this internal signal to the *Q* and *QN* outputs of the flip-flop. Because the flip-flop can change state in response to changes in *SN*, *RN*, and *CLK*, these three signals are in the sensitivity list of the always statement. Both *RN* and *SN* are active low signals. If *RN* = 0, the flip-flop is reset, and if *SN* = 0, the flip-slop is set. Since *RN* and *SN*, reset and set the flip-flop independently of the clock, they are tested first. If *RN* and *SN* are both 1, we test for the falling edge of the clock. In the **if** statement, both (~RN) and (RN == 1'b0) are acceptable.

FIGURE 2-22: J-K Flip-Flop Model

```
module JKFF (SN, RN, J, K, CLK, Q, QN);
input   SN, RN, J, K, CLK;
output  Q, QN;
reg Qint;
always @(negedge CLK or RN or SN)
begin
  if (~RN)
    #8  Qint <= 0;                                      // statement1
  else if (~SN)
    #8  Qint <= 1;                                      // statement2
  else
    Qint <= #10 ((J && ~Qint) || (~K && Qint));  // statement3
end
assign Q = Qint;                                        // statement4
assign QN = ~Qint;                                      // statement5
endmodule
```

The condition (negedge CLK) is TRUE only if *CLK* has just changed from 1 to 0. The next state of the flip-flop is determined by its characteristic equation:

$$Q^+ = JQ' + K'Q$$

The 8 ns delay represents the time it takes to set or clear the flip-flop output after *SN* or *RN* changes to 0. The 10 ns delay represents the time it takes for *Q* to change after the falling edge of the clock.

2.8 Always Blocks Using Event Control Statements

An alternative form for an always block uses wait or event control statements instead of a sensitivity list. If a sensitivity list is omitted at the always keyword, delays or time-controlled events must be specified inside the always block. For example,

```
always
begin
#10 clk <= ~clk;
end
```

will work as long as non-zero delay is specified.

An always block cannot have both wait statements and a sensitivity list. An always block with wait statements may have the form

```
always
begin
  sequential-statements
  wait-statement
  sequential-statements
  wait-statement
  . . .
end
```

Such an always block could look like

```
always
begin
  rst = 1; // sequential statements
  @(posedge CLK); //wait until posedge CLK
  // more sequential statements
end
```

This always block will execute the sequential-statements until a wait (event control) statement is encountered. Then it will wait until the specified condition is satisfied. It will then execute the next set of sequential-statements until another wait is encountered. It will continue in this manner until the end of the always block is reached. Then it will start over again at the beginning of the block.

2.8 Always Blocks Using Event Control Statements

The wait statement is used as a level-sensitive event control. The general syntax of the wait statement is

```
wait (Boolean-expression)
```

A procedural statement waits when the Boolean expression is FALSE. When the expression is TRUE, the statement is executed. The logic values 0, 'x', and 'z' are treated as FALSE. Logic 1 is TRUE. The following example will block the flow of the procedural block when the condition of the wait statement is FALSE. The wait statement can also be used to handshake or synchronize two concurrent processes, as illustrated in the following sequence:

```
always
begin
  wait (WR)
    MEM = DATA_IN;
  wait (~WR)
    DATA_OUT = MEM;
end
```

When the WR signal becomes true, DATA_IN gets written into MEM but as soon as the WR signal becomes false, the MEM value becomes available on DATA_OUT.

Example

For a half adder, sum and carry can be found using the equations sum = x XOR y ; carry = x AND y. What is wrong with the following code for a half adder that must add if add signal equals 1?

```
always @(*)
begin
if (add == 1)
    sum = x ^ y;
    carry = x & y;
end
```

(a) It will compile but not simulate correctly
(b) It will compile and simulate correctly but not synthesize correctly
(c) It will work correctly in simulation and synthesis
(d) It will not even compile

Answer: (a). This code will compile but will not simulate correctly. The if statement is missing begin and end. Currently only the sum is part of the if statement. The carry statement will get executed regardless of the add signal. This can be corrected by adding begin and end for the if statement. That will result in correct simulation. It can still lead to latches in synthesis. Latches can be avoided by adding else clause or by initializing sum and carry to 0 at the beginning of the always statement.

Example

What is wrong with the following code for a half adder that must add if add signal equals 1?

```
always @(*)
begin
if (add == 1)
   sum = x ^ y;
   carry = x & y;
 else
   sum = 0;
   carry = 0;
end
```

(a) It will compile but not simulate correctly
(b) It will compile and simulate correctly but not synthesize correctly
(c) It will work correctly in simulation and synthesis
(d) It will not even compile

Answer: (d). This code will not even compile due to the missing begin and end inside the if statement. When the compiler gets to the else, it finds that the corresponding if statement is missing. Both if and else clauses need begin and end. Once that is corrected, both simulation and synthesis will work correctly.

2.9 Delays in Verilog

In one of the initial examples in this chapter, we used the statement

```
assign #5 D = A && B;
```

to model an AND gate with a propagation delay of 5 ns (assuming its time unit is ns). The foregoing statement will model the AND gate's delay; however, it also introduces some complication, which many readers will not normally expect. If you simulate this AND gate with inputs that change very often in comparison to the gate delay (e.g., at 1 ns, 2 ns, 3 ns, etc.), the simulation output will not show the changes. This is due to the way Verilog delays work.

Basically, delays in Verilog can be categorized into two models: inertial delay and transport delay. The inertial delay for combinational blocks can be expressed in the following three ways:

```
// explicit continuous assignment
wire D;
assign #5 D = A && B;

// implicit continuous assignment
wire #5 D = A && B;

// net declaration
wire #5 D;
assign D = A && B;
```

2.9 Delays in Verilog

Any changes in A and B will result in a delay of 5 ns before the change in output is visible. If values in A or B are changed 5 ns before the evaluation of D output, the change in values will be propagated. However, an input pulse that is shorter than the delay of the assignment does not propagate to the output. This feature is called **inertial delay**. Inertial delay is intended to model gates and other devices that do not propagate short pulses from the input to the output. If a gate has an ideal inertial delay T, in addition to delaying the input signals by time T, any pulse with a width less than T is rejected. For example, if a gate has an inertial delay of 5 ns, a pulse of width 5 ns would pass through, but a pulse of width 4.999 ns would be rejected.

Transport delay is intended to model the delay introduced by wiring; it simply delays an input signal by the specified delay time. In order to model this delay, a delay value must be specified on the right-hand side of the statement. Figure 2-23 illustrates transport delay and inertial delay in Verilog. Consider the following code:

```
always @ (X)
begin
   Z1 <= #10 (X);      // transport delay
end
assign #10 Z2 = X;     // inertial delay
```

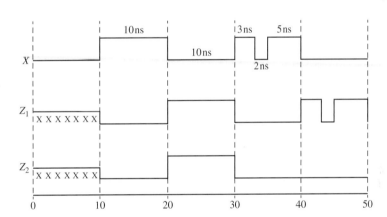

FIGURE 2-23: Inertial and Transport Delays

The first statement has transport delay while the second one has inertial delay. As shown in Figure 2-23, if the delay is shorter than 10 ns, the input signal will not be propagated to the output in the second statement. Only one pulse (between 10 ns and 20 ns) on input X is propagated to the output Z_2, since it has 10 ns pulse width. All other pulses are not propagated to the output Z_2. But Z_1 has transport delay and hence propagates all pulses. It is assumed that the output Z_1 and Z_2 are initialized to 0 at 0 ns. The delay in the statement Z_1 <= #10 X; is called **intra-assignment delay**. The expression on the right hand side is evaluated but not assigned to Z_1 until the delay has elapsed (also called **delayed assignment**). However, in a statement like #10 Z_1 <= X; the delay of #10 elapses first and then the expression is evaluated and assigned to Z_1 (also called **delayed evaluation**).

The placement of the delay on the right-hand side cannot be done with continuous assign statements. Hence the following statement is illegal. It will produce a compile-time error.

```
assign a = #10 b;
```

Verilog also has a type of delay called **net delay**. Consider the code

```
wire C1;
wire #10 C2;  // net delay on wire C2

assign #30 C1 = A || B;   // statement 1 - inertial delay
assign #20 C2 = A || B;   // statement 2 - inertial delay
                          //               will be
                          // added to net delay before being
                          // assigned to wire C2
```

The wire C_2 has a **net delay** of 10 ns associated with it, specified in its declaration whereas C_1 has no such net delay. Net delay refers to the time it takes from any driver on the net to change value to the time when the net value is updated and propagated further. There are inertial delays of 30 ns for C_1 in statement 1 and 20 ns for C_2 in statement 2, typically representative of gate delays. After statement 2 processes its delay of 20 ns, the net delay of 10 ns is added to it. Figure 2-24(a) indicates the difference between C_1 and C_2. C_1 rejects all narrow pulses less than 30 ns, whereas C_2 rejects only pulses less than 20 units.

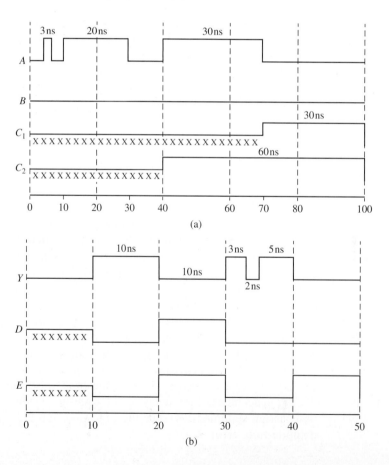

FIGURE 2-24: Example of Net Delays

Now consider the following two statement pairs with the Y waveform as shown in Figure 2-24(b).

```
wire #3 D; // net delay on wire D
assign #7 D = Y; // statement 1 - inertial delay

wire #7 E; // net delay on wire E
assign #3 E = Y; // statement 1 - inertial delay
```

The assign statement for D works with a 7 ns inertial delay and rejects any pulse below 7 ns. Hence D rejects the 3 ns, 2 ns and 5 ns pulses in Y. The 3 ns net delay from the wire statement is added to the signal that comes out from the assign statement. In the case of E, pulses below 3 ns are rejected. Hence the 3 ns pulse in Y passes through the assign statement for E, the 2 ns pulse is rejected and the 5 ns pulse is accepted. Hence the 3 ns and 5 ns pulses get combined in the absence of the 2 ns pulse to yield output on E appears as a big 10 ns pulse. The 7 ns net delay from the wire statement is added to the signal that comes out from the assign statement. If any pulses less than 7 ns are encountered at the net delay phase, they will be rejected. Figure 2-24(b) illustrates the waveforms for D and E.

Note that these delays are relevant only for simulation. Understanding how inertial delay works can remove a lot of frustration in your initial experience with Verilog simulation. The pulse rejection associated with inertial delay can inhibit many output changes. In simulations with basic gates and simple circuits, one should make sure that test sequences that you apply are wider than the inertial delays of the modeled devices. The focus of this book is synthesizeable Verilog where all specified delays are ignored. Hence we do not further dwell on simulator behavior for delays.

2.10 Compilation, Simulation, and Synthesis of Verilog Code

After describing a digital system in Verilog, simulation of the Verilog code is important for two reasons. First, we need to verify that the Verilog code correctly implements the intended design, and second, we need to verify that the design meets its specifications. We first simulate the design and then synthesize it to the target technology (FPGA or custom ASIC). In this section, first we describe steps in simulation and then introduce synthesis. As illustrated in Figure 2-25, there are three phases in the simulation of Verilog code: **analysis (compilation), elaboration, and simulation**.

FIGURE 2-25: Compilation, Elaboration, and Simulation of Verilog Code

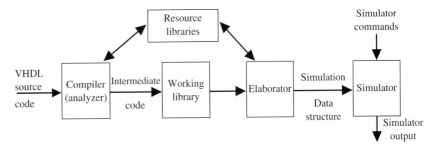

Before the Verilog model of a digital system can be simulated, the Verilog code must first be compiled. The Verilog compiler, also called an **analyzer**, first checks the Verilog source code to see that it conforms to the syntax and semantic rules of Verilog. If there is a syntax error, such as a missing semicolon, or if there is a semantic error, such as trying to add two signals of incompatible types, the compiler will output an error message. The compiler also checks to see that references to libraries are correct. If the Verilog code conforms to all of the rules, the compiler generates intermediate code, which can be used by a simulator or by a synthesizer.

After a Verilog design has been parsed but before simulation begins, the design must have the modules being instantiated linked to the modules being defined, the parameters propagated among the various modules, and hierarchical references resolved. This phase in understanding a Verilog description is referred to as **elaboration**. During elaboration, a *driver* is created for each signal. Each driver holds the current value of a signal and a queue of future signal values. Each time a signal is scheduled to change in the future, the new value is placed in the queue along with the time at which the change is scheduled. In addition, memory storage is allocated for the required signals; the interconnections among the port signals are specified; and a mechanism is established for executing the Verilog statements in the proper sequence. The resulting data structure represents the digital system being simulated.

The simulation process consists of an **initialization phase** and actual **simulation**. The simulator accepts simulation commands, which control the simulation of the digital system and which specify the desired simulator output. Verilog simulation uses what is known as **discrete event simulation**. The passage of time is simulated in discrete steps in this method of simulation. The initialization phase is used to give an initial value to the signal. To facilitate correct initialization, the initial value can be specified in the Verilog model. In the absence of any specifications of the initial values, some simulator packages may assign an initial value depending on the type of the signal. Please note that this initialization is only for simulation and not for synthesis.

A design consists of connected threads of execution or processes. Processes are objects that can be evaluated, that may have state, and that can respond to changes on their inputs to produce outputs. Processes include modules, initial and always procedural blocks, continuous assignments, procedural assignment statements, system tasks, and so forth.

Every change in value of a net or variable in the circuit being simulated is considered an **update event**. Processes are sensitive to update events. When an update event is executed, all the processes that are sensitive to that event are evaluated in an arbitrary order. The evaluation of a process is also an event, known as an **evaluation event**. The term **simulation time** is used to refer to the time value maintained by the simulator to model the actual time it would take for the circuit being simulated.

Events can occur at different times. In order to keep track of the events and to make sure they are processed in the correct order, the events are kept on an **event queue**, ordered by simulation time. Putting an event on the queue is called **scheduling an event**.

The Verilog event queue is logically segmented into five different regions:

> **i. Active event region:** Events that occur at the current simulation time are in this region. Events can be added to any of the five regions but can be removed only

from this region (i.e., the *active* region). Events can be processed in any order from within this region. (This freedom to choose any active event for immediate processing is an essential source of non-determinism in the Verilog HDL.)

ii. **Inactive event region:** Events that occur at the current simulation time but that shall be processed after all the active events are processed are in this region. Blocking assignments with zero delays are in this region until they get moved later to the active region.

iii. **Non-blocking assign update region:** Events that have been evaluated during some previous simulation time but that shall be assigned at this simulation time after all the active and inactive events are processed are in this region.

iv. **Monitor event region:** Events that shall be processed after all the active, inactive, and non-blocking assign update events are processed are in this region. These are the *monitor* events.

v. **Future event region:** Events that occur at some future simulation time are in this region. These are the *future* events. Future events are divided into *future inactive events* and *future non-blocking assignment update events*.

When each Verilog statement is processed, events are added to the various queue regions according to the following convention for each type of statement:

i. **Continuous assignment**—evaluate RHS and add to active region as an active update event.

ii. **Procedural continuous assign**—evaluate RHS and add to active region as an update event.

iii. **Blocking assignment with delay**—compute RHS and put into future event region for time after delay.

iv. **Blocking assignment with no delay**—compute RHS and put into inactive region for current time.

v. **Non-blocking assignment with no delay**—compute RHS and schedule as non-blocking assign update event for current time if zero delay.

vi. **Non-blocking assignment with delay**—compute RHS and schedule as non-blocking assign update event for future time if zero delay.

vii. **$monitor** and **$strobe** system tasks—create monitor events for these system tasks. (These events are continuously reenabled in every successive time step.)

The processing of all the active events is called a **simulation cycle**.

For each simulation time, the following actions are performed in order:

i. Process all active update events. (Whenever there is an active update event, the corresponding object is modified and new events are added to the various event queue regions for other processes sensitive to this update.)

ii. Then activate all inactive events for that time (and process them because now they are active).

iii. Then activate all non-blocking assign update events and process them.

iv. Then activate all monitor events and process them.

v. Advance time to the next event time and repeat from step i.

All of these five steps happen at the same time, but the events occur in the order active, inactive, non-blocking update, and monitor events.

> VHDL uses the concept of an infinitesimal delay called delta (Δ) delay to explicitly indicate the various update times within the same simulation time, however the Verilog Language Reference Model (LRM) does not mention delta delays. In the Verilog simulators we experimented with, there is a delta delay indicated in non-blocking procedural assignments with zero delay, but no delta delays were observed in blocking or continuous assign statements with zero delay. However, implicitly the simulation uses the ordering between the aforementioned 5 event queue regions to yield the correct ordering. Blocking assignments with zero delay are in the inactive queue first, and happen after continuous assignments but no delta delay is shown to indicate the delay. Non-determinism is usually avoided except that the freedom to choose any active event for immediate processing from the active queue region contributes to some non-determinism in the Verilog HDL.

Basically, the simulator works as follows with "<=": whenever a component input changes, the output is scheduled to change after the specified delay or after Δ if no delay is specified. When all events for the current time have been processed, simulated time is advanced to the next time at which an event is specified. When time is advanced by a finite amount (1 ns for example), the Δ counter is reset and simulation resumes. Real time does not advance again until all events associated with the current simulation time have been processed.

If two non-blocking updates are made to the same variable in the same time step, the second one dominates by the end of the time step. For example, in Figure 2-26(a) events are added to the event queue in source code order because of

FIGURE 2-26: Illustration of Non-Determinism

```
module determinate;
reg a;

initial a = 0;

always begin
a <= #5 0;
a <= #5 1;
end
// The assigned value of a is deterministic
// because of ordering from begin to end
endmodule
                                                          (a)
module nondeterminate;
reg a;
initial a = 0;

always a <= #5 0;
always a <= #5 1;
// The assigned value of a is non-deterministic
Endmodule
```

> The IEEE 1364 Standard Verilog Language Reference Manual (LRM) [1] provides definitions and interpretations for the various Verilog constructs, which all compliant simulators shall implement. However, there is a great deal of choice in the definitions, and some differences in the details of execution are to be expected between different simulators.
>
> Those who are accustomed to the elegant delta delay conventions in VHDL may be disappointed with Verilog simulation outputs. Verilog was created primarily with circuit synthesis in mind and Verilog simulation may not clearly indicate the precise ordering of multiple events happening at the same simulation time. The IEEE 1364 Standard Verilog Language Reference Manual (LRM) does not even use the word 'delta' in it.

the **begin...end**, and the two updates are performed in source order as well. Hence, the variable a will be assigned 0 first and then 1 in that order. There is no non-determinism in this code. However, for the code in Figure 2-26(b), the two **always** blocks are concurrent with respect to each other and there is no ordering between them. Hence the assigned value of a is non-deterministic.

2.10.1 Simulation with Multiple Processes (Initial or Always Blocks)

If a model contains more than one process, all processes execute concurrently with other processes. If there are concurrent statements outside always statements, they also execute concurrently. Statements inside of each always block execute sequentially. A process takes no time to execute unless it has wait statements in it. As an example of simulation of multiple processes, we trace execution of the Verilog code shown in Figure 2-27.

FIGURE 2-27: Verilog Code to Illustrate Process Simulation

```
module twoprocess
   reg A,B;
initial
begin
   A = 0;
   B = 0;
end
// process P1
always @(B)
begin
   A <= 1;
   A <= #5 0;
end
// process P2
always @(A)
begin
   if (A)
       B <= #10 ~B;
end
```

Figure 2-28 shows the drivers for the signals A and B as the simulation progresses. In the absence of an initial block, each driver would hold x, since this is the default initial value for a signal. When simulation begins, initialization takes place and each driver holds 0 since an initial block is included in the provided code. Both always statements wait until a signal on the sensitivity list changes. With the initial block here, the signal changes from initialization lead to the execution of the always statements. In the absence of the initial block, one can force input changes using simulation commands. When process *P1* executes at zero time, two changes in A are scheduled (A changes to 1 at time Δ and back to 0 at time = 5 ns). Meanwhile, process *P2* executes at zero time, but no change in B occurs since

FIGURE 2-28: Signal Drivers for Simulation Example

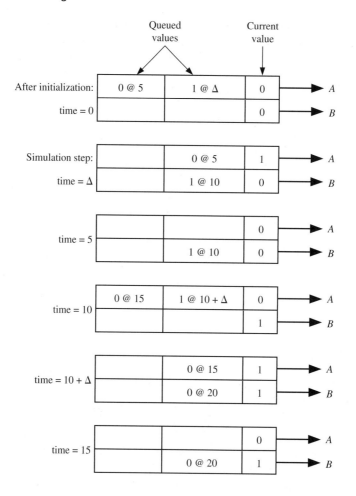

A is still 0 during execution at time 0 ns. Time advances to Δ, and A changes to 1. The change in A causes process $P2$ to execute, and since $A = 1$, B is scheduled to change to 1 at time 10 ns. The next scheduled change occurs at time = 5 ns, when A changes to 0. This change causes $P2$ to execute, but B does not change. B changes to 1 at time = 10 ns. The change in B causes $P1$ to execute, and two changes in A are scheduled. When A changes to 1 at time $10 + \Delta$, process $P2$ executes, and B is scheduled to change at time 20 ns. Then A changes at time 15 ns, and the simulation continues in this manner until the runtime limit is reached. It should be understood that A changes at 15 ns and not at $15 + \Delta$. The Δ delay comes into the picture only when no time delay is specified.

Verilog simulators use event-driven simulation, as illustrated in the preceding example. A change in a signal is referred to as an *event*. Each time an event occurs, any processes that have been waiting on the event are executed in zero time, and any resulting signal changes are queued up to occur at some future time. When all the active processes are finished executing, simulation time is advanced to the time for which the next event is scheduled, and the simulator processes that event. This continues until either no more events have been scheduled or the simulation time limit is reached.

Inertial delays can now be explained in the following manner. Each input change causes the simulator to schedule a change, which is scheduled to occur after the specified delay; however, if another input change happens before the specified delay has elapsed, the first change is dequeued from the simulation driver queue. Hence only pulses wider than the specified delay appear at the output.

One of the most important uses of Verilog is to synthesize or automatically create hardware from a Verilog description. The **synthesis** software for Verilog translates the Verilog code to a circuit description that specifies the needed components and the connections between the components. The initial steps (analysis and elaboration) in Figure 2-25 are common whether Verilog is used for simulation or synthesis. The simulation and synthesis processes are shown in Figure 2-29.

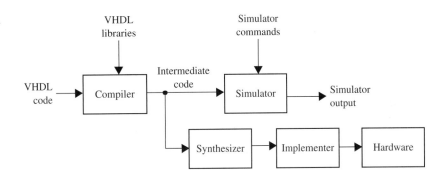

FIGURE 2-29: Compilation, Simulation, and Synthesis of Verilog Code

Although synthesis can be done in parallel to simulation, practically it follows simulation because designers would normally want to catch errors first before attempting to synthesize. After the Verilog code for a digital system has been simulated to verify that it works correctly, the Verilog code can be synthesized to produce a list of required components and their interconnections, typically called the **netlist**. The synthesizer output can then be used to implement the digital system using specific hardware, such as a CPLD or an FPGA or as an ASIC. The CAD software used for implementation generates the necessary information to program the CPLD or FPGA hardware. Synthesis and implementation of digital logic from Verilog code is discussed in more detail in Chapter 6.

2.11 Verilog Data Types and Operators

2.11.1 Data Types

Verilog has two main groups of data types: the variable data types and the net data types. These two groups differ in the way that they are assigned and hold values. They also represent different hardware structures.

The *net* data types can represent physical connections between structural entities, such as gates. Generally, it does not store values. Instead, its value is determined by the values of its drivers, such as a continuous assignment or a gate. A very popular *net* data type is the **wire**. There are also several other predefined data types that are part of nets. Examples are **tri** (for tristate), **wand** (for wired and), **wor** (for wired or).

The *variable* data type is an abstraction of a data storage element. A variable shall store a value from one assignment to the next. An assignment statement in a procedure acts as a trigger that changes the value in the data storage element. A very popular *variable* data type is the **reg**. There are also several other predefined data types that are part of *variables*. Examples are **reg**, **time**, **integer, real,** and **real-time**.

Unlike VHDL, all data types are predefined by the Verilog language and not by the user. Some of the popular predefined types are

nets	connections between hardware elements (declared with keywords such as **wir**e)
variables	data storage elements that can retain values (declared with the keywords such as **reg**)
integer	an integer is a variable data type (declared with the keyword **integer**)
real	real number constants and real variable data types for floating-point number (declared with the keyword **real**)
time	a special variable data type to store time information (declared with the keyword **time**)
vectors	wire or reg data types can be declared as vectors (multiple bits) (vectors can be declared with [range1 : range2])

In previous versions of the Verilog standard, the term **register** was used to encompass the **reg**, **integer**, **time**, **real**, and **realtime** types, but starting with the 2005 IEEE 1364Standard, that term is no longer used as a Verilog data type. A net or reg declaration without a range specification shall be considered 1 bit wide and is known as a **scalar**. Multiple bit net and reg data types shall be declared by specifying a range, which is known as a **vector**.

While VHDL is a strongly typed language where signals and variables of different types generally cannot be mixed in the same assignment statement, Verilog uses weak typing, which means some mixing of related data types is allowed.

2.11.2 Verilog Operators

Verilog operators are similar to the operators used in C language. Predefined Verilog operators can be grouped into several classes:

1. Unary sign and reduction operators:

+ −	Unary sign operators
&	Reduction and (unary operator to and bits in a vector and reduce to one bit)
~&	Reduction NAND
|	Reduction or (unary operator to or bits in a vector and reduce to one bit)
~|	Reduction NOR
^	Reduction XOR

	~^ or ^~	Reduction XNOR
	!	Logical negation
	~	Bit-wise negation

2. Arithmetic: Exponent

	**	Arithmetic (POWER)

3. Arithmetic: Multiplying, Modulus operators:

	*	Multiply
	/	Divide
	%	Modulus

4. Arithmetic: Addition:

	+	Add
	−	Subtract

5. Shift operators:

	<<	Logical left shift
	>>	Logical right shift
	<<<	Arithmetic left shift
	>>>	Arithmetic right shift

6. Relational operators:

	>	Greater than
	<	Less than
	>=	Greater than or equal
	<=	Less than or equal

7. Logical and bitwise operators: Equality and inequality

	==	Logical equality
	!=	Logical inequality
	===	Case equality
	!==	Case inequality

8. Bitwise operators:

	&	Bit-wise and (binary operator)

9. Logical and bitwise operators:

	^	Bit-wise exclusive or (binary operator)
	^~ or ~^	Bit-wise equivalence (binary operator)
	|	Bit-wise inclusive or (binary operator)

10. Logical and:

	&&	Logical and

11. Logical or:

	||	Logical or

12. Conditional

	? :	Conditional

13. Concatentation and replication

{}   Concatenation

{{}}   Replication

 When parentheses are not used, operators in class 1 have highest precedence and are applied first, followed by class 2, then class 3, and so forth. Class 13 operators have lowest precedence and are applied last, but if the expression with these operators is needed in order to perform another higher-precedence operation, it is evaluated before the other operator can be evaluated. Operators in the same class have the same precedence and are applied from left to right in an expression. The precedence order can be changed by using parentheses. The { } operator can be used to concatenate two vectors (or an element and a vector, or two elements) to form a longer vector. For example, {010, 1} is 0101 and {"ABC", "DEF"} is "ABCDEF". Consider the following expression where A, B, C, and D are vectors:

$$(\{A, \sim B\} \mid C \gg 2 \& D) == 110010$$

 One must note that this is a relational expression performing an *equality test*; it is not an assignment statement. To evaluate the expression inside ( ), the operator precedence shows the highest precedence for the following three operators in order:

$$\gg, \&, \mid$$

 In order to evaluate |, one of the operands of | has to be obtained by the concatentation, which forces the expression inside the concatenate to be evaluated and operators ~, { }, are applied before the | can be evaluated.

A Note on Operator Precedence
Different languages have differences in the order of precedence and hence those who program in many languages may write or interpret code erroneously, mixing the precedence between the different languages. Hence instead of worrying too much about the precedence order, a good strategy is to use parentheses and make code unambiguous.

 If $A = 110$, $B = 111$, $C = 011000$, and $D = 111011$, the computation proceeds as follows:

C >> 2	= 000110	(shift right 2 places)
C >> 2 & D	= 000010	(bit-wise and)
~ B = 000		(bit-wise negate)
{A, ~ B}	= 110000	(concatenation)
({A, ~ B}) | (C >> 2&D)	= 110010	(bit-wise or)
[({A, ~ B} | C >> 2) & D) == 110010]	= TRUE	(the parentheses force the equality) test to be done last and the result is TRUE)

The result of applying a relational operator is always a Boolean (FALSE or TRUE). Equals (==) and not equals (!=) can be applied to almost any type. The other relational operators can be applied to many numeric as well as to some array types. For example, if $A = 5$, $B = 4$, and $C = 3$, the expression (A >= B) && (B <= C) evaluates to FALSE. It is legal to use concatenate operator on the left side of the assignment, For example,

    {Carry, Sum} = A + B;

is legal. It adds A and B and the result goes into Sum and Carry. The most significant bit of the result is assigned to Carry. For the *logical equality* and *logical inequality* operators (== and !=), if, due to unknown or high-impedance bits in the operands, the relation is ambiguous, then the result shall be a 1-bit unknown value (x). For the *case equality* and *case inequality* operators (=== and !==), bits that are x or z shall be included in the comparison and shall match for the result to be considered equal. The result of these operators shall always be a known value, either 1 or 0.

The shift operators can be applied to signed and unsigned registers. One can declare a register to be signed in the following manner:

    reg signed [7:0] A = 8'hA5; //signed register A

In contrast, the register would have been unsigned if declared as follows:

    reg [7:0] B = 8'hA5; //unsigned register B

If the register is unsigned, arithmetic and logic shifts do the same operation. The following example illustrates the difference between signed and unsigned shifts on signed and unsigned data.

    reg signed [7:0] A = 8'hA5; // A is signed 1010 0101

    A >> 4 is 00001010   (shift right unsigned by 4, filled with 0).
    A >>> 4 is 11111010   (shift right signed by 4, filled with sign
                            bit).
    A << 4 is 01010000   (shift left unsigned, filled with 0).
    A <<< 4 is 01010000   (shift left signed, filled with 0
                            irrespective of rightmost bit).

    reg [7:0] B = 8'hA5; // B is unsigned 1010 0101

    B >> 4 is 00001010   (shift right unsigned by 4, filled with 0)
    B >>> 4 is 00001010   (shift right signed by 4, but B is unsigned,
                            filled with 0)
    B << 4 is 01010000   (shift left unsigned, filled with 0)
    B <<< 4 is 01010000   (shift left signed, filled with 0
                            irrespective of rightmost bit)

If A is declared as integer as in

    integer signed A = 8'hA5;

A >>> 4 yields 00001010 (shift right signed by 4, but integer type is 32 bits and the provided number has only eight bits, which does not include the sign and hence sign bit is).

But if A is initialized to 8'shA5 as in

integer A = 8'shA5;

A >>> 4 yields 11111010 (shift right signed by 4, A's sign bit is 1). The 'sh indicates that the value is signed hex.

However, in **reg** declarations, if a signed register is desired, it should be explicitly mentioned. For instance,

reg [7:0] A = 8'shA5

does not make the register signed. It should be declared as **reg signed [7:0] A = 8'hA5**

The + and − operators can be applied to any types, including integer or real numeric operands. When types are mixed, the expression self-evaluates to a type according to the types of the operands. If a and b are 16 bits each, (a + b) will evaluate to 16 bits. However, (a + b + 0) will evaluate to integer. If any operand is real, the result is real. **If any operand is unsigned, the result is unsigned, regardless of the operator**.

When expressions are evaluated, if the operands are of unequal bit lengths and if one or both operands are unsigned, the smaller operand shall be zero-extended to the size of the larger operand. If both operands are signed, the smaller operand shall be sign-extended to the size of the larger operand. If constants need to be extended, signed constants are sign-extended and unsigned constants are zero-extended.

The * and / operators perform multiplication and division on integer or floating-point operands. The ** operator raises an integer or floating-point number to an integer power. The % (modulus) operator calculates the remainder for integer operands.

> According to the Verilog Language Reference Manual (LRM), the result of the modulus operator takes the sign of the first operand. For example,
>
> −10 % 3
>
> results in –1 because the result takes the sign of the first operand. However, number theory books define **mod m** as a function from the set of integers to the set of {0,1,2,....,m−1} and hence –10 mod 3 is equal to 2. However, there is no reason for serious concern because –1 mod 3 is 2 according to the **mod m** definition. Hence -1 as given by Verilog and 2 as given by their definition are in fact the same number. VHDL has two separate operators, one for modulus and one for remainder. The VHDL **remainder** is signed according to the sign of the first operand whereas the modulus follows the **mod m** definition from number theory.

2.12 Simple Synthesis Examples

Synthesis tools try to infer the hardware components needed by "looking" at the Verilog code. In order for code to synthesize correctly, certain conventions must be followed. When writing Verilog code, you should always keep in mind that you are designing hardware, not simply writing a computer program. Each Verilog statement implies certain hardware requirements. Consequently, poorly written Verilog code may result in poorly designed hardware. Even if Verilog code gives the correct result when simulated, it may not result in hardware that works correctly when synthesized. Timing problems may prevent the hardware from working properly even though the simulation results are correct.

Consider the Verilog code in Figure 2-30. (Note that B is missing from the sensitivity list in the **always** statement.) This code will simulate as follows. Whenever A changes, it will cause the process to execute once. The value of C will reflect the values of A and B when the process began. If B changes now, that will not cause the process to execute.

FIGURE 2-30: Verilog Code Example Where Simulation and Synthesis Results in Different Outputs

```verilog
module Q1 (A, B, C);
input    A;
input    B;
output   C;

reg      C;

always @(A)
   C = #5 A | B;
endmodule
```

If this code is synthesized, most synthesizers will output an OR gate as in Figure 2-31. The synthesizer will warn you that B is missing from the sensitivity list in always statement, but will go ahead and synthesize the code properly. The synthesizer will also ignore the 5 ns delay on the preceding statement. If you want to model an exact 5 ns delay, you will have to use counters. The simulator output will not match the synthesizer's output since the always statement will not execute when B changes. This is an example of where the synthesizer guessed a little more than what you wrote; it assumed that you probably meant an OR gate and created that circuit (accompanied by a warning). But this circuit functions differently from what was simulated before synthesis. It is important that you always check for synthesizer warnings of missing signals in the sensitivity list. Perhaps the synthesizer helped you; perhaps it created hardware that you did not intend to.

FIGURE 2-31: Synthesize Output for Code in Figure 2-30

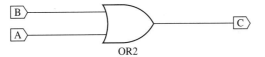

Now, consider the Verilog code in Figure 2-32. What hardware will you get if you synthesized this code?

Let us think about the block diagram of the circuit represented by this code without worrying about the details inside. The block diagram is as shown in

FIGURE 2-32: Example Verilog Code

```
module Q3 (A, B, F, CLK, G);
input    A;
input    B;
input    F;
input    CLK;
output   G;

reg      G;
reg      C;

always @(posedge CLK)
begin
  C <= A & B;    // statement 1
  G <= C | F;    // statement 2
end

endmodule
```

Figure 2-33. The ability to hide details and use abstractions is an important part of good system design.

FIGURE 2-33: Block Diagram for Verilog Code in Figure 2-32

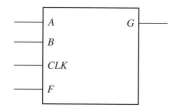

Note that *C* is an internal signal, and therefore, it does not show up in the block diagram.

Now, let us think about the details of the circuit inside this block. This circuit is not two cascaded gates; the signal assignment statements are in a process (an always statement). An edge-triggered clock is implied by the use of **posedge** or **negedge** in the clock statement preceding the signal assignment. Since the values of *C* and *G* need to be retained after the clock edge, flip-flops are required for both *C* and *G*. Please note that a change in the value of *C* from statement 1 will not be considered during the execution of statement 2 in that pass of the process. It will be considered only in the next pass, and the flip-flop for *C* makes this happen in the hardware also. Hence the code implies hardware shown in Figure 2-34.

FIGURE 2-34: Hardware Corresponding to Verilog Code in Figure 2-32

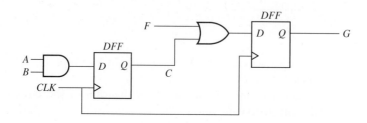

We saw earlier that the following code represents a D-latch:

```
always @(G or D)
begin
  if (G)  Q <= D;
end
```

Let us understand why this code does not represent an AND gate with G and D as inputs. If $G = 1$, an AND gate will result in the correct output to match the **if** statement. However, what happens if currently $Q = 1$ and then G changes to 0? When G changes to 0', an AND gate would propagate that to the output; however, the device we have modeled here should not. It is expected to make no changes to the output if G is not equal to 1. Hence, it is clear that this device has to be a D-latch and not an AND gate.

In order to infer flip-flops or registers that change state on the rising edge of a clock signal, most synthesizers require that the sensitivity list in an always statement should include an edge-triggered signal as in

always @(**posedge** CLK)

For every assignment statement in an always statement, a signal on the left side of the assignment will cause creation of a register or flip-flop. The moral to this story is that if you do not want to create unnecessary flip-flops, do not put the signal assignments in a clocked always statement. If `clock` is omitted in the sensitivity list of an always statement, the synthesizer may produce latches instead of flip-flops.

Now consider the Verilog code in Figure 2-35. If you attempt to synthesize this code, the synthesizer will generate an empty block diagram. This is because D, the output of the block shown in the Figure, is never assigned. The code assigns the new value to C, which is never brought out to the outside world. It will generate warnings that

```
Input <CLK> is never used.
Input <A> is never used.
Input <B> is never used.
Output <D> is never assigned.
```

FIGURE 2-35: Example Verilog Code That Will Not Synthesize

```
module no_syn (A, B, CLK, D);
  input    A;
  input    B;
  input    CLK;
  output   D;

  reg      C;

  always @(posedge CLK)
     C <= A & B;

endmodule
```

2.13 Verilog Models for Multiplexers

A multiplexer is a combinational circuit and can be modeled using concurrent statements only or using always statements. A conditional operator with **assign statement** can be used to model a multiplexer without always statements. A case statement or if-else statement can also be used to make a model for a multiplexer within an always statement.

2.13.1 Using Conditional Operator

Figure 2-36 shows a 2-to-1 multiplexer (MUX) with 2 data inputs and one control input. The MUX output is $F = A' \cdot I_0 + A \cdot I_1$. The corresponding Verilog statement is

assign F = (~A && I$_0$) || (A && I$_1$);

Here, the MUX can be modeled as a single concurrent signal assignment statement. Alternatively, we can represent the MUX by a conditional signal assignment statement as shown in Figure 2-36. This statement executes whenever A, I_0, or I_1 changes. The MUX output is I_0 when $A = 0$; otherwise it is I_1. In the conditional statement, I_0, I_1, and F can be one or more bits.

FIGURE 2-36: 2-to-1 Multiplexer

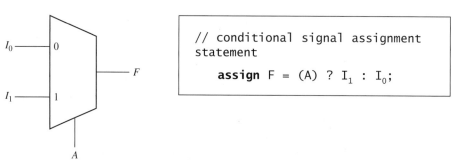

```
// conditional signal assignment
statement

    assign F = (A) ? I₁ : I₀;
```

The general form of a conditional signal assignment statement is

assign signal_name = condition ? expression_T : expression_F;

This concurrent statement is executed whenever a change occurs in a signal used in one of the expressions or conditions. If `condition` is true, `signal_name` is set equal to the value of `expression_T`. Otherwise if `condition` is false, `signal_name` is set equal to the value of `expression_F`. Figure 2-37 shows how two cascaded MUXes can be represented by a conditional signal assignment statement. The

FIGURE 2-37: Cascaded 2-to-1 MUXes Using Conditional Assignment

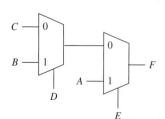

```
assign F = E ? A : ( D ? B : C );
// nested conditional assignment
```

output MUX selects A when the condition E is true; otherwise, it selects the output of the first MUX, which is B when the condition D is true, or it is C.

Figure 2-38 shows a 4-to-1 multiplexer (MUX) with four data inputs and two control inputs, A and B. The control inputs select which one of the data inputs is transmitted to the output. The logic equation for the 4-to-1 MUX is

$$F = A'B'I_0 + A'B\,I_1 + A\,B'I_2 + A\,B\,I_3$$

One way to model the MUX is with the Verilog statement

```
assign F = (~A && ~B && I0) || (~A && B && I1) ||
          A && ~B && I2) || (A && B && I3);
```

Another way to model the 4-to-1 MUX is to use a conditional assignment statement:

```
assign F = (A) ? (B ? I3 : I2) : ( B ? I1 : I0 );
```

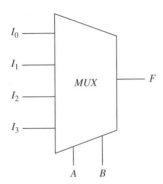

FIGURE 2-38: 4-to-1 Multiplexer

2.13.2 Using If-else or Case Statement in an Always Block

If a MUX model is used inside an always statement, a concurrent statement cannot be used. The MUX can be modeled using a **case** statement within an always block:

```
always @ (Sel or I_0 or I_1 or I_2 or I_3)
  case Sel
    2'b00 : F = I_0;
    2'b01 : F = I_1;
    2'b10 : F = I_2;
    2'b11 : F = I_3;
  endcase
```

Since this MUX has four input signals, the selection signal, Sel should be a 2-bit signal. The selection signals are represented as 2'b00, 2'b01, 2'b10, and 2'b11 in the form of <number of bits>'<base><value>. The b represents that the base is binary here. The case statement has the general form:

```
case expression
   choice1 : sequential statements1
   choice2 : sequential statements2
   . . .
   [default : sequential statements]
endcase;
```

The expression is evaluated first. If it is equal to choice1, then sequential statements1 are executed; if it is equal to choice2, then sequential statements2 are executed; and so forth. All possible values of the expression must be included in the choices. If all values are not explicitly given, a default clause is required in the case statement. As an alternative, the MUX can also be modeled using an if-else statement within an always block:

```
always @ (Sel or I0 or I1 or I2 or I3)
begin
   if      (Sel == 2'b00)    F = I0;
   else if (Sel == 2'b01)    F = I1;
   else if (Sel == 2'b10)    F = I2;
   else if (Sel == 2'b11)    F = I3;
end
```

One might notice that combinational circuits can be described using concurrent or sequential statements. Sequential circuits generally require an **always** statement. **Always** statements can be used to make sequential or combinational circuits.

The following are important coding practices while writing synthesizeable Verilog for combinational hardware:

(a) If possible use concurrent assignments (e.g., assign) to design combinational logic.
(b) When procedural assignments (always blocks) are used for combinational logic, use blocking assignments (e.g., "=").
(c) If Verilog 2001 or later is used, instead of specifying contents of sensitivity lists, use always@* to avoid accidental omission of inputs from sensitivity lists. The accidental omission results in incorrect hardware or deviation between simulation and synthesis.

2.14 Modeling Registers and Counters Using Verilog Always Statements

When several flip-flops change state on the same clock edge, statements representing these flip-flops can be placed in the same clocked always statements. Figure 2-39 shows three flip-flops connected as a **cyclic shift register** (or a **rotating shift register**). These flip-flops all change state following the rising edge of the clock. We have assumed a 5 ns propagation delay between the clock edge and the output change. Immediately following the clock edge, the three statements in the always statement execute in sequence with no delay. The new values of the Qs are then scheduled to change after 5 ns. If we omit the delay and replace the sequential statements with

```
Q1 <= Q3;    Q2 <= Q1;    Q3 <= Q2;
```

the operation is basically the same. The three statements execute in sequence in zero time, and then the Qs values change after a delta delay. In both cases, the old values of Q_1, Q_2, and Q_3 are used to compute the new values. This may seem strange

FIGURE 2-39: Cyclic Shift Register

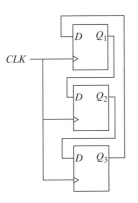

```
always @ (posedge CLK)
begin
  Q1 <= #5 Q3;
  Q2 <= #5 Q1;
  Q3 <= #5 Q2;
end
```

at first, but that is the way the hardware works. At the rising edge of the clock, all of the D inputs are loaded into the flip-flops, but the state change does not occur until after a propagation delay.

The order of the statements is not important when the non-blocking assignment operator "<=" is used. The same result is obtained even if the statements are in reverse order as shown here.

```
always @ (posedge CLK)
begin
  Q3 <= #5 Q2;
  Q2 <= #5 Q1;
  Q1 <= #5 Q3;
end
```

Example 1

What is the hardware obtained if the following code is synthesized?

```
module reg3 (Q1,Q2,Q3,A,CLK);
input    A;
input    CLK;
output   Q1,Q2,Q3;

reg      Q1,Q2,Q3;

always @(posedge CLK)
begin
  Q3 = Q2;   // statement 1
  Q2 = Q1;   // statement 2
  Q1 = A;    // statement 3
end

endmodule
```

Answer: A 3-bit shift register

Explanation: The list of statements executes from top to bottom in order. Note that the blocking operator is used. Therefore, the first statement finishes update before the second statement is executed. Synthesis results in a 3-bit shift register with serial input A, and outputs Q_1, Q_2, and Q_3.

Note: While a register can be modeled using the blocking operator "=" as in this example, it is generally advised to not do so. As mentioned previously, a good coding practice while writing synthesizeable code is to use non-blocking assignments (i.e., "<=") in always blocks intended to create sequential logic, and the blocking operator "=" in always blocks intended to create combinational logic.

Example 2

What is the hardware obtained if the following code is synthesized? Note that this is the same code as in the previous example, but with the statement order inside the always block reversed.

```
module reg31 (Q1,Q2,Q3,A,CLK);
input    A;
input    CLK;
output   Q1,Q2,Q3;

reg      Q1,Q2,Q3;

always @(posedge CLK)
begin
   Q1 = A;    // statement 1
   Q2 = Q1;   // statement 2
   Q3 = Q2;   // statement 3
end

endmodule
```

Answer: A single flip-flop

Explanation: The list of statements executes from top to bottom in order. Note that the blocking operator is used. So the first statement finishes update before the second statement is executed. Q_1 gets the value of the serial input A when statement 1 finishes. In statement 2, the same value propagates to Q_2. In statement 3, the same value propagates to Q_3. In effect, the input A has reached Q_3. Modern synthesis tools will generate a single flip-flop with input A when this code is synthesized. The outputs Q_1, Q_2, and Q_3 can all be connected to the output of the same flip-flop. If the synthesizer does not have good optimization algorithms, it might generate three parallel flip-flops, each with the same input A but with outputs Q_1, Q_2, and Q_3, respectively. As mentioned in the Note to Example 1, it is not a good practice to use the blocking operator "=" in always blocks intended to create sequential logic. If one were to use non-blocking statements, the order of the statements would not have mattered.

Figure 2-40 shows a simple register that can be loaded or cleared on the rising edge of the clock. If CLR is set to 1, the register is cleared, and if $Ld = 1$, the D inputs are loaded into the register. This register is fully synchronous so that the Q outputs change only in response to the clock edge and not in response to a change in Ld or CLR. In the Verilog code for the register, Q and D are 4-bit vectors dimensioned [3:>]. Since the register outputs can only change on the rising edge of the clock, *CLR* and *Ld* are not on the sensitivity list. The *CLR* and *Ld* signals are tested after the rising edge of the clock. If *CLR* = *Ld* = 0, no change of *Q* occurs. Since *CLR* is tested before *Ld*, if *CLR* = 1, the **else if** prevents *Ld* from being tested and *CLR* overrides *Ld*.

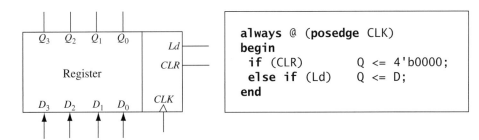

FIGURE 2-40: Register with Synchronous Clear and Load

Next, we will model a left shift register using a Verilog **always** statement. The register in Figure 2-40 is similar to that in Figure 2-41, except that we have added a left shift control input (*LS*). When *LS* is 1, the contents of the register are shifted left and the right-most bit is set equal to *Rin*. The shifting is accomplished by taking the rightmost 3 bits of *Q*, Q[2:0], and concatenating them with *Rin*. For example, if *Q* = 1101 and *Rin* = 0, then {Q[2:0], Rin} = 1010, and this value is loaded back into the *Q* register on the rising edge of *CLK*. The code implies that if *CLR* = *Ld* = *LS* = 0, then *Q* remains unchanged.

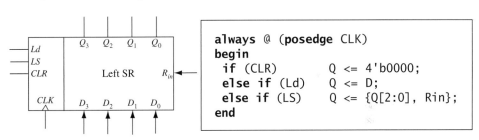

FIGURE 2-41: Left Shift Register with Synchronous Clear and Load

Figure 2-42 shows a simple synchronous counter. On the rising edge of the clock, the counter is cleared when *ClrN* = , and it is incremented when *ClrN* = *En* = 1. In this example, the signal *Q* represents the 4-bit value stored in the counter. The signal

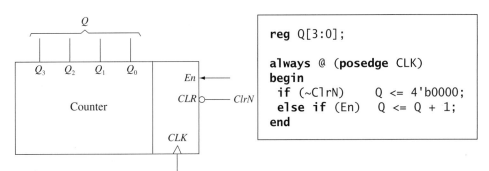

FIGURE 2-42: Verilog Code for a Simple Synchronous Counter

Q is declared to be of type **reg** with length of 4 bits. Then the statement `Q <= Q+1;` increments the counter. When the counter is in state 1111, the next increment takes it back to state 0000.

Now, let us create a Verilog model for a standard MSI counter, the 74163. It is a 4-bit fully synchronous binary counter, which is available in both TTL and CMOS logic families. Although rarely used in new designs at present, it represents a general type of counter that is found in many CAD libraries. In addition to performing the counting function, it can be cleared or loaded in parallel. All operations are synchronized by the clock, and all state changes take place following the rising edge of the clock input. A block diagram of the counter is provided in Figure 2-43.

FIGURE 2-43: 74163 Counter Operation

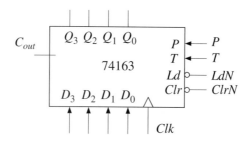

Control Signals			Next State				
ClrN	LdN	PT	Q_3^+	Q_2^+	Q_1^+	Q_0^+	
0	X	X	0	0	0	0	(clear)
1	0	X	D_3	D_2	D_1	D_0	(parallel load)
1	1	0	Q_3	Q_2	Q_1	Q_0	(no change)
1	1	1	present state + 1				(increment count)

This counter has four control inputs—*ClrN*, *LdN*, *P*, and *T*. Both *P* and *T* are used to enable the counting function. While *P* is an actual enable signal to the 4-bit generic counter, *T* is used for a carry connection signal when cascading multiple counters. Operation of the counter is as follows:

1. If *ClrN* = 0, all flip-flops are set to 0 following the rising clock edge.
2. If *ClrN* = 1 and *LdN* = 0, the *D* inputs are transferred (loaded) in parallel to the flip-flops following the rising clock edge.
3. If *ClrN* = *LdN* = 1 and *P* = *T* = 1, the count is enabled and the counter state will be incremented by 1 following the rising clock edge.

If *T* = 1, the counter generates a carry (C_{out}) in state 15; consequently

$$C_{out} = Q_3 Q_2 Q_1 Q_0 T$$

The truth table in Figure 2-43 summarizes the operation of the counter. Note that *ClrN* overrides the load and count functions in the sense that when *ClrN* = 0, clearing occurs regardless of the values of *LdN*, *P*, and *T*. Similarly *LdN* overrides the count function. The *ClrN* input on the 74163 is referred to as a *synchronous* clear input because it clears the counter in synchronization with the clock, and no clearing can occur if no clock pulse is present.

2.14 Modeling Registers and Counters Using Verilog Always Statements

The Verilog description of the counter is shown in Figure 2-44. Q represents the four flip-flops that comprise the counter. The counter output, $Qout$, changes whenever Q changes. The carry output is computed whenever Q or T changes. The always statement will be executed only at the rising edge of Clk. Since clear overrides load and count, the first **if** statement tests $ClrN$ first. Since load overrides count, LdN is tested next. Finally, the counter is incremented if both P and T are 1.

FIGURE 2-44: 74163 Counter Model

```verilog
// 74163 FULLY SYNCHRONOUS COUNTER
module c74163(LdN, ClrN, P, T, Clk, D, Cout, Qout);
input           LdN;
input           ClrN;
input           P;
input           T;
input           Clk;
input [3:0]     D;
output          Cout;
output [3:0]    Qout;

reg [3:0]       Q;

assign Qout = Q;
assign Cout = Q[3] & Q[2] & Q[1] & Q[0] & T;

always @(posedge Clk)
begin
  if (~ClrN)          Q <= 4'b0000;
  else if (~LdN)      Q <= D;
  else if (P & T)     Q <= Q + 1;
end

endmodule
```

To test the counter, we have cascaded two 74163s to form an 8-bit counter (Figure 2-45). When the counter on the right is in state 1111 and $T1 = 1$, $Carry1 = 1$. Then for the left counter, $PT = 1$ if $P = 1$. If $PT = 1$, on the next clock the right counter is incremented to 0000 at the same time the left counter

FIGURE 2-45: Two 74163 Counters Cascaded to Form an 8-Bit Counter

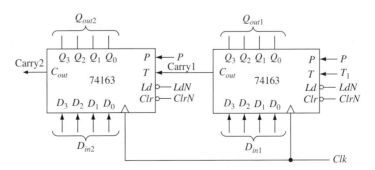

FIGURE 2-46: Verilog for 8-Bit Counter Using 4-Bit Counter Modules

```verilog
// 8-Bit counter using two 74163 counters using the model in Fig 2-44
module eight_bit_counter(ClrN, LdN, P, T1, Clk, Din1, Din2, Count, Carry2);
input         ClrN;
input         LdN;
input         P;
input         T1;
input         Clk;
input  [3:0]  Din1;
input  [3:0]  Din2;
output [7:0]  Count;
output        Carry2;

wire          Carry1;
wire   [3:0]  Qout1;
wire   [3:0]  Qout2;

c74163 ct1(LdN, ClrN, P, T1, Clk, Din1, Carry1, Qout1); // instance 1 (right)
c74163 ct2(LdN, ClrN, P, Carry1, Clk, Din2, Carry2, Qout2); //instance 2 (left)
assign Count 5 {Qout2, Qout1};
endmodule
```

is incremented. Figure 2-46 shows the Verilog code for the 8-bit counter. In this code we have used the c74163 model from Figure 2-44 as a component and have instantiated two copies of it. The module c74163 should be available for the module eight_bit_counter by defining it in same file or another file. The instantiation for the lower four bits is done using the statement

   c74163 ct1(LdN, ClrN, P, T1, Clk, Din1, Carry1, Qout1);

where c74163 is the previously defined component's module name and ct1 is the instance name. The inputs and outputs are mapped to Din1 and Qout1 and other control signals. VHDL uses a port-map keyword to accomplish this kind of instantiation whereas in Verilog, no keyword is required. This type of instantiation is required for **structural modeling** interconnecting previously defined modules. These instances should be outside always statements.

Let us now synthesize the Verilog code for a left shift register from Figure 2-41. Before synthesis is started, we must specify a target device (e.g., a particular FPGA or CPLD) so that the synthesizer knows what components are available. Let us assume that the target is a CPLD or FPGA that has D flip-flops with clock enable (D-CE flip-flops). Q and D are of length four bits. Because updates to Q follow on rising edge of the CLK, this suggests that Q must be a register composed of four flip-flops, which we will label Q_3, Q_2, Q_1, and Q_0. Since the flip-flops can change state when Clr, Ld, or Ls is 1, we connect the clock enables to an OR gate whose output is $Clr + Ld + Ls$. Then we connect gates to the D inputs to select the data to be loaded into the flip-flops. If $Clr =$ and $Ld = 1$, D is loaded into the register on the rising clock edge. If $Clr = Ld = 0$ and $Ls = 1$, then Q_2 is loaded into Q_3, Q_1 is loaded into Q_2, and so forth. Figure 2-47 shows the logic circuit for the first two flip-flops. If $Clr = 1$, the D flip-flop inputs are 0 and the register is cleared.

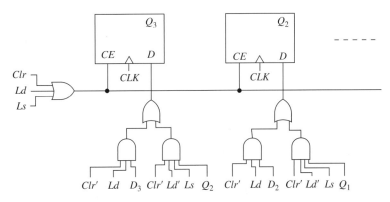

FIGURE 2-47: Synthesis of Verilog Code for Left Shift Register from Figure 2-41

> A Verilog synthesizer cannot synthesize delays. Clauses of the form "# time" will be ignored by most synthesizers, but some synthesizers may even require that the clauses be removed.
>
> Similarly, initial blocks are usually ignored by synthesis tools. Also, although initial values for signals may be specified in port and signal declarations, these initial values are usually ignored by the synthesizer. A reset signal should be provided if the hardware must be set to a specific initial state. Otherwise, the initial state of the hardware may be unknown and the hardware may malfunction. The only exception to this is that some synthesis tools for FPGAs may utilize the initial blocks and initial values in the initial bit stream that is downloaded into the FPGA. But for synthesis into custom hardware, initial values and initial blocks are typically ignored.

Avoiding Unwanted Latches

Verilog signals retain their current values until they are changed. This can result in the creation of unwanted latches when the Verilog code is synthesized. For example, in a combinational always block created using the statements

```
always @ (Sel or I₀ or I₁ or I₂)
begin
    if      (Sel == 2'b00)   F = I₀;
    else if (Sel == 2'b01)   F = I₁;
    else if (Sel == 2'b10)   F = I₂;
end
```

there would be latches to hold the value of F when Sel changes to 2'b11. Someone probably wrote this code intending a MUX. Circuits with latches are not combinational hardware any more. The latch creates a variety of timing problems and unexpected behavior. Instead of being 1 bit wide, if F is 8 bits wide, eight latches would be created.

One can avoid unwanted latches by assigning a value to combinational signals in every possible execution path in an always block intended to create combinational hardware. Hence one should include an **else** clause in every **if** statement or explicitly include all possible cases of the inputs. For example, the code

```
always @ (Sel or I0 or I1 or I2 or I3)
begin
    if        (Sel == 2'b00)   F = I0;
    else if   (Sel == 2'b01)   F = I1;
    else if   (Sel == 2'b10)   F = I2;
    else if   (Sel == 2'b11)   F = 0;
end
```

would not create any latches but would create a MUX. For **if** then **else** statements and **case** statements, it is important to have all cases specified.

Another method to avoid latches is by initializing at the beginning of the always statement. For instance, the following code is latch free, even though only three of the four possible cases are specified.

```
always @ (Sel or I0 or I1 or I2 or I3)
F = 0;
begin
    if        (Sel == 2'b00)   F = I0;
    else if   (Sel == 2'b01)   F = I1;
    else if   (Sel == 2'b10)   F = I2;
end
```

The following are important coding practices while writing synthesizable Verilog for sequential hardware:

(a) Use an edge-triggered clock in the sensitivity list using the **posedge** or **negedge** keywords.
(b) Use non-blocking assignments—that is, "<=" inside always blocks although it is possible to get sequential hardware by certain uses of the blocking "=" operator.
(c) Do not mix blocking and non-blocking statements in an always block.
(d) Do not make assignments to the same variable from more than one always block. This is not a compile-time error and hence may go unnoticed.
(e) Avoid unwanted latches by assigning a value to combinational output signals in every possible execution path in the always block. This can be done by
  i. including else clauses for if statements,
  ii. specifying all cases for case statements or have a **default** clause at the end, or
  iii. unconditionally assigning default values to all combinational output signals at the beginning of the always block.

2.15 Behavioral and Structural Verilog

Any circuit or device can be represented in multiple forms of abstraction. Consider the different representations for a NAND gate, as illustrated in Figure 2-48. When one hears the term NAND, different designers, depending on their domain of

FIGURE 2-48: Different Levels of Abstraction of a NAND Device

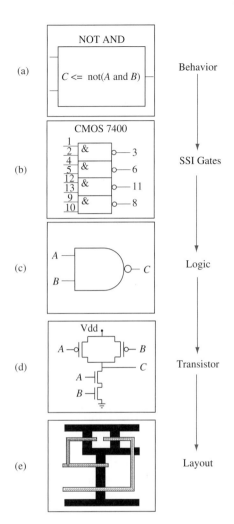

expertise, think of these different representations of the same NAND device. Some would think of just a block representing the behavior of a NAND operator, as illustrated in Fig 2-48(a). Some others might think of the four gates in a CMOS 7400 chip, as in Figure 2-48(b). Designers who work at the logic level think of the logic symbol for a NAND gate, as in Figure 2-48(c). Transistor-level circuit designers think of the transistor-level circuit to achieve the NAND functionality, as in Figure 2-48(d). What passes through the mind of a physical-level designer is the layout of a NAND gate, as in Figure 2-48(e). All of the figures represent the same device, but they differ in the amount of detail provided in the description.

Just as a NAND gate can be described in different ways, any logic circuit can be described with different levels of detail. Figure 2-49 indicates a behavioral level representation of the logic function $F = ab + bc$, whereas Figures 2-50(a) and 2-50(b) illustrate two equivalent structural representations. The functionality specified in the abstract description in Figure 2-49 can be achieved in different ways, two examples of which are by using two AND gates and one OR gate or three NAND gates.

A structural description gives different descriptions for Figures 2-50(a) and 2-50(b), whereas the same behavioral description could result in either of these two representations. A structural description specifies more details, whereas the behavioral level description specifies only the behavior at a higher level of abstraction.

FIGURE 2-49: A Block Diagram with A, B, C as Inputs and F = AB + BC as Output

FIGURE 2-50: Two Implementations of F = AB + BC

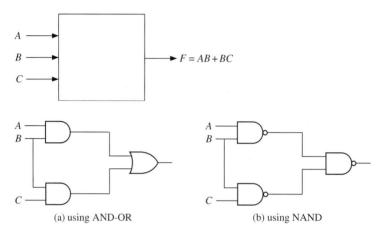

(a) using AND-OR   (b) using NAND

You probably noticed that the same circuit can be described in different ways. Similarly, Verilog allows you to create design descriptions at multiple levels of abstraction. The most common ones are **behavioral models**, **data flow (RTL) models**, **and structural models**. Behavioral Verilog models describe the circuit or system at a high level of abstraction without implying any particular structure or technology. Only the overall behavior is specified. In contrast, in structural models, the components used and the structure of the interconnection between the components are clearly specified. Structural models may be detailed enough to specify use of particular gates and flip-flops. The structural Verilog model is at a low level of abstraction. Verilog code can be written at an intermediate level of abstraction, at the data flow level or the RTL level, in addition to the pure behavioral level or structural level. RTL stands for Register Transfer Language. Register Transfer Languages have been used for decades to describe the behavior of synchronous systems where a system is viewed as registers plus the control logic required to perform loading and manipulation of registers. In the data flow model, data path and control signals are specified. The working of the system is described in terms of the data transfer between registers.

If designs are specified at higher levels of abstraction, they need to be converted to the lower levels in order to be implemented. In the early days of design automation, there were not enough automatic software tools to perform this conversion; hence, designs needed to be specified at the lower levels of abstraction. Designs were entered using schematic capture or lower levels of abstraction. Nowadays, synthesis tools perform very efficient conversion of behavioral level designs into target technologies.

Behavioral and structural design techniques are often combined. Different parts of the design are often done with different techniques. State-of-the-art design automation tools generate efficient hardware for logic and arithmetic circuits; hence, a large part of those designs are done at the behavioral level. However, memory structures often need manual optimizations and are done by custom design, as opposed to automatic synthesis.

2.15.1 Modeling a Sequential Machine

In this section, we discuss several ways of writing Verilog descriptions for sequential machines. Let us assume that we have to write a **behavioral** model for a Mealy sequential circuit represented by the state table in Figure 2-51 (one may note that this is the BCD to Excess-3 code converter designed in Chapter 1). A block diagram of this state machine is also shown in Figure 2-51. This view of the circuit can be used to write its entity description. Please note that the current state and next state are not visible externally.

FIGURE 2-51: State Table and Block Diagram of Sequential Machine

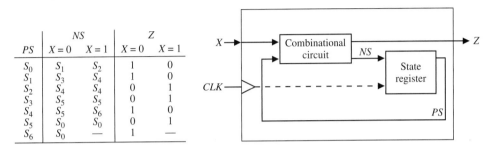

PS	NS X=0	NS X=1	Z X=0	Z X=1
S_0	S_1	S_2	1	0
S_1	S_3	S_4	1	0
S_2	S_4	S_4	0	1
S_3	S_5	S_5	0	1
S_4	S_5	S_6	1	0
S_5	S_0	S_0	0	1
S_6	S_0	—	1	—

There are several ways to model this sequential machine. One approach would be to use two **always** blocks to represent the two parts of the circuit. One **always** block models the combinational part of the circuit and generates the next state information and outputs. The other **always** block models the state register and updates the state at the appropriate edge of the clock. Figure 2-52 illustrates such a model for this Mealy machine. The first always block represents the combinational circuit. At the behavioral level, we will represent the state and next state of the circuit by integer signals initialized to 0. Please remember that this initialization is meaningful only for simulations. Since the circuit outputs, Z and *Nextstate*, can change when either the *State* or X changes, the sensitivity list includes both *State* and X. The case statement tests the value of *State*, and depending on the value of X, Z and *Nextstate* are assigned new values. The second always block represents the state register. Whenever the rising edge of the clock occurs, *State* is updated to the value of *Nextstate*, so *CLK* appears in the sensitivity list. The second always block will simulate correctly if written as

```
always @ (CLK)                  // State Register
begin                           // rising edge of clock (simulation)
   State <= Nextstate;
end
```

but in order to synthesize with edge-triggered flip-flops, the event expression (posedge) or negedge must be used, as in

```
always @ (posedge CLK)          // State Register
begin                           // (synthesis)
   State <= Nextstate;          // rising edge of clock
end
```

In Figure 2-52, *State* is defined as 3-bit register. If *State* is defined as an integer, some synthesis tools may synthesize it to be a 32-bit register, but most modern synthesis tools will figure out that only values less than 7 are used and will synthesize it to a 3-bit register. The statement **default** is not actually needed when the outputs

FIGURE 2-52: Behavioral Model for Excess-3 Code Converter

```verilog
// This is a behavioral model of a Mealy state machine (Figure 2-51)
// based on its state table. The output (Z) and next state are
// computed before the active edge of the clock. The state change
// occurs on the rising edge of the clock.
module Code_Converter(X, CLK, Z);

input X, CLK;
output Z;

reg     Z;
reg     [2:0] State;
reg     [2:0] Nextstate;

initial
begin
   State = 0;
   Nextstate = 0;
end

always @(State or X)
begin                                   // Combinational Circuit
case(State)
  0 : begin
          if(X == 1'b0)
          begin
              Z = 1'b1;
              Nextstate = 1;
          end
          else
          begin
              Z = 1'b0;
              Nextstate = 2;
          end
      end
  1 : begin
          if(X == 1'b0)
          begin
              Z = 1'b1;
              Nextstate = 3;
          end
          else
          begin
              Z = 1'b0;
              Nextstate = 4;
          end
      end
  2 : begin
          if(X == 1'b0)
          begin
              Z = 1'b0;
```

```verilog
                    Nextstate = 4;
            end
            else
            begin
                    Z = 1'b1;
                    Nextstate = 4;
            end
    end
3 : begin
            if(X == 1'b0)
            begin
                    Z = 1'b0;
                    Nextstate = 5;
            end
            else
            begin
                    Z = 1'b1;
                    Nextstate = 5;
            end
    end
4 : begin
            if(X == 1'b0)
            begin
                    Z = 1'b1;
                    Nextstate = 5;
            end
            else
            begin
                    Z = 1'b0;
                    Nextstate = 6;
            end
    end
5 : begin
            if(X == 1'b0)
            begin
                    Z = 1'b0;
                    Nextstate = 0;
            end
            else
            begin
                    Z = 1'b1;
                    Nextstate = 0;
            end
    end
6 : begin
            if(X == 1'b0)
            begin
                    Z = 1'b1;
                    Nextstate = 0;
            end
            else
```

```
              begin
                Z = 1'b0;
                Nextstate = 0;
              end
          end
      default : begin
      // should not occur
      end
    endcase
  end
  always @(posedge CLK)        // State Register
    begin                       // rising edge of clock
      State <= Nextstate;
    end
endmodule
```

and next states of all possible values of *State* are explicitly specified. In this case, however, it should be included because one of the eight states is not specified as a case. The empty statement implies no action, which is appropriate, since the other values of *State* should never occur. In case statements, the *State* variable should be bound to a specific number of bits, or results will be unpredictable.

A simulator command file that can be used to test Figure 2-52 is as follows:

```
add wave CLK X State Nextstate Z
force CLK 0 0, 1 100 -repeat 200
force X 0 0, 1 350, 0 550, 1 750, 0 950, 1 1350
run 1600
```

The first command specifies the signals that are to be included in the waveform output. The next command defines a clock with a period of 200 ns. *CLK* is 0 at time 0 ns, is 1 at time 100 ns, and repeats every 200 ns. In a command of the form

```
force signal_name v1 t1, v2 t2, ...
```

signal_name gets the value v1 at time t1, the value v2 at time t2, and so forth. *X* is 0 at time 0 ns, changes to 1 at time 350 ns, changes to 0 at time 550 ns, and so on. The *X* input corresponds to the sequence 0010 1001, and only the times at which *X* changes are specified. Execution of the preceding command file produces the waveforms shown in Figure 2-53.

FIGURE 2-53: Simulator Output for Excess-3 Code Converter

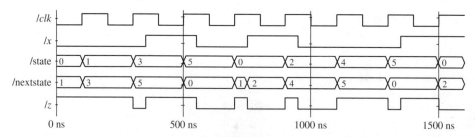

In Chapter 1, we manually designed this state machine (Figure 1-26). This circuitry contained three flip-flops, four 3-input NAND gates, two 3-input NAND gates, and one inverter. The behavioral model of Figure 2-52 may not result in exactly that circuit. In fact, when we synthesized it using Xilinx ISE tools, we obtained a circuit that contains seven D-flip-flops, fifteen 2-input AND gates, three 2-input OR gates, and one 7-input OR gate. Apparently, the Xilinx synthesis tool may be using one-hot design by default, instead of encoded design. One-hot design is a popular approach for FPGAs, where flip-flops are abundant.

Figure 2-54 shows an alternative behavioral model for the code converter that uses a single **always** block instead of two **always** blocks. The next state is not computed explicitly, but instead the state register is updated directly to the proper next state value on the rising edge of the clock. Since Z can change whenever State or X changes, Z should not be computed in the clocked block. Instead, we have used a conditional assignment statement to compute Z. If Z were updated in the clocked block, then a flip-flop would be created to store Z and Z would be updated at the wrong time. In general, the two always block model for a state machine is preferable

FIGURE 2-54: Behavioral Model for Code Converter Using a Single Process

```verilog
// This is a behavioral model of the Mealy state machine for BCD to
// Excess-3 Code Converter based on its state table. The state change
// occurs on the rising edge of the clock. The output is computed by a
// conditional assignment statement whenever State or Z changes.
module Code_Converter(X, CLK, Z);
input X, CLK;
output Z;
wire    X;              // Redundant declaration of X, CLK and Z.
wire    CLK;            // Input and output signals are wire data
wire    Z;              // type by default
reg     [2:0] State;
initial
begin
  State = 0;
end
always @(posedge CLK) begin
  case(State)
    0 : begin
      if(X == 1'b0)
        State <= 1;
      else
        State <= 2;
    end
    1 : begin
      if(X == 1'b0)
        State <= 3;
      else
        State <= 4;
    end
    2 :
      State <= 4;
    3 :
      State <= 5;
    4 : begin
```

```
            if(X == 1'b0)
               State <= 5;
            else
               State <= 6;
         end
         5 :
            State <= 0;
         6 :
            State <= 0;
         default : begin
         end
      endcase
   end
   assign Z = (State == 0 && X == 1'b0) || (State == 1 && X == 1'b0) ||
              (State == 2 && X == 1'b1) || (State == 3 && X == 1'b1) ||
              (State == 4 && X == 1'b0) || (State == 5 && X == 1'b1) ||
              State == 6 ? 1'b1 : 1'b0;
endmodule
```

to the one always block model, since the former corresponds more closely to the hardware implementation which uses a combinational circuit and a state register.

Another way to model this Mealy machine is using the **data flow** approach (i.e., using equations). The data flow Verilog model of Figure 2-55 is based on the next state and output equations, which are derived in Chapter 1 (Figure 1-25). The flip-flops are updated in an always block that is sensitive to CLK. When the rising edge of the clock occurs, Q_1, Q_2, and Q_3 are all assigned new values. A 10 ns delay is included

FIGURE 2-55: Sequential Machine Model Using Equations

```
// The following is a description of the sequential machine of
// the BCD to Excess-3 code converter in terms of its next state
// equations obtained as in Figure 1-25. The following state assignment was
// used: S0-->0; S1-->4; S2-->5; S3-->7; S4-->6; S5-->3; S6-->2
module Code_Converter(X, CLK, Z);
   input   X;
   input   CLK;
   output  Z;

   reg     Q1;
   reg     Q2;
   reg     Q3;

   always @(posedge CLK)

   begin
      Q1 <= #10 (~Q2);
      Q2 <= #10 Q1;
      Q3 <= #10 (Q1 & Q2 & Q3) | ((~X) & Q1 & (~Q3)) |
            (X & (~Q1) & (~Q2));
   end

   assign #20 Z = ((~X) & (~Q3)) | (X & Q3);
endmodule
```

to represent the propagation delay between the active edge of the clock and the change of the flip-flop outputs. Even though the statements in the always block are executed sequentially, Q_1, Q_2, and Q_3 are all scheduled to be updated at the same time, $T + \Delta$, where T is the time at which the rising edge of the clock occurred. Thus, the old value of Q_1 is used to compute Q_2^+, and the old values of Q_1, Q_2, and Q_3 are used to compute Q_3^+. The concurrent assignment statement for Z causes Z to be updated whenever a change in X or Q_3 occurs. The 20 ns delay represents two gate delays. One may note that in order to do Verilog modeling at this level, one needs to perform state assignments, derive next-state equations, etc. In contrast, at the behavioral level, the state table was sufficient to create the Verilog model.

Yet another approach to creating a Verilog model of the foregoing Mealy machine is to create a **structural** model describing the gates and flip-flops in the circuit. Figure 2-58 shows a structural Verilog representation of the circuit of Figure 1-26. One should note that the designer had to manually perform the design and obtain the gate level circuitry here in order to create a model as in Figure 2-58. Seven NAND gates, three D flip-flops, and one inverter are used in the design presented in Figure 1-26. When primitive components like gates and flip-flops are required, each of these components can be defined in a separate Verilog module. One could also use the built-in primitives predefined in Verilog. In this chapter, we are going to build our own building blocks. Predefined built-in primitives are described in Section 8.4. Depending on which CAD tools are used, the component modules must be included in the same file as the main Verilog description, or they must be inserted as separate files in a Verilog project. The code in Figure 2-58 requires component modules DFF, Nand3, Nand2, and Inverter. CAD tools might include modules with similar components. If such modules are used, one should use the exact component names and port-map statements that match the input-output signals of the component. The DFF module is as follows:

FIGURE 2-56: D Flipflop

```
// D Flip-Flop
module DFF(D, CLK, Q, QN);
    input    D;
    input    CLK;
    output   Q;
    output   QN;

    reg      Q;
    reg      QN;

    initial
    begin
       Q  = 1'b0;
       QN = 1'b1;
    end

    always @(posedge CLK)
        begin
           Q  <= #10 D;
           QN <= #10 (~D);
        end
endmodule
```

The Nand3 module is as follows:

FIGURE 2-57: 3-Input NAND Gate

```
// 3 input NAND gate
module Nand3(A1, A2, A3, Z);
    input    A1;
    input    A2;
    input    A3;
    output   Z;
    assign #10 Z = (~(A1 & A2 & A3));
endmodule
```

The Nand2 and Inverter modules are similar except for the number of inputs. We have assumed a 10-ns delay in each component, and this can easily be changed to reflect the actual delays in the hardware being used.

FIGURE 2-58: Structural Model of Sequential Machine

```
// The following is a STRUCTURAL Verilog description of
// the circuit to realize the BCD to Excess-3 code Converter.
// This circuit was illustrated in Figure 1-26.
// Uses components NAND3, NAND2, INVERTER and DFF
// The component modules can be included in the same file
// or they can be inserted as separate files.

module Code_Converter(X, CLK, Z);

input X, CLK;
output  Z;
wire    X;
wire    CLK;
wire    Z;
wire    A1;
wire    A2;
wire    A3;
wire    A5;
wire    A6;
wire    D3;
wire    Q1;
wire    Q2;
wire    Q3;
wire    Q1N;
wire    Q2N;
wire    Q3N;
wire    XN;

Inverter I1(X, XN);
Nand3 G1(Q1, Q2, Q3, A1);
```

```
Nand3 G2(Q1, Q3N, XN, A2);
Nand3 G3(X, Q1N, Q2N, A3);
Nand3 G4(A1, A2, A3, D3);
DFF FF1(Q2N, CLK, Q1, Q1N);
DFF FF2(Q1, CLK, Q2, Q2N);
DFF FF3(D3, CLK, Q3, Q3N);
Nand2 G5(X, Q3, A5);
Nand2 G6(XN, Q3N, A6);
Nand2 G7(A5, A6, Z);

endmodule
```

Since *Q1*, *Q2*, and *Q3* are initialized to 0, the complementary flip-flop outputs (*Q1N*, *Q2N*, and *Q3N*) are initialized to 1. *G1* is a 3-input NAND gate with inputs *Q1*, *Q2*, *Q3*, and output *A1*. *FF1* is a D flip-flop with the *D* input connected to *Q2N*. Executing the simulator command file given below produces the waveforms of Figure 2-59, which are very similar to Figure 1-39.

```
add wave CLK X Q1 Q2 Q3 Z
force CLK 0 0, 1 100 -repeat 200
force X 0 0, 1 350, 0 550, 1 750, 0 950, 1 1350
run 1600
```

FIGURE 2-59: Waveforms for Code Converter

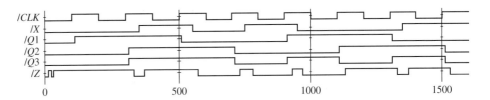

If you synthesize this structural description, you will certainly get exactly the same circuit that you had in mind. Now the circuit includes only three D-flip-flops, three 2-input NAND gates, and four 3-input NAND gates. Compare it with the seven D-flip-flops, fifteen 2-input AND gates, three 2-input OR gates, and one 7-input OR gate generated when Figure 2-52 was synthesized. When the designer specified all components and their interconnections, the synthesizer tool did not have to infer or "guess."

> **Synthesis with integers:** Integers are generally treated as 32-bit quantities. While writing synthesizeable code, it is a good idea to use the appropriate number of bits for your variables as needed as opposed to using integers. One should use integers only where they are not explicitly synthesized, eg: looping variables. If integers are used for state variables in the state machine design presented in these pages, the synthesis tools we used could figure out that the state variable is actually only 3-bits long. However, some tools may generate 32-bit registers.

Those who have developed C code with assembly inlining may feel some similarity to the phenomenon occurring here. By inlining the assembly code, one can precisely describe what microprocessor instruction sequence you want to be used, and the compiler gives you that. In a similar way, the synthesizer does not actually have to translate any structural descriptions that the designer wrote; it simply gives the hardware that the designer specified in a structural fashion. Some optimizing tools are capable of optimizing imperfect circuits that you might have specified. In general, you have more control over the generated circuitry when you use structural coding. However, it takes a lot more effort to produce a structural model, because one needs to perform state assignments, derive next state equations, and so forth. **Time to market** is an important criterion for success in the IC market; hence, designers often use behavioral design in order to achieve quick time to market. Additionally, CAD tools have matured significantly during the past decade, and most modern synthesis tools are capable of producing efficient hardware for arithmetic and logic circuits.

2.16 Constants

Verilog provides constants in addition to variables and nets. Verilog provides three different ways to define constant values, one of which is using `define as in

```
`define   constant_name   constant_value
```

The `define is one of the compiler directives in Verilog and is used to define a number or an expression for a meaningful string. The ` in `define is called *grave accent* (ASCII 0x60). It is different from the character ('), which is the *apostrophe* character (ASCII 0x27). More on compiler directives can be found in Chapter 8.

The `define compiler directive replaces 'constant_name with constant_value. For example:

```
`define wordsize 16
 reg [1:`wordsize] data;
```

causes the string wordsize to be replaced by 16. It then shows how data is declared to be a reg of width wordsize.

Another method to create constants is to use the parameter keyword as follows:

```
parameter constant_name = constant_value;
```

For example,

```
parameter msb = 15; // defines msb as a constant value 15
parameter [31:0] decim = 1'b1; // value converted to 32 bits
```

Another method to make constants is using localparam.

```
localparam constant_name = constant_value;
```

The localparam is similar to the parameter, but it cannot be directly changed. The localparam can be used to define constants that should not be changed.

Verilog can define constant values in a module using the parameter. The parameter can be used to customize the module instances. Typical uses of parameters are to specify delays and width of variables.

Verilog HDL parameters do not belong to either the variable or the net group. Parameters are not variables; they are constants. Since parameters represent constants, it is illegal to modify their values at run time. However, module parameters can be modified at compilation time to have values that are different from those specified in the declaration assignment. This allows customization of module instances. The parameter values can be changed at module instantiation or by using **defparam** statement. These are described in Chapter 8.

2.17 Arrays

A key feature of VLSI circuits is the repeated use of similar structures. Arrays in Verilog can be used while modeling the repetition. Digital systems often use memory arrays. Verilog arrays can be used to create memory arrays and specify the values to be stored in these arrays. In order to use an array in Verilog, we must declare the array upper and lower bound. There are two positions to declare the array bounds:

In one option, the array bounds are declared between the variable type (reg or net) and the variable name, and the array bound means the number of bits for the declared variable. If the array bound is defined as [7:0] as shown in the following example,

```
reg    [7:0]    eight_bit_register;
```

the register variable `eight_bit_register` can store one byte (eight bits) of information. The 8-bit register can be initialized to hold the value 00000001 using the following statement:

```
eight_bit_register = 8'b00000001;
```

As a second option, array bounds can be declared after the name of the array. In the declaration that follows, **rega** is an array of n 1-bit registers while **regb** is a single n-bit register.

```
reg    rega [1:n]; // This is an array of n 1-bit registers
reg    [1:n] regb; // This is an n-bit register
```

We can define multiple 8-bit registers in one array declaration. In this case, additional upper and lower bound(s) must be declared after the name of the array. In the example that follows, 16 registers are declared; each register can store one-byte (8-bit) vector information.

```
reg    [7:0]    eight_bit_register_array [15:0];
```

The foregoing declaration means that each of the 16 variables in the array can have 8-bit vector information. This array can be initialized as follows:

```
eight_bit_register_array[15] = 8'b00001100;
eight_bit_register_array[14] = 8'b00000000;
           . . . . . . .
eight_bit_register_array[1]  = 8'b11001100;
eight_bit_register_array[0]  = 8'b00010001;
```

Arrays can be created of various data types. Arrays of wires and integers can be declared as follows:

```
wire wire_array[5:0]; // declares an array of 6 wires
integer inta[1:64]; // declares an array of 64 integer values
```

Matrices

Multidimensional array types may also be defined with two or more dimensions. The following example defines a 2-dimensional array variable in an initial statement, which is a matrix of integers with four rows and three columns with 8-bit elements:

```
reg  [7:0]  matrixA [0:3][0:2] = { { 1,  2,  3},
                                   { 4,  5,  6},
                                   { 7,  8,  9},
                                   {10, 11, 12}};
```

The array element *matrixA[3][1]* references the element in the fourth row and second column, which has a value of 11.

Look-Up Table Method Using Arrays and Parameters

The array construct together with parameter can be used to create look-up tables which can be used to create combinational circuits using the ROM or Look-up Table (LUT) method.

Example

Parity bits are often used in digital communication for error detection and correction. The simplest of these involve transmitting one additional bit with the data, a parity bit. Use Verilog arrays to represent a parity generator that generates a 5-bit-odd-parity generation for a 4-bit input number using the look-up table (LUT) method.

Answer: The input word is a 4-bit binary number. A 5-bit odd-parity representation will contain exactly an odd number of 1s in the output word. This can be accomplished by the ROM or LUT method using a look-up table of size 16 entries × 5 bits. The look-up table is indicated in Figure 2-60.

FIGURE 2-60: LUT Contents for a Parity Code Generator

Input (LUT Address)				Output (LUT Data)				
A	B	C	D	P	Q	R	S	T
0	0	0	0	0	0	0	0	1
0	0	0	1	0	0	0	1	0
0	0	1	0	0	0	1	0	0
0	0	1	1	0	0	1	1	1
0	1	0	0	0	1	0	0	0
0	1	0	1	0	1	0	1	1
0	1	1	0	0	1	1	0	1
0	1	1	1	0	1	1	1	0
1	0	0	0	1	0	0	0	0
1	0	0	1	1	0	0	1	1
1	0	1	0	1	0	1	0	1
1	0	1	1	1	0	1	1	0
1	1	0	0	1	1	0	0	1
1	1	0	1	1	1	0	1	0
1	1	1	0	1	1	1	0	0
1	1	1	1	1	1	1	1	1

The Verilog code for the parity generator is illustrated in Figure 2-61. The first 4 bits of the output are identical to the input. Hence, instead of storing all 5 bits of the output, one might store only the parity bit and then concatenate it to the input bits. The parameter construct is used to define the ParityBit which is 1-bit constant, and the 4-bit input data and 1-bit ParityBit are concatenated to make a parity code as an output.

FIGURE 2-61: Parity Code Generator Using the LUT Method

```
module parity_gen(X, Y);
input     [3:0] X;
output    [4:0] Y;

wire            ParityBit;

parameter [0:15] OT = {1'b1, 1'b0, 1'b0, 1'b1, 1'b0, 1'b1, 1'b1, 1'b0, 1'b0, 1'b1,
1'b1, 1'b0, 1'b1, 1'b0, 1'b0, 1'b1};

assign ParityBit = OT[X];
assign Y = {X, ParityBit};

endmodule
```

2.18 Loops in Verilog

Often, one has systems where some activity is happening in a repetitive fashion. Verilog loop statements can be used to express this behavior. A loop statement is a sequential statement. Verilog has several kinds of loop statements including **for** loops and **while** loops. There is also a **repeat** loop.

Forever Loop (Infinite Loop)

Infinite loops are undesirable in common computer languages, but they can be useful in hardware modeling where a device works continuously and continues to work until the power is off. Below is an example for a forever loop:

```
begin
   clk = 1'b0;
   forever #10 clk = ~clk;
end
```

For Loop

One way to augment the basic loop is to use the **for** loop, where the number of invocations of the loop can be specified. Syntax is similar to C language except that begin and end are used instead of { } for more than one statement. Note that, unlike C language, we do not have the operators ++ and − in Verilog. Instead of using those operators as in C language, we need to use its full operational statement, i = i + 1. The general form of a **for** loop is as follows:

```
for (initial_statement; expression; incremental_statement)
begin

    sequential statement(s);

end
```

The following example is of initializing an array variable, eight_bit_register_array, using a **for** loop.

```
reg  [7:0]  eight_bit_register_array [15:0];

for (i=0; i<16; i=i+1)
begin

  register A[i] = 8'b00000000;

end
```

One could use this type of loop in behavioral models. The following excerpt models a 4-bit adder. The loop index (*i*) will be initialized to 0 when the **for** loop is entered, and the sequential statements will be executed. Execution will be repeated for $i = 1$, $i = 2$, and $i = 3$; then the loop will terminate. The carry out from one iteration (C_{out}) is copied to the carry in (C_{in}) before the end of the loop. Since variables are used for the sum and carry bits, the update of carry out happens instantaneously. Code like this often appears in Verilog tasks and functions (described in Chapter 8).

```
for (i=0; i<4; i=i+1)
begin

  Cout = (A[i] && B[i]) || (A[i] && Cin) || (B[i] && Cin);
    sum[i] = A[i] ^ B[i] ^ Cin;
    Cin = Cout;

end
```

You could also use the **for** loop to create multiple copies of a basic cell. When the foregoing code is synthesized, the synthesizer typically provides four copies of a 1-bit adder connected in a ripple carry fashion. If you actually desire to create just one copy of the cell and simply use it multiple times as a serial adder, then you have to design a sequential circuit. The loop construct will not synthesize into that behavior.

While Loop

As in **while** loops in most languages, a condition is tested before each iteration. The loop is terminated if the condition is false. The general form of a while loop is

```
while condition
begin

    sequential statements;

end
```

A while loop is used primarily for simulation.

Figure 2-62 illustrates a while loop that models a down counter. We use the **while** statement to continue the decrementing process until the stop is encountered or the counter reaches 0. The counter is decremented on every rising edge of clk until either the count is 0 or stop is 1.

FIGURE 2-62: Use of While Loop

```verilog
`define MAX 100
  module counter_100;
  integer count;
  initial begin
    count = 0;
    while (count < `MAX) begin
      count = count + 1;
    end // while

    $display("number = %d ", count);
  end // initial begin
endmodule
```

Repeat Loop

The repeat loop repeats the sequential statement(s) for specified times. The number of repetitions is set by a constant value or a logical expression.

```verilog
repeat( 8 )
  begin

    x = x + 1;
    y = y + 2;

  end
```

2.19 Testing a Verilog Model

Once a Verilog model for a system has been made, the next step is to test it. A model has to be tested and validated before it can be successfully used.

A test bench is a piece of Verilog code that can provide input combinations to test a Verilog model for the system under test. It provides stimuli to the system or circuit under test. Test benches are frequently used during simulation to provide sequences of inputs to the circuit or Verilog model under test. Figure 2-63 shows a test bench for testing the 4-bit binary adder that we created earlier in this chapter. The adder we are testing will be treated as a component and will be embedded in the test-bench program. The signals generated within the test bench are interfaced to the adder, as shown in Figure 2-63. The module on the right side is the module under test.

FIGURE 2-63: Interfacing of Signals While Using a Test Bench to Test a 4-Bit Adder

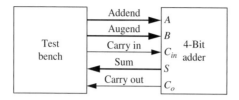

Figure 2-64 shows an example of a test bench for a 4-bit adder without using assertion monitors.

Assertion statements available in languages such as System Verilog provide an easy mechanism to specify properties that the design must adhere to and to verify them. Verilog does not have assertion statements. Hence in this test bench, we simply provide test inputs and check whether the module outputs match the expected outputs. It may be noticed that the test bench module does not have external inputs and outputs; hence, the port list is empty in the module declaration in statement 1. One can see use of Verilog parameter construct in statement 2.

FIGURE 2-64: Test Bench for 4-Bit Adder

```verilog
module TestAdder_v2;  //Statement 1
parameter  N = 11;    //Statement 2
reg   [3:0] addend;
reg   [3:0] augend;
reg         cin;
wire  [3:0] sum;
wire        cout;

reg [3:0] addend_array[1:N];
reg [1:N] cin_array;
reg [3:0] augend_array[1:N];
reg [3:0] sum_array[1:N];
reg [1:N] cout_array;

initial
begin
  //initialization of addend_array
  addend_array[1]  = 4'b0111;
  addend_array[2]  = 4'b1101;
  addend_array[3]  = 4'b0101;
  addend_array[4]  = 4'b1101;
  addend_array[5]  = 4'b0111;
  addend_array[6]  = 4'b1000;
  addend_array[7]  = 4'b0111;
  addend_array[8]  = 4'b1000;
  addend_array[9]  = 4'b0000;
  addend_array[10] = 4'b1111;
  addend_array[11] = 4'b0000;
```

```verilog
      //initialization of cin_array
      cin_array[1]  = 1'b0;
      cin_array[2]  = 1'b0;
      cin_array[3]  = 1'b0;
      cin_array[4]  = 1'b0;
      cin_array[5]  = 1'b1;
      cin_array[6]  = 1'b0;
      cin_array[7]  = 1'b0;
      cin_array[8]  = 1'b0;
      cin_array[9]  = 1'b1;
      cin_array[10] = 1'b1;
      cin_array[11] = 1'b0;

      //initialization of augend_array
      augend_array[1]  = 4'b0101;
      augend_array[2]  = 4'b0101;
      augend_array[3]  = 4'b1101;
      augend_array[4]  = 4'b1101;
      augend_array[5]  = 4'b0111;
      augend_array[6]  = 4'b0111;
      augend_array[7]  = 4'b1000;
      augend_array[8]  = 4'b1000;
      augend_array[9]  = 4'b1101;
      augend_array[10] = 4'b1111;
      augend_array[11] = 4'b0000;

      //initialization of sum_array (expected sum outputs)
      sum_array[1]  = 4'b1100;
      sum_array[2]  = 4'b0010;
      sum_array[3]  = 4'b0010;
      sum_array[4]  = 4'b1010;
      sum_array[5]  = 4'b1111;
      sum_array[6]  = 4'b1111;
      sum_array[7]  = 4'b1111;
      sum_array[8]  = 4'b0000;
      sum_array[9]  = 4'b1110;
      sum_array[10] = 4'b1111;
      sum_array[11] = 4'b0000;

      //initialization of cout_array (expected carry output)
      cout_array[1]  = 1'b0;
      cout_array[2]  = 1'b1;
      cout_array[3]  = 1'b1;
      cout_array[4]  = 1'b1;
      cout_array[5]  = 1'b0;
      cout_array[6]  = 1'b0;
      cout_array[7]  = 1'b0;
      cout_array[8]  = 1'b1;
      cout_array[9]  = 1'b0;
      cout_array[10] = 1'b1;
      cout_array[11] = 1'b0;
end
```

```verilog
integer i;
always
begin
  for(i = 1 ; i <= N ; i = i + 1)
  begin
    $display(i);
    addend <= addend_array[i];//apply an addend test vector
    augend <= augend_array[i];//apply an augend test vector
    cin <= cin_array[i];//apply a carry in

    #(40);//adder expected to take 40 time units

    if(!(sum == sum_array[i] & cout == cout_array[i]))
    begin
      $write("ERROR: ");
      $display("Wrong Answer ");
    end
    else begin
      $display("Correct!!");
    end
  end
  $display("Test Finished");
end
Adder4 add1(addend, augend, cin, sum, cout); //module under test instantiated
endmodule
```

Input and output vectors are hard-coded into the test bench. An exhaustive test of the adder module requires 512 tests, since there are 9 input bits (4 addend; 4 augend and 1 carryin). We have a random set of 11 tests.

The **$display** and **$write** statements are used to print test results. The two sets of tasks are identical except that **$display** automatically adds a new line character to the end of its output, whereas the **$write** task does not. In Verilog, a name following the $ is interpreted as a *system task* or a *system function*. The dollar sign (**$**) essentially introduces a language construct that enables the development of user-defined system tasks and functions. These system tasks are not design semantics, but refer to simulator functionality.

The test module has to be instantiated into the test bench outside the always statement. In this example, Adder4 is the adder module from Figure 2-12. It is instantiated into the test bench and the inputs mapped to test vectors. Use of appropriate delays becomes important in test benches. A **#(40)** delay is specified after applying a set of inputs so that the simulation model has time to evaluate the results. This delay value is computed based on the fact that each bit of the ripple carry adder takes 10ns in the model in Figure 2-10. Use of incorrect values of delays may give misleading test results. An improperly created test bench may incorrectly indicate that a properly working circuit is faulty.

Similar to **$display**, there are several other Verilog statements to observe outputs—for example, **$strobe**, and **$monitor**. The format string is like that in C/C++ and may contain format characters, as in the following:

```
$display ("Value is %d", para1);
$strobe  ("Value is %d", para1);
$monitor ("Value is %d", para1);
```

The **$display** and **$strobe** display once every time they are executed (i.e., every time the statement is encountered), whereas **$monitor** displays every time one of its parameters changes. The difference between **$display** and **$strobe** is that **$strobe** displays the parameters at the very end of the current simulation time unit whereas **$display** outputs them exactly where it is executed.

2.20 A Few Things to Remember

Verilog has different kinds of signal assignments and constructs. Which construct is to be used depends on the purpose of the Verilog model being created. Verilog is typically written for the following three reasons:

i. to design hardware (i.e., to model and synthesize hardware)
ii. to model hardware (i.e., to create simulation models that are not necessarily synthesizeable)
iii. to verify hardware (i.e., to test designs)

The following are illegal in all models:

```
assign a <= b;      // statement 1
assign a <= #10 b;  // statement 2
assign a =  #10 b;  // statement 3
```

The assign statement is to be used with '=', not with '<='. Hence the first two statements are illegal. Statement 3 uses delayed assignment, which cannot be done with the assign statement. No delay can be specified on the right side of a continuous assign statement. Hence statement 3 is illegal.

The following are legal.

```
assign a = b;       \\ Statement 4
assign #10 a = b;   \\ Statement 5
```

Statement 4 is a concurrent statement. It executes concurrently to all other concurrent assign statements and always blocks. Statement 5 behaves with the inertial type of delay.

How we use assign, blocking, and non-blocking concepts depends on what we are trying to write Verilog for. Some strategies can be followed to result in appropriate models for each purpose.

1. **To design hardware (i.e., to model and synthesize hardware)** This is a major use of Verilog, and the following guidelines are important

   (a) Do not use initial blocks. Initial blocks are usually ignored during synthesis, except in some FPGA synthesis tools.

- (b) Do not use delays (either delayed assignment or delayed evaluation). Delays are ignored during synthesis.
- (c) If possible, use concurrent assignments (assign) to design combinational logic.
- (d) It is possible to use procedural assignments (always blocks) to design either combinational logic or sequential logic.
- (e) When procedural assignments (always blocks) are used for combinational logic, use blocking assignments (e.g., '='). In Verilog 2001 or later, use always@* to avoid accidental omission of signals from sensitivity lists.
- (f) When procedural assignments (always block) are used for sequential logic, use non-blocking assignments (e.g., '<=').
- (g) Do not mix blocking and non-blocking statements in an always block.
- (h) Do not make assignments to the same variable from more than one always block. This is not a compile-time error, but it leads to timing problems, which are very difficult to debug.
- (i) Avoid unwanted latches by assigning a value to combinational output signals in every possible execution path in the always block. This can be done by
    - i. including else clauses for if statements,
    - ii. specifying all cases for case statements or have a **default** clause at the end, or
    - iii. unconditionally assigning default values to all combinational outputs at the beginning of the always block.

2. **To model hardware (i.e., to create simulation models which are not necessarily synthesizable)** There are several choices in the types of statements used when synthesis is not required. The following are important points to remember:

- (a) If delays are not to be modeled, use blocking assignments (e.g., A = B) for combinational logic and non-blocking assignments (e.g., A <= B) for sequential logic.
- (b) In blocking assignments with no delay specification (e.g., A = B), the new values are assigned immediately without any delta delays. However, in non-blocking assignments with no delay specification (e.g., A <= B), the change is scheduled to occur after a delta time.
- (c) To model combinational logic with inertial delays, use delayed evaluation blocking statements (e.g., #10 A = B;)
- (d) To model combinational logic with transport delays, use delayed assignment non-blocking assignments (e.g., A <= #10 B;). This has to be inside an always statement, because non-blocking assignments cannot be outside always statements.
- (e) To model sequential logic with delays, use delayed assignment non-blocking assignments (e.g., A <= #10 B;). This has to be inside an always statement because non-blocking assignments cannot be outside always statements.
- (f) Use inertial delays if pulse-rejection behavior is required. If inertial delays are used, remember this fact when checking simulation outputs. If input pulses are narrower than the inertial delay values, output changes will not

occur and not paying attention to this fact may make a designer think that the model is wrong, even when the model is fine.

(g) Do not make assignments to the same variable from more than one always block. Since this is not a compile-time error, it can lead to hard debugging challenges. It may appear that the circuit is working at times and not working at times.

Verilog Language Reference Model (LRM) outlines the possibility of race conditions during the simulation since execution of expression evaluation and net update events may be intermingled. For instance, consider the following code:

```
assign p = q;
initial begin
q = 1;
#1 q = 0;
$display(p);
end
```

The Verilog LRM mentions that a simulator is correct in displaying either a 1 or a 0. The assignment of 0 to q enables an update event for p. The simulator may either continue and execute the **$display** task or execute the update for p, followed by the **$display** task.

3. **To verify hardware.** Verification models and test benches mostly use initial blocks with blocking assignments and delayed assignments (see Figure 2-64 for an example). All types of assignment statements can be used in verification models. Delays are often used to create test stimuli arrive to unit under test at precise times and results are compared against expected results after the delays for the circuit to work. Good strategies for simulation apply here as well, but the following are especially important:

(a) Although all types of assignment statements can be used in verification models, use blocking assignments if possible.
(b) When delays are used, pay attention to inertial behavior.
(c) Use initial blocks to hard code test stimulus values.
(d) Use parameters for creating constants so test benches can easily be modified.
(e) Be aware of the differences between **$display**, **$strobe** and **$monitor** so that wrong conclusions are not made about correctly working circuits.

  i. **$monitor** displays every time one of its parameters changes.
  ii. **$strobe** displays at the very end of the current simulation time unit.
  iii. **$display** outputs values exactly where it is executed.

In this chapter, we have covered the basics of Verilog. We have shown how to use Verilog to model combinational logic and sequential machines. Since Verilog is a hardware description language, it differs from an ordinary programming language in several ways. Most importantly, Verilog statements execute concurrently, since they must model real hardware in which the components are all in operation at the same time. Statements within an always block execute sequentially, but the always blocks themselves operate concurrently. We will cover more advanced features of Verilog in Chapter 8.

Problems

2.1 In the following Verilog code, $A, B, C,$ and D are 0 at time 10 ns. If D changes to 1 at 20 ns, specify the times at which $A, B,$ and C will change and the values they will take.

(a)
```
always @(D)
begin
        #5 A <= 1;
        B <= A + 1;
        #10 C <= B;
end
```

(b)
```
always @(D)
begin
        A <= #5 1;
        B <= A + 1;
        C <= #10 B;
end
```

2.2 (a) What device does the following Verilog code represent?
```
always @(CLK, Clr, Set)
begin
   if(Clr == 1'b1)
      Q <= 1'b0;
   else if(Set == 1'b1)
      Q <= 1'b1;
   else if(CLK == 1'b0)
      Q <= D;
   else begin
   end
end
```

(b) What happens if $Clr = Set = 1$ in the device in part a?

2.3 Write a Verilog description of an S-R latch using an always block.

2.4 An M-N flip-flop responds to the falling clock edge as follows:

If $M = N = 0$, the flip-flop changes state.

If $M = 0$ and $N = 1$, the flip-flop output is set to 1.

If $M = 1$ and $N = 0$, the flip-flop output is set to 0.

If $M = N = 1$, no change of flip-flop state occurs.

The flip-flop is cleared asynchronously if $CLRn = 0$.

Write a complete Verilog module that implements an M-N flip-flop.

2.5 A DD flip-flop is similar to a D flip-flop, except that the flip-flop can change state ($Q^+ = D$) on both the rising edge and the falling edge of the clock input. The flip-flop has a direct reset input, R, and $R = 0$ resets the flip-flop to $Q = 0$ independent of the clock. Similarly, it has a direct set input, S, that sets the flip-flop to 1 independent of the clock. Write a Verilog description of a DD flip-flop.

2.6 **(a)** Assume $D_1=0$, $D_2=5$, and D_1 changes to 1 at time=10ns. What are the values of D_1 and D_2 after the following code has been executed once? Do the values of D_1 and D_2 swap?

```
always @ (D1)
begin
    D2 <= D1;
    D1 <= D2;
End
```

(b) Assume $D_1=0$, $D_2=5$, and D_1 changes to 1 at time=10ns. What are the values of D_1 and D_2 after the following code has been executed once? Do the values of D_1 and D_2 swap?

```
always @ (D1)
begin
    D2 = D1;
    D1 = D2;
end
```

(c) How many latches will result when the following code is synthesized? Assume B is 3 bits long.

```
always @ (State) begin
    case(State)
        2'b00: B = 5;
        2'b01: B = 3;
        2'b10: B = 0;
    endcase
end
```

Circle the correct choice:

  i. 1 latch because 1 case is missing
  ii. 2 latches because state is 2 bits
  iii. 3 latches because B is 3 bits
  iv. 5 latches because B is 3 bits and state is 2 bits
  v. None of the above. But it results in _____ latches.

2.7 What is wrong with the following code for a half adder that must add if add signal equals 1?

```
always @(x)
begin
```

```
            if (add == 1)
            begin
                sum = x ^ y;
                carry = x & y;
            end
            else
            begin
                sum = 0;
                carry = 0;
            end
        end
```

(a) It will compile but will not simulate correctly.
(b) It will compile and simulate correctly but will not synthesize correctly.
(c) It will work correctly in simulation and in synthesis.
(d) It will not even compile.

2.8 For the following Verilog code, assume that D changes to 1 at time 5 ns. Give the values of A, B, C, D, E, and F each time a change occurs. That is, give the values at time 5 ns, $5 + \Delta, 5 + 2\Delta$, and so forth. Carry thisced out until 20 steps have occurred, until no further change occurs, or until a repetitive pattern emerges.

```
module prob(D);
inout D;
wire A, C;
reg B,E,F,temp_D;

initial
begin
    B = 1'b0;
    E = 1'b0;
    F = 1'b0;
    temp_D = 1'b0;
end

assign C = A;
assign A = (B & !E) | D;
assign D = temp_D;

always @(A)
begin
    B = A;
end

always
begin
    wait(A)
    E <= #5 B;
    temp_D <= 1'b0;
    F <= E;
end
endmodule
```

2.9 Assuming B is driven by the simulator command:

   force B 0 0, 1 10, 0 15, 1 20, 0 30, 1 35

Draw a timing diagram illustrating A, B, and C if the following concurrent statements are executed:

```
always @(B)
begin
  A = #5 B;
end
assign #8 C = B;
```

2.10 Assuming B is driven by the simulator command:

   force B 0 0, 1 4, 0 10, 1 15, 0 20, 1 30, 0 40

Draw a timing diagram illustrating A, B, and C if the following concurrent statements are executed:

```
always @(B)
begin
  A <= #5 B;
end
assign #5 C <= B;
```

2.11 (a) Write a conditional signal assignment statement to represent the 4-to-1 MUX shown subsequently. Assume that there is an inherent delay in the MUX that causes the change in output to occur 10 ns after a change in input.
   (b) Repeat (a) using an if-else statement.
   (c) Repeat (a) using a case statement.

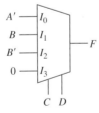

2.12 Draw the circuit represented by the following Verilog process:
```
always @(clk,clr)
begin
    if(clr == 1'b1)
        Q <= 1'b0;
    else if(clk == 1'b0 && CE == 1'b1)
        begin
            if(C == 1'b0)
                Q <= A & B;
```

```
                else
                    Q <= A | B;
            end
    end
```

Why is *clr* on the sensitivity list whereas *C* is not?

2.13 Implement the following Verilog code using these components: D flip-flops with clock enable, a multiplexer, an adder, and any necessary gates. Assume that *Ad* and *Ora* will never be 1 at the same time and enable the flip-flops only when *Ad* or *Ora* is 1.

```
module module1(A,B,Ad,Ora,clk,C);
input Ad,Ora,clk;
input [2:0]A,B;
output reg[2:0]C;

initial
begin
    C = 3'd0;
end

always @(posedge clk)
begin
    if(Ad == 1'b1)
            C <= A + B;
    if(Ora == 1'b1)
            C <= A | B;
end
endmodule
```

2.14 Consider the following Verilog code:

```
module Q3(A,B,C,F,Clk,E);
input A,B,C,F,Clk;
output reg E;

reg D,G;

initial
begin
    E = 1'b0;
    D = 1'b0;
    G = 1'b0;
end

always @(posedge Clk)
begin
    D <= A & B & C;
    G <= ~A & ~B;
    E <= D | G | F;
end
endmodule
```

(a) Draw a block diagram for the circuit (no gates and at block level only).
(b) Give the circuit generated by the preceding code (at the gate level).

2.15 If A = 101, B = 011, and C = 010, what are the values of the following statements? Assume A, B, and C are of reg type. Assume as many bits as necessary for the result.

(a) {A,B} | {B,C}
(b) A >> 2
(c) A >>> 2
(d) {A,(~B)} == 111110
(e) A | B & C

2.16 In the following Verilog Code, A, B, C, and D are registers that are 0 at time = 4 ns. If A changes to 1 at time 5 ns, make a table showing the values of A, B, C, and D as a function of time until time = 18 ns. Include deltas. Indicate the times at which each process begins executing.

```
always @(A)
begin
    B <= #5 A;
    C <= #2 B;
end

always
begin
    wait(B);
    A <= ~B;
    D <= ~A ^ B;
end
```

2.17 Given the following timing waveform for Y, draw D and E.

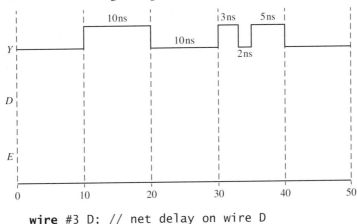

```
wire #3 D; // net delay on wire D
assign #5 D = Y; // statement 1 - inertial delay

wire #5 E; // net delay on wire E
assign #3 E = Y; // statement 1 - inertial delay
```

2.18 Given the following timing waveform for Y, draw D and E.

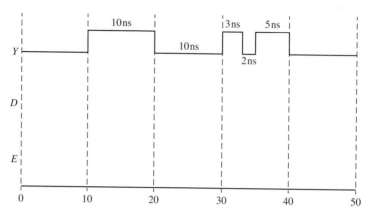

```
wire #4 D; // net delay on wire D
assign #6 D = Y; // statement 1 - inertial delay

wire #6 E; // net delay on wire E
assign #4 E = Y; // statement 1 - inertial delay
```

2.19 Assuming B is driven by the simulator command:

force B 0 0, 1 4, 0 10, 1 15, 0 20, 1 30, 0 40

Draw a timing diagram illustrating A, B, and C if the following statement is executed:

(a)
```
always @(A,B)
begin
    #5 C <= A && B;
end
```

(b)
```
always @(A,B)
begin
    #5 C = A && B;
end
```

(c)
```
always @(A,B)
begin
    C <= #5 A && B;
end
```

(d)
```
always @(A,B)
begin
    C = #5 A && B;
end
```

2.20 An inhibited toggle flip-flop has inputs I_0, I_1, T, and *Reset* and outputs Q and QN. *Reset* is active high and overrides the action of the other inputs. The flip-flop works as follows: If $I_0 = 1$, the flip-flop changes state on the rising edge of T; if $I_1 = 1$, the flip-flop changes state on the falling edge of T. If $I_0 = I_1 = 0$, no state change occurs (except on reset). Assume the propagation delay from T to output is 8 ns and from reset to output is 5 ns.

(a) Write a complete Verilog description of this flip-flop.
(b) Write a sequence of simulator commands that will test the flip-flop for the input sequence $I_1 = 1$, toggle T twice, $I_1 = 0$, $I_0 = 1$, toggle T twice.

2.21 Evaluate

$$A \gg 4$$
$$A \ggg 4$$
$$A \ll 4$$
$$A \lll 4$$

given

(a) reg signed [7:0] A = 8'hC7;
(b) reg [7:0] A = 8'hA5;
(c) reg signed [7:0] A = 8'hC7;
(d) integer signed A = 8'hC7;
(e) integer A = 8'hC7;
(f) integer A = 8'shC7;
(g) integer signed A = 8'shC7;
(h) integer signed A = 32'hFFFFFFC7;

2.22 In the following Verilog process $A, B, C,$ and D are all registers that have a value of 0 at time = 10 ns. If E changes from 0 to 1 at time = 20 ns, specify the time(s) at which each signal will change and the value to which it will change. List these changes in chronological order.

```
always
begin
    wait(E);
    A <= #5 1;
    B <= A + 1;
    C <= #10 B;
    D <= #3 B;
    A <= #15 A + 5;
    B <= B + 7;
end
```

2.23 Given

```
reg [7:0] C;
reg signed [7:0] D;
reg [7:0] A = 8'hD5;
```

evaluate

  i. C = A >> 4
 ii. C = A >>> 4
iii. C = A << 4
 iv. C = A <<< 4
  v. D = A >> 4
 vi. D = A >>> 4

vii. D = A << 4
viii. D = A <<< 4

2.24 In the following Verilog process A, B, C, and D are all registers that have a value of 0 at time = 10 ns. If E changes from 0 to 1 at time = 20 ns, specify the time(s) at which each signal will change and the value to which it will change. List these changes in chronological order.

```
always @(E)
begin
    A <= #5 1;
    B <= A + 1;
    C <= #10 B;

    D <= #3 B;
    A <= #15 A + 5;
    B <= B 1 7;
end
```

2.25 **Given**

```
reg [7:0] C;
reg signed [7:0] D;
reg [7:0] A = 8'shD5;
```

evaluate

i. C = A >> 4
ii. C = A >>> 4
iii. C = A << 4
iv. C = A <<< 4
v. D = A >> 4
vi. D = A >>> 4
vii. D = A << 4
viii. D = A <<< 4

2.26 **Given**

```
wire a = 1'b1;
wire [1:0] b = 2'b10;
wire [2:0] c = 3'b101;
```

evaluate

i. {b, c}
ii. {a, b, c, 2'b01}
iii. {a, b[0], c[1]}

2.27 Draw the hardware obtained if the following code is synthesized:

```
module  reg3 (Q1,Q2,Q3,Q4, A,CLK);
input    A;
input    CLK;
output   Q1,Q2,Q3,Q4;
```

```
      reg      Q1,Q2,Q3,Q4;
      always @(posedge CLK)
      begin
        Q4 = Q3;
        Q3 = Q2;
        Q2 = Q1;
        Q1 = A;
      end

    endmodule
```

2.28 **Given**

```
      wire a = 1'b0;
      wire [1:0] b = 2'b10;
      wire [2:0] c = 3'b101;
```

evaluate
  i. { 4{a} }
  ii. { 4{a}, 2{b} }
  iii. { 4{a}, c}

2.29 Draw the hardware obtained if the following code is synthesized:

```
      module   reg3 (Q1,Q2,Q3,Q4, A,CLK);
      input    A;
      input    CLK;
      output   Q1,Q2,Q3,Q4;
      reg      Q1,Q2,Q3,Q4;
      always @(posedge CLK)
      begin
        Q1 = A;
        Q2 = Q1;
        Q3 = Q2;
        Q4 = Q3;
      end

    endmodule
```

2.30 **Given**

```
      reg [7:0] C;
      reg signed [7:0] D;
      reg signed [7:0] A = 8'hD5;
```

evaluate
  i. C = A >> 4
  ii. C = A >>> 4
  iii. C = A << 4
  iv. C = A <<< 4
  v. D = A >> 4

vi. D = A >>> 4
vii. D = A << 4
viii. D = A <<< 4

2.31 Draw the hardware obtained if the following two modules are synthesized and describe the differences.

```
module   reg3 (Q1,Q2,Q3,Q4,A,CLK);
input    A;
input    CLK;
output   Q1,Q2,Q3,Q4;

reg      Q1,Q2,Q3,Q4;

// first
always @(posedge CLK)
begin
  Q1 <= A;
  Q2 <= Q1;
  Q3 <= Q2;
  Q4 <= Q3;
end

endmodule
```

```
module   reg3 (Q1,Q2,Q3,Q4,A,CLK);
input    A;
input    CLK;
output   Q1,Q2,Q3,Q4;

reg      Q1,Q2,Q3,Q4;

// first -> second
always @(posedge CLK)
begin
  Q4 <= Q3;
  Q3 <= Q2;
  Q2 <= Q1;
  Q1 <= A;
end

endmodule
```

2.32 A Moore sequential machine with two inputs (X_1 and X_2) and one output (Z) has the following state table:

Present State	Next State X_1X_2 = 00	01	10	11	Output (Z)
1	1	2	2	1	0
2	2	1	2	1	1

Write Verilog code that describes the machine at the behavioral level. Assume that state changes occur 10 ns after the falling edge of the clock and output changes occur 10 ns after the state changes.

2.33 Complete the following Verilog code to implement a counter that counts in the following sequence: $Q = 1000, 0111, 0110, 0101, 0100, 0011, 1000, 0111, 0110, 0101, 0100, 0011, \ldots$ (repeats). The counter is synchronously loaded with 1000 when $Ld8 = 1$. It goes through the prescribed sequence when $Enable = 1$. The counter outputs $S_5 = 1$ whenever it is in state 0101. Do not change the provided structure of the following module in any way. Your code must be synthesizable.

```
module countQ1(clk,Ld8,Enable,S5,Q);
input clk,Ld8,Enable;
output reg S5;
output reg[3:0] Q;
     .
     .
     .
endmodule
```

2.34 Write Verilog code to implement the following state table. Use two always blocks. State changes should occur on the falling edge of the clock. Implement the Z_1 and Z_2 outputs using concurrent conditional statements. Assume that the combinational part of the sequential circuit has a propagation delay of 10 ns and the propagation delay between the rising-edge of the clock and the state register output is 5 ns.

Present State	Next State $X_1X_2 = 00$	01	11	Output (Z_1Z_2)
1	3	2	1	00
2	2	1	3	10
3	1	2	3	01

2.35 A synchronous 4-bit UP/DOWN binary counter has a synchronous clear signal CLR and a synchronous load signal LD. CLR has higher priority than LD. Both CLR and LD are active high. D is a 4-bit input to the counter, and Q is the 4-bit output from the counter. UP is a signal that controls the direction of counting. If CLR and LD are not active and $UP = 1$, the counter increments. If CLR and LD are not active and UP 0, the counter decrements. All changes occur on the falling edge of the clock.

(a) Write a behavioral Verilog description of the counter.
(b) Use the foregoing UP/DOWN counter to implement a synchronous modulo 6 counter that counts from 1 to 6. This modulo 6 counter has an external reset, which, if applied, makes the count = 1. A count enable signal CNT makes it count in the sequence 1, 2, 3, 4, 5, 6, 1, 2, ..., incrementing once for each clock pulse. You should use any necessary logic to make the counter go to count = 1 after count = 6. The modulo 6 counter counts only in the UP sequence. Provide a textual/pictorial description of your approach.
(c) Write a behavioral Verilog description for the modulo-6 counter in part (b).

2.36 In the following code, *state* and *nextstate* are integers with a range of 0 to 2.

```verilog
always @(state,X)
begin
    case(state)
    0:begin
        if(X == 1'b1)
            nextstate = 2'b01;
    end
    1:begin
        if(X == 1'b0)
            nextstate = 2'b10;
    end
    2:begin
        if(X == 1'b1)
            nextstate = 2'b00;
    end
    endcase
end
```

(a) Explain why a latch would be created when the code is synthesized.
(b) What signal would appear at the latch output?
(c) Make changes in the case statement that would eliminate the latch.
(d) Can you make changes outside the case statement that would eliminate the latch? If yes, illustrate. If no, explain.

2.37 Examine the following Verilog code and answer the following questions

```verilog
module Problem(X,CLK,Z1,Z2);
input X,CLK;
output Z1,Z2;

reg [1:0]State,Nextstate;

initial
begin
    State = 2'b00;
    Nextstate = 2'b00;
end

always @(State,X)
begin
    case(State)
    0:begin
        if(X == 1'b0)begin
            Z1 = 1'b1;
            Z2 = 1'b0;
            Nextstate = 2'b00;
        end
        else begin
            Z1 = 1'b0;
            Z2 = 1'b0;
            Nextstate = 2'b01;
        end
```

```verilog
            end
    1:begin
            if(X == 1'b0)begin
                    Z1 = 1'b0;
                    Z2 = 1'b1;
                    Nextstate = 2'b01;
            end
            else begin
                    Z1 = 1'b0;
                    Z2 = 1'b1;
                    Nextstate = 2'b10;
            end
    end
    2:begin
            if(X == 1'b0)begin
                    Z1 = 1'b0;
                    Z2 = 1'b1;
                    Nextstate = 2'b10;
            end
            else begin
                    Z1 = 1'b0;
                    Z2 = 1'b1;
                    Nextstate = 2'b11;
            end
    end
    3:begin
            if(X == 1'b0)begin
                    Z1 = 1'b0;
                    Z2 = 1'b0;
                    Nextstate = 2'b00;
            end
            else begin
                    Z1 = 1'b1;
                    Z2 = 1'b0;
                    Nextstate = 2'b01;
            end
    end
        endcase
end
always @(posedge CLK)
begin
    State <= Nextstate;
end
endmodule
```

(a) Draw a block diagram of the circuit implemented by this code.

(b) Write the state table that is implemented by this code.

2.38 What is wrong with the following model of a 4-to-1 MUX? (It is not a syntax error.)

```verilog
reg [1:0]sel;
always @(A,B,I0,I1,I2,I3)
```

```
            begin
                sel = 0;
                if(A == 1'b1)
                    sel = sel + 1;
                else
                begin
                end
                if(B == 1'b1)
                    sel = sel + 2;
                else
                begin
                end
                case(sel)
                    0:begin
                        F = I0;
                    end
                    1:begin
                        F = I1;
                    end
                    2:begin
                        F = I2;
                    end
                    3:begin
                        F = I3;
                    end
                endcase
            end
```

2.39 **(a)** Write a behavioral Verilog description of the state machine you designed in Problem 1.30.

Assume that state changes occur on the falling edge of the clock pulse. Instead of using if-then-else statements, represent the state table and output table by arrays. Compile and simulate your code using the following test sequence:

$$X = 1101\ 1110\ 1111$$

X should change 1/4 clock period after the rising edge of the clock.

(b) Write a data flow Verilog description using the next-state and output equations to describe the state machine. Indicate on your simulation output at which times S and V are to be read.

(c) Write a structural model of the state machine in Verilog that contains the interconnection of the gates and D flip-flops.

2.40 When the following Verilog code is simulated, A is changed to 1 at time 5 ns. Make a table that shows all changes in A, B, and D and the times at which they occur through time = 40 ns.

```
module Q1F00(A)
inout A;
reg B,C;
wire D;
```

```
        assign #10 D = A ^ B;
        always @(D)
        begin
            C = ~D;
            if(C == 1'b1)
                    A <= #15 ~A;
            B <= D;
        end
        endmodule
```

2.41 **(a)** Write a behavioral Verilog description of the state machine that you designed in Problem 1.29. Assume that state changes occur on the falling edge of the clock pulse. Use a case statement together with if-then-else statements to represent the state table. Compile and simulate your code using the following test sequence:

$$X = 1011\ 0111\ 1000$$

X should change 1/4 clock period after the falling edge of the clock.

(b) Write a data flow Verilog description using the next-state and output equations to describe the state machine. Indicate on your simulation output at which times D and B should be read.

(c) Write a structural model of the state machine in Verilog that contains the interconnection of the gates and J-K flip-flops.

2.42 What device does the following Verilog code represent?

```
reg Qtmp;

always @(CLK,RST)
begin
    if(RST == 1'b1)
            Qtmp = 1'b0;
    else if(CLK == 1'b1 && T == 1'b1)
            Qtmp = ~Qtmp;
    else begin
    end
    Q <= Qtmp;
end
```

2.43 **(a)** Write a Verilog module for a LUT with four inputs and three outputs. The 3-bit output should be a binary number equal to the number of 1s in the LUT input.

(b) Write a Verilog module for a circuit that counts the number of 1s in a 12-bit number. Use three of the modules from **(a)** along with overloaded addition operators.

(c) Simulate your code and test it for the following data inputs:

111111111111, 010110101101, 100001011100

2.44 Implement a 3-to-8 decoder using a LUT. Give the LUT truth table and write the Verilog code. The inputs should be A, B, and C, and the output should be an 8-bit unsigned vector.

2.45 A is an array of 20 4-bit registers. Write Verilog code that finds the largest register in the array,

   (a) Using a for loop.
   (b) Using a while loop.

2.46 Write Verilog code to test a Mealy sequential circuit with one input (X) and one output (Z). The code should include the Mealy circuit as a component. Assume the Mealy circuit changes state on the rising edge of CLK. Your test code should generate a clock with a 100-ns period. The code should apply the following test sequence:

$$X = 0, 1, 1, 0, 1, 1, 0, 1, 1, 1, 0, 0$$

X should change 10 ns after the rising edge of CLK. Your test code should read Z at an appropriate time and should verify that the following output sequence was generated:

$$Z = 1, 0, 0, 1, 1, 0, 1, 1, 0, 1, 1, 0$$

Report an error if the output sequence from the Mealy circuit is incorrect; otherwise, report "sequence correct." Complete the following architecture for the tester:

```
module tester;
reg CLK;reg[11:0]X;
reg[11:0]Z;
initial
begin
    X = 12'b011011011100;
    Z = 12'b100110110110;
    CLK = 1;
end
    .
    .
    .
    .
endmodule
```

2.47 Write a Verilog test bench that will test the Verilog code for the sequential circuit of Figure 2-58. Your test bench should generate all 10 possible input sequences (0000, 0001, 0010, ..., 1001) and should verify that the output sequences are correct. Remember that the components have a 10-ns delay. The input should be changed 1/4 of a clock period after the rising edge of the clock, and the output should be read at the appropriate time. Report "Pass" if all sequences are correct; otherwise, report "Fail."

2.48 Write a test bench to test the counter of Problem 2.33. The test bench should generate a clock with a 100-ns period. The counter should be loaded on the first clock, then it should count for 5 clocks, then it should do nothing for 2 clocks, then it should continue counting for 10 clocks. The test bench port should output the current time (in time units, not the count) whenever $S_5 = 1$. Use only concurrent statements in your test bench.

2.49 Complete the following Verilog code to implement a test bench for the sequential circuit SMQ1. Assume that the Verilog code for the SMQ1 sequential circuit module is already available. Use a clock with a 50-ns half period. Your test bench should test the circuit for the input sequence $X = 1, 0, 0, 1, 1$. Assume that the correct output sequence for this input sequence is $1, 1, 0, 1, 0$. Use a single concurrent statement to generate the X sequence. The test bench should read the values of output Z at the proper times and should compare them with the correct values of Z. The correct answer is initialized as a constant:

$$answer = 11010;$$

The port signal *correct* should be set to TRUE if the answer is correct; otherwise, it should be set to FALSE. Make sure that you read Z at the correct time. Use wait statements in your test bench.

```
module testSMQ1(correct);
output reg correct;
reg CLK;
reg [4:0] X;
reg [4:0] Z;
reg [4:0] answer;

initial
begin
 correct = 1;
 X = 5'b10011;
 Z = 5'b11010;
 answer = 5'b11010;
end
   .
   .
   .
   .
endmodule
```

2.50 (a) Write at least two different Verilog modules that are equivalent to the following pseudo code:

A = B1 **when** C = 1 **else** B2 **when** C = 2 **else** B3 **when** C = 3 **else** 0;

(b) Draw a circuit to implement the following statement,

A = B1 **when** C1 = 1 **else** B2 **when** C2 = 1 **else** B3 **when** C3 = 1 **else** 0;

2.51 Write a Verilog module that describes a 16-bit serial-in, serial-out shift register with inputs *SI* (serial input), *EN* (enable), *CK* (clock, shifts on rising edge), and a serial output (*SO*).

2.52 Write a Verilog description of an SR latch.

(a) Use a conditional assignment statement (i.e., a behavioral description).
(b) Use the characteristic equation in the Verilog description.
(c) Use the logic gate level structure of an SR latch in the model.

2.53 A description of a 74194 4-bit bidirectional shift register follows.
The *CLRb* input is asynchronous and active low and overrides all the other control inputs. All other state changes occur following the rising edge of the clock. If the control inputs $S_1 = S_0 = 1$, the register is loaded in parallel. If $S_1 = 1$ and $S_0 = 0$, the register is shifted right and *SDR* (serial data right) is shifted into Q_3. If $S_1 = 0$ and $S_0 = 1$, the register is shifted left and *SDL* is shifted into Q_0. If $S_1 = S_0 = 0$, no action occurs.

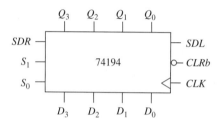

(a) Write a behavioral-level Verilog model for the 74194.
(b) Draw a block diagram and write a Verilog description of an 8-bit bidirectional shift register that uses two 74194s as components. The parallel inputs and outputs to the 8-bit register should be *X*([7:0]) and *Y*([7:0]). The serial inputs should be *RSD* and *LSD*.

2.54 Given

```
reg [3:0] A = 4'b1101;
reg [2:0] B = 3'b111;
reg signed [3:0] C = 4'b1101;
reg signed [2:0] D = 3'b111;
reg signed [7:0] S;
```

evaluate

    i.   S = A + B;
    ii. S = A + B + 0;
    iii. S = C + D;
    iv. S = C + D + 0;
    v. S = A + D;
    vi. S = A + D + 0;

2.55 A synchronous (4-bit) up/down decade counter with output *Q* works as follows: All state changes occur on the rising edge of the *CLK* input, except the asynchronous clear (*CLR*). When *CLR* = 0, the counter is reset regardless of the values of the other inputs.

If the *LOAD* input is 0, the data input *D* is loaded into the counter.
If *LOAD* = *ENT* = *ENP* = *UP* = 1, the counter is incremented.
If *LOAD* = *ENT* = *ENP* = 1 and *UP* = 0, the counter is decremented.
If *ENT* = *UP* = 1, the carry output (*CO*) = 1 when the counter is in state 9.
If *ENT* = 1 and *UP* = 0, the carry output (*CO*) = 1 when the counter is in state 0.

(a) Write a Verilog description of the counter.
(b) Draw a block diagram and write a Verilog description of a decimal counter that uses two of the previously specified counters to form a 2-decade decimal up/down counter that counts up from 00 to 99 or down from 99 to 00.
(c) Simulate for the following sequence: load counter with 98, increment three times, do nothing for two clocks, decrement four times, and clear.

2.56 Given

```
integer A = 8'shA5;
integer B = 8'shB6;
integer C = 8'hA5;
integer D= 8'hB6;
reg signed [31:0] S;
```

evaluate

    i.   S = A + B;
    ii.   S = A + B + 0;
    iii.   S = C + D;
    iv.   S = C + D + 0;
    v.   S = A + D;
    vi.   S = A + D +0;

2.57 Write a Verilog model for a 74HC192 synchronous 4-bit up/down counter. Ignore all timing data. Your code should contain a statement of the form `always @(DOWN, UP, CLR, LOADB)`.

2.58 A 4-bit magnitude comparator chip (e.g., 74LS85) compares two 4-bit numbers *A* and *B* and produces outputs to indicate whether $A < B$, $A = B$, or $A > B$. There are three output signals to indicate each of the foregoing conditions. Note that exactly one of the output lines will be high and the other two lines will be low at any time. The chip is a cascadable chip and has three inputs—*A>B.IN*, *A=B.IN*, and *A<B.IN*—in order to allow cascading the chip to make 8-bit or larger magnitude comparators.

(a) Draw a block diagram of a 4-bit magnitude comparator.
(b) Draw a block diagram to indicate how you can construct an 8-bit magnitude comparator using two 4-bit magnitude comparators.
(c) Write a behavioral Verilog description for the 4-bit comparator.
(d) Write Verilog code for the 8-bit comparator using two 4-bit comparators as components.

2.59 Consider the following 8-bit bidirectional synchronous shift register with parallel load capability. The notation used to represent the input/output pins is explained as follows:

   CLR   Asynchronous Clear, which overrides all other inputs
   Q(7:0) 8-bit output
   D(7:0) 8-bit input
   S_0, S_1 mode control inputs
   LSI   serial input for left shift
   RSI   serial input for right shift

The mode control inputs work as follows:

S_0	S_1	Action
0	0	No action
0	1	Right shift
1	0	Left shift
1	1	Load parallel data (i.e., Q = D)

   (a) Write a Verilog module for this shift register.
   (b) Draw a block diagram illustrating how two of these can be connected to form a 16-bit cyclic shift register, which is controlled by signals L and R. If $L = 1$ and $R = 0$, the 16-bit register is cycled left. If $L = 0$ and $R = 1$, the register is cycled right. If $L = R = 1$, the 16-bit register is loaded from $X(15:0)$. If $L = R = 0$, the register is unchanged.
   (c) Write a Verilog module for the module in part (b), using the module built in (a).

2.60 (a) What do the acronyms HDL and FPGA stand for?
   (b) How does a hardware description language like Verilog differ from an ordinary programming language?
   (c) What are the advantages of using a hardware description language as compared with schematic capture in the design process?

2.61 (a) Which of the following are legal Verilog identifiers? `123A`, `A_123`, `_A123`, `$A123_`, `c1__c2`, `and`, `and1`
   (b) Which of the following Verilog identifiers are equivalent? `aBC`, `ABC`, `Abc`, `abc`

2.62 Given the concurrent Verilog statements:

```
assign #3 B = A && C;
assign #2 C = !B;
```

   (a) Draw the circuit the statements represent.
   (b) Draw a timing diagram if initially $A = B = 0$ and $C = 1$, and A changes to 1 at time 5 ns.

2.63 Write a Verilog description of the following combinational circuit using concurrent statements. Each gate has a 5-ns delay, excluding the inverter, which has a 2-ns delay.

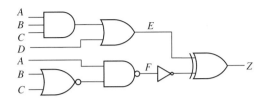

2.64 (a) Write Verilog code for a full subtracter using logic equations.
(b) Write Verilog code for a 4-bit subtracter using the module defined in (a) as a component.

2.65 Write Verilog code for the following circuit. Assume that the gate delays arenegligible.

(a) Using concurrent statements.
(b) Using an always block with sequential statements. No latches should be generated.

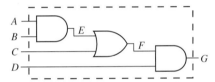

Introduction to Programmable Logic Devices

Chapter 1 illustrated how the same digital circuit can be implemented using a variety of standard building blocks. If one can put several of these building blocks into an integrated circuit (IC) and provide the user with mechanisms to modify the configuration, one can implement almost any circuit within a chip. This is the general principle of programmable logic devices.

This chapter introduces the use of programmable logic devices in digital design. Read-only memories (ROMs), programmable logic arrays (PLAs), and programmable array logic (PAL) devices are discussed first. Then complex programmable logic devices (CPLDs) and field-programmable gate arrays (FPGAs) are introduced. Use of these devices allows us to implement complex logic functions, which require many gates and flip-flops, with a single IC chip. Although FPGAs are introduced, only an overview is provided in this chapter. A detailed treatment of FPGAs can be found in Chapter 6.

3.1 Brief Overview of Programmable Logic Devices

Designers have always liked programmable logic devices such as PALs and FPGAs for implementation of digital circuits. First, there is reasonable integration ability, allowing implementation of a significant amount of functionality into one physical chip. Programmable logic devices remove the need for multiple off-the-shelf devices along with the inconvenience and unreliability associated with external wires. Second, there is the increased ability to change designs. Many of the programmable devices allow easy reprogramming. In general, it is easier to change the design in case of errors or changes in design specifications. Currently, programmable logic comes in different types—(1) devices that can be programmed only once and (2) those that can be reprogrammed many times.

Figure 3-1 illustrates a classification of popular programmable logic devices. Programmable logic can be considered to fall into field-programmable logic and factory-programmable logic. The term **field** indicates that this type of device is

3.1 Brief Overview of Programmable Logic Devices

FIGURE 3-1: Major Programmable Logic Devices

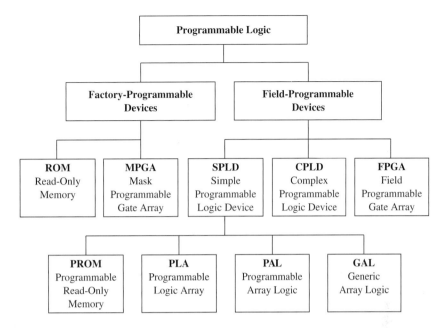

programmed in the user's "field" rather than in a semiconductor fab. Often, many may refer to programmable logic to mean devices that are field programmable. However, there are factory-programmable devices as well. These are generic devices that can be programmed at the factory to meet customers' requirements. The programming technology uses an irreversible process; hence, programming can be done only once. Examples of factory-programmable logic are mask-programmable gate arrays (MPGAs) and read-only memories (ROMs). The earliest generations of many programmable devices were programmable only at the factory.

Read-only memories (ROMs) can be considered an early form of programmable logic. While primarily meant for use as memory, ROMs can be used to implement any combinational circuitry. This is illustrated subsequently in the section on ROMs. MPGAs are traditional gate arrays, which require a mask to be designed. MPGAs are often simply called **gate arrays** and have been a popular technology for creating application-specific integrated circuits (ASICs).

User programmable logic in the form of AND-OR circuits was developed at the beginning of the 1970s. By 1972–73, one-time field-programmable logic arrays that permitted instant customizations by designers were available. Some referred to these devices as field-programmable logic arrays or FPLAs. Monolithic Memories Inc. (MMI), a company that was bought by Advanced Micro Devices (AMD), created integrated circuits called programmable logic arrays (PLAs) in 20–24 pin packages that could yield the same functionality as 5 to 20 off-the-shelf chips. A similar device is the programmable array logic or PAL.

PALs and PLAs contain arrays of gates. In the PLA, there is a programmable AND array and a programmable OR array, allowing users to implement combinational functions in two levels of gates. The PAL is a special case of a PLA, in that the OR array is fixed and only the AND array is programmable. Many PALs also contain flip-flops.

In the 1970s and 1980s, PALs and PLAs were very popular. Part of the popularity was due to the ease of design. MMI and Advanced Micro Devices created a simple programming language, called PALASM, to easily convert Boolean equations into PLA configurations. PALASM made programming PALs and PLAs relatively simple.

The early programmable devices allowed only one-time programming. The next technological innovation that helped programmable logic was advancement in erasure of programmable devices. In the early days, erasure of programmable logic used ultraviolet light. With ultraviolet light, erasing the configuration of a device meant removing the device from the circuit and placing it in an ultraviolet environment. Hence, in-circuit erasure was not possible. Ultraviolet erasers were slow; typically 10 or 15 minutes were required to perform erasures. Then electrically erasable technology came along. This led to the creation of field-programmable logic arrays that can be easily and quickly erased and reprogrammed without removing the chip from the board.

The early PALs and PLAs were soon followed by CMOS electrically erasable programmable logic devices (PLDs). While the term PLDs can be used to refer to any programmable logic devices, there are a set of devices, including the popular PALCE22V10, that are often referred to as PLDs. PLDs contain macroblocks with arrays of gates, multiplexers, flip-flops, or other standard building blocks. Several of these macroblocks appear in a PLD. Lattice Semiconductor created similar devices with easy reprogrammability and called their line of devices **GALs (generic array logic)**.

Now, many refer to PLAs, PALs, GALs, PLDs, and PROMs collectively as simple PLDs (SPLDs) in contrast to another product that has come into the market, complex PLDs (CPLDs). As the name suggests, CPLDs have more integration capability than SPLDs. They come in sizes ranging from 500 to 16,000 gates. CPLDs essentially put multiple PLDs into the same chip with some kind of an interconnection circuit, typically a crossbar switch.

During the late 1980s, Xilinx started using static RAM storage elements to hold configuration information for programmable devices and created devices called field-programmable gate arrays (FPGAs), which can integrate a fairly large amount of logic. Contrary to their names, the basic building blocks in these devices were not arrays of gates but were bigger and complex blocks containing static RAMs and multiplexers. Several PLD vendors and gate array companies soon jumped into the market creating a variety of FPGA architectures, some of which used reprogrammable technologies while others used one-time programmable fuse technologies. The FPGA technology has continually improved during the past 15 years. Now, there are FPGAs that can contain more than 5 million gates.

Programmable logic devices basically contain an array of basic building blocks that can be used to implement whatever functionality one desires. Different programmable devices differ in the building blocks or the amount of programmability they provide. Table 3-1 illustrates a comparison of various programmable logic devices. FPGAs are bigger and more complex than CPLDs. The routing resources in FPGAs are more complex than those in simple programmable devices. The variety of alternate routes that can be taken cause the paths taken by signals to be unpredictable. FPGAs are more expensive than CPLDs and SPLDs. They contain more overhead for programming. In this chapter, we describe various programmable devices, including SPLDs, CPLDs, and FPGAs.

TABLE 3-1: A Comparison of Programmable Devices

	SPLD	CPLD	FPGA
Density	Low Few hundred gates	Low to Medium 500 to 12,000 gates	Medium to High 3,000 to 5,000,000 gates
Timing	Predictable	Predictable	Unpredictable
Cost	Low	Low to Medium	Medium to High
Major Vendors	Lattice Cypress AMD	Xilinx Altera	Xilinx Altera Lattice Microsemi
Example Device Families	**Lattice** GAL16LV8 GAL22V10 **Cypress** PALCE16V8 **AMD** 22V10	**Xilinx** CoolRunner XC9500 **Altera** MAX	**Xilinx** Kintex Artix Virtex Spartan **Altera** Stratix Cyclone Arria **Lattice** Mach ECP **Microsemi** Axcelerator Fusion

> Many names and abbreviations in this field have historically been used to refer to specific types of programmable devices, although, one may not find the name to be meaningful. Consider PALs and PLAs; both are arrays of logic. The fact that PLAs contain programmable AND and OR arrays and PALs contain only programmable AND arrays is due to nothing but historical reasons. They could very well be named the other way around. But it is important for students to understand what these names popularly refer to because they will need to communicate with fellow designers and other design teams. Conventions are important in facilitating communication.

3.2 Simple Programmable Logic Devices (SPLDs)

With the advent of CPLDs and FPGAs, the early-generation programmable logic devices, such as ROMs, PALs, PLAs, and PLDs, can be collectively called simple programmable logic devices. In this section, we describe the implementation of digital circuits in simple PLDs.

3.2.1 Read-Only Memories (ROM)

A read-only memory (ROM) consists of an array of semiconductor devices that are interconnected to store an array of binary data. Once binary data is stored in the ROM, it can be read out whenever desired, but the data that is stored cannot be changed under normal operating conditions. Figure 3-2(a) shows a ROM that has three input lines and four output lines. Figure 3-2(b) shows a typical truth table, which relates the ROM inputs and outputs. For each combination of input values on the three input lines, the corresponding pattern of 0s and 1s appears on the ROM output lines. For example, if the combination $ABC = 010$ is applied to the input lines, the pattern $F_0F_1F_2F_3 = 0111$ appears on the output lines. Each of the output patterns that is stored in the ROM is called a *word*. Since the ROM has three input lines, we have $2^3 = 8$ different combinations of input values. Each input combination serves as an *address*, which can select one of the eight words stored in the memory. Since there are four output lines, each word is 4 bits long, and the size of this ROM is 8 words $\times$ 4 bits.

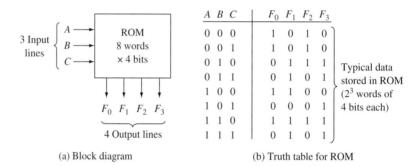

FIGURE 3-2: An 8-Word $\times$ 4-Bit ROM

(a) Block diagram

(b) Truth table for ROM

A ROM that has n input lines and m output lines (Figure 3-3) contains an array of 2^n words, and each word is m bits long. The input lines serve as an address to select one of the 2^n words. When an input combination is applied to the ROM, the pattern of 0s and 1s stored in the corresponding word in the memory appears at the output lines. For the example in Figure 3-3, if 00 ... 11 is applied to the input (address lines) of the ROM, the word 110 ... 010 will be selected and transferred to the output lines. A $2^n \times m$ ROM can realize m functions of n variables since it can store a truth table with 2^n rows and m columns. Typical sizes for commercially available ROMs range from 32 words $\times$ 4 bits to 512K words $\times$ 8 bits, or larger.

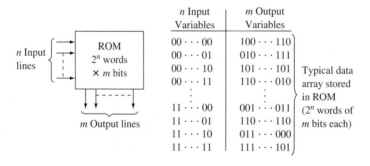

FIGURE 3-3: Read-Only Memory with n Inputs and m Outputs

A ROM basically consists of a decoder and a memory array. When a pattern of n 0s and 1s is applied to the decoder inputs, exactly one of the 2^n decoder outputs is 1. This decoder output line selects one of the words in the memory array, and the bit pattern stored in this word is transferred to the memory output lines.

Basic types of ROMs include mask-programmable ROMs, user-programmable ROMs (PROMs), erasable programmable ROMs (usually called EPROMs), electrically erasable and programmable ROMs (EEPROMs), and flash memories. In the mask-programmable ROM, the data array is permanently stored at the time of manufacture. This is accomplished by selectively including or omitting the switching elements at the row-column intersections of the memory array. This requires preparation of a special "mask," which is used during fabrication of the integrated circuit. Preparation of this mask is expensive, so use of mask-programmable ROMs is economically feasible only if a large quantity (typically several thousand or more) is required with the same data array. There are also one-time user-programmable ROMs or PROMs.

Because modification of the data stored in a ROM is often necessary during the developmental phases of a digital system, EPROMs are used instead of mask-programmable ROMs. EPROMs use a special charge-storage mechanism to enable or disable the switching elements in the memory array. An EPROM programmer is used to provide appropriate voltage pulses to store electronic charges in the memory array locations. The data stored in this manner is generally permanent until erased using ultraviolet light. After erasure, a new set of data can be stored in the EPROM.

The electrically erasable PROM (or EEPROM) is similar to an EPROM, except that erasure is accomplished using electrical pulses instead of ultraviolet light. A traditional EEPROM can be erased and reprogrammed only a limited number of times, typically 100 to 1000 times. **Flash memories** are similar to EEPROMs, except that they use a different charge-storage mechanism. They usually have built-in programming and erasure capability so that data can be written to the flash memory while it is in a circuit without the need for a separate programmer.

A ROM can implement any combinational circuit. Essentially, if the outputs for all combinations of inputs are stored in the ROM, the outputs can be "looked up" in the table stored in the ROM. The ROM method is also called the **Look-Up Table (LUT)** method for this reason.

Consider the implementation of a 2-bit adder in a ROM. This adder must add two 2-bit numbers. Since the maximum value of a 2-bit number is 3, the maximum sum is 6, necessitating 3 bits for the sum. The truth table for such an adder is illustrated in Figure 3-4. One could also design a 2-bit full adder assuming a carry input in addition to the two 2-bit numbers.

This 2-bit adder can be implemented with a 16×3 ROM. The input numbers (X and Y) must be connected to the four address lines, and the three data lines will produce the sum bits.

Figure 3-5 illustrates the ROM implementation of this 2-bit full adder. Assuming the connections that are shown, the contents of the ROM in its 16 locations should be 0, 1, 2, 3, 1, 2, 3, 4, 2, 3, 4, 5, 3, 4, 5, and 6 respectively (representing the digits in decimal). The LSB of the sum will come from the LSB of the data bus.

FIGURE 3-4: Block Diagram and Truth Table of a 2-Bit Adder

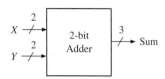

X_1	X_0	Y_1	Y_0	S_2	S_1	S_0
0	0	0	0	0	0	0
0	0	0	1	0	0	1
0	0	1	0	0	1	0
0	0	1	1	0	1	1
0	1	0	0	0	0	1
0	1	0	1	0	1	0
0	1	1	0	0	1	1
0	1	1	1	1	0	0
1	0	0	0	0	1	0
1	0	0	1	0	1	1
1	0	1	0	1	0	0
1	0	1	1	1	0	1
1	1	0	0	0	1	1
1	1	0	1	1	0	0
1	1	1	0	1	0	1
1	1	1	1	1	1	0

FIGURE 3-5: ROM Implementation of a 2-Bit Full Adder

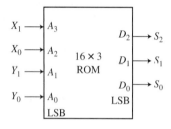

Example

Compute the size of the ROM required to implement an 8-to-3 priority encoder.

Answer: An encoder performs the inverse function of a decoder. An 8-to-3 priority encoder is illustrated in Figure 3-6. If input y_i is 1 and the other inputs are 0, then the *abc* outputs represent a binary number equal to i. An additional output d is used to indicate invalid outputs. A value of 1 on bit d indicates that the output bits a, b, and c are valid. If more than one input is 1 in a priority encoder, the highest numbered input determines the output. The truth table in Figure 3-6 illustrates the output combinations for each input combination. The Xs in the truth table indicate "don't cares." As illustrated, the 8-to-3 priority encoder has 8 inputs and 4 outputs. Hence, it needs a $2^8 \times 4$ bit ROM.

Comment: There will be 256 entries in this ROM. When all the "don't cares" in the truth table in Figure 3-6 are expanded, it does result in 256 entries.

FIGURE 3-6: 8-to-3 Priority Encoder

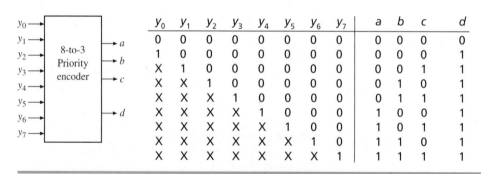

Example

Implement, in ROM, a sequential machine whose state table is given in Figure 3-7: You may note that this is the BCD to Excess-3 code converter that we designed in Chapter 1.

FIGURE 3-7: State Table for a Sequential Circuit

PS	NS X=0	NS X=1	Z X=0	Z X=1
S_0	S_1	S_2	1	0
S_1	S_3	S_4	1	0
S_2	S_4	S_4	0	1
S_3	S_5	S_5	0	1
S_4	S_5	S_6	1	0
S_5	S_0	S_0	0	1
S_6	S_0	—	1	—

Answer: A sequential circuit can easily be designed using a ROM and flip-flops. The combinational part of the sequential circuit can be realized using the ROM. The ROM can be used to realize the output functions and the next state functions. The state of the circuit can then be stored in a register of D flip-flops and fed back to the input of the ROM. Use of D flip-flops is preferable to J-K flip-flops since using 2-input flip-flops would require increasing the number of inputs for the flip-flops (which are outputs from the ROM). The fact that the D flip-flop input equations would generally require more gates than the J-K equations is of no consequence since the size of the ROM depends only on the number of inputs and outputs and not on the complexity of the equations being realized. For this reason, the state assignment used is also of little importance, and generally, a state assignment in straight binary order is as good as any.

In order to realize this sequential machine, a ROM and three D flip-flops are necessary. The ROM will generate the next-state equations and output Z from the present states and input X. Hence, the ROM needs four address lines (three coming from flip-flops and one for X), and it should provide four outputs (three next state bits and output Z). Figure 3-8 illustrates the general organization of the implementation. Since the ROM has four inputs, it contains $2^4 = 16$ words. In general, a Mealy sequential circuit with i inputs, j outputs, and k state variables can be realized using k D flip-flops and a ROM with $i + k$ inputs (2^{i+k} words) and $j + k$ outputs.

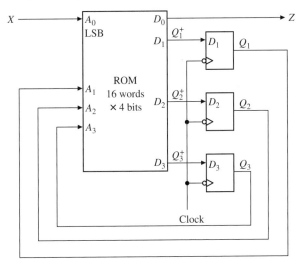

FIGURE 3-8: Realization of a Mealy Sequential Circuit with a ROM

Now, let us derive the contents of the ROM. Table 3-2 gives the truth table for the sequential circuit, which implements the state table of Figure 3-7 with the "don't cares" replaced by 0s, and using a straight binary state assignment.

TABLE 3-2: ROM Truth Table

Q_1	Q_2	Q_3	X	Q_1^+	Q_2^+	Q_3^+	Z
0	0	0	0	0	0	1	1
0	0	0	1	0	1	0	0
0	0	1	0	0	1	1	1
0	0	1	1	1	0	0	0
0	1	0	0	1	0	0	0
0	1	0	1	1	0	0	1
0	1	1	0	1	0	1	0
0	1	1	1	1	0	1	1
1	0	0	0	1	0	1	1
1	0	0	1	1	1	0	0
1	0	1	0	0	0	0	0
1	0	1	1	0	0	0	1
1	1	0	0	0	0	0	1
1	1	0	1	0	0	0	0
1	1	1	0	0	0	0	0
1	1	1	1	0	0	0	0

Assuming that Q_1, Q_2, Q_3, and X are connected to the address lines in that order, with X connected to the LSB, the contents of the ROM to implement this sequential machine are 3, 4, 6, 8, 9, 8, A, B, B, C, 0, 1, 1, 0, 0, and 0 (in hexadecimal representation). The hexadecimal (hex) representation is a concise and convenient way to represent the outputs. The output Z will come from the LSB of the data lines. The next-state information will be available from the three MSBs of the ROM data lines.

3.2.2 Programmable Logic Arrays (PLAs)

A programmable logic array (PLA) performs the same basic function as a ROM. A PLA with n inputs and m outputs (Figure 3-9) can realize m functions of n variables. The internal organization of the PLA is different from that of the ROM. The

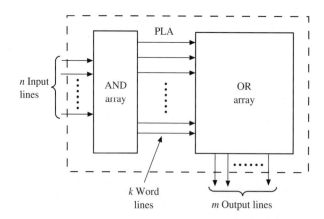

FIGURE 3-9:
Programmable Logic Array Structure

decoder is replaced with an AND array that realizes selected product terms of the input variables. The OR array OR's together the product terms needed to form the output functions.

Figure 3-10 shows a PLA that realizes the following functions:

$$F_0 = \Sigma m(0, 1, 4, 6) = A'B' + AC \tag{3-1}$$

$$F_1 = \Sigma m(2, 3, 4, 6, 7) = B + AC'$$

$$F_2 = \Sigma m(0, 1, 2, 6) = A'B' + BC'$$

$$F_3 = \Sigma m(2, 3, 5, 6, 7) = AC + B$$

The foregoing logic functions contain three variables. In a PLA implementation, each product term in the equation is created first, and then the required product terms are OR'ed using the OR gate. Hence, product terms can be shared while using the PLA. Instead of minimizing each function separately, we want to minimize the total number of product terms. There are five distinct product terms in the preceding four equations. Figure 3-10 illustrates a PLA with three inputs, five product terms, and four outputs, implementing the four equations. It should be noted that the number of terms in each equation is not important, as long as there are AND gates to generate all the product terms required for all outputs together.

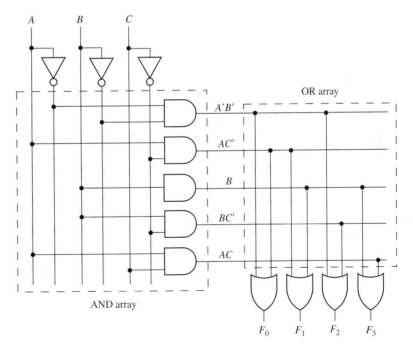

FIGURE 3-10: PLA with Three Inputs, Five Product Terms, and Four Outputs (Logic Level)

Internally, the PLA may use NOR-NOR logic instead of AND-OR logic. The array shown in Figure 3-10 is thus equivalent to the nMOS PLA structure shown in Figure 3-11. Logic gates are formed in the array by connecting nMOS switching transistors between the column lines and the row lines.

FIGURE 3-11: PLA with Three Inputs, Five Product Terms, and Four Outputs (Transistor Level)

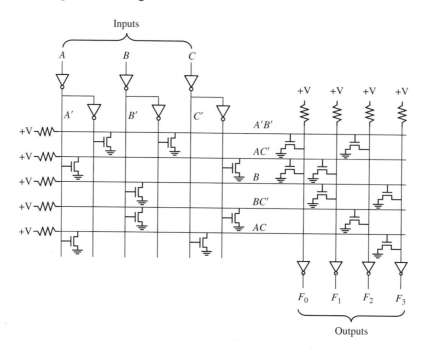

Figure 3-12 shows the implementation of a 2-input NOR gate using nMOS transistors. The transistors act as switches, so if the gate input is a logic 0, the transistor is off. If the gate input is a logic 1, the transistor provides a conducting path to ground. If $X_1 = X_2 = 0$, both transistors are off, and the pull-up resistor brings the Z output to a logic 1 level (+V). If either X_1 or X_2 is 1, the corresponding transistor is turned on, and $Z = 0$. Thus, $Z = (X_1 + X_2)' = X_1'X_2'$, which corresponds to a NOR gate. The part of the PLA array that realizes F_0 is equivalent to the NOR-NOR gate structure shown in Figure 3-13. After canceling the extra inversions, this reduces to an AND-OR structure.

FIGURE 3-12: nMOS NOR Gate

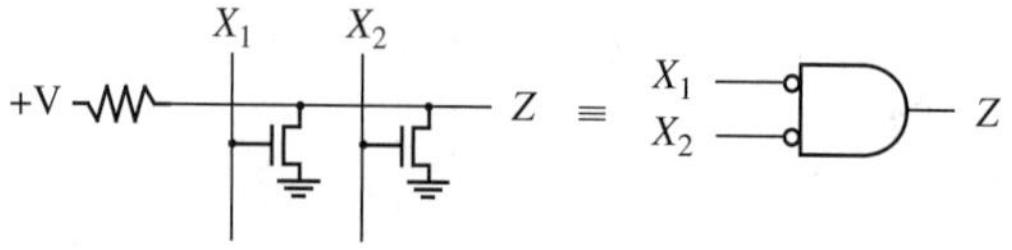

FIGURE 3-13: Conversion of NOR-NOR to AND-OR

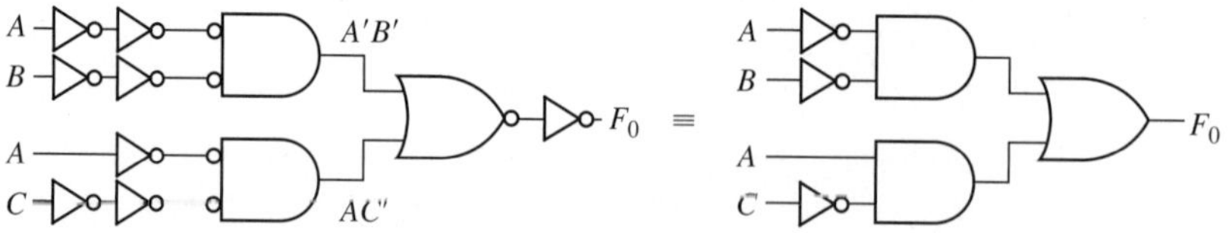

> Source, drain, and gate are the names of the three terminals of the metal oxide semiconductor (MOS) transistor. The gate is the one that is used to control the ON/OFF action. There are two types of MOS transistors, nMOS and

pMOS. The illustrations in this section use nMOS transistors. A popular technology since the 1990s is complementary MOS (CMOS), where nMOS and pMOS transistors are used together in a complementary fashion.

The contents of a PLA can be specified by a modified truth table. Table 3-3 specifies the PLA in Figure 3-10. The input side of the table specifies the product terms. The symbols 0, 1, and – indicate whether a variable is complemented, not complemented, or not present in the corresponding product term. The output side of the table specifies which product terms appear in each output function. A 1 or 0 indicates whether a given product term is present or not present in the corresponding output function. Thus, the first row of Table 3-3 indicates that the term $A'B'$ is present in output functions F_0 and F_2, and the second row indicates that AC' is present in F_0 and F_1.

TABLE 3-3: PLA Table for Equations 3-1

Product Term	Inputs			Outputs			
	A	B	C	F_0	F_1	F_2	F_3
$A'B'$	0	0	–	1	0	1	0
AC'	1	–	0	1	1	0	0
B	–	1	–	0	1	0	1
BC'	–	1	0	0	0	1	0
AC	1	–	1	0	0	0	1

Next, we will realize the following functions using a PLA:

$$F_1 = \Sigma m(2, 3, 5, 7, 8, 9, 10, 11, 13, 15) \quad (3\text{-}2)$$
$$F_2 = \Sigma m(2, 3, 5, 6, 7, 10, 11, 14, 15)$$
$$F_3 = \Sigma m(6, 7, 8, 9, 13, 14, 15)$$

If we minimize each function separately, the result is

$$F_1 = bd + b'c + ab' \quad (3\text{-}3)$$
$$F_2 = c + a'bd$$
$$F_3 = bc + ab'c' + abd$$

If we implement these reduced equations in a PLA, a total of eight different product terms (including c) is required.

Instead of minimizing each function separately, we want to minimize the total number of rows in the PLA table. In this case, the number of terms in each equation is not important, since the size of the PLA does not depend on the number of terms within an equation. Equations (3-3) are plotted on the Karnaugh maps shown in Figure 3-14. Since the term $ab'c'$ is already needed for F_3, we can use it in F_1 instead of ab'. The other two 1s in ab' are covered by the $b'c$ term. This eliminates the need to use a row of the PLA table for ab'. Since the terms $a'bd$ and abd are needed in F_2 and F_3, respectively, we can replace bd in F_1 with $a'bd + abd$. This eliminates the need for a row to implement bd. Since $b'c$ and bc are used in F_1 and F_3, respectively, we can replace c in F_2 with $b'c + bc$. The resulting equations (3-4) correspond to the reduced

FIGURE 3-14: Multiple-Output Karnaugh Maps

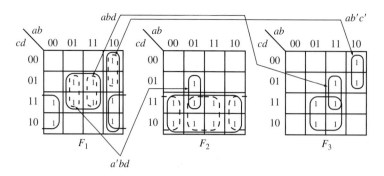

PLA table (Table 3-4). Instead of using Karnaugh maps to reduce the number of rows in the PLA, the Espresso algorithm can be used. This complex algorithm is described in *Logic Minimization Algorithms for VLSI Synthesis* by Brayton [10].

TABLE 3-4: Reduced PLA Table

a	b	c	d	F_1	F_2	F_3
0	1	–	1	1	1	0
1	1	–	1	1	0	1
1	0	0	–	1	0	1
–	0	1	–	1	1	0
–	1	1	–	0	1	1

$$F_1 = a'bd + abd + ab'c' + b'c \qquad (3\text{-}4)$$
$$F_2 = a'bd + b'c + bc$$
$$F_3 = abd + ab'c' + bc$$

Because equations 3-4 have only five different product terms, the PLA table has only five rows. This is a significant improvement over Equations 3-3, which require eight product terms. Figure 3-15 shows the corresponding PLA structure, which has

FIGURE 3-15: PLA Realization of Equations 3-4

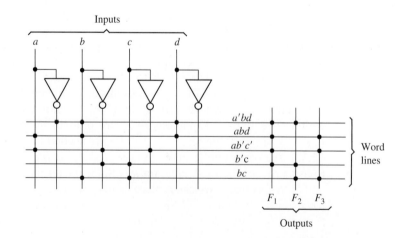

four inputs, five product terms, and three outputs. A dot at the intersection of a word line and an input or output line indicates the presence of a switching element in the array.

A PLA table is significantly different from a truth table for a ROM. In a truth table, each row represents a minterm; therefore, exactly one row will be selected by each combination of input values. The 0s and 1s of the output portion of the selected row determine the corresponding output values. On the other hand, each row in a PLA table represents a general product term. Therefore, zero, one, or more rows may be selected by each combination of input values. To determine the value of F for a given input combination, the values of F in the selected rows of the PLA table must be OR'ed together. The following examples refer to the PLA table of Table 3-4. If $abcd = 0001$, no rows are selected, and all F_is are 0. If $abcd = 1001$, only the third row is selected, and $F_1 F_2 F_3 = 101$. If $abcd = 0111$, the first and fifth rows are selected. Therefore, $F_1 = 1 + 0 = 1$, $F_2 = 1 + 1 = 1$, and $F_3 = 0 + 1 = 1$.

Next, we realize the sequential machine BCD to Excess-3 Code Converter of Figure 1-23 using a PLA and three D flip-flops. The circuit structure is the same as Figure 3-8, except that the ROM is replaced by a PLA. The required PLA table, based on the equations given in Figure 1-25, is Table 3-5.

TABLE 3-5: PLA Table

Product Term	Q_1	Q_2	Q_3	X	Q_1^+	Q_2^+	Q_3^+	Z
Q_2'	—	0	—	—	1	0	0	0
Q_1	1	—	—	—	0	1	0	0
$Q_1 Q_2 Q_3$	1	1	1	—	0	0	1	0
$Q_1 Q_3' X'$	1	—	0	0	0	0	1	0
$Q_1' Q_2' X$	0	0	—	1	0	0	1	0
$Q_3' X'$	—	—	0	0	0	0	0	1
$Q_3 X$	—	—	1	1	0	0	0	1

3.2.3 Programmable Array Logic (PAL)

The PAL (programmable array logic) is a special case of the programmable logic array in which the AND array is programmable and the OR array is fixed. The basic structure of the PAL is the same as the PLA shown in Figure 3-9. Because only the AND array is programmable, the PAL is less expensive than the more general PLA, and the PAL is easier to program. For this reason, logic designers frequently use PALs to replace individual logic gates when several logic functions must be realized.

Figure 3-16(a) represents a segment of an unprogrammed PAL. The symbol

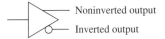

represents an input buffer, which is logically equivalent to

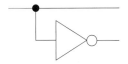

A buffer is used since each PAL input must drive many AND gate inputs. When the PAL is programmed, some of the interconnection points are programmed to make the desired connections to the AND gate inputs. Connections to the AND gate inputs in a PAL are represented by Xs as shown here

$$\begin{array}{c} A \\ B \\ C \end{array} \Rightarrow ABC \equiv \overset{A\ B\ C}{\underset{}{\text{---}}} ABC$$

As an example, we will use the PAL segment of Figure 3-16(a) to realize the function $I_1 I'_2 + I'_1 I_2$. The Xs in Figure 3-16(b) indicate that I_1 and I'_2 lines are connected to the first AND gate, and the I'_1 and I_2 lines are connected to the other gate.

FIGURE 3-16: PAL Segment

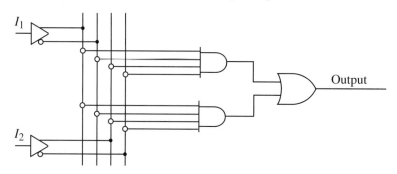

(a) Unprogrammed

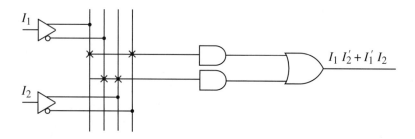

(b) Programmed

When designing with PALs, we must simplify our logic equations and try to fit them into one (or more) of the available PALs. Unlike the more general PLA, the AND terms cannot be shared among two or more OR gates; therefore, each function to be realized can be simplified by itself without regard to common terms. For a given type of PAL, the number of AND terms that feed each output OR gate is fixed and limited. If the number of AND terms in a simplified function is too large, we may be forced to choose a PAL with more gate inputs and fewer outputs.

As an example of programming a PAL, we will implement a full adder. The logic equations for the full adder are as follows:

$$Sum = X'Y'C_{in} + X'YC_{in}' + XY'C_{in}' + XYC_{in}$$
$$C_{out} = XC_{in} + YC_{in} + XY$$

Figure 3-17 shows a section of a PAL where each OR gate is driven by four AND gates. The Xs on the diagram show the connections that are programmed into the PAL to implement the full adder equations. For example, the first row of Xs implements the product term $X'Y'C_{in}$.

FIGURE 3-17:
Implementation of a Full Adder Using a PAL

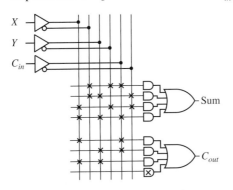

Typical combinational PALs have from 10 to 20 inputs and from 2 to 10 outputs, with 2 to 8 AND gates driving each OR gate. PALs are also available that contain D flip-flops with inputs driven from the programmable array logic. Such PALs are called sequential PALs. They provide a convenient way of realizing sequential circuits. Figure 3-18 shows a segment of a sequential PAL. The D flip-flop is driven from an OR gate, which is fed by two AND gates. The flip-flop output is fed back to the programmable AND array through a buffer. Thus, the AND gate inputs can be connected to A, A', B, B', Q, or Q'. The diagram shows the realization of the next-state equation:

$$Q^+ = D = A'BQ' + AB'Q$$

The flip-flop output is connected to an inverting tristate buffer, which is enabled when $EN = 1$.

FIGURE 3-18: Segment of a Sequential PAL

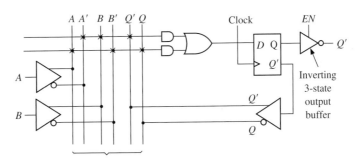

A few decades ago, PALs were very popular among digital system designers. A very popular PAL was the 16R4. This PAL has an AND gate array with 16

input variables, and it has four D flip-flops. Currently, several other programmable devices such as GALs (described in the following section), CPLDs, and FPGAs have arrived. PALs have practically disappeared; hence, we do not further describe any of the traditional PAL devices.

3.2.4 Programmable Logic Devices (PLDs)/Generic Array Logic (GAL)

PALs and PLAs have been very popular for implementing small circuitry and interface logic often needed by designers. As integrated circuit technology has improved, a wide variety of other programmable logic devices have become available. Traditional PALs are not reprogrammable. However, there are flash erasable/reprogrammable PALs now. Often, these are referred to as PLDs or GALs.

The 22CEV10 (Figure 3-19) is a CMOS electrically erasable PLD that can be used to realize both combinational and sequential circuits. The abbreviation PLD has been used as a generic term for all programmable logic devices and to refer to specific devices such as the 22CEV10. In addition to the AND-OR arrays that the PALs have, most PLDs have some type of a macroblock that contains some multiplexers and some additional programmability. These PLDs are named with reference to their input and output capability. For instance, the 22CEV10 has 12 dedicated input pins and 10 pins that can be programmed as either inputs or outputs. It contains 10 D flip-flops and 10 OR gates. The number of AND gates that

FIGURE 3-19: Block Diagram for 22V10

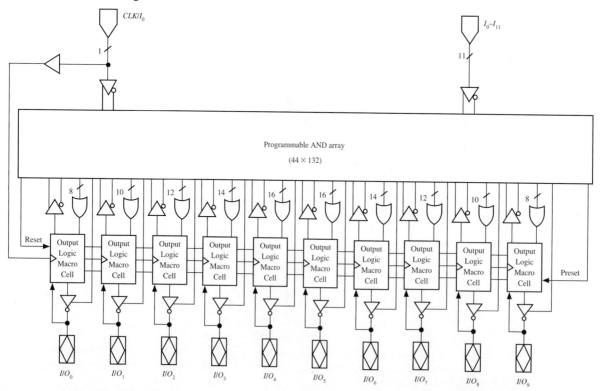

feed each OR gate ranges from 8 through 16. Each OR gate drives an *output logic macrocell*. Each macrocell contains one of the 10 D flip-flops. The flip-flops have a common clock, a common asynchronous reset (AR) input, and a common synchronous preset (SP) input. The name 22V10 indicates a versatile PAL with a total of 22 input and output pins, 10 of which are bidirectional I/O (input/output) pins.

Figure 3-20 shows the details of a 22CEV10 output macrocell. The connections to the output pins are controlled by programming this macrocell. The output MUX control inputs S_1 and S_0 select one of the data inputs. For example, $S_1 S_0 = 10$ selects data input 2. Each macrocell has two programmable interconnect bits. S_1 or S_0 is connected to ground (logic 0) when the corresponding bit is programmed. Erasing a bit disconnects the control line (S_1 or S_0) from ground and allows it to float to logic 1. When $S_1 = 1$, the flip-flop is bypassed, and the output is from the OR gate. The OR gate output is connected to the I/O pin through the multiplexer and the output buffer. The OR gate is also fed back so that it can be used as an input to the AND gate array. If $S_1 = 0$, then the flip-flop output is connected to the output pin, and it is also fed back so that it can

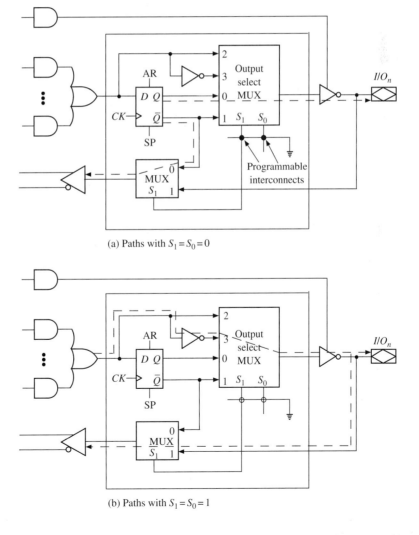

FIGURE 3-20: PLD Output Macrocell

be used for AND gate inputs. When $S_0 = 1$, the output is not inverted, so it is an active high. When $S_0 = 0$, the output is inverted, so it is an active low. The output pin is driven by a tristate inverting buffer. When the buffer output is in a high-impedance state, the OR gate and flip-flop are disconnected from the output pin, and the pin can be used as an input. The dashed lines on Figure 3-20(a) show the path when both S_1 and S_0 are 0, and the dashed lines on Figure 3-20(b) show the path when both S_1 and S_0 are 1. Note that in the first case, the flip-flop output Q is inverted by the output buffer, and in the second case, the OR gate output is inverted twice, so there is no net inversion.

Several PLDs similar to the 22V10 have been popular. Typically these PLDs had 8 to 12 I/O pins. Each output pin is typically connected to an output macrocell, and each macrocell has a D flip-flop. The I/O pins can be programmed so that they act as inputs or as combinational or flip-flop outputs. Some of the PLDs have a dedicated clock input and the others have a dual-purpose pin that can be used either as a clock or as an input. All the PLDs typically have tristate buffers at the outputs, and some of them have a dedicated output enable ($\overline{OE}$).

Lattice Semiconductor created similar devices that are in-circuit programmable and called them generic array logic (**GAL**). GALs are perfect for implementing small amounts of interface logic, often called "glue" logic. Most of the common PLDs, like the PALCE22V10, PALCE20V8, and so on, have GAL equivalents, called GAL22V10, GAL20V8, and so on.

Design Flow for PLDs

Computer-aided design programs for PALs and PLDs are widely available. Such programs accept logic equations, truth tables, state graphs, or state tables as inputs and automatically generate the required bit patterns. These patterns can then be downloaded into a PLD programmer, which will create the necessary connections and verify the operation of the PAL. Many of the newer types of PLD are erasable and reprogrammable in a manner similar to EPROMs and EEPROMs. Hence, in these newer devices, bit patterns corresponding to the required EEPROM content will be generated by the software.

PALASM and **ABEL** are two languages that were popularly used with PALs and PLDs. PALASM is a PLD design language from MMI and AMD. ABEL is a PLD design language from DATA I/O. Intel used to manufacture PLDs and had a PLD language called PLDShell. While PALASM and ABEL can still be used, designs for GALs can now be done using hardware description languages such as Verilog or VHDL.

3.3 Complex Programmable Logic Devices (CPLDs)

Improvements in integrated circuit technology have made it possible to create programmable ICs equivalent to several PLDs in the same chip. These chips are called complex programmable logic devices (CPLDs). When storage elements such as flip-flops are also included on the same IC, a small digital system can be implemented with a single CPLD.

3.3 Complex Programmable Logic Devices (CPLDs)

CPLDs are an extension of the PAL concept. In general, a CPLD is an IC that consists of a number of PAL-like logic blocks together with a programmable interconnect matrix. CPLDs typically contain 500 to 10,000 logic gates. Essentially, several PLDs are interconnected using a crossbar-like switch and fabricated inside the same IC. An $N \times M$ crossbar switch is one in which each of the N input lines can be connected to any of the M output lines simultaneously. It is expensive to build these switches; however, use of such a switch results in predictable timing. Many CPLDs are electronically erasable and reprogrammable and are sometimes referred to as EPLDs (erasable PLDs).

A typical CPLD contains a number of macrocells that are grouped into function blocks. Connections between the function blocks are made through an interconnection array. Each macrocell contains a flip-flop and an OR gate, which has its inputs connected to an AND gate array. Some CPLDs are based on PALs, in which case each OR gate has a fixed set of AND gates associated with it. Other CPLDs are based on PLAs, in which case any AND gate output within a function block can be connected to any OR gate input in that block.

Xilinx, Altera, Lattice Semiconductor, Cypress, and Atmel are the major CPLD manufacturers in the market today. The major products available in the market are listed in Table 3-6. Some vendors specify their gate capacities in usable gates and some specify it in terms of logic elements.

TABLE 3-6: Major CPLDs and Their Approximate Capacity

Vendor	CPLD Family	Gate Count
Xilinx	CoolRunner-II	750 to 12K
	CoolRunner XPLA3	750 to 12K
	XC9500XV	800 to 6400
	XC9500	800 to 6400
	XC9500XL	800 to 6400
Atmel	CPLD ATF15	750 to 3000 usable gates
	CPLD-2 22V10	500 usable gates
	CPLD-Proprietary	2500
Cypress	Delta39K	30K to 200K
	Flash370i	800 to 3200
	Quantum38K	30K to 100K
	Ultra37000	960 to 7700
	MAX340 high-density EPLDs	600 to 3750
Lattice	ispXPLD 5000MX	75K to 300K
	ispMACH 4000B/C/V/Z	640 to 10,240
Altera	MAX II	240 to 2,210 logic elements
	MAX3000	600 to 10K usable gates
	MAX7000	600 to 10K usable gates

3.3.1 An Example CPLD: The Xilinx CoolRunner

Xilinx has two major series of CPLDs—the CoolRunner and the XC9500. Figure 3-21 shows the basic architecture of a CoolRunner family CPLD, the Xilinx XCR3064XL. This CPLD has 4 function blocks, and each block has 16 associated **macrocells**

178 Chapter 3 Introduction to Programmable Logic Devices

FIGURE 3-21: Architecture of Xilinx CoolRunner XCR3064XL CPLD

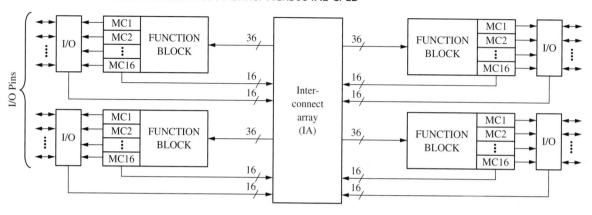

(MC1, MC2,...). Each function block is a programmable AND-OR array that is configured as a PLA. Each macrocell contains a flip-flop and multiplexers that route signals from the function block to the input/output (I/O) block or to the **interconnect array (IA).** The interconnect array selects signals from the macrocell outputs or I/O blocks and connects them back to function block inputs. Thus, a signal generated in one function block can be used as an input to any other function block. The I/O blocks provide an interface between the bidirectional I/O pins on the IC and the interior of the CPLD.

Figure 3-22 shows how a signal generated in the PLA (function block) is routed to an I/O pin through a macrocell. Any of the 36 inputs from the IA (or their complements) can be connected to any inputs of the 48 AND gates. Each OR gate can accept up to 48 product term inputs from the AND array. The macrocell logic in this diagram is a simplified version of the actual logic. The first MUX (1) can be programmed to select the OR gate output or its complement. The MUX (2) at the output of the macrocell can be programmed to select either the combinational output (G) or the flip-flop output (Q). This output goes to the interconnect array and to the output cell. The output cell includes a 3-state buffer (3) to drive the I/O pin. The buffer enable input can be programmed from several sources. When the I/O pin is used as an input, the buffer must be disabled.

FIGURE 3-22: CPLD Function Block and Macrocell (Simplified Version of XCR3064XL)

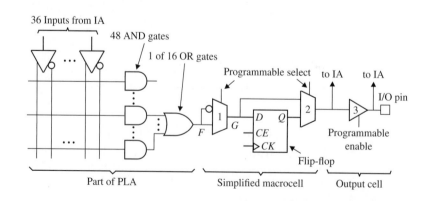

Figure 3-23 shows how a Mealy sequential machine with two inputs, two outputs, and two flip-flops can be implemented by a CPLD. Four macrocells are required, two to generate the *D* inputs to the flip-flops and two to generate the *Z* outputs. The flip-flop outputs are fed back to the AND array inputs via the interconnection matrix (not shown). The number of product terms required depends on the complexity of the equations for the *D*s and the *Z*s.

FIGURE 3-23: CPLD Implementation of a Mealy Machine

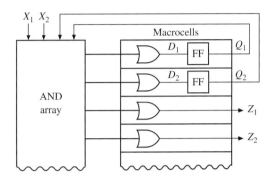

CPLD Implementation of a Parallel Adder with Accumulator

Assume that we need to implement an adder with an accumulator, as in Figure 3-24, in a CPLD. The accumulator register needs one flip-flop for each bit. Each bit also needs to generate the sum and carry bits corresponding to that bit.

FIGURE 3-24: N-Bit Parallel Adder with Accumulator

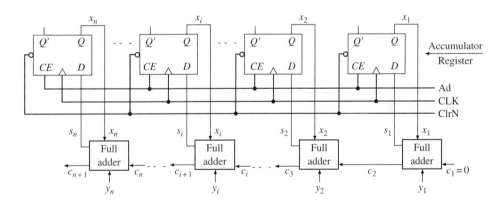

Figure 3-25 shows how three bits of such a parallel adder with an accumulator can be implemented using a CPLD. Each bit of the adder requires two macrocells. One of the macrocells implements the sum function and an accumulator flip-flop. The other macrocell implements the carry, which is fed back into the AND array. The *Ad* signal can be connected to the enable input (*CE*) of each flip-flop via an AND gate (not shown). Each bit of the adder requires eight product terms (four for the sum, three for the carry, and one for *CE*). For each accumulator flip-flop

$$D_i = X_i^+ = S_i = X_i \oplus Y_i \oplus C_i$$

FIGURE 3-25: CPLD Implementation of a Parallel Adder with Accumulator

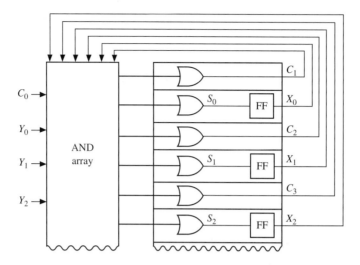

If the flip-flops are programmed as T flip-flops, then the logic for the sum can be simplified. For each accumulator flip-flop

$$X_i^+ = X_i \oplus Y_i \oplus C_i$$

Therefore, the T input is

$$T_i = X_i^+ \oplus X_i = Y_i \oplus C_i$$

The add signal can be AND'ed with the T_i input so that the flip-flop state can change only when $Ad = 1$

$$T_i = Ad(Y_i \oplus C_i) = Ad\ Y_i\ C_i' + Ad\ Y_i'\ C_i$$

The equation for carry is

$$C_{i+1} = X_i Y_i + X_i C_i + Y_i C_i$$

• • • • • • • • •

3.4 Field-Programmable Gate Arrays (FPGAs)

In this section, we introduce field-programmable gate arrays (FPGAs). FPGAs are ICs that contain an array of identical logic blocks with programmable interconnections. The user can program the functions realized by each logic block and the connections between the blocks. FPGAs have revolutionized the way prototyping and designing is done in the world. The flexibility offered by reprogrammable FPGAs has enhanced the design process. While different kinds of programmable devices had been around, when Xilinx used SRAM storage elements to create programmable logic blocks and introduced its family of XC2000 devices in 1985, the world received a totally new and powerful technology. There are a variety of FPGA products available in the market now. Xilinx, Altera, Lattice Semiconductor, and Microsemi, are examples of companies that design and sell FPGAs.

FPGAs provide several advantages over traditional gate arrays or mask-programmable gate arrays (**MPGAs**). A traditional gate array can be used to implement any circuit, but it is programmable only in the factory. A specific mask to match the particular circuit is created in order to fabricate the gate array. The design time of a gate-array–based IC is a few months. FPGAs are standard off-the-shelf products. Manufacturing time reduces from months to hours as one adopts FPGAs instead of MPGAs. Design iterations become easier with FPGAs. This is a tremendous advantage when it comes to time to market. It becomes easy to correct mistakes that creep into designs. Mistakes and design specification changes become less costly. Prototyping cost is reduced. At low volumes, FPGAs are cheaper than MPGAs.

FPGAs have disadvantages as well. FPGAs are less dense than traditional gate arrays (MPGAs). In FPGAs, a lot of resources are spent merely to achieve the needed programmability. MPGAs have better performance than FPGAs. Programmable points have resistance and capacitance. They slow down signals, so FPGAs are slower than traditional gate arrays. In addition, interconnection delays are unpredictable in FPGAs. PLDs such as PALs and GALs are simple and inexpensive. CPLDs are faster and cheaper than FPGAs. The overhead for programmability is fairly low in PALs and CPLDs. The main advantage of CPLDs over FPGAs is the lower cost and predictability in timing.

Several commercial FPGAs are listed in Table 3-7. As one can notice, some of these chips contain logic equivalent to 5 million gates. The capacities of some FPGAs are specified in number of look-up tables (LUTs). Due to the large capacity,

TABLE 3-7: Examples of Commercial FPGAs

Vendor	FPGA Product	Capacity (Approx) in Gates/LUTs
Xilinx	Kintex 7	41,000 to 298,600 LUTs
	Artix 7	63,400 to 134,600 LUTs
	Virtex-6	46,560 to 474,240 LUTs
	Spartan-3	50K to 5M
	Virtex-5	19,200 to 207,360 LUTs
Altera	Arria V	76,800 to 516,096 LUTs
	Arria II	45,125 to 256,500 LUTs
	ACEX 1K	56K to 257K
	APEX II	1.9M to 5.25M
	FLEX 10K	10K to 50K
	Stratix/Stratix II	10,570 to 132,540 logic elements
Lattice	LatticeECP2	6K to 68K LUTs
	Lattice SC	15.2K to 115.2K LUTs
	ispXPGA	139K to 1.25M
	MachXO	256 to 2280 LUTs
	LatticeECP	6.1K to 32.8K LUTs
Microsemi	Fusion	90K to 1.5M system gates
	IGLOO	15K to 3M system gates
	Axcelerator	125K To 2M
	eX	3K to 12K
	ProASIC3	30K to 3M
	MX	3K to 54K
Quick Logic	Eclipse/EclipsePlus	248K to 662K
	Quick RAM	45K to 176K
	pASIC 3	5K to 75K
Atmel	AT40K	5K to 40K
	AT40KAL	5K to 50K

it is possible to prototype or even manufacture large systems in a single FPGA. In this chapter, we describe the basic organization of FPGAs. Design examples with FPGAs are presented in Chapter 6.

3.4.1 Organization of FPGAs

Figure 3-26 shows the layout of a typical FPGA. The interior of FPGAs typically contains three elements that are programmable:

- Programmable logic blocks
- Programmable input/output blocks
- Programmable routing resources

FIGURE 3-26: Layout of a Typical FPGA

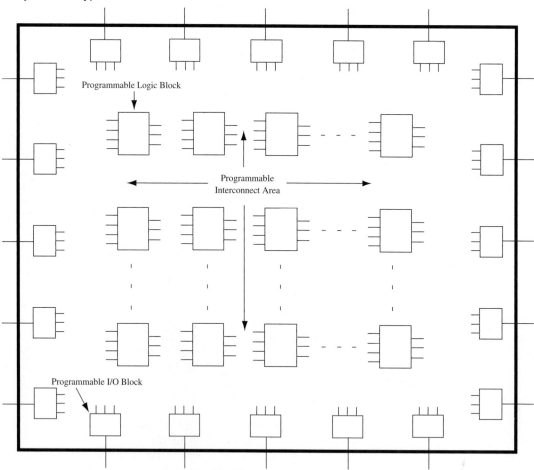

Arrays of programmable logic blocks are distributed within the FPGA. These logic blocks are surrounded by input/output (I/O) interface blocks. These I/O blocks can be considered to be on the periphery of the chip. They connect the logic signals to FPGA pins. The space between the logic blocks is used to route connections between the logic blocks.

The "field" programmability in FPGAs is achieved by reconfigurable elements, which can be programmed or reconfigured by the user. As mentioned, there are three major programmable elements in FPGAs: the logic block, the interconnect, and the input/output block. Programmable logic blocks are created by using multiplexers, look-up tables, and AND-OR or NAND-NAND arrays. "Programming" them means changing the input or control signals to the multiplexers, changing the look-up table contents, or selecting or not selecting particular gates in AND-OR gate blocks. For a programmable interconnect, "programming" means making or breaking specific connections. This is required to interconnect various blocks in the chip and to connect specific I/O pins to specific logic blocks. Programmable I/O blocks denote blocks that can be programmed to be input, output, or bidirectional lines. Typically, they can also be "programmed" to adjust the properties of their buffers such as inverting/non-inverting, tristate, passive pull-up, or even to adjust the **slew rate**, which is the rate of change of signals on that pin.

What makes an FPGA distinct from a CPLD is the flexible general-purpose interconnect. In a CPLD, the interconnect is fairly restricted. The general-purpose interconnect in an FPGA gives it a lot of flexibility, but it also has the disadvantage of being slow. A connection from one part of the chip to another part might have to travel through several programmable interconnect points, resulting in large and unpredictable signal delays.

Although Figure 3-26 was used to illustrate the general structure of an FPGA, not all FPGAs look like that. Commercial FPGAs use a variety of architectures. The FPGA architecture or organization refers to the manner or topology in which the logic blocks and interconnect resources are distributed inside the FPGA. The organization that is presented in Figure 3-26 is often referred to as *symmetrical array architecture*. If one examines the various FPGAs that have been on the market since their inception in the late 1980s, one could classify them into four different basic architectures or topologies:

Matrix-based (symmetrical array) architectures

Row-based architectures

Hierarchical PLD architectures

Sea-of-gates architecture

These architectures are illustrated in Figure 3-27. This classification is based on the layout of the general purpose logic region in the FPGAs. Modern FPGAs contain special-purpose blocks including a microprocessor, and the special-purpose blocks are usually embedded into the center or on the peripheries.

Matrix-Based (Symmetrical Array) Architectures

The logic blocks in this type of FPGA are organized in a matrix-like fashion, as illustrated in Figure 3-27(a). Most Xilinx FPGAs belong to this category. The logic blocks in these architectures are typically of a large granularity (capable of implementing 4-variable functions or more). These architectures typically contain 8×8 arrays in the smaller chips and 100×100 or larger arrays in the bigger chips. The routing resources are interspersed between the logic blocks. The routing in these architectures is often called two-dimensional channeled routing since routing resources are generally available in horizontal and vertical directions.

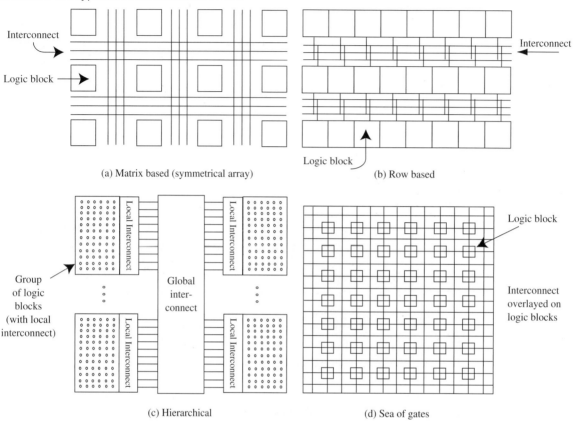

FIGURE 3-27: Typical Architectures for FPGAs

Row-Based Architectures

These architectures were inspired by traditional gate arrays. The logic blocks in this architecture are organized into rows as illustrated in Figure 3-27(b). Thus, there are rows of logic blocks and routing resources. The routing resources interspersed between the rows can be used to interconnect the various logic blocks. Traditional mask-programmable gate arrays use very similar architectures. The routing in these architectures is often called one-dimensional channeled routing, because the routing resources are located as a channel in between rows of logic resources. Some Microsemi FPGAs employ this architecture.

Hierarchical Architectures

In some FPGAs, blocks of logic cells are grouped together by a local interconnect, and several such groups are interconnected by another level of interconnect. For instance, in Altera APEX20 and APEX II FPGAs, 10 or so logic elements are connected to form what Altera calls a Logic Array Block (LAB), and then several LABs are connected to form a MEGALAB. Thus, there is a hierarchy in the organization of these FPGAs. These FPGAs contain clusters of logic blocks with localized resources for interconnection. The global interconnect network is used for the interconnections between the clusters of logic blocks in these FPGAs.

Sea-of-Gates Architecture

The sea-of-gates architecture is yet another manner to organize the logic blocks and interconnect in an FPGA. The general FPGA fabric consists of a large number of gates, and then there is an interconnect superimposed on the sea of gates as illustrated in Figure 3-27(d). Plessey, a manufacturer that was in the FPGA market in the mid-1990s, made FPGAs of this architecture. The basic cell used was a NAND gate, in contrast to the larger basic cells used by manufacturers such as Xilinx. While the terminology sea of gates is the most popular, there are also the terminologies **sea of cells** and **sea of tiles** to indicate the topology of FPGAs with a large number of fine-grain logic cells. The Microsemi Fusion FPGAs contain a sea of tiles, where each tile can be configured as a 3-input logic function or a flip-flop/latch.

3.4.2 FPGA Programming Technologies

FPGAs consist of a large number of logic blocks interspersed with a programmable interconnect. The logic block is programmable in the sense that the same building block can be "programmed" or "configured" to create any desired circuitry. There is also programmability in the interconnections between the logic blocks.

Several techniques have been used to achieve the programmable interconnections between FPGAs. The term "**programming technology**" is used here to denote the technology by which the programmability in an FPGA is achieved. In some devices, the reconfigurability is achieved by changing the contents of static RAM cells. In some devices, it is achieved by using flash memory cells. In others, it is achieved by fusing metal links. In general, FPGAs use one of the following programming methods:

StaticRAM programming technology

EPROM/EEPROM/flash programming technology

AntiFuse programming technology

The SRAM Programming Technology

The SRAM programming technology involves creating reconfigurability by bits stored in static RAM (SRAM) cells. The logic blocks, I/O blocks, and interconnects can be made programmable by using configuration bits stored in SRAM. Reconfigurable logic blocks can easily be implemented as look-up tables (LUTs), which is the same approach as the ROM method described in Section 3.2.1. Sixteen SRAM cells can implement any function of four variables. The programmable interconnect can also be achieved by SRAM. The key idea is to use pass transistors to create switches and then control them using the SRAM content. Consider the arrangement in Figure 3-28(a). The SRAM cell is connected to the gate of the pass transistor. When the SRAM cell content is 0, the pass transistor is OFF; hence, no connection exists between points A and B. A closed path can be achieved by turning the pass transistor ON by making the SRAM cell content 1. SRAM bits can be used to construct routing matrices by using multiplexers as shown in Figure 3-28(b). Changing the contents of the SRAM in the arrangement in Figure 3-28(b) will allow the designer to change what is connected to point X. The bits that are stored in the SRAM for deciding the LUT functionality or interconnection are called **configuration bits**.

FIGURE 3-28: Routing with Static RAM Programming

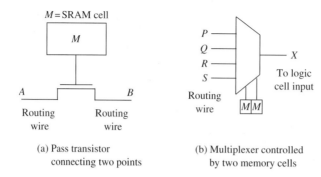

(a) Pass transistor connecting two points

(b) Multiplexer controlled by two memory cells

An SRAM cell usually takes six transistors, as illustrated in Figure 3-29. Four cross-coupled transistors are required to create a latch, and two additional transistors are used to control passing data bits into the latch. When the *Word Line* is set to high, the values on the *Bit Line* will be latched into the cell. This is the write operation. The read operation is performed by precharging the *Bit Line* and *Bit Line'* to a logic 1 and then setting *Word Line* to high. The contents stored in the cell will then appear on the *Bit Line*. Some SRAM cell implementations use only five transistors. One advantage of using static RAM is that it is volatile and you can write new contents again and again. This provides flexibility during prototyping and development. Another advantage is that the fabrication steps for making SRAM cells are not different from the steps for making logic. The major disadvantage of the SRAM programming technology is that five or six transistors are used for every SRAM cell. This adds a tremendous cost to the chip. For example, if an FPGA has 1 million programmable points, it means that approximately 5 or 6 million transistors are employed in achieving this programmability.

FIGURE 3-29: Typical 6-Transistor SRAM Cell

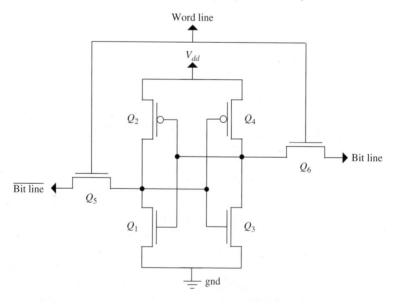

Being volatile can become a disadvantage when an FPGA is used in the final product. Hence, when SRAM FPGAs are used, a nonvolatile device such as an

EPROM should be used to permanently store the configuration bits. Typically, what is done is to use the EPROM as a "boot ROM," The EPROM contents are transferred to the SRAM when power comes up.

Xilinx FPGAs were the first FPGAs to use SRAM as the programming technology. In fact, it is the flexibility and reprogrammability of SRAM FPGAs that caused FPGAs to become widely popular. Now, many companies use the SRAM programming technology for their FPGAs.

EPROM/EEPROM Programming Technology

In the EPROM/EEPROM programming technology, EPROM cells are used to control programmable connections. Assume that EPROM/EEPROM cells are used instead of the SRAM cells in Figure 3-28. A transistor with two gates—a floating gate and a control gate—is used to create an EPROM cell. Figure 3-30 illustrates an EPROM cell. The pull-up resistor connects the drain of the transistor to the power supply (labeled V_{DD} in the figure). To turn the transistor off, charge can be injected on the floating gate using a high voltage between the control gate and the drain of the transistor. This charge increases the threshold voltage of the transistor and turns it off. The charge can be removed by exposing the floating gate to ultraviolet light. This lowers the threshold voltage of the transistor and makes it function normally.

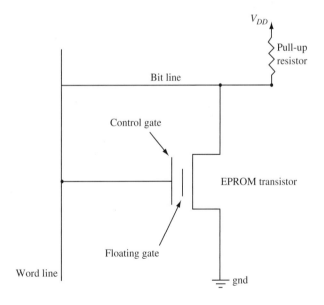

FIGURE 3-30: The EPROM Programming Technology

EPROMs are slower than SRAM; hence, SRAM-based FPGAs can be programmed faster. EPROMs also require more processing steps than SRAM. EPROM-based switches have high ON resistance and high static power consumption. The EEPROM is similar to EPROM, but removal of the gate charge can be done electrically.

Flash memory is a form of EEPROM that allows multiple locations to be erased in one operation. Flash memory stores information in floating gate transistors as in traditional EPROM. The floating gate is isolated by an insulating oxide layer; hence, any electrons placed there are trapped. The cell is read by placing a specific voltage on the control gate. When the voltage to read is placed, electrical current will or will not flow depending on the threshold voltage of the cell, which is controlled by the number of electrons trapped in the floating gate. In some devices, the information is stored as absence or presence of current. In some advanced devices, the amount of current flow is sensed; hence, multiple bits of information can be stored in a cell. To erase, a large voltage differential is placed between the control gate and the source, which pulls electrons off. Flash memory is erased in segments/sectors; all cells in a block are erased at the same time.

The Antifuse Programming Technology

In some FPGAs, the programmable connections between different points are achieved by what is called an "antifuse." Contrary to fuse wires that blow open when high current passes through them, the "antifuse" programming element changes from high resistance (open) to low resistance (closed) when a high voltage is applied to it. Antifuses are often built using dielectric layers between N+ diffusion and polysilicon layers or by amorphous silicon between metal layers. Antifuses are normally OFF; permanently connected links are created when they are programmed. The process is irreversible; hence, antifuse FPGAs are only one-time programmable. Programming an antifuse requires applying a high voltage and currents in excess of normal currents. Special programming transistors larger than normal transistors are incorporated into the device in order to accomplish the programming. There are various antifuse technologies; a popular one is the Via antifuse technology.

Antifuse technology has the advantage that the area consumed by the programmable switch is small. Another advantage is that antifuse-based connections are faster than SRAM- and EEPROM-based switches. The disadvantage of the antifuse technology is that it is not reprogrammable. It is a permanent connection; if an error or design change necessitates reprogramming, a new device is required.

Comparison of Programming Technologies

Table 3-8 compares the characteristics of the major programming technologies used by FPGAs. Only the SRAM and EEPROM programming technologies allow in-circuit programmability. In-circuit programmability means that an FPGA

TABLE 3-8: Characteristics of the Major FPGA Programming Technologies

Programming Technology	Volatile/Nonvolatile	Reprogrammable	Area Overhead	Resistance	Capacitance
SRAM	Volatile	In-circuit reprogrammable	Large	Medium-high	High
EPROM	Nonvolatile	Out-of-circuit reprogrammable	Small	High	High
EEPROM/Flash	Nonvolatile	In-circuit reprogrammable	Medium to high	High	High
Antifuse	Nonvolatile	Not reprogrammable	Small	Small	Small

can be reprogrammed without removing it from the board in which it is used. In-circuit programmability is not possible in traditional EPROM-based devices, but EEPROM and flash technologies allow in-circuit reprogrammability.

SRAM FPGAs have several disadvantages: high area overhead, large delays, volatility, and others. However, the in-circuit programmability and fast programmability have made them very popular. SRAM FPGAs are more expensive than other types of FPGAs because each programmable point uses six transistors. This extra hardware contributes only to the reprogrammability but not to the actual circuitry realized with the FPGA. EEPROM- and Flash-based FPGAs are comparable to SRAM FPGAs in many respects; however, they are not as fast as SRAM FPGAs.

3.4.3 Programmable Logic Block Architectures

FPGAs in the past have employed different kinds of programmable logic blocks as the basic building block. In this section, we present some generalized versions of typical building blocks in commercial FPGAs.

The logic blocks vary in the basic components they use. For instance, some FPGAs use Look-Up Table (LUT) based logic blocks, while others use multiplexers and logic gates to build their logic blocks. There also have been FPGAs where logic blocks simply consisted of transistor pairs (e.g., Crosspoint FPGAs). Logic building blocks in early Altera FPGAs were PLD blocks. There were also FPGAs that used NAND gates as the building block (e.g., Plessey).

The logic blocks also vary in their architecture and size. Some FPGAs use large basic blocks, which can implement large functions (several 5-variable or 4-variable functions), and have several flip-flops in each basic block. In contrast, there are FPGA building blocks which only allow a 3-variable function or a flip-flop in one block. Some FPGAs allow choices as to whether latched/unlatched or both kinds of outputs can be brought out. Some FPGAs allow one to control the type of flip-flop that is realized. Some allow positive edge/negative edge clock, direct set/reset inputs to the flip-flop, etc.

Different FPGA manufacturers use different names (often trademarked) to denote their logic blocks. In the Xilinx literature, a programmable logic block is called a **Configurable Logic Block (CLB)**. Altera calls their basic blocks **Logic Elements (LE)** and a collection of 8 or 10 of them **Logic Array Blocks (LABs)**. The basic cells in Microsemi Fusion FPGAs are referred to as "**VersaTiles**."

Look-Up-Table–Based Programmable Logic Blocks

Many look-up-table–based FPGAs use a 4-variable look-up table plus a flip-flop as the basic element and then combine several of them in various topologies. Consider the structure in Figure 3-31. There are two 4-variable look-up tables (often denoted by the short form **LUT4**) and two flip-flops in this programmable logic block. The LUT4 can also be called a **4-variable function generator** since it can generate any function of four variables. The two LUT4s can generate any two functions of four variables. The inputs to the X-function generator are called X_1, X_2, X_3, and X_4, and the inputs to the Y-function generator are called Y_1, Y_2, Y_3, and Y_4. The functions can be steered to the output of the block (X and Y) in combinational or latched form. There are two D flip-flops in the logic block. The D flip-flops are versatile in the sense that they have clock enable, direct set, and direct reset inputs. A multiplexer selects between the combinatorial output and the latched version of the output. The little box with "M" in it (beneath the multiplexer) indicates a memory cell that is required to provide appropriate select signals to select between the latched and unlatched form of the function. An early Xilinx FPGA, the XC3000, used building blocks very similar to this structure.

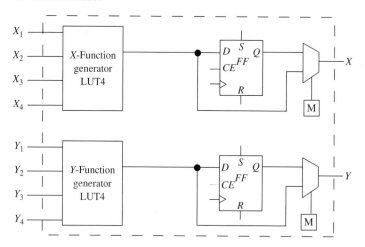

FIGURE 3-31: A Look-Up-Table–Based Programmable Logic Block

Let us assume that we want to implement the function $F_1 = A'B'C + A'BC' + AB$ using an FPGA with programmable logic blocks as shown in Figure 3-31. Since this is a 3-variable function, a 4-input LUT is more than sufficient to implement the function. The path highlighted in Figure 3-32 assumes that the X-function generator (top LUT) is used. Let us assume that X_1 is the LSB and X_4 is the MSB to the LUT. Since function F_1 uses only three variables, the X_4 input is not used. A truth table can be constructed to represent the function, and the LUT contents can be derived.

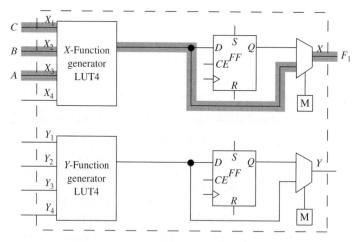

FIGURE 3-32: Highlighting Paths for Function F_1

The LUT contents to implement function F_1 will be 0, 1, 1, 0, 0, 0, 1, 1, 0, 1, 1, 0, 0, 0, 1, 1. The first 8 bits in the LUT reflect the truth table outputs when the function is represented in a truth table form. Since input X_4 is not grounded, the first 8 bits are repeated to take care of the possibility that the X_4 input might stay at a logic 1 when it is unused. Since the functions are stored in look-up table (LUT) form, the number of terms in the function is not important. Common minimizations to reduce the number of terms are not relevant. The number of variables is what is important.

Many commercial FPGAs use LUTs. Examples are the Xilinx Spartan/Virtex, Altera Cyclone II/APEX II, QuickLogic Eclipse/PolarPro, and Lattice Semiconductor ECP. Many of these FPGAs put two or more 4-input LUTs into a block in various topologies. Some FPGAs also provide multiplexers in addition to look-up tables.

Logic Blocks Based on Multiplexers and Gates

Some FPGAs use multiplexers as the basic building block. As you know, any combinational function can be implemented using multiplexers alone. In the most naïve method, a 4-to-1 multiplexer can generate any 2-input function. If inverted inputs can be provided, a 4-to-1 multiplexer can generate any 3-input function. Examples of multiplexer-based basic blocks are given in Figure 3-33. Logic blocks similar to these were used in early Microsemi FPGAs such as the ACT I and ACT II.

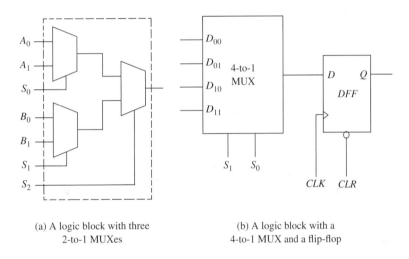

FIGURE 3-33: Multiplexer-Based Logic Blocks in FPGAs

(a) A logic block with three 2-to-1 MUXes

(b) A logic block with a 4-to-1 MUX and a flip-flop

Let us assume that we want to implement the function $F_1 = A'B'C + A'BC' + AB$ using an FPGA with programmable logic blocks consisting of 4-to-1 multiplexers. Two of the 3-input variables can be connected to the multiplexer select lines. Then, we have to provide appropriate signals to the multiplexer data input lines in order to realize the function. To derive these inputs, we will first construct a truth table of the function as shown below:

A	B	C	F	MUX input in terms of $\{0,1,C,C'\}$
0	0	0	0	} C
0	0	1	1	
0	1	0	1	} C'
0	1	1	0	
1	0	0	0	} 0
1	0	1	0	
1	1	0	1	} 1
1	1	1	1	

Let us assume that A and B are connected to the select inputs of the multiplexer. Next, we will derive values of inputs to provide to the multiplexer input lines in terms of the third variable in the function. The third variable is C, and by providing one of the four values $\{C, C', 0, 1\}$, any 3-variable function can be expressed. Considering the first two rows of the truth table, it can be seen that $F = C$ when $AB = 00$. Similarly, considering the third and fourth rows of the truth table, $F = C'$ when $AB = 01$. When $AB = 10$, $F = 0$ irrespective of the value of C. Similarly, when $AB = 11$, the value of the function equals 1. The last column in the truth table presents the required multiplexer inputs. Hence, one 4-to-1 multiplexer with the connections shown in Figure 3-34 can implement function F_1.

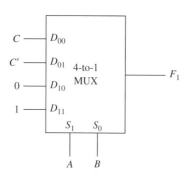

FIGURE 3-34: Multiplexer Implementing Function F_1

In the preceding three sections, we have provided an overview of the general architecture, logic block types and programming technologies that can be used to build FPGAs. The general architecture, programming technology, and logic block types of several example commercial FPGAs are summarized in Table 3-9. LUT-based FPGAs appear to be very common, especially for Xilinx and Altera. Microsemi is the manufacturer of multiplexer based FPGAs. SRAM programming technology, while expensive, also appears to be common.

3.4.4 Programmable Interconnects

A key element of an FPGA is the general-purpose programmable interconnect interspersed between the programmable logic blocks. There are different types of interconnection resources in all commercial FPGAs. Every vendor has its own specific names for the different types of interconnects in their FPGA.

Interconnects in Symmetric Array FPGAs

In this section, we discuss some of the basic elements used for interconnection in symmetric array FPGAs.

General-Purpose Interconnect: Many FPGAs use switch matrices that provide interconnections between routing wires connected to the switch matrix. Figure 3-35(a) illustrates interconnecting logic blocks in an FPGA using switch matrices. Many FPGAs use this type of interconnect. A typical switch matrix is illustrated in Figure 3-35(b), where there is a switch at each intersection (i.e., wherever

TABLE 3-9: Architecture, Technology, and Logic Block Types of Commercial FPGAs

Company	Device Names	General Architecture	Logic Block Type	Programming Technology
Microsemi	IGLOO	Sea of Tiles	LUT	Flash
	ProASIC/ ProASIC3/ ProASIC^{plus}	Sea of Tiles	Multiplexers & Basic Gates	Flash, SRAM
	SX/SXA/eX/MX	Sea of Modules	Multiplexers & Basic Gates	Antifuse
	Axcelerator	Sea of Modules	Multiplexers & Basic Gates	SRAM
	Fusion	Sea of Tiles	Multiplexers & Basic Gates	Flash, SRAM
Xilinx	Kintex	Symmetrical Array	LUT	SRAM
	Virtex	Symmetrical Array	LUT	SRAM
	Spartan	Symmetrical Array	LUT	SRAM
Atmel	AT40KAL	Cell-Based	Multiplexers & Basic Gates	SRAM
QuickLogic	Eclipse II	Flexible Clock	LUT	SRAM
	PolarPro	Cell-Based	LUT	SRAM
Altera	Cyclone II	Two-Dimensional Row and Column Based	LUT	SRAM
	Stratix II	Two-Dimensional Row and Column Based	LUT	SRAM
	APEX II	Row and Column, But Hierarchical Interconnect	LUT	SRAM

the lines cross). A switch matrix that supports every possible connection from every wire to every other wire is very expensive. The connectivity is often limited to some subset of a full crossbar connection; moreover, not all connections might be possible simultaneously. In the switch matrix illustrated in Figure 3-35(b), each wire from a side of the switch can be routed to other wires using some combination of the switches. In order to support this type of connection, each cross point in the switch matrix must support six possible interconnections, as shown in Figure 3-35(c).

Depending on the programming technology, SRAM cells, flash memory cells, or antifuse connections control the configuration of the switches. The switch matrices interspersed between the logic blocks in an FPGA allow general-purpose interconnectivity between arbitrary points in the chip. However, the switch matrices are expensive in area and time (delay). If a signal passes through several of these switch matrices, it could contribute to a significant signal delay. Moreover, the delays are variable and unpredictable depending on the number of the switch matrices involved in each signal. In contrast, the interconnection resources in a CPLD are more restricted. However, interconnections in CPLDs result in smaller and more predictable delays.

FIGURE 3-35: Routing Matrix for General-Purpose Interconnection in an FPGA. Based on Xilinx.

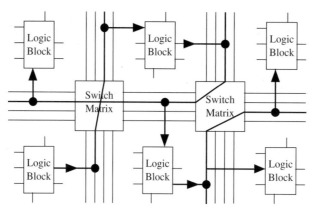

(a) Logic blocks interconnected with switch matrices

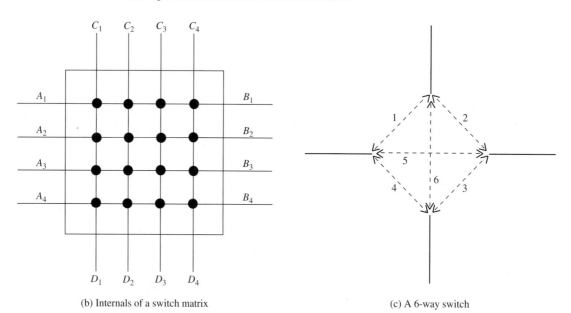

(b) Internals of a switch matrix

(c) A 6-way switch

Direct Interconnects: Many FPGAs provide special connections between adjacent logic blocks. These interconnects are fast because they do not go through the routing matrix. Many FPGAs provide direct interconnections to the four nearest neighbors: top, bottom, left, and right. Figure 3-36 illustrates examples of direct connections. In some cases, there are special interconnections to 8 neighboring blocks, including the diagonally located logic blocks (Figure 3-36(b)). The direct interconnections do not go through the switch matrix but are implemented with dedicated switches resulting in smaller delays. These types of direct interconnects are used in some Xilinx FPGAs.

Global lines: For purposes such as high fan-out and low-skew clock distribution, most FPGAs provide routing lines that span the entire width/height of device. A limited number (two or four) of such global lines are provided by many FPGAs in

FIGURE 3-36: Direct Interconnects between Neighboring CLBs. Based on Xilinx.

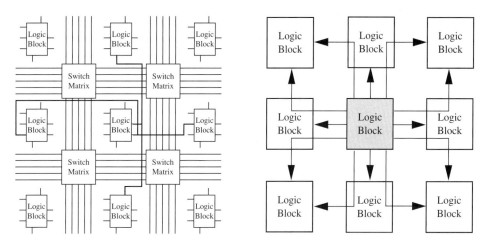

(a) Direct interconnects to four near neighbors

(b) Special connections to 8 neighbors

the horizontal and vertical directions. Figure 3-37 illustrates **horizontal long lines** (global lines) in an example FPGA. The logic blocks often have tristate buffers to connect to the global lines.

FIGURE 3-37: Global Lines. Based on Xilinx.

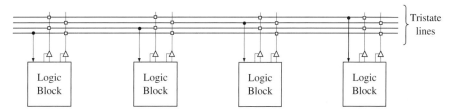

Clock Skew

There are several million gates in modern FPGA chips. When a clock is distributed to various parts of such a large chip, the delays in the wire carrying the clock can result in the clock edge arriving at different times at different parts. This difference in the actual edge of the clock as it arrives at different flip-flops or other devices is called clock skew. Clock skew is a problem in large systems including modern microprocessors. Carefully planned clock distribution circuits are implemented in most systems in order to minimize the effect of clock skew. Modern FPGAs provide specialized clock distribution circuitry in order to create a clock of sufficient strength and low skew.

Interconnects in Row-Based FPGAs

Many of the interconnect resources mentioned previously are very characteristic of symmetric array devices with a two-dimensional array of logic blocks (e.g., Xilinx). In devices that are row based, there are rows of logic blocks and there are channels of switches to enable connections between the logic blocks. Several switches are used to route a signal from a logic block in one row to another logic block elsewhere in the chip. There are arrays of switches in the routing channel between the rows of logic. The routing resources in these FPGAs are very similar to routing in traditional gate arrays.

The interconnects in row-based channeled architecture can be classified into two categories—non-segmented routing and segmented routing. In order to understand different types of channel routing, consider the connections *x*, *y*, and *z* in Figure 3-38(a). Figure 3-38(b) indicates what is called as a **non-segmented channel routing** architecture. There are three horizontal rows or tracks in this figure. There are several vertical wires and switches at the crosspoints. The switches technically can use any programming technology (SRAM, EPROM, or antifuse), although FPGAs that use this type of routing are typically antifuse FPGAs. Desired connectivity is obtained by programming the appropriate switches. Connectivity between the points marked "x" is obtained by the two switches at row 1, columns 1 and 4. Typically this is called net "x". Net 'x' simply means a wire that is named "x". The connectivity for net 'y' is obtained by programming the switches at row 2, columns 2 and 7. It may be noticed that row 1 cannot be used for any other connections other than net "x". Similarly row 2 is exclusively used for net "y". Thus, a problem with this type

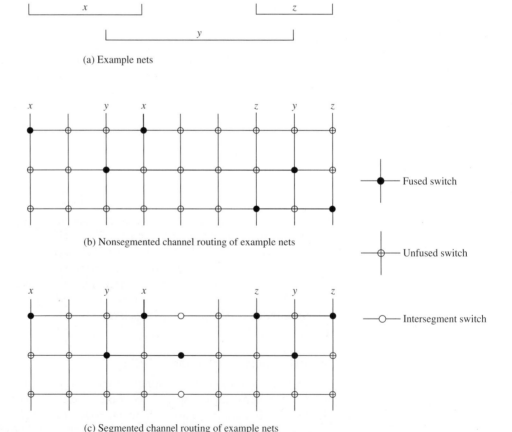

FIGURE 3-38: Typical Routing Resources in a Row-Based FPGA

of interconnect resource is that a full-length track (i.e., an entire row) is used even for a short net. The area overhead of this type of routing is very high for this reason.

In order to reduce the area overhead associated with using full-length tracks for each net, we can use **segmented tracks**, as shown in Figure 3-38(c). Instead of being full length, a track is divided into segments. If a track in row 1 is segmented into two segments, one could use the same track for one more net. For example, nets "x" and "z" can both be routed on row 1 in Figure 3-38(c). That is the principle of segmented track routing. More nets can be routed using the same number of tracks; however, when long nets are desired, intersegment switches must be used to join the segments. These switches introduce more resistance and capacitance into the net. However, the overall routing resource area will reduce with segmented routing.

3.4.5 Programmable I/O Blocks in FPGAs

The I/O pads on an FPGA are connected to programmable input/output blocks, which facilitate connecting the signals from FPGA logic blocks to the external world in desired forms and formats. I/O blocks on modern FPGAs allow use of the pin as input and/or output, in direct (combinational) or latched forms, in tristate true or inverted forms, and with a variety of I/O standards.

Figure 3-39 shows an example configurable input/output block (I/OB). Each I/OB has a number of I/O options, which can be selected by configuration memory cells, indicated by boxes with an "M." The I/O pad can be programmed to be an output or an input. To use the cell as an output, the tristate buffer must be enabled. To use the cell as an input, the tristate control must be set to place the tristate buffer, which drives the output pin, in the high-impedance state.

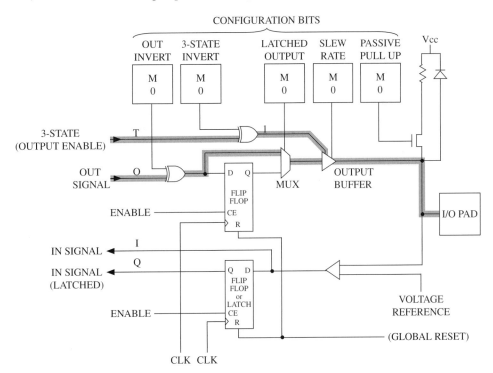

FIGURE 3-39: Programmable I/O Block for an FPGA. Based on Xilinx.

Flip-flops are provided so that input and output values can be stored within the I/O block. The flip-flops are bypassed when direct input or output is desired. The input flip-flop on many FPGAs can be programmed to act as an edge-triggered D flip-flop or as a transparent latch. Even if the I/O pin is not used, the I/O flip-flops can still be used to store data.

The configuration memory cells (marked "M") allow control of various aspects associated with the I/O block. An output signal can be inverted by the I/O block if desired. The inversion is done using an XOR gate. The output signal goes through an exclusive-OR gate, where it is either complemented or not, depending on the contents of the configuration bit in the *OUT-INVERT* cell. The 3-STATE INVERT configuration bit allows one to create an active high or active low tristate control signal. If the *3-STATE* signal is 1 and the *3-STATE INVERT* bit is 0 (or if the *3-STATE* signal is 0 and the *3-STATE INVERT* bit is 1), the buffer drives the output signal to the I/O pad. When the I/O pad is used as an input, the output buffer must be in the high-impedance state. An external signal coming into the I/O pad goes through a buffer and then to the input of a D flip-flop. The buffer output provides a *DIRECT IN* signal to the logic array. Alternatively, the input signal can be stored in the D flip-flop, which provides the *LATCHED IN* signal to the logic array.

The *LATCHED OUTPUT* configuration bit allows one to provide the output in latched or combinational form. Depending on how the *LATCHED OUTPUT* bit is programmed, either the *OUT* signal or the flip-flop output goes to the output buffer. The *SLEW RATE* bit controls the rate at which the output signal can change. When the output drives an external device, reduction of the slew rate is desirable to reduce the induced noise that can occur when the output changes rapidly. When the *PASSIVE PULL-UP* bit is set, a pull-up resistor is connected to the I/O pad. This internal pull-up resistor can be used to avoid floating inputs. The highlighted path indicates the I/O block in an output configuration, with tristate enabled.

I/O Standards

Early FPGAs provided TTL and CMOS signal compatibility, but currently there are many more standards for input/output signals. I/O blocks on modern FPGAs allow transforming signals to a variety of I/O signal standards, some of which appear in the following list:

- LVTTL: low-voltage transistor-transistor logic
- PCI: peripheral component interconnect
- LVCMOS: low-voltage complementary metal-oxide semiconductor
- LVPECL: low-voltage positive emitter-coupled logic
- SSTL: stub-series terminated logic
- AGP: advanced graphics port
- CTT: center tap terminated
- GTL: gunning transceiver logic
- HSTL: high-speed transceiver logic;

Some of these standards use 5 volts whereas some use 3.3 volts or even 1.5 volts. The LVTTL is an example of a 3.3V standard that can tolerate 5V signals. The LVCMOS2 is a 2.5V signal standard that can tolerate 5V signals. The PCI standard has 5V and 3.3 V versions. Some standards need an input voltage reference.

3.4.6 Dedicated Specialized Components in FPGAs

In the early days, FPGAs were simply logic blocks of medium or low complexity, integrated with programmable I/O and interconnect. More recently, FPGA vendors have incorporated embedded processors, DSP processors, dedicated multipliers, dedicated memory, analog-to-digital (A/D) converters, and the like into FPGAs. These specialized components help to efficiently achieve the provided special-purpose functionality. For instance, if dedicated multipliers are not provided, one will have to implement them using general-purpose logic blocks, albeit in an inefficient manner.

Dedicated Memory

A key feature of modern FPGAs is the embedding of dedicated memory blocks (RAM) onto the chip. The embedded RAM can be used to implement the memory needs of the circuit being designed. It could be a table storing constants/coefficients during processing, or it could be implementing memory for an embedded processor that you are designing using the FPGA. Modern FPGAs include 16K to 10M bits of memory. The width of the embedded RAM often can be adjusted. Let us assume that there are 32K of SRAM bits provided as blocks of RAM. This RAM can be used as 32K × 1, 16K × 2, 8K × 4, or 4K × 8. Essentially there are several tiles or blocks of memory. They can be placed in different ways to achieve different aspect ratios. The number of address lines and data lines get adjusted according to the aspect ratio, as illustrated in Table 3-10.

TABLE 3-10: Variable-Width RAM Aspect Ratios

Width	Depth	Addr Bus	Data Bus
1	32 K	15 bits	1 bit
2	16 K	14 bits	2 bits
4	8 K	13 bits	4 bits
8	4 K	12 bits	8 bits
16	2 K	11 bits	16 bits

Dedicated Arithmetic Units

Many users of FPGAs use them to implement arithmetic logic. When logic is implemented in FPGA logic blocks, the implementation generally takes more area and power and is slower than custom implementations. Hence, if most of the target users use arithmetic units such as adders and multipliers, it is beneficial to provide support for such dedicated operations inside the chip. Most FPGAs provide dedicated fast-carry logic to create fast adders. Nowadays, many FPGAs also contain

TABLE 3-11: Examples of FPGAs with Dedicated Multipliers

FPGA	Dedicated Multipliers
Xilinx Virtex-4, Virtex-II Pro/X, Spartan-3E, Spartan 3/3L	18 × 18 multipliers
Xilinx Virtex-5 and Virtex-6	25 × 18 multipliers
Altera Stratix II, III, IV Cyclone II, III, IV	18 × 18 multipliers

dedicated multipliers (See Table 3-11). Thus, instead of mapping a multiplier into several logic blocks, dedicated multipliers provided on the FPGA fabric can be used. These dedicated multipliers are more efficient than a multiplier one could implement using the programmable logic in the FPGA. As indicated in Table 3-11, many Xilinx and Altera FPGAs provide 18-bit × 18-bit multipliers. Xilinx Virtex-5 and Virtex-6 series contain 25 × 18 multipliers, which are touted to be highly useful for digital signal processing (DSP) applications.

Digital Signal Processing (DSP) Blocks

Multiplication is a common operation in DSP. Hence the dedicated multipliers help DSP applications. Xilinx Virtex 5 FPGAs provide DSP slices which contain 25 × 18 multipliers, 48-bit addition, and so forth. Most FPGAs contain carry chains to facilitate addition. The Altera Stratix IV FPGAs contain built-in adders in each logic module. As with adders and multipliers, an FPGA vendor can provide DSP building blocks such as hardware for fast Fourier transforms (FFTs), finite impulse response (FIR) filters, infinite impulse response (IIR) filters, and so forth. Encryption/decryption, compression/decompression, and security functions can also be provided. Once a large amount of specialized components have been provided, a large part of an FPGA may be unused in applications that do not warrant such specialized components. In some FPGAs, DSP support is limited to the dedicated multipliers.

TABLE 3-12: Examples of FPGAs with Embedded Microprocessors

FPGA	Embedded Processor
Xilinx Virtex-4, Virtex-II Pro/X	IBM PowerPC
Xilinx Spartan-3E, Spartan 3/3l	MicroBlaze PicoBlaze
Xilinx Kintex	MicroBlaze
Altera Stratix II, Cyclone II	Nios II
Altera APEX, APEX II	ARM, MIPS, Nios
Altera Excalibur	ARM 9
Altera Arria	ARM 9
Microsemi Fusion	ARM7

Embedded Processors

Many modern FPGAs contain an entire processor core. This is extremely useful when designers use hybrid solutions, where part of a system is in a programmable processor but part of the system is implemented in hardware. Circuitry that needs a large amount of flexibility can be implemented in the microprocessor, but circuit parts that need better performance than that of a programmable processor can be implemented in the FPGA logic blocks. Some FPGAs include the core of a small MIPS processor such as the MIPS R 4000, and some include an embedded version of the IBM PowerPC processor. Some FPGAs include custom processors designed by the FPGA vendors such as the MicroBlaze from Xilinx and the Nios processor from Altera.

Content Addressable Memories (CAM)

In some FPGAs, the memory blocks can be used as **content addressable memories** (CAMs). The general concept of a memory is that the user provides a memory address and the memory unit responds with the content. A CAM is a special kind of memory in which the content, not the address, is used to search the memory. A user provides a data element and the CAM responds with addresses where that data was found. CAMs contain more logic than RAMs because all locations of the memory have to be searched simultaneously to see whether the particular content is in any of the locations. Some FPGAs allow embedded CAM (e.g., Altera APEX II).

> The **Microsemi Fusion** architecture shown in Figure 3-40 provides several specialized components, including embedded RAM, decryption, and A/D converters. At the core of the chip are tiles of logic blocks (VersaTiles in Microsemi terminology). The embedded RAM is in the form of rows of SRAM blocks above and below the tiles of logic blocks. Several specialized components appear below the SRAM blocks in the bottom. There is a dedicated decryption unit that implements the AES decryption algorithm. (AES stands for advanced encryption standard and has been the cryptographic standard for the U.S. government since 2001.) There is an analog-to-digital converter (ADC) that accepts inputs from several analog quads, which are circuitry to condition analog signals received by the FPGA. The analog quads contain circuitry to monitor and condition signals according to voltage, current, and temperature.

3.4.7 Applications of FPGAs

FPGAs have become a popular mode of circuit implementation for various applications.

Rapid Prototyping

FPGAs are very useful for building rapid prototypes of large systems. A designer can build proof-of-concept systems very quickly using field-programmable gate arrays. Since FPGAs are large enough to contain 5 million or more gates, many large real-world systems can be prototyped using a single FPGA. If a single FPGA will not suffice, multiple FPGAs can be interconnected to realize large systems.

FIGURE 3-40: Overview of the Microsemi Fusion Chip. Based on Xilinx.

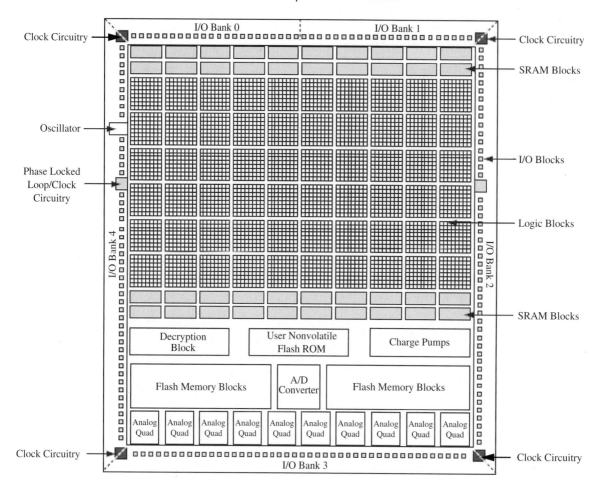

Rapid prototyping of large systems is done by using boards with multiple FPGAs and plugging multiple boards into a backplane (motherboard).

FPGAs As Final Product in Medium-Speed Systems

Circuits realized using FPGAs typically operate in the 150–200 MHz clock rate. For applications where this speed is sufficient, FPGAs can be used for the final product itself as opposed to the prototype. When an FPGA is used as the final product, enhancements to the system can be done as software updates rather than as hardware changes. Modern FPGA speeds are adequate for many applications.

Reconfigurable Circuits and Systems

The reprogrammability of FPGAs lends itself to building dynamically reconfigurable circuits and systems. SRAM-based FPGAs make it possible to implement "soft" hardware. FPGAs have been used to design circuits and systems that need multiple functionalities at various times.

As an example, consider a reprogrammable Tomahawk missile that the U.S. Navy designed using FPGAs. [TomaHawk reference]. The conventional Tomahawk is a long-range Navy cruise missile designed to perform a variety of missions. The Navy designed a reconfigurable Tomahawk, which can operate in one of two modes, depending on the mission at hand. Rather than designing separate logic for each mode, the missile designers used FPGAs so that the configuration for each mode can be kept on-board in ROM. Depending on the mode of operation, the FPGA could be configured in mid-flight.

Glue Logic

FPGAs have become the medium of choice for implementing interface or glue logic between modules and components. Small changes in interface protocols or formats would conventionally necessitate building new interface logic. With SRAM FPGAs, the new interface logic can be implemented on the same FPGA as in a software update.

Hardware Accelerators/Coprocessors

A software application running on a conventional system can be accelerated if a coprocessor/accelerator can implement some key routines/kernels from the application in hardware. An FPGA can be used to implement the key kernel. An SRAM-based, reconfigurable FPGA is well suited for this use, because depending on the application running, different kernels can dynamically be programmed into the FPGA. This approach has been demonstrated for applications such as pattern matching. FPGA-based hardware is used for several applications, including computer architecture simulator acceleration, emulation boards, hardware test/verification, among others.

3.4.8 Design Flow for FPGAs

Sophisticated CAD tools are available to assist with the design of systems using programmable gate arrays. Designs can be entered in many ways.

In the early days of FPGAs, designs were entered using schematic entry or even lower levels of design entry tools. Low-level design entry means less abstraction, whereas high-level means entering designs at a higher level of abstraction (e.g., behavioral Verilog/VHDL description). Early FPGA tools allowed low-level utilities to enter logic equations, Karnaugh maps, and other objects into specific logic blocks in the FPGA. Schematic capture technique means that the designer develops a schematic of the design. Schematic diagrams utilizing standard hardware components are created and entered into the CAD software.

Currently, automatic synthesis tools are available that will take a Verilog description of the system as an input and generate an interconnection of gates and flip-flops to realize the system. Behavioral models can be translated into design implementations reasonably efficiently. Synthesis tools have advanced significantly in the last decade. The most common method of designing a digital system with an FPGA uses the following steps:

1. Create a behavioral, RTL, or structural model of the design in a hardware description language such as Verilog or VHDL.
2. Simulate and debug the design.
3. Synthesize the design targeting the desired device.
4. Run a mapping/partitioning program. This program will break the logic diagram into pieces that will fit into the configurable logic blocks.
5. Run an automatic place-and-route program. This will place the logic blocks in appropriate places in the FPGA and will then route the interconnections between the logic blocks.
6. Run a program that will generate the bit pattern necessary to program the FPGA.
7. Download the bit pattern into the internal configuration cells in the FPGA and test the operation of the FPGA.

Steps 3, 4, and 5 are often integrated in modern CAD tools. However, the processes mentioned in the steps are happening whether presented as one step or several steps. This is analogous to how general-purpose compilers have integrated compiling and assembling steps. In the early days of high-level language compilers, the term compiling meant only translation into an assembly language format. Converting from assembly language to machine language code was considered the assembler's job. Currently, the steps are integrated in most high-level language compilation environments.

In SRAM-based FPGAs, when the final system is built, the bit pattern for programming the FPGA is normally stored in an EPROM and automatically loaded into the FPGA when the power is turned on. The EPROM is connected to the FPGA, as shown in Figure 3-41. The FPGA resets itself after the power has been applied. Then it reads the configuration data from the EPROM by supplying a sequence of addresses to the EPROM inputs and storing the EPROM output data in the FPGA internal configuration memory cells. This is not required in flash memory based FPGAs because the flash technology is non-volatile. In antifuse FPGAs, the configuration bits permanently alter the switches.

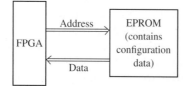

FIGURE 3-41: EPROM Connections for SRAM FPGA Initialization

In this chapter we introduced several different types of programmable logic devices and used them for designing circuits. The technology underlying early programmable logic devices, such as ROMs, PALs and PLAs, was presented first. Simple PLDs and GALs were presented next. Examples were presented to illustrate implementations of simple logic functions in these devices. CPLDs and FPGAs were presented next. The discussion of FPGAs was limited to an overview of the general technology underlying this class of devices. General organization of FPGAs, general structure of logic blocks, typical programming techniques, and so forth were discussed. More details on FPGAs are presented in Chapter 6.

Problems

3.1 Write VHDL code that describes the output macrocell of a 22V10 (the part enclosed by dashed lines on Figure 3-20). The entity should include S_1 and S_0. Note that the flip-flop has an asynchronous reset (AR) and a synchronous preset (SP).

3.2 An N-bit bidirectional shift register has N parallel data inputs, N outputs, a left serial input (LSI), a right serial input (RSI), a clock input, and the following control signals:

> *Load*: Load the parallel data into the register (load overrides shift).
> *Rsh*: Shift the register right (LSI goes into the left end).
> *Lsh*: Shift the register left (RSI goes into the right end).

(a) If the register is implemented using a 22V10, what is the maximum value of N?
(b) Give equations for the rightmost two cells.

3.3 **(a)** In which applications should a designer use a CPLD rather than an FPGA?
(b) In which applications should a designer use an MPGA rather than an FPGA?
(c) In which applications should a designer use an FPGA rather than an MPGA?
(d) A company is designing an experimental product that is in version 1 now. It is expected that the product will undergo several revisions. The company's plan is to use an FPGA for the actual design. What type of FPGA (SRAM or antifuse) should be used?
(e) A company is designing a product using an FPGA. The company's plan is to use an FPGA for the actual design. The product has undergone several revisions and is fairly stable. Minimizing area, power, and cost is important for the company. Which type of FPGA (SRAM or antifuse) should be used?
(f) A company is designing a product. It expects to sell 1000 copies of the product. Should the company use an MPGA or an FPGA for this product?
(g) A company is designing a product. It expects to sell 100 million copies of the product. Should the company use an MPGA or an FPGA for this product?

3.4 **(a)** Implement the function $F_1 = A'BC + B'C + ABC$ using an FPGA with programmable logic blocks consisting of 4-to-1 multiplexers. Assume inputs and their complements are available as shown in Figure 3-34.
(b) Implement the function $F_1 = A'B + AB' + AC' + A'C$ using a multiplexer. What is the size of the smallest multiplexer needed, assuming inputs and their complements are available?

3.5 Show how the left shift register of Figure 2-41 could be implemented using a CPLD. Draw a diagram similar to Figure 3-25. Give the equations for the flip-flop D inputs.

3.6 A Mealy sequential circuit with four output variables is realized using a 22V10. What is the maximum number of input variables it can have? What is the maximum number of states? Can any Mealy circuit with these numbers of inputs and outputs be realized with a 22V10? Explain.

3.7 **(a)** What is the difference between a traditional gate array and an FPGA?
(b) What are the different types of FPGAs based on architecture (organization)?

(c) What are the different programming technologies for FPGAs?
(d) What is the main advantage of SRAM FPGAs?
(e) What is the main advantage of antifuse FPGAs?
(f) What are the major programmable elements in an FPGA?
(g) What are the disadvantages of SRAM FPGAs?
(h) What are the disadvantages of antifuse FPGAs?
(i) How many transistors are typically required to make an SRAM cell?
(j) What is an MPGA?
(k) What is difference between a CPLD and an FPGA?
(l) What is an advantage of a CPLD over an FPGA?
(m) What is the advantage of an FPGA over a CPLD?
(n) Name three vendors of CPLDs.
(o) Name three vendors of FPGAs.

3.8 (a) Route the "w", "x", "y", and "z" nets on the non-segmented tracks shown in the diagram that follows. Use the minimum number of tracks possible.
(b) Route the "w", "x", "y", and "z" nets on the segmented tracks shown in the diagram that follows. Use the minimum number of tracks possible.

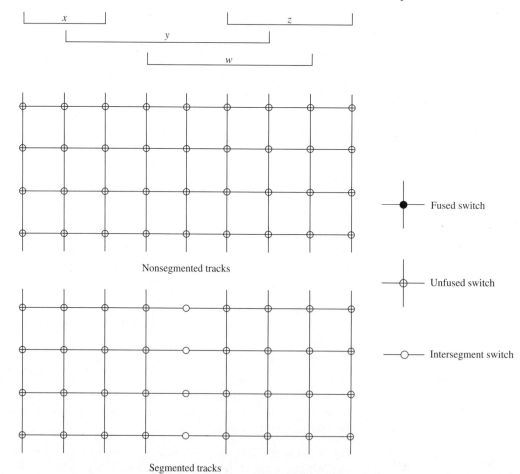

3.9 Consider the following programmable I/O block:

Highlight the connections to configure this I/O block as an INPUT pin. Specify the five configuration bits.

3.10 What is the size of the smallest ROM that is needed to implement the following?
  (a) An 8-bit full adder (assume carry-in and carry-out)
  (b) A BCD-to-binary converter (2 BCD digits)
  (c) A 4-to-1 MUX
  (d) A 32-bit adder (adds two 32-bit numbers to give a 33-bit sum)
  (e) A 3-to-8 decoder
  (f) A 32-bit adder (no carry in or carry out)
  (g) A 16 × 16 bit multiplier
  (h) A 16-bit full adder (with carry-in and carry-out)
  (i) An 8-to-3 priority encoder
  (j) A 10-to-4 priority encoder
  (k) An 8-to-1 multiplexer

3.11 Find a minimum-row PLA to implement the following three functions:

$$f(A, B, C, D) = \Sigma m(3, 6, 7, 11, 15)$$
$$g(A, B, C, D) = \Sigma m(1, 3, 4, 7, 9, 13)$$
$$h(A, B, C, D) = \Sigma m(4, 6, 8, 10, 11, 12, 14, 15)$$

(a) Use Karnaugh maps to find common terms. Give the logic equations with common terms underlined, the PLA table, and also a PLA diagram similar to Figure 3-15.

(b) Use the Espresso multiple output simplification routine that is in *LogicAid*. Compare the *LogicAid* results with part (a). They might not be exactly the same since *LogicAid* Espresso finds only minimum row tables; it does not necessarily minimize the number of variables in each AND term. *Note*: enter the variable names A, B, C, D, F, G, and H in *LogicAid*. Printouts with variable names X_1, X_2, X_3, X_4, and so forth are not acceptable.

3.12 Given $F = A'B' + BC'$ and $G = AC + B'$, write a complete Verilog module that realizes the functions F and G using an 8-word × 2-bit ROM. Include the array type declaration and the constant declaration that defines the contents of the ROM.

3.13 Find a minimum-row PLA table to implement the following sets of functions:
(a) $f_1(A, B, C, D) = \Sigma m(0, 2, 3, 6, 7, 8, 9, 11, 13)$,
$f_2(A, B, C, D) = \Sigma m(3, 7, 8, 9, 13)$,
$f_3(A, B, C, D) = \Sigma m(0, 2, 4, 6, 8, 12, 13)$

(b) $f_1(A, B, C, D) = cd + ad + a'bc'd'$
$f_2(A, B, C, D) = bc'd' + ac' + ad'$

3.14 Implement the following state table using a ROM and two D flip-flops. Use a straight binary state assignment.
(a) Show the block diagram and the ROM truth table. Truth table column headings should be in the order $Q_1 \, Q_0 \, X \, D_1 \, D_0 \, Z$
(b) Write Verilog code for the implementation. Use an array to represent the ROM table, and use two processes.

Present State	Next State		Output (Z)	
	X = 0	X = 1	X = 0	X = 1
S_0	S_0	S_1	0	1
S_1	S_2	S_3	1	0
S_2	S_1	S_3	1	0
S_3	S_3	S_2	0	1

3.15 (a) Find a minimum-row PLA table to implement the following equations:
$$x(A, B, C, D) = \Sigma m(0, 1, 4, 5, 6, 7, 8, 9, 11, 12, 14, 15)$$
$$y(A, B, C, D) = \Sigma m(0, 1, 4, 5, 8, 10, 11, 12, 14, 15)$$
$$z(A, B, C, D) = \Sigma m(0, 1, 3, 4, 5, 7, 9, 11, 15)$$

(b) Indicate the connections that will be made to program a PLA to implement your solution to part (a) on a diagram similar to Figure 3-15.

3.16 The following state table is implemented using a ROM and two D flip-flops (falling edge triggered):

Q_1Q_2	$Q_1^+Q_2^+$		Z	
	$X=0$	$X=1$	$X=0$	$X=1$
00	01	10	0	1
01	10	00	1	1
10	00	01	1	0

(a) Draw the block diagram.
(b) Write Verilog code that describes the system. Assume that the ROM has a delay of 10 ns and each flip-flop has a propagation delay of 15 ns.

Design Examples

CHAPTER 4

In this chapter, we present several Verilog design examples to illustrate the design of small digital systems. We present the concept of dividing a design into a controller and a data path and using the control circuit to control the sequence of operations in a digital system. We use Verilog to describe a digital system at the behavioral level so that we can simulate the system to test the algorithms used. We also show how designs have to be coded structurally if specific hardware structures are to be generated.

In any design, first, one should understand the problem and the design specifications clearly. If the problem has not been stated clearly, try to get the specification of the design clarified. In real-world designs, if another team or a client company is providing your team with the specifications, getting the design specifications clarified properly can save you a lot of grief later. Good design starts with a clear specification document.

Once the problem has been stated clearly, often designers start thinking about the basic blocks necessary to accomplish what is specified. Designers often think of standard building blocks, such as adders, shift-registers, counters, and the like. Traditional design methodology splits a design into a "data path" and a "controller." The term "data path" refers to the hardware that actually performs the data processing. The controller sends control signals or commands to the data path, as shown in Figure 4-1. The controller can obtain feedback in the form of status signals from the data path.

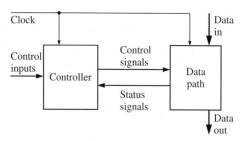

FIGURE 4-1: Separation of a Design into Data Path and Controller

In the context of a microprocessor, the data path is the ALU (Arithmetic and Logic Unit) that performs the core of the processing. The controller is the control logic that sends appropriate control signals to the data path, instructing it to perform addition, multiplication, shifting, or whatever action is called for by the instruction. Many users have a tendency to mistakenly consider the term "data path" to be synonymous with the data bus, but "data path" in traditional design terminology refers to the actual data processing unit.

Maintaining a distinction between data path and controller helps in debugging (i.e., finding errors in the design). It also helps while modifying the design. Many modifications can be accomplished by changing only the control path because the same data path can still support the new requirements. The controller can generate the new sequence of control signals to accomplish the functionality of the modified design. Design often involves refining the data path and controller in iterations.

In this chapter, we will discuss various design examples. Several arithmetic and non-arithmetic examples are presented. Non-arithmetic examples include a 7-segment decoder, a traffic light, a scoreboard, and a keypad scanner. Arithmetic circuits such as adders, multipliers, and dividers are also presented.

4.1 BCD to 7-Segment Display Decoder

Seven segment displays are often used to display digits in digital counters, watches, and clocks. A digital watch displays time by turning on a combination of the segments on a 7-segment display. For this example, the segments are labeled as follows, and the digits have the forms as indicated in Figure 4-2.

FIGURE 4-2: 7-Segment Display

Let us design a BCD to 7-segment display decoder. BCD stands for "binary coded decimal." In this format, each digit of a decimal number is encoded into 4-bit binary representation. This decoder is a purely combinational circuit; hence, no state machine is involved here. A block diagram of the decoder is shown in Figure 4-3. The decoder for one BCD digit is presented.

FIGURE 4-3: Block Diagram of a BCD to 7-Segment Display Decoder

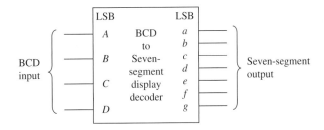

We will create a behavioral Verilog architectural description of this BCD to 7-segment decoder by using a single process with a case statement to model this

combinational circuit, as in Figure 4-4. The sensitivity list of the process consists of the BCD number (4 bits).

FIGURE 4-4: Behavioral Verilog Code for BCD to 7-Segment Decoder

```verilog
module bcd_seven (bcd, seven);
   input [3:0] bcd;
   output[7:1] seven;

   reg    [7:1] seven;

   always @(bcd)
   begin
     case (bcd)
       4'b0000 : seven = 7'b0111111 ;
       4'b0001 : seven = 7'b0000110 ;
       4'b0010 : seven = 7'b1011011 ;
       4'b0011 : seven = 7'b1001111 ;
       4'b0100 : seven = 7'b1100110 ;
       4'b0101 : seven = 7'b1101101 ;
       4'b0110 : seven = 7'b1111101 ;
       4'b0111 : seven = 7'b0000111 ;
       4'b1000 : seven = 7'b1111111 ;
       4'b1001 : seven = 7'b1101111 ;
       default : seven = 7'b0000000 ;
     endcase
   end
endmodule
```

4.2 A BCD Adder

In this example, we design a 2-digit BCD adder, which will add two BCD numbers and produce the sum in BCD format. In BCD representation, each decimal digit is encoded into binary. For instance, decimal number 97 will be represented as 1001 0111 in the BCD format, where the first 4 bits represent digit 9 and the next 4 bits represent digit 7. One may note that the BCD representation is different from the binary representation of 97, which is 110001. It takes 8 bits to represent 97 in BCD, whereas the binary representation of 97 (110001) only requires 6 bits. The 4-bit binary combinations 1010, 1011, 1100, 1101, 1110, and 1111 corresponding to hexadecimal numbers A to F are not used in the BCD representation. Since 6 out of 16 representations possible with 4 binary bits are skipped, a BCD number will take more bits than the corresponding binary representation.

When BCD numbers are added, each sum digit should be adjusted to skip the six unused codes. For instance, if 6 is added with 8, the sum is 14 in decimal form. A binary adder would yield 1110, but the lowest digit of the BCD sum should read 4. In order to obtain the correct BCD digit, 6 should be added to the sum whenever it is greater than 9. Figure 4-5 illustrates the hardware that will be required to

FIGURE 4-5: Addition of Two BCD Numbers

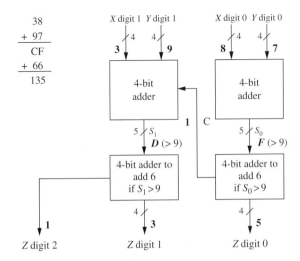

```
   38
 + 97
   CF
 + 66
  135
```

perform the addition of 2 BCD digits. A binary adder adds the least significant digits. If the sum is greater than 9, an adder adds 6 to yield the correct sum digit along with a carry digit to be added with the next digit. The addition of the higher digits is performed in a similar fashion.

The Verilog code for the BCD adder is shown in Figure 4-6. The input BCD numbers are represented by *X* and *Y*. The BCD sum of two 2-digit BCD numbers can exceed two digits and hence three BCD digits are provided for the sum, which is represented by *Z*. The compiler directive 'define' can be used to denote each digit of each BCD number. For example, the upper digit of *X* can be denoted by *Xdig1* by using the Verilog statement:

`` `define      Xdig1      X[7:4] ``

This statement allows us to use the name *Xdig1* whenever we wish to refer to the upper digit of *X*. If BCD numbers 97 and 38 are added, the sum is 135; hence, *Zdig2*

FIGURE 4-6: Verilog Code for BCD Adder

```verilog
`define Xdig1 X[7:4]
`define Xdig0 X[3:0]
`define Ydig1 Y[7:4]
`define Ydig0 Y[3:0]
`define Zdig2 Z[11:8]
`define Zdig1 Z[7:4]
`define Zdig0 Z[3:0]

module BCD_Adder (X, Y, Z);

  input[7:0] X;
  input[7:0] Y;
  output[11:0] Z;
```

```
    wire[4:0] S0;
    wire[4:0] S1;
    wire C;

assign S0 = `Xdig0 + `Ydig0 ;
assign `Zdig0 = (S0 > 9) ? S0[3:0] + 6 : S0[3:0] ;
assign C = (S0 > 9) ? 1'b1 : 1'b0 ;

assign S1 = `Xdig1 + `Ydig1 + C ;
assign `Zdig1 = (S1 > 9) ? S1[3:0] + 6 : S1[3:0] ;
assign `Zdig2 = (S1 > 9) ? 4'b0001 : 4'b0000 ;

endmodule
```

equals 1, *Zdig1* equals 3 and *Zdig0* equals 5. In Verilog code, the defined name should be used with ` (e.g., `Xdig1 or `Zdig2).

During the addition of the second digit, the carry digit from the addition of the *XDig0* and *Ydig0* is also added. Thus, the addition of the second digit is accomplished by the statement:

```
assign S1 = `Xdig1 + `Ydig1 + C ;
```

4.3 32-Bit Adders

Let us assume that we have to design a 32-bit adder. A simple way to construct an adder is to build a **ripple-carry adder**, as shown in Figure 4-7. In this type of adder, 32 copies of a one-bit full adder are connected in succession to create the 32-bit adder. The carry "ripples" from the least significant bit to the most significant bit. If gate delays are t_g, a one-bit adder delay is $2*t_g$ (assuming a sum-of-products expression for sum and carry, and ignoring delay for inverters), and a 32-bit ripple-carry adder will take approximately 64 gate delays. For instance, if gate delays are 1 ns, the maximum frequency at which the 32-bit ripple-carry adder can operate is approximately 16 MHz! This is inadequate for many applications. Hence, designers often resort to faster adders.

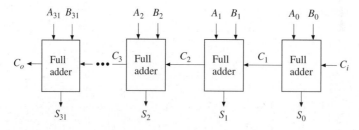

FIGURE 4-7: A 32-Bit Ripple-Carry Adder

Carry Look-Ahead Adders

A popular fast-addition technique is carry look-ahead (CLA) addition. In the carry look-ahead adder, the carry signals are calculated in advance, based on the input signals. For any bit position i, one can see that a carry will be generated if the corresponding input bits (i.e., A_i, B_i) are 1 or if there was a carry-in to that bit and at least one of the input bits are 1. In other words, bit i has carry-out if A_i and B_i are 1 (irrespective of carry-in to bit i); bit i also has a carry-out if $C_i = 1$ and either A_i or B_i is 1. Thus, for any stage i, the carry-out is

$$C_{i+1} = A_i B_i + (A_i \oplus B_i) \cdot C_i \tag{4-1}$$

The "$\oplus$" stands for the exclusive OR operation. Equation 4-1 simply expresses that there is a carry out from a bit position if it **generated** a carry by itself (e.g., $A_i B_i = 1$) or it simply **propagated** the carry from the lower bit forwarded to it (i.e., $(A_i \oplus B_i) \cdot C_i$).

Since $A_i B_i = 1$ indicates that a stage generated a carry, a general **generate (Gi) function** may be written as

$$G_i = A_i B_i \tag{4-2}$$

Similarly, since $(A_i \oplus B_i)$ indicates whether a stage should propagate the carry it receives from the lower stage, a general **propagate (P_i) function** may be written as

$$P_i = A_i \oplus B_i \tag{4-3}$$

Notice that the propagate and generate functions depend only on the input bits and can be realized with one or two gate delays. Since there will be a carry whether one of A_i or B_i is 1 or both are 1, one can also write the propagate expression as

$$P_i = A_i + B_i \tag{4-4}$$

where the OR operation is substituted for the XOR operation. Logically this propagate function also results in the correct carry-out; however, traditionally it has been customary to define the propagate function as the XOR; that is, the bit position simply propagates a carry (without generating a carry by itself). Also, typically, the sum signal is expressed as

$$S_i = A_i \oplus B_i \oplus C_i = P_i \oplus C_i \tag{4-5}$$

The expression $P_i \oplus C_i$ can be used for sum only if P_i is defined as $A_i \oplus B_i$.

The carry-out equation can be rewritten by substituting (4-2) and (4-3) in (4-1) for G_i and P_i as:

$$C_{i+1} = G_i + P_i C_i \tag{4-6}$$

In a 4-bit adder, the C_is can be generated by repeatedly applying Equation 4-6 as shown here

$$C_1 = G_0 + P_0 C_0 \tag{4-7}$$

$$C_2 = G_1 + P_1 C_1 = G_1 + P_1 G_0 + P_1 P_0 C_0 \tag{4-8}$$

$$C_3 = G_2 + P_2 C_2 = G_2 + P_2 G_1 + P_2 P_1 G_0 + P_2 P_1 P_0 C_0 \tag{4-9}$$

$$C_4 = G_3 + P_3 C_3 = G_3 + P_3 G_2 + P_3 P_2 G_1 + P_3 P_2 P_1 G_0 + P_3 P_2 P_1 P_0 C_0 \tag{4-10}$$

These carry bits are the look-ahead carry bits. They are expressed in terms of P_is, G_is, and C_0. Thus, the sum and carry from any stage can be calculated without waiting for the carry to ripple through all the previous stages. Since G_is and P_is can be generated with one or two gate delays, the C_is will be available in three or four gate delays. The advantage is that these delays will be the same, independent of the number of bits one needs to add, in contrast to the ripple counter. Of course this is achieved with the extra gates to generate the look-ahead carry bits. A 4-bit carry look-ahead adder can now be built, as illustrated in Figure 4-8. The output of partial full adder will be **propagate** (P) and **generate** (G) in addition to **sum** (S), and **carry** (C) will be generated by carry look-ahead logic rather than the partial full adder. C_4 can also be generated with the following equation:

$$C_4 = G_G + P_G C_0 \tag{4-11}$$

This is accomplished by computing a **group propagate** (P_G) and **group generate** (G_G) signal, which is produced by **carry look-ahead logic**:

$$P_G = P_3 P_2 P_1 P_0 \tag{4-12}$$

$$G_G = G_3 + P_3 G_2 + P_3 P_2 G_1 + P_3 P_2 P_1 G_0 \tag{4-13}$$

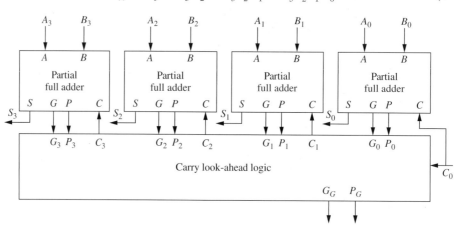

FIGURE 4-8: Block Diagram of a 4-Bit CLA

The disadvantage of the carry look-ahead adder is that the look-ahead carry logic as shown in Equations 4-7 through 4-13, is not simple. It gets quite complicated for more than 4 bits. For that reason, carry look-ahead adders are usually implemented as 4-bit modules and are used in a hierarchical structure to realize adders that have multiples of 4 bits. Figure 4-9 shows the block diagram for a 16-bit carry look-ahead adder. Four carry look-ahead adders, similar to the ones introduced previously, are used. Instead of relying on each 4-bit adder to send its carry-out to the next 4-bit adder, the **block carry look-ahead logic** generates input carry bits to be fed to each 4-bit adder using a group propagate (P_G) and group generate (G_G) signal, which is produced by each 4-bit adder. The next level of carry look-ahead logic uses these group propagates/generates and generates the required carry bits in parallel. The propagate for a group is true if all the propagates in that group are true. The generate for a group is true if the MSB generated a carry or if a lower bit generated a carry and every higher bit in the group propagated it.

The group propagate P_G and generate G_G will be available after three and four gate delays, respectively (one or two additional delays than the P_i and G_i signals,

FIGURE 4-9: Block Diagram of a 16-Bit CLA

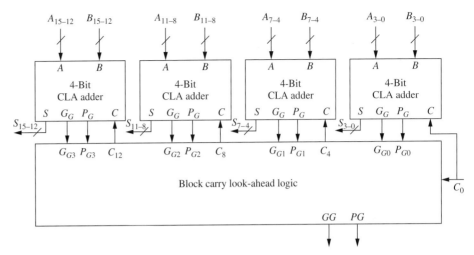

respectively). The carry equations for the block carry look-ahead logic are as follows:

$$C_4 = G_{G0} + P_{G0}C_0 \qquad (4\text{-}14)$$

$$C_8 = G_{G1} + P_{G1}G_{G0} + P_{G1}P_{G0}C_0 \qquad (4\text{-}15)$$

$$C_{12} = G_{G2} + P_{G2}G_{G1} + P_{G2}P_{G1}G_{G0} + P_{G2}P_{G1}P_{G0}C_0 \qquad (4\text{-}16)$$

C_{16}, which is a final carry of 16-bit CLA, will be

$$C_{16} = GG + PG\,C_0 \qquad (4\text{-}17)$$

One can derive the propagate (PG) and generate (GG) equation for block carry look-ahead logic in a manner similar to equation 4-12 and 4-13.

Figure 4-10 illustrates the Verilog description of a 4-bit carry look-ahead adder.

FIGURE 4-10: Verilog Description of a 4-Bit Carry Look-Ahead Adder

```
module CLA4 (A, B, Ci, S, Co, PG, GG);
   input[3:0] A;
   input[3:0] B;
   input Ci;
   output[3:0] S;
   output Co;
   output PG;
   output GG;

   wire[3:0] G;
   wire[3:0] P;
   wire[3:1] C;
   CLALogic CarryLogic (G, P, Ci, C, Co, PG, GG);
   GPFullAdder FA0 (A[0], B[0], Ci, G[0], P[0], S[0]);
   GPFullAdder FA1 (A[1], B[1], C[1], G[1], P[1], S[1]);
   GPFullAdder FA2 (A[2], B[2], C[2], G[2], P[2], S[2]);
   GPFullAdder FA3 (A[3], B[3], C[3], G[3], P[3], S[3]);
endmodule
```

```verilog
module CLALogic (G, P, Ci, C, Co, PG, GG);
    input[3:0] G;
    input[3:0] P;
    input Ci;
    output[3:1] C;
    output Co;
    output PG;
    output GG;

    wire GG_int;
    wire PG_int;

    assign C[1] = G[0] | (P[0] & Ci) ;
    assign C[2] = G[1] | (P[1] & G[0]) | (P[1] & P[0] & Ci) ;
    assign C[3] = G[2] | (P[2] & G[1]) | (P[2] & P[1] & G[0]) | (P[2] & P[1] &
                  P[0] & Ci) ;
    assign PG_int = P[3] & P[2] & P[1] & P[0] ;
    assign GG_int = G[3] | (P[3] & G[2]) | (P[3] & P[2] & G[1]) | (P[3] & P[2] &
                    P[1] & G[0]) ;
    assign Co = GG_int | (PG_int & Ci) ;
    assign PG = PG_int ;
    assign GG = GG_int ;
endmodule

module GPFullAdder (X, Y, Cin, G, P, Sum);
    input X;
    input Y;
    input Cin;
    output G;
    output P;
    output Sum;

    wire P_int;

    assign G = X & Y ;
    assign P = P_int ;
    assign P_int = X ^ Y ;
    assign Sum = P_int ^ Cin ;
endmodule
```

Verilog code for a 16-bit carry look-ahead adder can be developed by instantiating four copies of the 4-bit carry look-ahead adder and one additional copy of the carry look-ahead logic. A 64-bit adder can be built by one more level of block carry look-ahead logic. The delay increases by only two gate delays when the adder size increases from 16 bits to 64 bits. Developing Verilog code for 16- and 64-bit carry look-ahead logic is left as subjects of exercise problems.

Figure 4-11 illustrates behavioral Verilog code for a 32-bit adder using the "+" operator. If this code is synthesized, depending on the tools used and the target technology, an adder with characteristics in between a ripple-carry adder and a fast 2-level adder will be obtained. The various topologies result in different area, power, and delay characteristics.

FIGURE 4-11: Behavioral Model for a 32-Bit Adder

```
module Adder32 (A, B, Ci, S, Co);

   input[31:0] A;
   input[31:0] B;
   input Ci;
   output[31:0] S;
   output Co;

   wire[32:0] Sum33;

   assign Sum33 = A + B + Ci ;
   assign S = Sum33[31:0] ;
   assign Co = Sum33[32] ;
endmodule
```

Example

If gate delays are t_g, what is the delay of the fastest 32-bit adder? Assume that the amount of hardware consumed is not a constraint. Only speed is important.

Answer: One can express each sum bit of a 32-bit adder as a sum-of-products expression of the input bits. There will be 33 such equations, including one for the carry out bit. These equations will be very long, and some of them could include 60+ variables in the product term. Nevertheless, if gates with any number of inputs are available, theoretically a 2-level adder can be made. Although it is not very practical, theoretically, the delay of the fastest adder will be $2t_g$ if gate delays are t_g.

Example

Is ripple-carry adder the smallest 32-bit adder?

Answer: A 32-bit ripple-carry adder uses 32 1-bit adders. One could design a 32-bit serial adder using a single 1-bit full adder. The input numbers are shifted into the adder, one bit at a time, and carry output from addition of each pair of bits is saved in a flip-flop and fed back to the next addition. The hardware illustrated in Figure 4-12 accomplishes this. The delay

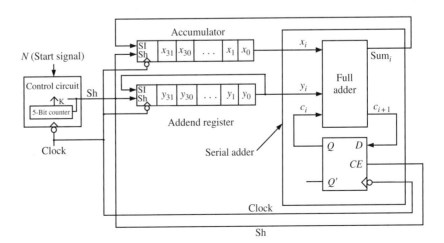

FIGURE 4-12: A 32-Bit Serial Adder Built from a Single 1-Bit Adder

of adder will be 32 * ($2t_g + t_{ff}$), where $2t_g$ is the delay of the one-bit full adder and t_{ff} is the delay of the flip-flop (including set up time). If a flip-flop delay is at least two gate delays, the delay of the 32-bit serial adder will be at least $128t_g$. The adder hardware is simple; however, there is also the control circuitry needed to generate 32 shift signals. The registers storing the operands must have shift capability as well.

Even if you write Verilog code based on data flow equations, as shown in Figure 4-10, that does not guarantee that the synthesizer will produce a carry look-ahead adder with the delay characteristics we have discussed. The software might optimize the synthesis output depending on the specific hardware components available in the target technology. For instance, if you are using an FPGA with fast adder support, the software may map some of the functions into the fast adder circuitry. Depending on the number of FPGA logic blocks and interconnects used, the delays will be different from the manual calculations. The delays of a ripple-carry, carry look-ahead, and serial adder for a gate-based implementation are presented in Table 4-1 for various adder sizes. One can see that the carry look-ahead adder is very attractive for large adders.

TABLE 4-1: Comparison of Ripple-Carry and Carry Look-Ahead Adders

Adder size	Ripple-carry adder delay	CLA delay	Serial adder delay
4 bit	$8\ t_g$	$5\text{-}6\ t_g$	$16\ t_g$
16 bit	$32\ t_g$	$7\text{-}8\ t_g$	$64\ t_g$
32 bit	$64\ t_g$	$9\text{-}10\ t_g$	$128\ t_g$
64 bit	$128\ t_g$	$9\text{-}10\ t_g$	$256\ t_g$

4.4 Traffic Light Controller

Let us design a sequential traffic light controller for the intersection of street "A" and street "B." Each street has traffic sensors, which detect the presence of vehicles approaching or stopped at the intersection. $Sa = 1$ means a vehicle is approaching on street "A," and $Sb = 1$ means a vehicle is approaching on street "B." Street "A" is a main street and has a green light until a car approaches on "B." Then the lights change, and "B" has a green light. At the end of 50 seconds, the lights change back unless there is a car on street "B" and none on "A," in which case the "B" cycle is extended for 10 additional seconds. If cars continue to arrive on street "B" and no car appears on street "A," "B" continues to have a green light. When "A" is green, it remains green at least 60 seconds, and then the lights change only when a car approaches on "B." Figure 4-13 shows the external connections to the controller. Three of the outputs (Ga, Ya, and Ra) drive the green, yellow, and red lights on

FIGURE 4-13: Block Diagram of Traffic Light Controller

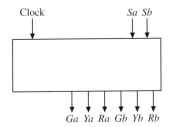

street "A." The other three (*Gb*, *Yb*, and *Rb*) drive the corresponding lights on street "B."

Figure 4-14 shows a Moore state graph for the controller. For timing purposes, the sequential circuit is driven by a clock with a 10-second period. Thus, a state change can occur at most once every 10 seconds. The following notation is used: *GaRb* in a state means that $Ga = Rb = 1$ and all the other output variables are 0. *Sa'Sb* on an arc implies that $Sa = 0$ and $Sb = 1$ will cause a transition along that arc. An arc without a label implies that a state transition will occur when the clock occurs, independent of the input variables. Thus, the green "A" light will stay on for six clock cycles (60 seconds) and then change to yellow if a car is waiting on street "B."

FIGURE 4-14: State Graph for Traffic Light Controller

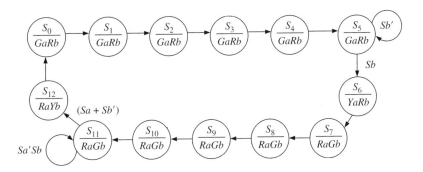

The Verilog code for the traffic light controller (Figure 4-15) represents the state machine with two always statements. Whenever the state—*Sa* or *Sb*—changes, the first always statement updates the outputs and *nextstate*. When the rising edge of the clock occurs, the second always statement updates the state register. The case statement illustrates the use of multiple case numbers. Since states S_0 through S_4 have the same outputs and the next states are in numeric sequence, we use multiple case numbers together instead of five separate case behaviors.

```
                    0, 1, 2, 3, 4 :
                        begin
                            Ga_tmp = 1'b1 ;
                            Rb_tmp = 1'b1 ;
                            nextstate = state + 1 ;
                        end
```

FIGURE 4-15: Verilog Code for Traffic-Light Controller

```verilog
module traffic_light (clk, Sa, Sb, Ra, Rb, Ga, Gb, Ya, Yb);
input clk;
input Sa;
input Sb;
inout Ra;
inout Rb;
inout Ga;
inout Gb;
inout Ya;
inout Yb;

reg Ra_tmp;
reg Rb_tmp;
reg Ga_tmp;
reg Gb_tmp;
reg Ya_tmp;
reg Yb_tmp;

reg[3:0] state;
reg[3:0] nextstate;
parameter[1:0] R = 0;
parameter[1:0] Y = 1;
parameter[1:0] G = 2;
wire[1:0] lightA;
wire[1:0] lightB;

assign Ra = Ra_tmp;
assign Rb = Rb_tmp;
assign Ga = Ga_tmp;
assign Gb = Gb_tmp;
assign Ya = Ya_tmp;
assign Yb = Yb_tmp;

initial
begin
    state = 0;
end

always @(state or Sa or Sb)
begin
    Ra_tmp = 1'b0 ;
    Rb_tmp = 1'b0 ;
    Ga_tmp = 1'b0 ;
    Gb_tmp = 1'b0 ;
    Ya_tmp = 1'b0 ;
    Yb_tmp = 1'b0 ;
    nextstate = 0;

    case (state)
        0, 1, 2, 3, 4 :
                begin
```

4.4 Traffic Light Controller

```verilog
                        Ga_tmp = 1'b1 ;
                        Rb_tmp = 1'b1 ;
                        nextstate = state + 1 ;
                    end
        5 :
                    begin
                        Ga_tmp = 1'b1 ;
                        Rb_tmp = 1'b1 ;
                        if (Sb == 1'b1)
                        begin
                            nextstate = 6 ;
                        end
                        else
                        begin
                            nextstate = 5 ;
                        end
                    end
        6 :
                    begin
                        Ya_tmp = 1'b1 ;
                        Rb_tmp = 1'b1 ;
                        nextstate = 7 ;
                    end
        7, 8, 9, 10 :
                    begin
                        Ra_tmp = 1'b1 ;
                        Gb_tmp = 1'b1 ;
                        nextstate 5 state + 1 ;
                    end
        11 :
                    begin
                        Ra_tmp = 1'b1 ;
                        Gb_tmp = 1'b1 ;
                        if (Sa == 1'b1 | Sb == 1'b0)
                        begin
                            nextstate = 12 ;
                        end
                        else
                        begin
                            nextstate = 11 ;
                        end
                    end
        12 :
                    begin
                        Ra_tmp = 1'b1 ;
                        Yb_tmp = 1'b1 ;
                        nextstate = 0 ;
                    end
    endcase
end
```

```
always @(posedge clk)
begin
    state <= nextstate ;
end

assign lightA = (Ra==1'b1) ? R : (Ya==1'b1) ? Y : (Ga==1'b1) ? G : lightA;
assign lightB = (Rb==1'b1) ? R : (Yb==1'b1) ? Y : (Gb==1'b1) ? G : lightB;

endmodule
```

For each state, only the signals that are 1 are listed within the case statement. Since in Verilog a signal will hold its value until it is changed, we should turn off each signal when the next state is reached. In state 6 we should set *Ga* to 0, in state 7 we should set *Ya* to 0, and so forth. This could be accomplished by inserting appropriate statements in each case. For example, we could insert Ga <= '0' in the **case of state=6 -> case 6.** An easier way to turn off the outputs is to set them all to 0 before the case statement, as shown in Figure 4-15. At first, it seems that a glitch might occur in the output when we set a signal to 0 that should remain 1. However, this is not a problem, because the sequential statements within an always statement execute instantaneously. For example, suppose that at time = 20 ns a state change from S_2 to S_3 occurs. *Ga* and *Rb* are 1, but as soon as the always statement starts executing, the first line of code is executed and *Ga* and *Rb* are scheduled to change to 0 at time 20 + D. The case statement then executes, and *Ga* and *Rb* are scheduled to change to 1 at time 20 + D. Since this is the same time as before, the new value (1) preempts the previously scheduled value (0), and the signals never change to 0.

One may also notice the statement

$$\text{nextstate = 0;}$$

before the case statement. This statement helps to avoid latches for `nextstate` even if all cases are not specified. In this example, only 13 of the 16 cases are specified and latches will be generated for `nextstate` unless a default clause is created where `nextstate` is set to 0. The basic thing to remember is that `nextstate` is a combinational signal and `state` is the actual signal that holds the state value in flip-flops.

Before completing the design of the traffic controller, we will test the Verilog code to see that it meets specifications. As a minimum, our test sequence should cause all of the arcs on the state graph to be traversed at least once. We may want to perform additional tests to check the timing for various traffic conditions, such as heavy traffic on both "A" and "B," light traffic on both, heavy traffic on "A" only, heavy traffic on "B" only, and special cases such as a car failing to move when the light is green, a car going through the intersection when the light is red and so forth.

To make it easier to interpret the simulator output, we define a type named light with the values R, Y, and G and two signals, *lightA* and *lightB*, which can assume these values. Then we add code to set *lightA* to R when the light is red, to Y when the light is yellow, and to G when the light is green. The following simulator command file first tests the case where both self-loops on the graph are traversed and then the case where neither self-loop is traversed:

```
add wave clk Sa Sb state lightA lightB
force clk 0 0, 1 5 sec -repeat 10 sec
force Sa 1 0, 0 40, 1 170, 0 230, 1 250 sec
force Sb 0 0, 1 70, 0 100, 1 120, 0 150, 1 210, 0 250, 1 270 sec
run 300 sec
```

The test results in Figure 4-16 verify that the traffic lights change at the specified times.

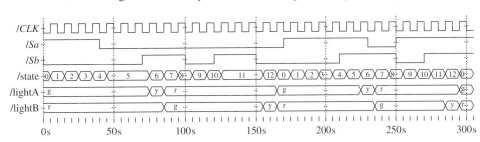

FIGURE 4-16: Test Results for Traffic Light Controller

4.5 State Graphs for Control Circuits

Before continuing with additional examples, we describe the notation we use on control state graphs and then state the conditions that must be satisfied to have a proper state graph. We usually label control state graphs using variable names instead of 0s and 1s. This makes the graph easier to read, especially when the number of inputs and outputs is large. If we label an arc on a Mealy state graph $X_i X_j / Z_p Z_q$, this means if inputs X_i and X_j are 1 (we don't care what the other input values are), the outputs Z_p and Z_q are 1 (and the other outputs are 0), and we will traverse this arc to go to the next state. For example, for a circuit with four inputs (X_1, X_2, X_3, X_4) and four outputs (Z_1, Z_2, Z_3, Z_4), the label $X_1 X_4' / Z_2 Z_3$ is equivalent to 1–0/0110. In general, if we label an arc with an input expression, I, we will traverse the arc when $I = 1$. For example, if the input label is $AB + C'$, we will traverse the arc when $AB + C' = 1$.

In order to have a completely specified proper state graph in which the next state is always uniquely defined for every input combination, we must place the following constraints on the input labels for every state S_k:

1. If I_i and I_j are any pair of input labels on arcs exiting state S_k, then $I_i I_j = 0$ if $i \neq j$.
2. If n arcs exit state S_k and the n arcs have input labels $I_1, I_2, \ldots, I_n$, respectively, then $I_1 + I_2 + \cdots + I_n = 1$.

Condition 1 assures us that at most one input label can be 1 at any given time, and condition 2 assures us that at least one input label will be 1 at any given time. Therefore, exactly one label will be 1, and the next state will be uniquely defined for every input combination. For example, consider the partial state graph in Figure 4-17 where $I_1 = X_1$, $I_2 = X_1' X_2'$, and $I_3 = X_1' X_2$.

FIGURE 4-17: Example Partial State Graph

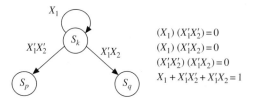

$(X_1)(X_1'X_2') = 0$
$(X_1)(X_1'X_2) = 0$
$(X_1'X_2')(X_1'X_2) = 0$
$X_1 + X_1'X_2' + X_1'X_2 = 1$

Conditions 1 and 2 are satisfied for S_k.

An incompletely specified proper state graph must always satisfy condition 2, and it must satisfy condition 1 for all combinations of values of input variables that can occur for each state S_k. Thus, the partial state graph in Figure 4-18 represents part of a proper state graph only if input combination $X_1 = X_2 = 1$ cannot occur in state S_k.

FIGURE 4-18: Example Partial State Graph

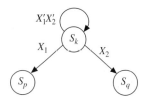

If there are three input variables (X_1, X_2, X_3), the preceding partial state graph represents the following state table row:

000	001	010	011	100	101	110	111
S_k	S_k	S_k	S_q	S_q	S_p	S_p	–	–

4.6 Scoreboard and Controller

In this example, we will design a simple scoreboard, which can display scores from 0 to 99 (decimal). The input to the system should consist of a reset signal and control signals to increment or decrement the score. The 2-digit decimal count gets incremented by 1 if increment signal is true and is decremented by 1 if decrement signal is true. If increment and decrement are true simultaneously, no action occurs.

The current count is displayed on 7-segment displays. In order to prevent accidental erasure, the reset button must be pressed for five consecutive cycles in order to erase the scoreboard. Scoreboard should allow down counts to correct a mistake (in case of accidentally incrementing more than required).

Data Path

At the core of the design will be a 2-digit BCD counter to perform the counting. Two 7-segment displays will be needed to display the current score. One will also require BCD to 7-segment decoders to facilitate the display of each BCD digit. Figure 4-19 illustrates a block diagram of the system. Since true reset should happen

4.6 Scoreboard and Controller

FIGURE 4-19: Overview of Simple Scoreboard

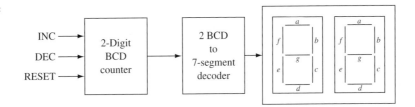

only after pressing reset for five clock cycles, we will also use a 3-bit reset counter called *rstcnt*.

Controller

The controller for this circuit works as follows. There are two states in this finite state machine (FSM), as indicated in Figure 4-20. In the initial state (S_0), the BCD counter is cleared. The reset counter is also made equal to 0. Essentially, S_0 is an initialization state where all the counters are cleared. After the initial start state, the FSM moves to the next state (S_1), which is where counting gets done. In this state, in every clock cycle, incrementing or decrementing is done according to the input signals. If reset signal *rst* arrives, the *rstcnt* is incremented. If reset count has already reached 4, and reset command is still persisting in the fifth clock cycle, a transition to state S_0 is made. If the *inc* signal is present and *dec* is not present, the BCD counter is incremented. The notation *add1* on the arc on the top-right is used to indicate that the BCD counter is incremented. If the *dec* signal is present and *inc* is not present, the BCD counter is decremented. The notation *sub1* on the arc on the bottom-right is used to indicate that the BCD counter is decremented. In any cycle that the reset signal is not present, the *rstcnt* is cleared. If both the *inc* and *dec* signals are true, or neither are true, the rest counter (*rstcnt*) is cleared and the BCD counter is left unchanged.

FIGURE 4-20: State Graph for Scoreboard

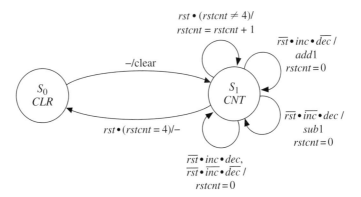

Verilog Model

The Verilog code for the scoreboard is given in Figure 4-21. The two 7-segment displays, *seg7disp1* and *seg7disp2*, are declared as unsigned 7-bit vectors. The segments of the 7-segment display are labeled *a* through *g*, as in Figure 4-19. The unsigned

FIGURE 4-21: Verilog Code for Scoreboard

```verilog
module Scoreboard (clk, rst, inc, dec, seg7disp1, seg7disp0);

    input clk;
    input rst;
    input inc;
    input dec;
    output[6:0] seg7disp1;
    output[6:0] seg7disp0;

    reg[0:0] State; //Only 2 states; So 1 bit sufficient
    reg[3:0] BCD1;
    reg[3:0] BCD0;
    reg[2:0] rstcnt;
    reg[6:0] seg7Rom[0:9];

    initial
    begin
        BCD1 = 4'b0000;
        BCD0 = 4'b0000;
        rstcnt = 0;
        State = 0;

        seg7Rom[0] = 7'b0111111;
        seg7Rom[1] = 7'b0000110;
        seg7Rom[2] = 7'b1011011;
        seg7Rom[3] = 7'b1001111;
        seg7Rom[4] = 7'b1100110;
        seg7Rom[5] = 7'b1101101;
        seg7Rom[6] = 7'b1111100;
        seg7Rom[7] = 7'b0000111;
        seg7Rom[8] = 7'b1111111;
        seg7Rom[9] = 7'b1100111;
    end

    always @(posedge clk)
    begin
        case (State)
            0 :
                    begin
                        BCD1 <=4'b0000 ;
                        BCD0 <= 4'b0000 ;
                        rstcnt <= 0 ;
                        State <= 1 ;
                    end
            1 :
                    begin
                        if (rst == 1'b1)
                        begin
                            if (rstcnt == 4)
                            begin
```

```verilog
                                    State <= 0 ;
                                end
                            else
                            begin
                                rstcnt <= rstcnt + 1 ;
                            end
                        end
                        else if (inc == 1'b1 && dec == 1'b0)
                        begin
                            rstcnt <= 0 ;
                            if (BCD0 < 4'b1001)
                            begin
                                BCD0 <= BCD0 + 1 ;
                            end
                            else if (BCD1 < 4'b1001)
                            begin
                                BCD1 <= BCD1 + 1 ;
                                BCD0 <= 4'b0000 ;
                            end
                        end
                        else if (dec == 1'b1 && inc == 1'b0)
                        begin
                            rstcnt <= 0 ;
                            if (BCD0 > 4'b0000)
                            begin
                                BCD0 <= BCD0 - 1 ;
                            end
                            else if (BCD1 > 4'b0000)
                            begin
                                BCD1 <= BCD1 - 1 ;
                                BCD0 <= 4'b1001 ;
                            end
                        end
                        else if (inc == 1'b1 && dec == 1'b1)
                        begin
                            rstcnt <= 0 ;
                        end
                    end
            endcase
        end
    assign seg7disp0 = seg7Rom[BCD0] ;
    assign seg7disp1 = seg7Rom[BCD1] ;

endmodule
```

type is used so that the overloaded "+" operator can be used for incrementing the counter by one. The decoder for the 7-segment display can be implemented as an array or look-up table. The look-up table consists of 10 7-bit vectors. A reg datatype *seg7Rom* is declared for the array of the 7-segment values corresponding to each BCD digit. It is a two-dimensional array with 10 elements, each of which is a 7-bit unsigned vector. The look-up table must be addressed with a BCD digit, and the statement

$$\textbf{assign}\ seg7disp0\ =\ seg7rom[BCD0];$$

accesses the appropriate element from the array *seg7rom* to convert the decimal digit to the 7-segment form. BCD addition is accomplished with adjustment of the sum if the sum is greater than 9. If the current count is less than 9, it is incremented. If it is 9, adding 1 results in a 0, but the next digit should be incremented. Similarly, decrementing from 0 is performed by borrowing a 1 from the next higher digit.

4.7 Synchronization and Debouncing

The *inc*, *dec*, and *rst* signals to the scoreboard in the previous design are external inputs. An issue in systems involving external inputs is synchronization. Outputs from a keypad or push button switches are not synchronous to the system clock signal. Since they will be used as inputs to a synchronous sequential circuit, they should be synchronized.

Another issue in systems involving external inputs is switch bounce. When a mechanical switch is closed or opened, the switch contact will bounce, causing noise in the switch output, as shown in Figure 4-22(a). The contact may bounce for several milliseconds before it settles down to its final position. After a switch closure has been detected, we must wait for the bounce to settle before reading the key. In any circuit involving mechanical switches, one should debounce the switches. Debouncing means removing the transients in the switch output.

FIGURE 4-22:
Debouncing
Mechanical Switches

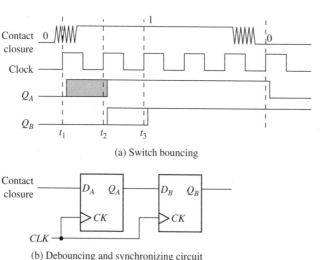

(a) Switch bouncing

(b) Debouncing and synchronizing circuit

Flip-flops are very useful devices when contacts must be synchronized and debounced. Figure 4-22(b) shows a proposed debouncing and synchronizing circuit. In this design, the clock period is greater than the bounce time. If the rising edge of the clock occurs during the bounce, either a 0 or 1 will be clocked into the flip-flop at t_1. If a 0 was clocked in, a 1 will be clocked in at the next active clock edge (t_2). Consequently, it appears that Q_A will be a debounced and synchronized version of the contact closure. However, a possibility of failure exists if *the switch* changes very close to the clock edge such that the setup or hold time is violated. In this case the flip-flop output Q_A may oscillate or otherwise malfunction. Although this situation will occur very infrequently, it is best to guard against it by adding a second flip-flop. We will choose the clock period so that any oscillation at the output of Q_A will have died out before the next active edge of the clock so that the input D_B will always be stable at the active clock edge. The debounced signal, Q_B, will always be clean and synchronized with the clock, although it may be delayed up to two clock cycles after the switch is pressed.

Single Pulser

One assumption in the scoreboard design is that each time the *inc* and *dec* signals are provided, they last only for one clock cycle. Digital systems generally run at speeds higher than actions by humans, and practically, it is very difficult for humans to produce a signal that lasts only for a clock pulse. If the pressing of the button lasted longer than a clock cycle, the counters will continue to be incremented in the foregoing design. A solution to the problem is to develop a circuit that generates a single pulse for a human action of pressing a button or switch. Such a circuit can be used in a variety of applications involving humans, push buttons, and switches.

Now, let us design a single pulser circuit that delivers a synchronized pulse that is a single clock cycle long when a button is pressed. The circuit must sense the pressing of a button and assert an output signal for one clock cycle. Then the output stays inactive until the button is released.

Let us create a state diagram for the single pulser. The single pulser circuit must have two states: one in which it will detect the pressing of the key and one in which it will detect the release of the key. Let us call the first state S_0 and the second state S_1. Let us use the symbol SYNCPRESS to denote the synchronized key press. When the circuit is in state S_0 and the button is pressed, the system produces the single pulse and moves to state S_1. The single pulse is a Mealy output as the state changes from S_0 to S_1. Once the system is in state S_1, it waits for the button to be released. As soon as it is released, it moves to the start state S_0 waiting for the next button press. The single pulse output is true only during the transition from S_0 to S_1. The state diagram is illustrated in Figure 4-23.

FIGURE 4-23: State Diagram of Single Pulse

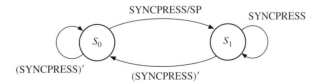

Since there are only two states for this circuit, it can be implemented using one flip-flop. A single pulser can be implemented as in Figure 4-24. The first block consists of the circuitry in Figure 4-22 and generates a synchronized button press, SYNCPRESS. The flip-flop implements the two states of the state machine. Let us

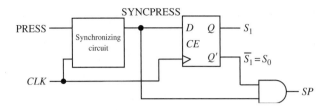

FIGURE 4-24: Single Pulser and Synchronizer Circuit

assume the state assignments are $S_0 = 0$ and $S_1 = 1$. In such a case, the Q output of the second flip-flop is synonymous with S_1, and the Q' output of the second flip-flop is synonymous with S_0. The equation for the single pulse SP is

$$SP = S_0 \cdot \text{SYNCPRESS}$$

It may also be noted that $S_0 = S_1'$. Including the 2 flip-flops inside the synchronizing block, three flip-flops can provide debouncing, synchronization, and single-pulsing. If button pushes can be passed through such a circuit, a single pulse that is debounced and synchronized, with respect to the system clock, can be obtained. It is a good practice to feed external push-button signals through such a circuit in order to obtain controlled and predictable operation.

4.8 A Shift-and-Add Multiplier

In this section, we will design a multiplier for unsigned binary numbers. When we form the product $A \times B$, the first operand (A) is called the *multiplicand* and the second operand (B) is called the *multiplier*. As illustrated here, binary multiplication requires only shifting and adding. In the following example, we multiply 13_{10} by 11_{10} in binary

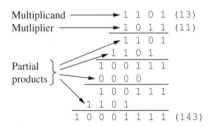

Note that each partial product is either the multiplicand (1101) shifted over by the appropriate number of places or zero. Instead of forming all the partial products first and then adding, each new partial product is added in as soon as it is formed, which eliminates the need for adding more than two binary numbers at a time.

Multiplication of two 4-bit numbers requires a 4-bit multiplicand register, a 4-bit multiplier register, a 4-bit full adder, and an 8-bit register for the product. The product register serves as an accumulator to accumulate the sum of the partial products. If the multiplicand were shifted left each time before it was added to the accumulator, as was done in the previous example, an 8-bit adder would be needed. Therefore, it is better to shift the contents of the product register to the right each time, as shown in the block diagram of Figure 4-25.

4.8 A Shift-and-Add Multiplier

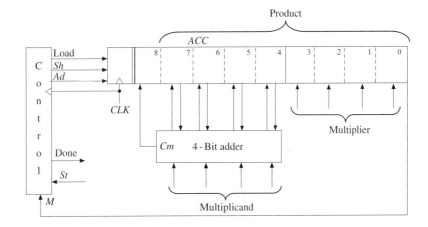

FIGURE 4-25: Block Diagram for Binary Multiplier

This type of multiplier is sometimes referred to as a serial-parallel multiplier, since the multiplier bits are processed serially, but the addition takes place in parallel. As indicated by the arrows on the diagram, 4 bits from the accumulator (ACC) and 4 bits from the multiplicand register are connected to the adder inputs; the 4 sum bits and the carry output from the adder are connected back to the accumulator. When an add signal (Ad) occurs, the adder outputs are transferred to the accumulator by the next clock pulse, thus causing the multiplicand to be added to the accumulator. An extra bit at the left end of the product register temporarily stores any carry that is generated when the multiplicand is added to the accumulator. When a shift signal (Sh) occurs, all 9 bits of ACC are shifted right by the next clock pulse.

Since the lower 4 bits of the product register are initially unused, we will store the multiplier in this location instead of in a separate register. As each multiplier bit is used, it is shifted out the right end of the register to make room for additional product bits. A shift signal (Sh) causes the contents of the product register (including the multiplier) to be shifted right one place when the next clock pulse occurs. The control circuit puts out the proper sequence of add and shift signals after a start signal ($St = 1$) has been received. If the current multiplier bit (M) is 1, the multiplicand is added to the accumulator followed by a right shift; if the multiplier bit is 0, the addition is skipped and only the right shift occurs. The multiplication example (13×11) is reworked as follows showing the location of the bits in the registers at each clock time:

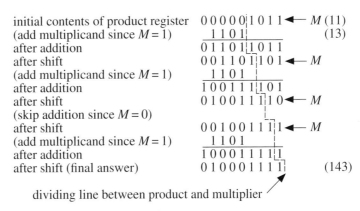

dividing line between product and multiplier

The control circuit must be designed to output the proper sequence of add and shift signals. Figure 4-26 shows a state graph for the control circuit. In Figure 4-26, S_0 is the reset state, and the circuit stays in S_0 until a start signal ($St = 1$) is received. This generates a *Load* signal, which causes the multiplier to be loaded into the lower 4 bits of the accumulator (ACC) and the upper 5 bits of the accumulator to be cleared. In state S_1, the low-order bit of the multiplier (M) is tested. If $M = 1$, an add signal is generated, and if $M = 0$, a shift signal is generated. Similarly, in states S_3, S_5, and S_7, the current multiplier bit (M) is tested to determine whether to generate an add or shift signal. A shift signal is always generated at the next clock time following an add signal (states S_2, S_4, S_6, and S_8). After four shifts have been generated, the control network goes to S_9 and a done signal is generated before returning to S_0.

FIGURE 4-26: State Graph for Binary Multiplier Control

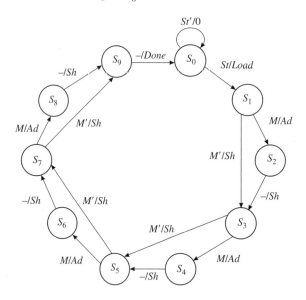

The behavioral Verilog model (Figure 4-27) corresponds directly to the state graph. Since there are 10 states, we have declared an integer ranging from 0 to 9 for the state signal. The signal ACC represents the 9-bit accumulator output. The statement

`` `define M ACC[0] ``

allows us to use the name M in place of $ACC(0)$. The notation 1, 3, 5, 7 : means when the state is 1 or 3 or 5 or 7, the action that follows occurs. All register operations and state changes take place on the rising edge of the clock. For example, in state 0, if St is 1, the multiplier is loaded into the accumulator at the same time the state changes to 1. The expression {1'b0, ACC[7:4]} + Mcand is used to compute the sum of two 4-bit unsigned vectors to give a 5-bit result. This represents the adder output, which is loaded into ACC at the same time the state counter is incremented. The right shift on ACC is accomplished by loading ACC with 0 concatenated with the upper 8 bits of ACC. The expression {1'b0, ACC[8:1]} could be replaced with ACC >> 1.

4.8 A Shift-and-Add Multiplier

FIGURE 4-27: Behavioral Model for 4 × 4 Binary Multiplier

```verilog
// This is a behavioral model of a multiplier for unsigned
// binary numbers.  It multiplies a 4-bit multiplicand
// by a 4-bit multiplier to give an 8-bit product.

// The maximum number of clock cycles needed for a
// multiply is 10.

`define M ACC[0]

module mult4X4 (Clk, St, Mplier, Mcand, Done, Result);

    input Clk;
    input St;
    input[3:0] Mplier;
    input[3:0] Mcand;
    output Done;
    output[7:0] Result;

    reg[3:0] State;
    reg[8:0] ACC;

    initial
    begin
        State = 0;
        ACC   = 0;
    end

    always @(posedge Clk)
    begin
        case (State)
            0 :
                    begin
                        if (St ==1'b1)
                        begin
                            ACC[8:4] <= 5'b00000 ;
                            ACC[3:0] <= Mplier ;
                            State <= 1 ;
                        end
                    end
            1, 3, 5, 7 :
                    begin
                      if (`M == 1'b1)
                      begin
                         ACC[8:4] <= {1'b0, ACC[7:4]} + Mcand ;
                          State <= State + 1 ;
                      end
                      else
                      begin
                         ACC <= {1'b0, ACC[8:1]} ;
```

```verilog
                            State <= State + 2 ;
                    end
                end
            2, 4, 6, 8 :
                begin
                    ACC <= {1'b0, ACC[8:1]} ;
                    State <= State + 1 ;
                end
            9 :
                begin
                    State <= 0 ;
                end
        endcase
    end

    assign Done = (State == 9) ? 1'b1 : 1'b0 ;
    assign Result = (State == 9) ? ACC[7:0] : 8'b01010101 ;

endmodule
```

The *Done* signal should be turned on only in state 9. If we had used the statement State <= 0; Done <= '1'; for the behavior of State = 9, *Done* would be turned on at the same time *State* changes to 0. This is too late, since we want *Done* to turn on when *State* becomes 9. Therefore, we used a separate concurrent assignment statement. This statement is placed outside the process so that *Done* will be updated whenever *State* changes.

As the state graph for the multiplier indicates, the control performs two functions—generating add or shift signals as needed and counting the number of shifts. If the number of bits is large, it is convenient to divide the control circuit into a counter and an add-shift control, as shown in Figure 4-28(a). First, we will derive a state graph for the add-shift control that tests St and M and outputs the proper sequence of add and shift signals (Figure 4-28(b)). Then we will add a completion signal (K) from the counter that stops the multiplier after the proper number of shifts have been completed. Starting in S_0 in Figure 4-28(b), when a start signal $St = 1$ is received, a load signal is generated and the circuit goes to state S_1. Then if $M = 1$, an add signal is generated and the circuit goes to state S_2; if $M = 0$, a shift signal is generated and the circuit stays in S_1. In S_2, a shift signal is generated since a shift always follows an add. The graph of Figure 4-28(b) will generate the proper sequence of add and shift signals, but it has no provision for stopping the multiplier.

To determine when the multiplication is completed, the counter is incremented each time a shift signal is generated. If the multiplier is n bits, n shifts are required. We will design the counter so that a completion signal (K) is generated after $n - 1$ shifts have occurred. When $K = 1$, the circuit should perform one more addition, if necessary, and then do the final shift. The control operation in Figure 4-28(c) is the same as Figure 4-28(b) as long as $K = 0$. In state S_1, if $K = 1$, we test M as usual. If $M = 0$, we output the final shift signal and go to the done state (S_3); however, if $M = 1$, we add before shifting and go to state S_2. In state S_2, if $K = 1$, we output one

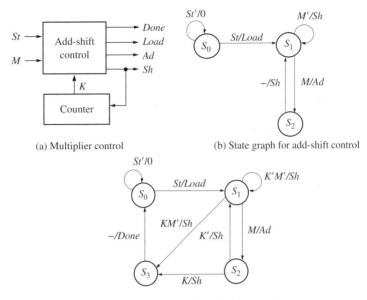

FIGURE 4-28: Multiplier Control with Counter

(a) Multiplier control

(b) State graph for add-shift control

(c) Final state graph for add-shift control

more shift signal and then go to S_3. The last shift signal will increment the counter to 0 at the same time the add-shift control goes to the done state.

As an example, consider the multiplier of Figure 4-25, but replace the control circuit with Figure 4-28(a). Since $n = 4$, a 2-bit counter is needed to count the four shifts, and $K = 1$ when the counter is in state 3 (11_2). Table 4-2 shows the operation of the multiplier when 1101 is multiplied by 1011. S_0, S_1, S_2, and S_3 represent states of the control circuit (Figure 4-28(c)). The contents of the product register at each step are the same as given on page 233 of this chapter.

TABLE 4-2: Operation of Multiplier Using a Counter

Time	State	Counter	Product Register	St	M	K	Load	Ad	Sh	Done
t_0	S_0	00	000000000	0	0	0	0	0	0	0
t_1	S_0	00	000000000	1	0	0	1	0	0	0
t_2	S_1	00	000001011	0	1	0	0	1	0	0
t_3	S_2	00	011011011	0	1	0	0	0	1	0
t_4	S_1	01	001101101	0	1	0	0	1	0	0
t_5	S_2	01	100111101	0	1	0	0	0	1	0
t_6	S_1	10	010011110	0	0	0	0	0	1	0
t_7	S_1	11	001001111	0	1	1	0	1	0	0
t_8	S_2	11	100011111	0	1	1	0	0	1	0
t_9	S_3	00	010001111	0	1	0	0	0	0	1

At time t_0, the control is reset and waits for a start signal. At time t_1, the start signal St is 1, and a *Load* signal is generated. At time t_2, $M = 1$, so an *Ad* signal is generated. When the next clock occurs, the output of the adder is loaded into the accumulator and the control goes to S_2. At t_3, an *Sh* signal is generated, so at the next clock shifting occurs and the counter is incremented. At t_4, $M = 1$ so $Ad = 1$, and the adder output is loaded into the accumulator at the next clock. At t_5 and t_6, shifting and counting occur. At t_7, three shifts have occurred and the counter state is 11, so $K = 1$. Since $M = 1$, addition occurs and control goes to S_2. At t_8, $Sh = K = 1$, so at the next clock the final shift occurs and the counter is incremented back to state 00. At t_9, a *Done* signal is generated.

The multiplier design given here can easily be expanded to 8, 16, or more bits simply by increasing the register size and the number of bits in the counter. The add-shift control would remain unchanged.

4.9 Array Multiplier

An array multiplier is a parallel multiplier that generates the partial products in a parallel fashion. The various partial products are added as soon as they are available. Consider the process of multiplication as illustrated in Table 4-3. Two 4-bit unsigned numbers, $X_3X_2X_1X_0$ and $Y_3Y_2Y_1Y_0$, are multiplied to generate a product that is possibly 8 bits. Each of the X_iY_j product bits can be generated by an AND gate. Each partial product can be added to the previous sum of partial products using a row of adders. The sum output of the first row of adders, which adds the first two partial products, is $S_{13}S_{12}S_{11}S_{10}$, and the carry output is $C_{13}C_{12}C_{11}C_{10}$. Similar results occur for the other two rows of adders. (We have used the notation S_{ij} and C_{ij} to represent the sums and carries from the ith row of adders.)

TABLE 4-3: 4-Bit Multiplier Partial Products

					X_3	X_2	X_1	X_0	Multiplicand
					Y_3	Y_2	Y_1	Y_0	Multiplier
					X_3Y_0	X_2Y_0	X_1Y_0	X_0Y_0	partial product 0
				X_3Y_1	X_2Y_1	X_1Y_1	X_0Y_1		partial product 1
				C_{12}	C_{11}	C_{10}			1st row carries
			C_{13}	S_{13}	S_{12}	S_{11}	S_{10}		1st row sums
			X_3Y_2	X_2Y_2	X_1Y_2	X_0Y_2			partial product 2
			C_{22}	C_{21}	C_{20}				2nd row carries
		C_{23}	S_{23}	S_{22}	S_{21}	S_{20}			2nd row sums
		X_3Y_3	X_2Y_3	X_1Y_3	X_0Y_3				partial product 3
		C_{32}	C_{31}	C_{30}					3rd row carries
	C_{33}	S_{33}	S_{32}	S_{31}	S_{30}				3rd row sums
	P_7	P_6	P_5	P_4	P_3	P_2	P_1	P_0	final product

Figure 4-29 shows the array of AND gates and adders to perform this multiplication. If an adder has three inputs, a full adder (FA) is used, but if an adder has only two inputs, a half-adder (HA) is used. A half-adder is the same as a full adder with one of the inputs set to 0. This multiplier requires 16 AND gates, 8 full adders,

4.9 Array Multiplier

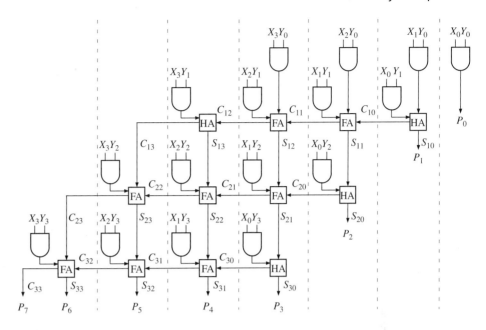

FIGURE 4-29: Block Diagram of 4 × 4 Array Multiplier

and 4 half-adders. After the X and Y inputs have been applied, the carry must propagate along each row of cells, and the sum must propagate from row to row. The time required to complete the multiplication depends primarily on the propagation delay in the adders. The longest path from input to output goes through 8 adders. If t_{ad} is the worst-case (longest possible) delay through an adder, and t_g is the longest AND gate delay, then the worst-case time to complete the multiplication is $8t_{ad} + t_g$.

In general, an n-bit-by-n-bit array multiplier would require n^2 AND gates, $n(n - 2)$ full adders, and n half-adders. So the number of components required increases quadratically. For the serial-parallel multiplier previously designed, the amount of hardware required in addition to the control circuit increases linearly with n.

For an $n \times n$ array multiplier, the longest path from input to output goes through n adders in the top row, n-1 adders in the bottom row and n-3 adders in the middle rows. The corresponding worst-case multiply time is $(3n - 4)t_{ad} + t_g$. The longest delay in a circuit is called a critical path. The worst-case delay can be improved to $2nt_{ad} + t_g$ by forwarding carry from each adder to the diagonally lower adder rather than the adder on the left side. When $n = 4$, both expressions are the same; however, for larger values of n, it is beneficial to pass carry diagonally as opposed to rippling it to the left. One may note that this multiplier has no sequential logic or registers.

The shift-and-add multiplier that we previously designed requires $2n$ clocks to complete the multiply in the worst case, although this can be reduced to n clocks using a technique discussed in the following section. The minimum clock period depends on the propagation delay through the n-bit adder as well as the propagation delay and setup time for the accumulator flip-flops.

Verilog Coding

If the topology has to be exactly what the designer wants, one needs to do structural coding. If one made a behavioral model of a multiplier without specifying the topology, the topology generated by the synthesizer will depend on the synthesis tool. Here, we present a structural model for an array multiplier in Figure 4-30. Full-adder and half-adder modules are created and used as components for the array multiplier. The full adders and half adders are interconnected according to the array multiplier topology. Several instantiation statements are used for this purpose.

FIGURE 4-30: Verilog Code for 4 × 4 Array Multiplier

```verilog
module Array_Mult (X, Y, P);

   input[3:0] X;
   input[3:0] Y;
   output[7:0] P;

   wire[3:0] C1;
   wire[3:0] C2;
   wire[3:0] C3;
   wire[3:0] S1;
   wire[3:0] S2;
   wire[3:0] S3;
   wire[3:0] XY0;
   wire[3:0] XY1;
   wire[3:0] XY2;
   wire[3:0] XY3;

   assign XY0[0] = X[0] & Y[0] ;
   assign XY1[0] = X[0] & Y[1] ;
   assign XY0[1] = X[1] & Y[0] ;
   assign XY1[1] = X[1] & Y[1] ;
   assign XY0[2] = X[2] & Y[0] ;
   assign XY1[2] = X[2] & Y[1] ;
   assign XY0[3] = X[3] & Y[0] ;
   assign XY1[3] = X[3] & Y[1] ;
   assign XY2[0] = X[0] & Y[2] ;
   assign XY3[0] = X[0] & Y[3] ;
   assign XY2[1] = X[1] & Y[2] ;
   assign XY3[1] = X[1] & Y[3] ;
   assign XY2[2] = X[2] & Y[2] ;
   assign XY3[2] = X[2] & Y[3] ;
   assign XY2[3] = X[3] & Y[2] ;
   assign XY3[3] = X[3] & Y[3] ;

   FullAdder FA1 (XY0[2], XY1[1], C1[0], C1[1], S1[1]);
   FullAdder FA2 (XY0[3], XY1[2], C1[1], C1[2], S1[2]);
   FullAdder FA3 (S1[2], XY2[1], C2[0], C2[1], S2[1]);
   FullAdder FA4 (S1[3], XY2[2], C2[1], C2[2], S2[2]);
   FullAdder FA5 (C1[3], XY2[3], C2[2], C2[3], S2[3]);
   FullAdder FA6 (S2[2], XY3[1], C3[0], C3[1], S3[1]);
   FullAdder FA7 (S2[3], XY3[2], C3[1], C3[2], S3[2]);
   FullAdder FA8 (C2[3], XY3[3], C3[2], C3[3], S3[3]);
   HalfAdder HA1 (XY0[1], XY1[0], C1[0], S1[0]);
```

```
    HalfAdder HA2 (XY1[3], C1[2], C1[3], S1[3]);
    HalfAdder HA3 (S1[1], XY2[0], C2[0], S2[0]);
    HalfAdder HA4 (S2[1], XY3[0], C3[0], S3[0]);

    assign P[0] = XY0[0] ;
    assign P[1] = S1[0] ;
    assign P[2] = S2[0] ;
    assign P[3] = S3[0] ;
    assign P[4] = S3[1] ;
    assign P[5] = S3[2] ;
    assign P[6] = S3[3] ;
    assign P[7] = C3[3] ;
endmodule

// Full Adder and half adder modules
// should be in the project

module FullAdder (X, Y, Cin, Cout, Sum);

    input X;
    input Y;
    input Cin;
    output Cout;
    output Sum;

    assign Sum = X ^ Y ^ Cin ;
    assign Cout = (X & Y) | (X & Cin) | (Y & Cin) ;
endmodule

module HalfAdder (X, Y, Cout, Sum);

    input X;
    input Y;
    output Cout;
    output Sum;

    assign Sum = X ^ Y ;
    assign Cout = X & Y ;
endmodule
```

4.10 A Signed Integer/Fraction Multiplier

Several algorithms are available for multiplication of signed binary numbers. The following procedure is a straightforward way to carry out the multiplication:

1. Complement the multiplier if negative.
2. Complement the multiplicand if negative.
3. Multiply the two positive binary numbers.
4. Complement the product if it should be negative.

Although this method is conceptually simple, it requires more hardware and computation time than some of the other available methods.

The next method we describe requires only the ability to complement the multiplicand. Complementation of the multiplier or product is not necessary. Although the method works equally well with integers or fractions, we illustrate the method with fractions, since we will later use this multiplier as part of a multiplier for floating-point numbers. Using 2's complement for negative numbers, we will represent signed binary fractions in the following form:

$$0.101 \quad +5/8 \quad 1.011 \quad -5/8$$

The digit to the left of the binary point is the sign bit, which is 0 for positive fractions and 1 for negative fractions. In general, the 2's complement of a binary fraction F is $F^* = 2 - F$. Thus, $-5/8$ is represented by $10.000 - 0.101 = 1.011$. (This method of defining 2's complement fractions is consistent with the integer case ($N^* = 2^n - N$), since moving the binary point $n - 1$ places to the left is equivalent to dividing by 2^{n-1}.) The 2's complement of a fraction can be found by starting at the right end and complementing all the digits to the left of the first 1, the same as for the integer case. The 2's complement fraction $1.000\ldots$ is a special case. It actually represents the number -1, since the sign bit is negative and the 2's complement of $1.000\ldots$ is $2 - 1 = 1$. We cannot represent $+1$ in this 2's complement fraction system, since $0.111\ldots$ is the largest positive fraction.

Binary Fixed Point Fractions

Fixed point numbers are number formats in which the decimal or binary point is at a fixed location. One can have a fixed-point 8-bit number format where the binary point is assumed to be after 4 bits (i.e., 4 bits for the fractional part and 4 bits for the integral part). If the binary point is assumed to be located 2 more bits to the right, there will be 6 bits for the integral part and 2 bits for the fraction. The range and precision of the numbers that can be represented in the different formats depend on the location of the binary point. For instance, if there are 4 bits for the fractional part and 4 bits for the integer, the range, assuming unsigned numbers, is 0.00 to 15.925. If only 2 bits are allowed for the fractional part and 6 bits for the integer, the range increases but the precision reduces. Now, the range would be 0.00 to 63.75, but the fractional part can be specified only as a multiple of 0.25.

Let us say we need to represent -13.45 in a 2's complement fixed-point number representation with 4 fractional bits. To convert any decimal fraction into the binary fraction, one technique is to repeatedly multiply the fractional part (only the fractional part in each intermediate step) by 2. So, starting with 0.45, the repeated multiplication results in

$$\mathbf{0}.90$$
$$\mathbf{1}.80$$
$$\mathbf{1}.60$$

4.10 A Signed Integer/Fraction Multiplier

> **1.**20
> **0.**40
> **0.**80
> **1.**60
> **1.**20
>
> Now, the binary representation can be obtained by considering the digits in bold. An appropriate representation can be obtained depending on the number of bits available (e.g., 0111 if 4 bits are available, 01110011 if 8 bits are available, and so on). The representation for decimal number 13.45 in the fixed point format with 4 binary places will be as follows:
>
> 13.45:   1101.0111
>
> One may note that the represented number is only an approximation of the actual number. The represented number can be converted back to decimal and seen to be 13.4375 (slightly off from the number we started with). The representation approaches the actual number as more and more binary places are added to the representation.
>
> Negative fractions can be represented in 2's complement form. Let us represent -13.45 in 2's complement form. This cannot be done if we have only four places for the integer. We need to have at least 5 bits for the integer in order to handle the sign. Assuming 5 bits are available for the integer, in a 9 bit format:
>
> | 13.45: | 01101.0111 |
> |----------------|------------|
> | 1's complement | 10010.1000 |
> | 2's complement | 10010.1001 |
>
> Hence, -13.45 = 10010.1001 in this representation.

When multiplying signed binary numbers, we must consider four cases:

Multiplicand	Multiplier
+	+
−	+
+	−
−	−

When both the multiplicand and the multiplier are positive, standard binary multiplication is used. For example,

$$
\begin{array}{r}
0.1\;1\;1 \\
\times\;0.1\;0\;1 \\
\hline
(0.\;0\;0)0\;1\;1\;1 \\
(0.)0\;1\;1\;1 \\
\hline
0.\;1\;0\;0\;0\;1\;1
\end{array}
\quad
\begin{array}{l}
(+7/8) \quad \leftarrow \quad \text{Multiplicand} \\
(+5/8) \quad \leftarrow \quad \text{Multiplier} \\
\\
(+7/64) \quad \leftarrow \quad \text{\textit{Note}: The proper representation of the} \\
(+7/16) \quad \leftarrow \quad \text{fractional partial products requires extension} \\
\qquad\qquad\qquad\quad \text{of the sign bit past the binary point, as} \\
(+35/64) \qquad\quad \text{indicated in parentheses. (Such extension} \\
\qquad\qquad\qquad\quad \text{is not necessary in the hardware.)}
\end{array}
$$

When the multiplicand is negative and the multiplier is positive, the procedure is the same as in the previous case, except that we must extend the sign bit of the multiplicand so that the partial products and final product will have the proper negative sign. For example,

$$
\begin{array}{r}
1.1\;0\;1 \\
\times\;0.1\;0\;1 \\
\hline
(1.\;1\;1)1\;1\;0\;1 \\
(1.)1\;1\;0\;1 \\
\hline
1.\;1\;1\;0\;0\;0\;1
\end{array}
\quad
\begin{array}{l}
(-3/8) \\
(+5/8) \\
\\
(-3/64) \quad \leftarrow \quad \text{\textit{Note}: The extension of the} \\
(-3/16) \quad \leftarrow \quad \text{sign bit provides proper} \\
\qquad\qquad\qquad\quad \text{representation of the} \\
(-15/64) \qquad\quad \text{negative products.}
\end{array}
$$

When the multiplier is negative and the multiplicand is positive, we must make a slight change in the multiplication procedure. A negative fraction of the form $1.g$ has a numeric value $-1 + 0.g$; for example, $1.011 = -1 + 0.011 = -(1 - 0.011) = -0.101 = -5/8$. Thus, when multiplying by a negative fraction of the form $1.g$, we treat the fraction part $(.g)$ as a positive fraction, but the sign bit is treated as -1. Hence, multiplication proceeds in the normal way as we multiply by each bit of the fraction and accumulate the partial products. However, when we reach the negative sign bit, we must add in the 2's complement of the multiplicand instead of the multiplicand itself. The following example illustrates this:

$$
\begin{array}{r}
0.1\;0\;1 \\
\times\;1.1\;0\;1 \\
\hline
(0.0\;0)0\;1\;0\;1 \\
(0.)0\;1\;0\;1 \\
\hline
(0.)0\;1\;1\;0\;0\;1 \\
1.\;0\;1\;1 \\
\hline
1.\;1\;1\;0\;0\;0\;1
\end{array}
\quad
\begin{array}{l}
(+5/8) \\
(-3/8) \\
\\
(+5/64) \\
(+5/16) \\
\\
\\
(-5/8) \quad \leftarrow \quad \text{\textit{Note}: The 2's complement of} \\
(-15/64) \qquad\quad \text{the multiplicand is added at} \\
\qquad\qquad\qquad\quad \text{this point.}
\end{array}
$$

When both the multiplicand and the multiplier are negative, the procedure is the same as before. At each step, we must be careful to extend the sign bit of the partial product to preserve the proper negative sign, and at the final step we must add in the 2's complement of the multiplicand, since the sign bit of the multiplier is negative. For example,

$$
\begin{array}{r}
1.1\,0\,1 \\
\times\,1.1\,0\,1 \\
\hline
\end{array}
\quad
\begin{array}{l}
(-3/8) \\
(-3/8)
\end{array}
$$

$$
\begin{array}{r}
(1.\;1\,1)\,1\,1\,0\,1 \\
(1.)1\,1\;\;0\,1 \\
\hline
1.\,1\,1\;\;0\,0\,0\,1 \\
0.\,0\,1\,1 \\
\hline
0.\,0\,0\;\;1\,0\,0\,1
\end{array}
\quad
\begin{array}{l}
(-3/64) \quad \leftarrow \text{ \textit{Note}: Extend sign bit} \\
(-3/16) \\
\\
(+3/8) \quad \leftarrow \text{ Add the 2's complement of the} \\
(+9/64) \quad\quad\;\; \text{multiplicand.}
\end{array}
$$

In summary, the procedure for multiplying signed 2's complement binary fractions is the same as for multiplying positive binary fractions, except that we must be careful to preserve the sign of the partial product at each step, and if the sign of the multiplier is negative, we must complement the multiplicand before adding it in at the last step. The hardware is almost identical to that used for multiplication of positive numbers, except a complementer must be added for the multiplicand.

Figure 4-31 shows the hardware required to multiply two 4-bit fractions (including the sign bit). A 5-bit adder is used so the sign of the sum is not lost due to a carry into the sign bit position. The M input to the control circuit is the currently active bit of the multiplier. Control signal Sh causes the accumulator to shift right one place with sign extension. Ad causes the ADDER output to be loaded into the left 5 bits of the accumulator. The carry out from the last bit of the adder is discarded, since we are doing 2's complement addition. Cm causes the multiplicand (Mcand) to be complemented (1's complement) before it enters the adder inputs. Cm is also connected to the carry input of the adder so that when $Cm = 1$, the adder adds 1 plus the 1's complement of Mcand to the accumulator, which is equivalent to adding the

FIGURE 4-31: Block Diagram for 2's Complement Multiplier

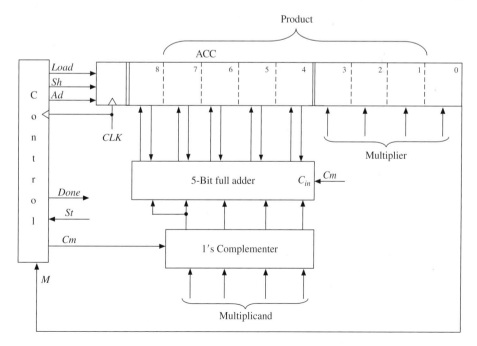

2's complement of Mcand. Figure 4-32 shows a state graph for the control circuit. Each multiplier bit (M) is tested to determine whether to add and shift or whether to just shift. In state S_7, M is the sign bit, and if $M = 1$, the complement of the multiplicand is added to the accumulator.

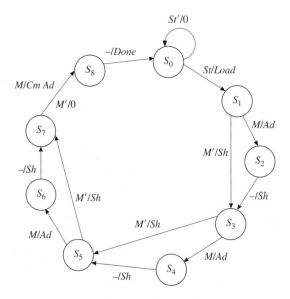

FIGURE 4-32: State Graph for 2's Complement Multiplier

When the hardware in Figure 4-31 is used, the add and shift operations must be done at two separate clock times. We can speed up the operation of the multiplier by moving the wires from the adder output one position to the right (Figure 4-33) so that the adder output is already shifted over one position when it is loaded into the accumulator. With this arrangement, the add and shift operations can occur at the same clock time, which leads to the control state graph of Figure 4-34. When the multiplication is complete, the product (6 bits plus sign) is in the lower 3 bits

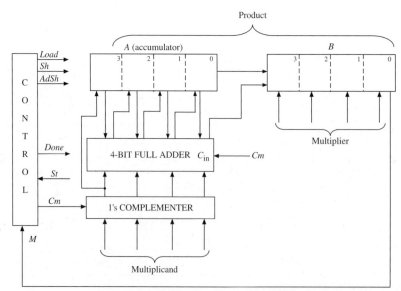

FIGURE 4-33: Block Diagram for Faster Multiplier

4.10 A Signed Integer/Fraction Multiplier

FIGURE 4-34: State Graph for Faster Multiplier

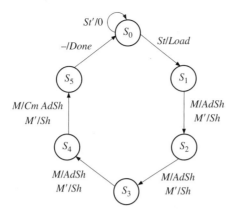

of A followed by B. The binary point then is in the middle of the A register. If we wanted it between the left two bits, we would have to shift A and B left one place.

A behavioral Verilog model for this multiplier is shown in Figure 4-35. Shifting the A and B registers together is accomplished by the sequential statements

```
A <= {A[3], A[3:1]} ;
B <= {A[0], B[3:1]} ;
```

FIGURE 4-35: Behavioral Model for 2's Complement Multiplier

```
`define M B[0]
module mult2C (CLK, St, Mplier, Mcand, Product, Done);
    input CLK;
    input St;
    input[3:0] Mplier;
    input[3:0] Mcand;
    output[6:0] Product;
    output Done;

    reg[2:0] State;
    reg[3:0] A;
    reg[3:0] B;
    reg[3:0] addout;

    initial
    begin
        State = 0;
    end

    always @(posedge CLK)
    begin
        case (State)
            0 :
                begin
                    if (St == 1'b1)
                    begin
                        A <= 4'b0000 ;
                        B <= Mplier ;
```

```verilog
                                                State <= 1 ;
                                            end
                                        else
                                                State <= 0;
                                    end
                    1, 2, 3 :
                                    begin
                                        if (`M == 1'b1)
                                        begin
                                            addout = A + Mcand;
                                            A <= {Mcand[3], addout[3:1]} ;
                                            B <= {addout[0], B[3:1]} ;
                                        end
                                        else
                                        begin
                                            A <= {A[3], A[3:1]} ;
                                            B <= {A[0], B[3:1]} ;
                                        end
                                            State <= State + 1 ;
                                    end
                    4 :
                                    begin
                                        if (`M == 1'b1)
                                        begin
                                            addout = A + ~Mcand + 1;
                                            A <= {~Mcand[3], addout[3:1]} ;
                                            B <= {addout[0], B[3:1]} ;
                                        end
                                        else
                                        begin
                                            A <= {A[3], A[3:1]} ;
                                            B <= {A[0], B[3:1]} ;
                                        end
                                            State <= 5 ;
                                    end
                    5 :
                                    begin
                                        State <= 0 ;
                                    end

            default :
                                    begin
                                        State <= 0 ;
                                    end

            endcase
        end
    assign Done = (State == 5) ? 1'b1 : 1'b0 ;
    assign Product = {A[2:0], B} ;
endmodule
```

Although these statements are executed sequentially, A and B are both scheduled to be updated at the same delta time. Therefore, the old value of A[0] is used when computing the new value of B.

A register *addout* has been defined to represent the 5-bit output of the adder. In states 1 through 4, if the current multiplier bit M is 1, then the sign bit of the multiplicand followed by 3 bits of *addout* are loaded into A. At the same time, the low-order bit of *addout* is loaded into B along with the high-order 3 bits of B. The *Done* signal is turned on when control goes to state 5, and then the new value of the product is outputted.

Before continuing with the design, we will test the behavioral level Verilog code to make sure that the algorithm is correct and consistent with the hardware block diagram. At early stages of testing, we will want a step-by-step printout to verify the internal operations of the multiplier and to aid in debugging, if required. When we think that the multiplier is functioning properly, then we will only want to look at the final product's output so that we can quickly test a large number of cases.

Figure 4-36 shows the command file and test results for multiplying +5/8 by −3/8. A clock is defined with a 20-ns period. The *St* signal is turned on at 2 ns and turned off one clock period later. By inspection of the state graph, the multiplication requires six clocks, so the run time is set at 120 ns.

FIGURE 4-36: Command File and Simulation Results for (+5/8 by −3/8)

```
// command file to test signed multiplier
add list CLK St State A B Done Product
force St 1 2, 0 22
force CLK 1 0, 0 10 - repeat 20
// (5/8 * -3/8)
force Mcand  0101
force Mplier 1101
run 120

ns    delta CLK St State  A    B    Done Product
0     +1    1   0   0     0000 0000  0   0000000
2     +0    1   1   0     0000 0000  0   0000000
10    +0    0   1   0     0000 0000  0   0000000
20    +1    1   1   1     0000 1101  0   0000000
22    +0    1   0   1     0000 1101  0   0000000
30    +0    0   0   1     0000 1101  0   0000000
40    +1    1   0   2     0010 1110  0   0000000
50    +0    0   0   2     0010 1110  0   0000000
60    +1    1   0   3     0001 0111  0   0000000
70    +0    0   0   3     0001 0111  0   0000000
80    +1    1   0   4     0011 0011  0   0000000
90    +0    0   0   4     0011 0011  0   0000000
100   +2    1   0   5     1111 0001  1   1110001
110   +0    0   0   5     1111 0001  1   1110001
120   +1    1   0   0     1111 0001  0   1110001
```

To thoroughly test the multiplier, we need to test not only the four standard cases (+ +, + −, − +, and − −) but also special cases and limiting cases. Test values for the multiplicand and multiplier should include 0, the largest positive fraction, the most negative fraction, and all 1s. We will write a Verilog test bench to test the multiplier. The **test bench** will provide a sequence of values for the multiplicand and the multiplier. Thus, it provides stimuli to the system under test, the multiplier. The test bench can also check for the correctness of the multiplier output. The multiplier we are testing will be treated as a component and embedded in the test bench program. The signals generated within the test bench are interfaced to the multiplier as shown in Figure 4-37.

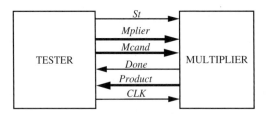

FIGURE 4-37: Interface between Multiplier and Its Test Bench

Figure 4-38 shows the Verilog code for the multiplier test bench. The test sequence consists of 11 sets of multiplicands and multipliers, provided in the *Mcandarr* and *Mplierarr* arrays. The expected outputs from the multiplier

FIGURE 4-38: Test Bench for Signed Multiplier

```
module testmult ();

   parameter N = 11;
   reg[3:0] Mcandarr[1:N];
   reg[3:0] Mplierarr[1:N];
   reg[6:0] Productarr[1:N];
   reg CLK;
   reg St;
   wire Done;
   reg[3:0] Mplier;
   reg[3:0] Mcand;
   wire[6:0] Product;
   integer i;

   initial
   begin
      CLK = 1'b1;
      Mcandarr[1] = 4'b0111;
      Mcandarr[2] = 4'b1101;
      Mcandarr[3] = 4'b0101;
      Mcandarr[4] = 4'b1101;
      Mcandarr[5] = 4'b0111;
      Mcandarr[6] = 4'b1000;
```

```
            Mcandarr[7]  = 4'b0111;
            Mcandarr[8]  = 4'b1000;
            Mcandarr[9]  = 4'b0000;
            Mcandarr[10] = 4'b1111;
            Mcandarr[11] = 4'b1011;
            Mplierarr[1]  = 4'b0101;
            Mplierarr[2]  = 4'b0101;
            Mplierarr[3]  = 4'b1101;
            Mplierarr[4]  = 4'b1101;
            Mplierarr[5]  = 4'b0111;
            Mplierarr[6]  = 4'b0111;
            Mplierarr[7]  = 4'b1000;
            Mplierarr[8]  = 4'b1000;
            Mplierarr[9]  = 4'b1101;
            Mplierarr[10] = 4'b1111;
            Mplierarr[11] = 4'b0000;
            Productarr[1]  = 7'b0100011;
            Productarr[2]  = 7'b1110001;
            Productarr[3]  = 7'b1110001;
            Productarr[4]  = 7'b0001001;
            Productarr[5]  = 7'b0110001;
            Productarr[6]  = 7'b1001000;
            Productarr[7]  = 7'b1001000;
            Productarr[8]  = 7'b1000000;
            Productarr[9]  = 7'b0000000;
            Productarr[10] = 7'b0000001;
            Productarr[11] = 7'b0000000;
        end
    always
    begin
        #10 CLK <= ~CLK ;
    end
    always @(posedge CLK)
    begin
        for(i = 1; i <= N; i = i + 1)
        begin
            Mcand  <= Mcandarr[i] ;
            Mplier <= Mplierarr[i] ;
            St <= 1'b1 ;
            @(posedge CLK);
            St <= 1'b0 ;
            @(negedge Done);
            if (~(Product == Productarr[i])) //compare with expected answer
                $display("Incorrect Product (error)");
        end
        $display("TEST COMPLETED (ERROR)");
    end
    mult2C mult1 (CLK, St, Mplier, Mcand, Product, Done);
endmodule
```

are provided in another array, the *Productarr*, in order to test the correctness of the multiplier outputs. The test values and results are placed in constant arrays in the Verilog code. The multiplier is instantiated and all signals are mapped with the test sequences.

```
mult2C mult1 (CLK, St, Mplier, Mcand, Product, Done);
```

The tester also generates the clock and start signal. The for loop reads values from the *Mcandarr* and *Mplierarr* arrays and then sets the start signal to 1. After the next clock, the start signal is turned off. Then, the test bench waits for the *Done* signal. When the trailing edge of *Done* arrives, the multiplier output is compared against the expected output in the array *Productarr*. An error is reported if the answers do not match. Since the *Done* signal is turned off at the same time the multiplier control goes back to S_0, the process waits for the falling edge of *Done* before looping back to supply new values of *Mcand* and *Mplier*. One may note that the **multiplier instatiation** is outside the always statement which generates the stimulus. The multiplier constantly receives some set of inputs and generates the corresponding set of outputs.

Figure 4-39 shows the command file and simulator output. We have annotated the simulator output to interpret the test results. The −NOtrigger together with the −Trigger done in the list statement causes the output to be displayed only when the *Done* signal changes. Without the −NOtrigger and −Trigger, the output would be displayed every time any signal on the list changed. All the product outputs are correct, except for the special case of −1 × −1 (1.000 × 1.000), which gives 1.000000 (−1) instead of +1. This occurs because no representation of +1 is possible without adding another bit.

FIGURE 4-39: Command File and Simulation of Signed Multiplier

```
// Command file to test results of signed multiplier
add list -NOtrigger Mplier Mcand Product -Trigger Done
run 1320

  ns    delta  Mplier Mcand  Product  Done
   0     +0    xxxx   xxxx   xxxxxxx   x
   0     +2    0101   0111   xxxxxxx   0
 100     +2    0101   0111   0100011   1    5/8 *  7/8 =  35/64
 120     +2    0101   1101   0100011   0
 220     +2    0101   1101   1110001   1    5/8 * -3/8 = -15/64
 240     +2    1101   0101   1110001   0
 340     +2    1101   0101   1110001   1   -3/8 *  5/8 = -15/64
 360     +2    1101   1101   1110001   0
 460     +2    1101   1101   0001001   1   -3/8 * -3/8 =   9/64
 480     +2    0111   0111   0001001   0
 580     +2    0111   0111   0110001   1    7/8 *  7/8 =  49/64
 600     +2    0111   1000   0110001   0
 700     +2    0111   1000   1001000   1    7/8 *  -1  =  -7/8
 720     +2    1000   0111   1001000   0
```

```
820    +2    1000  0111 1001000    1   -1 * 7/8 = -7/8
840    +2    1000  1000 1001000    0
940    +2    1000  1000 1000000    1   -1 * -1 = -1 (error)
960    +2    1101  0000 1000000    0
1060   +2    1101  0000 0000000    1   -3/8 * 0 = 0
1080   +2    1111  1111 0000000    0
1180   +2    1111  1111 0000001    1   -1/8 * -1/8 = 1/64
1200   +2    0000  1011 0000001    0
1300   +2    0000  1011 0000000    1   0 * -3/8 = 0
1320   +2    0000  1011 0000000    0
```

Next, we refine the Verilog model for the signed multiplier by explicitly defining the control signals and the actions that occur when each control signal is asserted. The Verilog code (Figure 4-40) is organized in a manner similar to the Mealy machine model of Figure 1-17. In the first process, the *Nextstate* and output control signals are defined for each present *State*. In the second process, after waiting for the rising edge of the clock, the appropriate registers are updated and the *State* is updated. We can test the Verilog code of Figure 4-40 using the same test file we used previously and verify that we get the same product outputs.

FIGURE 4-40: Model for 2's Complement Multiplier with Control Signals

```verilog
`define M B[0]

// This Verilog model explicitly defines control signals.
module mult2C2 (CLK, St, Mplier, Mcand, Product, Done);

    input CLK;
    input St;
    input[3:0] Mplier;
    input[3:0] Mcand;
    output[6:0] Product;
    output Done;

    reg Done;
    reg[2:0] State;
    reg[2:0] Nextstate;
    reg[3:0] A;
    reg[3:0] B;
    wire[3:0] compout;
    wire[3:0] addout;
    reg AdSh;
    reg Sh;
    reg Load;
    reg Cm;

    always @(State or St or `M)
    begin
        Load = 1'b0 ;
```

```verilog
                AdSh = 1'b0 ;
                Sh = 1'b0 ;
                Cm = 1'b0 ;
                Done = 1'b0 ;
                Nextstate = 1'b0;
                case (State)
                    0 :
                                    begin
                                        if (St == 1'b1)
                                        begin
                                            Load = 1'b1 ;
                                            Nextstate = 1 ;
                                        end
                                        else
                                        begin
                                            Load = 1'b0 ;
                                            Nextstate = 0 ;
                                        end
                                    end
                    1, 2, 3 :
                                    begin
                                        if (`M == 1'b1)
                                        begin
                                          AdSh = 1'b1 ;
                                        end
                                        else
                                        begin
                                            Sh = 1'b1 ;
                                        end
                                        Nextstate = State + 1 ;
                                    end
                    4 :
                                    begin
                                        if (`M == 1'b1)
                                        begin
                                            Cm = 1'b1 ;
                                            AdSh = 1'b1 ;
                                        end
                                        else
                                        begin
                                            Sh = 1'b1 ;
                                        end
                                        Nextstate = 5 ;
                                    end
                    5 :
                                    begin
                                        Done = 1'b1 ;
                                        Nextstate = 0 ;
                                    end
                    default:
```

```verilog
                    begin
                        Done = 1'b0 ;
                        Nextstate = 0 ;
                    end
            endcase
    end

    assign compout = (Cm == 1'b1) ? ~Mcand : Mcand ;
    assign addout = A + compout + Cm ;

    always @(posedge CLK)
    begin
        if (Load == 1'b1)
        begin
            A <= 4'b0000 ;
            B <= Mplier ;
        end
        if (AdSh == 1'b1)
        begin
            A <= {compout[3], addout[3:1]} ;
            B <= {addout[0], B[3:1]} ;
        end
        if (Sh == 1'b1)
        begin
            A <= {A[3], A[3:1]} ;
            B <= {A[0], B[3:1]} ;
        end
        State <= Nextstate ;
    end

    assign Product = {A[2:0], B} ;
endmodule
```

4.11 Keypad Scanner

In this example, we design a scanner for a keypad with three columns and four rows as in Figure 4-41. The keypad is wired in matrix form with a switch at the intersection of each row and column. Pressing a key establishes a connection between a row and column. The purpose of the scanner is to determine which key has been pressed and to output a binary number $N = N_3N_2N_1N_0$, which corresponds to the key number. For example, pressing key 5 must output 0101, pressing the * key must output 1010, and pressing the # key must output 1011. When a valid key has been detected, the scanner should output a signal V for one clock time. Assume that only one key

FIGURE 4-41: Keypad with Three Columns and Four Rows

1	2	3
4	5	6
7	8	9
*	0	#

is pressed at a time. The design must include hardware to protect the circuitry from malfunction due to keypad bounces.

The overall block diagram of the circuit is presented in Figure 4-42. The keypad contains resistors that are connected to ground. When a switch is pressed, a path is established from the corresponding column line to the ground. If a voltage can be applied on the column lines C_0, C_1, and C_2, then the voltage can be obtained on the row line corresponding to the key that is pressed. One among the rows R_0, R_1, R_2, or R_3 will have an active signal.

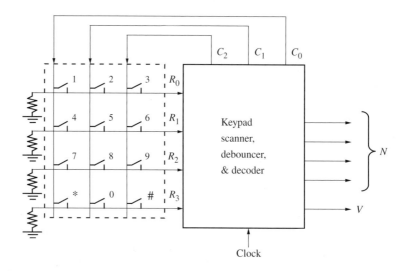

FIGURE 4-42: Block Diagram for Keypad Scanner

We will divide the design into several modules, as shown in Figure 4-43. The first part of the design will be a scanner that scans the rows and columns of the keypad. The keyscan module generates the column signals to scan the keyboard. The debounce module generates a signal K when a key has been pressed and a signal Kd after it has been debounced. When a valid key is detected, the decoder determines the key number from the row and column numbers.

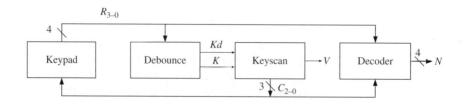

FIGURE 4-43: Scanner Modules

Scanner

We will use the following procedure to scan the keyboard: First apply logic 1s to columns C_0, C_1, and C_2 and wait. If any key is pressed, a 1 will appear on R_0, R_1, R_2, or R_3. Then apply a 1 to column C_0 only. If any of the R_is is 1, a valid key is detected. If R_0 is received, one knows that switch 1 was pressed. If R_1, R_2, or R_3 is received, it indicates switch 4, 7, or * was pressed. If so, set $V = 1$ and output the corresponding N. If no key is detected in the first column, apply a 1 to C_1 and repeat. If no key is detected in the second column, repeat for C_2. When a valid key is detected, apply 1_s to C_0, C_1, and C_2 and wait until no key is pressed. This last step is necessary so that only one valid signal is generated each time a key is pressed.

Debouncer

As discussed in the scoreboard example, we need to debounce the keys to avoid malfunctions due to switch bounce. Figure 4-44 shows a proposed debouncing and synchronizing circuit. The four row signals are connected to an OR gate to form signal K, which turns on when a key is pressed and a column scan signal is applied. The debounced signal Kd will be fed to the sequential circuit.

FIGURE 4-44: Debouncing and Synchronizing Circuit

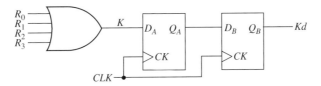

Decoder

The decoder determines the key number from the row and column numbers using the truth table given in Table 4-4. The truth table has one row for each of the 12 keys. The remaining rows have don't care outputs since we have assumed that only one key is pressed at a time. Since the decoder is a combinational circuit, its output will

TABLE 4-4: Truth Table for Decoder

R_3	R_2	R_1	R_0	C_0	C_1	C_2	N_3	N_2	N_1	N_0	
0	0	0	1	1	0	0	0	0	0	1	
0	0	0	1	0	1	0	0	0	1	0	
0	0	0	1	0	0	1	0	0	1	1	
0	0	1	0	1	0	0	0	1	0	0	
0	0	1	0	0	1	0	0	1	0	1	
0	0	1	0	0	0	1	0	1	1	0	
0	1	0	0	1	0	0	0	1	1	1	
0	1	0	0	0	1	0	1	0	0	0	
0	1	0	0	0	0	1	1	0	0	1	
1	0	0	0	1	0	0	1	0	1	0	(*)
1	0	0	0	0	1	0	0	0	0	0	
1	0	0	0	0	0	1	1	0	1	1	(#)

Logic Equations for Decoder

$N_3 = R_2 C_0' + R_3 C_1'$

$N_2 = R_1 + R_2 C_0$

$N_1 = R_0 C_0' + R_2' C_2 + R_1' R_0' C_0$

$N_0 = R_1 C_1 + R_1' C_2 + R_3' R_1' C_1'$

change as the keypad is scanned. At the time a valid key is detected ($K = 1$ and $V = 1$), its output will have the correct value and this value can be saved in a register at the same time the circuit goes to S_5.

Controller

Figure 4-45 shows the state diagram of the controller for the keypad scanner. It waits in S_1 with outputs $C_0 = C_1 = C_2 = 1$ until a key is pressed. In S_2, $C_0 = 1$, so if the key that was pressed is in column 0, $K = 1$ and the circuit outputs a valid signal and goes to S_5. Signal K is used instead of Kd, since the key press is already debounced. If no key press is found in column 0, column 1 is checked in S_3, and if necessary, column 2 is checked in S_4. In S_5, the circuit waits until all keys are released and Kd goes to 0 before resetting.

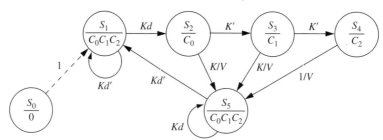

FIGURE 4-45: State Graph for Keypad Scanner

The foregoing state diagram as presented works for many cases, but it does have some timing problems. Let us analyze the following situations.

1. **Is K true whenever a button is pressed?**
   No. Although K is true if any one of the row signals R_1, R_2, R_3, or R_4 is true, if the column scan signals are not active, none of R_1–R_4 can be true, although the button is pressed.

2. **Can Kd be false when a button is continuing to be pressed?**
   Yes. Signal Kd is nothing but K delayed by two clock cycles. K can go to 0 during the scan process even when the button is being pressed. For instance, consider the case when a key in the rightmost column is pressed. During scan of the first two columns, K goes to 0. If K goes to 0 at any time, Kd will go to zero two cycles later. Hence, neither K nor Kd is the same as pressing the button.

3. **Can you go from S_5 to S_1 when a button is still pressed?**
   In the state diagram in Figure 4-45, the S_4-to-S_5 transition could happen when Kd is false. Kd might have become false while scanning C_0 and C_1. Hence, it is possible that one reaches back to S_1 when the key is still being pressed. As an example, let us assume that a button is pressed in column C_2. This is to be detected in S_4. However, during the scanning process in S_2 and S_3, K is 0; hence, two cycles later Kd will be 0 even if the button stays pressed. During the scan in S_4, the correct key can be found; however, the system can reach S_5 when Kd is still 0 and a malfunction can happen. S_5 is intended to sense the release of the key. However, Kd is not synonymous to pressing the button and Kd' does not truly indicate that the button got released. Since Kd' can appear when the button is still pressed, if you reach S_5 when Kd' is true due to scanning activity in a previous state, the system

4.11 Keypad Scanner

can go from S_5 to S_1 without a key release. In such a case, the same key may be read multiple times.

4. **What if a key is pressed for only one or two clock cycles?**
   If the key is pressed and released very quickly, there would be problems especially if the key is in the third column. By the time the scanner reaches state S_4, the key might have been released already. The key should be pressed long enough for the scanner to go through the longest path in the state graph from S_0 to S_5. This may not be a serious problem, because usually, the digital system clock is much faster than any mechanical switch.

   These problems can be fixed by assuring that one can reach S_5 only if Kd is true. A modified state diagram is presented in Figure 4-46. Before transitioning to state S_5, this circuit waits in state S_2, S_3, and S_4 until Kd also becomes 1.

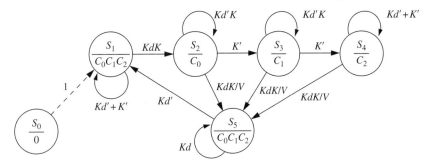

FIGURE 4-46: Modified State Graph for Keypad Scanner

Verilog Code

The Verilog code used to implement the design is shown in Figure 4-47. The decoder equations as well as the equations for K and V are implemented by concurrent statements. The process implements the next state equations for the keyscan and debounce flip-flops.

FIGURE 4-47: Verilog Code for Scanner

```verilog
module scanner (R0, R1, R2, R3, CLK, C0, C1, C2, N0, N1, N2, N3, V);
    input R0;
    input R1;
    input R2;
    input R3;
    input CLK;
    inout C0;
    inout C1;
    inout C2;
    output N0;
    output N1;
    output N2;
    output N3;
    output V;

    reg V;
    reg C0_tmp, C1_tmp, C2_tmp;
```

260 Chapter 4 Design Examples

```verilog
    reg QA;
    wire K;
    reg Kd;
    reg[2:0] state;
    reg[2:0] nextstate;
assign C0 = C0_tmp;
assign C1 = C1_tmp;
assign C2 = C2_tmp;

assign K  = R0 | R1 | R2 | R3 ;
assign N3 = (R2 & ~C0) | (R3 & ~C1) ;
assign N2 = R1 | (R2 & C0) ;
assign N1 = (R0 & ~C0) | (~R2 & C2) | (~R1 & ~R0 & C0) ;
assign N0 = (R1 & C1) | (~R1 & C2) | (~R3 & ~R1 & ~C1) ;
initial
begin
   state = 0;
   nextstate = 0;
end

always @(state or R0 or R1 or R2 or R3 or C0 or C1 or C2 or K or Kd or QA)
begin
  C0_tmp = 1'b0 ;
  C1_tmp = 1'b0 ;
  C2_tmp = 1'b0 ;
  V = 1'b0 ;
  case (state)
     0 :
                 begin
                     nextstate = 1 ;
                 end
       1 :
                 begin
                    C0_tmp = 1'b1 ;
                    C1_tmp = 1'b1 ;
                    C2_tmp = 1'b1 ;
                     if ((Kd & K) == 1'b1)
                     begin
                         nextstate = 2 ;
                     end
                     else
                     begin
                         nextstate = 1 ;
                     end
                 end
       2 :
                 begin
                    C0_tmp = 1'b1 ;
                     if ((Kd & K) == 1'b1)
                     begin
                         V = 1'b1 ;
```

```verilog
                        nextstate = 5 ;
                    end
                else if (K == 1'b0)
                    begin
                        nextstate = 3 ;
                    end
                else
                begin
                    nextstate = 2 ;
                end
            end
3 :
            begin
                C1_tmp = 1'b1 ;
                if ((Kd & K) == 1'b1)
                begin
                    V = 1'b1 ;
                    nextstate = 5 ;
                end
                else if (K == 1'b0)
                begin
                    nextstate = 4 ;
                end
                else
                begin
                    nextstate = 3 ;
                end
            end
4 :
            begin
                C2_tmp <= 1'b1 ;
                if ((Kd & K) == 1'b1)
                begin
                    V <= 1'b1 ;
                    nextstate = 5 ;
                end
                else
                begin
                    nextstate = 4 ;
                end
            end
5 :
            begin
                C0_tmp = 1'b1 ;
                C1_tmp = 1'b1 ;
                C2_tmp = 1'b1 ;
                if (Kd == 1'b0)
                 begin
                    nextstate = 1 ;
                end
                else
```

```
                        begin
                            nextstate = 5 ;
                        end
                end
    endcase
end
always @(posedge CLK)
begin
        state <= nextstate ;
        QA <= K ;
        Kd <= QA ;
end
endmodule
```

Test Bench for Keypad Scanner

This Verilog code would be very difficult to test by supplying waveforms for the inputs R_0, R_1, R_2, and R_3, since these inputs depend on the column outputs (C_0, C_1, C_2). A much better way to test the scanner is by using a test bench in Verilog. The scanner we are testing will be treated as a component and embedded in the test bench program. The signals generated within the test bench are interfaced to the scanner as shown in Figure 4-48. The test bench simulates a key press by supplying the appropriate R signals in response to the C signals from the scanner. When test bench receives $V = 1$ from the scanner, it checks to see whether the value of N corresponds to the key that was pressed.

FIGURE 4-48: Interface for Test Bench

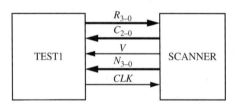

The Verilog code for the keypad test bench is shown in Figure 4-49. A copy of the scanner is instantiated within the TEST1 module, and connections to the scanner are made by the input/output mapping. The sequence of key numbers used for testing is stored in the array *KARRAY*. The tester simulates the keypad operation using concurrent statements for R_0, R_1, R_2, and R_3. Whenever C_0, C_1, C_2, or the key number (KN) changes, new values for the Rs are computed. For example, if $KN = 5$ (to simulate pressing key 5), then $R_0R_1R_2R_3 = 0100$ is sent to the scanner when $C_0C_1C_2 = 010$. The test process is as follows:

1. Read a key number from the array to simulate pressing a key.
2. Wait until $V = 1$ and the rising edge of the clock occurs.
3. Verify that the N output from the scanner matches the key number.
4. Set $KN = 15$ to simulate no key pressed. (Since 15 is not a valid key number, all Rs will go to 0.)
5. Wait until $Kd = 0$ before selecting a new key.

FIGURE 4-49: Verilog for Scanner Test Bench

```verilog
module scantest ();

   integer KARRAY[0:23];
   wire C0;
   wire C1;
   wire C2;
   wire V;
   reg CLK;
   wire R0;
   wire R1;
   wire R2;
   wire R3;
   wire[3:0] N;
   integer KN;
   integer i;

   initial
   begin
       CLK = 0;
      KARRAY[0] = 2;
      KARRAY[1] = 5;
      KARRAY[2] = 8;
      KARRAY[3] = 0;
      KARRAY[4] = 3;
      KARRAY[5] = 6;
      KARRAY[6] = 9;
      KARRAY[7] = 11;
      KARRAY[8] = 1;
      KARRAY[9] = 4;
      KARRAY[10] = 7;
      KARRAY[11] = 10;
      KARRAY[12] = 1;
      KARRAY[13] = 2;
      KARRAY[14] = 3;
      KARRAY[15] = 4;
      KARRAY[16] = 5;
      KARRAY[17] = 6;
      KARRAY[18] = 7;
      KARRAY[19] = 8;
      KARRAY[20] = 9;
      KARRAY[21] = 10;
      KARRAY[22] = 11;
      KARRAY[23] = 0;
   end

   always #20 CLK = ~CLK ;
   assign R0 = ((C0 == 1'b1 & KN == 1) | (C1 == 1'b1 & KN == 2)
             | (C2 == 1'b1 & KN == 3)) ? 1'b1 : 1'b0 ;
   assign R1 = ((C0 == 1'b1 & KN == 4) | (C1 == 1'b1 & KN == 5)
             | (C2 == 1'b1 & KN == 6)) ? 1'b1 : 1'b0 ;
```

```verilog
    assign R2 = ((C0 == 1'b1 & KN == 7) | (C1 == 1'b1 & KN == 8)
                  | (C2 == 1'b1 & KN == 9)) ? 1'b1 : 1'b0 ;
    assign R3 = ((C0 == 1'b1 & KN == 10) | (C1 == 1'b1 & KN == 0)
                  | (C2 == 1'b1 & KN == 11)) ? 1'b1 : 1'b0 ;
    always @(posedge CLK)
    begin
        for(i = 0; i <= 23; i = i + 1)
        begin
            KN <= KARRAY[i] ;
            @(posedge CLK);
            if (V == 1'b1)
                if (~(N == KN)) $display("Numbers don't match.");
            KN <= 15 ;
            @(posedge CLK);
            @(posedge CLK);
            @(posedge CLK);
        end
        $display("Test Complete.");
    end
    scanner scanner1(R0, R1, R2, R3, CLK, C0, C1, C2, N[0],N[1],
    N[2],N[3],V);
endmodule
```

Key presses in row order and column order are tried using the various numbers in KARRAY. The test bench will test whether the reported number matches the key pressed. The `display` statement will report "Numbers don't match" if the scanner generates the wrong key number, and it will report "Testing Complete." when all keys have been tested.

4.12 Binary Dividers

Unsigned Divider

We will consider the design of a parallel divider for positive binary numbers. As an example, we will design a circuit to divide an 8-bit dividend by a 4-bit divisor to obtain a 4-bit quotient. The following example illustrates the division process:

```
                              1010       quotient
            Divisor   1101 / 10000111    dividend
                              1101
    (135 ÷ 13 = 10 with        0111
      a remainder of 5)        0000
                               1111
                               1101
                                0101
                                0000
                                0101    Remainder
```

4.12 Binary Dividers

Just as binary multiplication can be carried out as a series of add and shift operations, division can be carried out by a series of subtract and shift operations. To construct the divider, we will use a 9-bit dividend register and a 4-bit divisor register, as shown in Figure 4-50. During the division process, instead of shifting the divisor right before each subtraction, we will shift the dividend to the left. Note that an extra bit is required on the left end of the dividend register so that a bit is not lost when the dividend is shifted left. Instead of using a separate register to store the quotient, we will enter the quotient bit by bit into the right end of the dividend register as the dividend is shifted left.

FIGURE 4-50: Block Diagram for Parallel Binary Divider

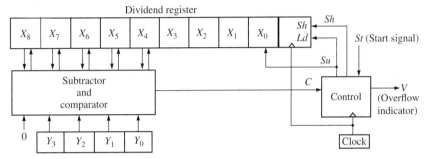

The preceding division example (135 divided by 13) is reworked next, showing the location of the bits in the registers at each clock time. Initially, the dividend and divisor are entered as follows:

0	1	0	0	0	0	1	1	1

1	1	0	1

Subtraction cannot be carried out without a negative result, so we will shift before we subtract. Instead of shifting the divisor one place to the right, we will shift the dividend one place to the left.

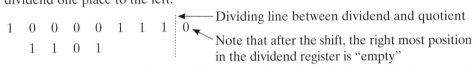

Subtraction is now carried out and the first quotient digit of 1 is stored in the unused position of the dividend register:

0 0 0 1 1 1 1 ┊ 1 ◄─ first quotient digit

Next we shift the dividend one place to the left.

0 0 1 1 1 1 1 ┊ 1 0
         1 1 0 1 ┊

Since subtraction would yield a negative result, we shift the dividend to the left again, and the second quotient bit remains zero.

0 1 1 1 1 1 ┊ 1 0 0
         1 1 0 1 ┊

Subtraction is now carried out, and the third quotient digit of 1 is stored in the unused position of the dividend register.

0 0 0 1 0 1 ┊ 1 0 1 ◄─ third quotient digit

A final shift is carried out and the fourth quotient bit is set to 0.

$$\underbrace{0 \quad 0 \quad 1 \quad 0 \quad 1}_{\text{remainder}} \quad | \quad \underbrace{1 \quad 0 \quad 1 \quad 0}_{\text{quotient}}$$

The final result agrees with that obtained in the first example.

If, as a result of a division operation, the quotient contains more bits than are available for storing the quotient, we say that an *overflow* has occurred. For the divider of Figure 4-50, an overflow would occur if the quotient is greater than 15, since only 4 bits are provided to store the quotient. It is not actually necessary to carry out the division to determine whether an overflow condition exists, since an initial comparison of the dividend and divisor will tell if the quotient will be too large. For example, if we attempt to divide 135 by 7, the initial contents of the registers are

$$0 \quad 1 \quad 0 \quad 0 \quad 0 \quad 0 \quad 1 \quad 1 \quad 1$$
$$ 0 \quad 1 \quad 1 \quad 1$$

Since subtraction can be carried out with a nonnegative result, we should subtract the divisor from the dividend and enter a quotient bit of 1 in the rightmost place in the dividend register. However, we cannot do this because the rightmost place contains the least significant bit of the dividend, and entering a quotient bit here would destroy that dividend bit. Therefore, the quotient would be too large to store in the 4 bits we have allocated for it, and we have detected an overflow condition. In general, for Figure 4-50, if initially $X_8X_7X_6X_5X_4 \geq Y_3Y_2Y_1Y_0$ (i.e., if the left 5 bits of the dividend register exceed or equal the divisor), the quotient will be greater than 15 and an overflow occurs. Note that if $X_8X_7X_6X_5X_4 \geq Y_3Y_2Y_1Y_0$, the quotient is

$$\frac{X_8X_7X_6X_5X_4X_3X_2X_1X_0}{Y_3Y_2Y_1Y_0} \geq \frac{X_8X_7X_6X_5X_4 0000}{Y_3Y_2Y_1Y_0} = \frac{X_8X_7X_6X_5X_4 \times 16}{Y_3Y_2Y_1Y_0} \geq 16$$

The operation of the divider can be explained in terms of the block diagram of Figure 4-50. A shift signal (Sh) will shift the dividend one place to the left. A subtract signal (Su) will subtract the divisor from the 5 leftmost bits in the dividend register and set the quotient bit (the rightmost bit in the dividend register) to 1. If the divisor is greater than the 5 leftmost dividend bits, the comparator output is $C = 0$; otherwise, $C = 1$. Whenever $C = 0$, subtraction cannot occur without a negative result, so a shift signal is generated. Whenever $C = 1$, a subtract signal is generated, and the quotient bit is set to 1. The control circuit generates the required sequence of shift and subtract signals.

Figure 4-51 shows the state diagram for the control circuit. When a start signal (St) occurs, the 8-bit dividend and 4-bit divisor are loaded into the appropriate registers. If C is 1, the quotient would require 5 or more bits. Since space is provided only for a 4-bit quotient, this condition constitutes an overflow, so the divider is stopped and the overflow indicator is set by the V output. Normally, the initial value of C is 0, so a shift will occur first and the control circuit will go to state S_2. Then, if $C = 1$, subtraction occurs. After the subtraction is completed, C will always be 0, so the next clock pulse will produce a shift. This process continues until four shifts

FIGURE 4-51: State Diagram for Divider Control Circuit

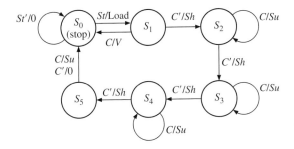

have occurred and the control is in state S_5. Then a final subtraction occurs, if necessary, and the control returns to the stop state. For this example, we will assume that when the start signal (St) occurs, it will be 1 for one clock time and then it will remain 0 until the control circuit is back in state S_0. Therefore, St will always be 0 in states S_1 through S_5.

Table 4-5 gives the state table for the control circuit. Since we assumed that $St = 0$ in states S_1, S_2, S_3, and S_4, the next states and outputs are "don't cares" for these states when $St = 1$. The entries in the output table indicate which outputs are 1. For example, the entry Sh means $Sh = 1$ and the other outputs are 0.

TABLE 4-5: State Table for Divider Control Circuit

State	StC 00	01	11	10	StC 00	01	11	10
S_0	S_0	S_0	S_1	S_1	0	0	Load	Load
S_1	S_2	S_0	–	–	Sh	V	–	–
S_2	S_3	S_2	–	–	Sh	Su	–	–
S_3	S_4	S_3	–	–	Sh	Su	–	–
S_4	S_5	S_4	–	–	Sh	Su	–	–
S_5	S_0	S_0	–	–	0	Su	–	–

This example illustrates a general method for designing a divider for unsigned binary numbers, and the design can easily be extended to larger numbers such as 16 bits divided by 8 bits or 32 bits divided by 16 bits. Using a separate counter to count the number of shifts is recommended if more than four shifts are required.

Signed Divider

We now design a divider for signed (2's complement) binary numbers that divides a 32-bit dividend by a 16-bit divisor to give a 16-bit quotient. Although algorithms exist to divide the signed numbers directly, such algorithms are rather complex. So we take the easy way out and complement the dividend and divisor if they are negative; when division is complete, we complement the quotient if it should be negative.

Figure 4-52 shows a block diagram for the divider. We use a 16-bit bus to load the registers. Since the dividend is 32 bits, two clocks are required to load the upper and lower halves of the dividend register, and one clock is needed to load

FIGURE 4-52: Block Diagram for Signed Divider

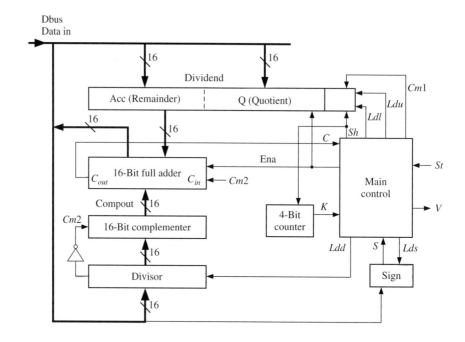

the divisor. An extra sign flip-flop is used to store the sign of the dividend. We will use a dividend register with a built-in 2's complementer. The subtractor consists of an adder and a complementer, so subtraction can be accomplished by adding the 2's complement of the divisor to the dividend register. If the divisor is negative, using a separate step to complement it is unnecessary; we can simply disable the complementer and add the negative divisor instead of subtracting its complement. The control circuit is divided into two parts—a main control, which determines the sequence of shifts and subtracts, and a counter, which counts the number of shifts. The counter outputs a signal $K = 1$ when 15 shifts have occurred. Control signals are defined as follows:

LdU	Load upper half of dividend from bus.
LdL	Load lower half of dividend from bus.
Lds	Load sign of dividend into sign flip-flop.
S	Sign of dividend.
Cm1	Complement dividend register (2's complement).
Ldd	Load divisor from bus.
Su	Enable adder output onto bus (*Ena*), and load upper half of dividend from bus.
Cm2	Enable complementer. (*Cm2* equals the complement of the sign bit of the divisor, so a positive divisor is complemented and a negative divisor is not.)
Sh	Shift the dividend register left one place and increment the counter.

C	Carry output from adder. (If $C = 1$, the divisor can be subtracted from the upper dividend.)
St	Start.
V	Overflow.
Qneg	Quotient will be negative. ($Qneg = 1$ when the sign of the dividend and divisor are different.)

The procedure for carrying out the signed division is as follows:

1. Load the upper half of the dividend from the bus, and copy the sign of the dividend into the sign flip-flop.
2. Load the lower half of the dividend from the bus.
3. Load the divisor from the bus.
4. Complement the dividend if it is negative.
5. If an overflow condition is present, go to the done state.
6. Else carry out the division by a series of shifts and subtracts.
7. When division is complete, complement the quotient if necessary and go to the done state.

Testing for overflow is slightly more complicated than for the case of unsigned division. First, consider the case of all positive numbers. Since the divisor and quotient are each 15 bits plus sign, their maximum value is 7FFFh. Since the remainder must be less than the divisor, its maximum value is 7FFEh. Therefore, the maximum dividend for no overflow is

$$\text{divisor} \times \text{quotient} + \text{remainder} = 7FFFh \times 7FFFh + 7FFEh = 3FFF7FFFh$$

If the dividend is 1 larger (3FFF8000h), division by 7FFFh (or anything smaller) will give an overflow. We can test for the overflow condition by shifting the dividend left one place and then comparing the upper half of the dividend (divu) with the divisor. If divu $\geq$ divisor, the quotient would be greater than the maximum value, which is an overflow condition. For the preceding example, shifting 3FFF8000h left once gives 7FFF0000h. Since 7FFFh equals the divisor, there is an overflow. On the other hand, shifting 3FFF7FFFh left gives 7FFEFFFEh, and since 7FFEh < 7FFFh, no overflow occurs when dividing by 7FFFh.

Another way of verifying that we must shift the dividend left before testing for overflow is as follows. If we shift the dividend left one place and then divu $\geq$ divisor, we could subtract and generate a quotient bit of 1. However, this bit would have to go in the sign bit position of the quotient. This would make the quotient negative, which is incorrect. After testing for overflow, we must shift the dividend left again, which gives a place to store the first quotient bit after the sign bit. Since we work with the complement of a negative dividend or a negative divisor, this method for detecting overflow will work for negative numbers, except for the special case where the dividend is 80000000h (the largest negative value). Modifying the design to detect overflow in this case is left as an exercise.

Figure 4-53 shows the state graph for the control circuit. When $St = 1$, the registers are loaded. In S_2, if the sign of the dividend (S) is 1, the dividend is

FIGURE 4-53: State Graph for Signed Divider Control Circuit

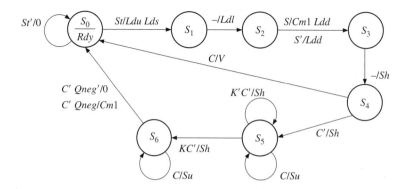

complemented. In S_3, we shift the dividend left one place and then we test for overflow in S_4. If $C = 1$, subtraction is possible, which implies an overflow, and the circuit goes to the done state. Otherwise, the dividend is shifted left. In S_5, C is tested. If $C = 1$, then $Su = 1$, which implies Ldu and Ena, so the adder output is enabled onto the bus and loaded into the upper dividend register to accomplish the subtraction. Otherwise, $Sh = 1$ and the dividend register is shifted. This continues until $K = 1$, at which time the last shift occurs if $C = 0$, and the circuit goes to S_6. Then if the sign of the divisor and the saved sign of the dividend are different, the dividend register is complemented so that the quotient will have the correct sign.

The Verilog code for the signed divider is shown in Figure 4-54. Since the 1's complementer and adder are combinational circuits, we have represented their operation by concurrent statements. All the signals that represent register outputs are updated on the rising edge of the clock, so these signals are updated in the process after waiting for CLK to change to 1. The counter is simulated by a signal, *count*. For convenience in listing the simulator output, we have added a ready signal (Rdy), which is turned on in S_0 to indicate that the division is completed.

FIGURE 4-54: Verilog Model of 32-Bit Signed Divider

```
`define Acc Dividend[31:16]

module sdiv (CLK, St, Dbus, Quotient, Remainder, V, Rdy);

    input CLK;
    input St;
    input[15:0] Dbus;
    output[15:0] Quotient;
    output[15:0] Remainder;
    output V;
    output Rdy;

    reg V;
```

4.12 Binary Dividers

```verilog
reg[2:0] State;
reg[3:0] Count;
reg Sign;
wire C;
wire Cm2;
reg[15:0] Divisor;
wire[15:0] Sum;
wire[15:0] Compout;
reg[31:0] Dividend;

assign Cm2 = ~Divisor[15] ;
assign Compout = (Cm2 == 1'b0) ? Divisor : ~Divisor ;
assign Sum = `Acc + Compout + Cm2;
assign C = ~Sum[15] ;
assign Quotient = Dividend[15:0] ;
assign Remainder = Dividend[31:16];
assign Rdy = (State == 0) ? 1'b1 : 1'b0 ;

initial
begin
   State = 0;
end

always @(posedge CLK)
begin
     case (State)
        0 :
                  begin
                     if (St == 1'b1)
                     begin
                         `Acc <= Dbus ;
                         Sign <= Dbus[15] ;
                         State <= 1 ;
                         V <= 1'b0 ;
                         Count <= 4'b0000 ;
                     end
                  end
        1 :
                  begin
                     Dividend[15:0] <= Dbus ;
                     State <= 2 ;
                  end
        2 :
                  begin
                     Divisor <= Dbus ;
                     if (Sign == 1'b1)
                     begin
                         Dividend <= ~Dividend + 1 ;
                     end
                     State <= 3 ;
                  end
```

```verilog
            3 :
                    begin
                        Dividend <= {Dividend[30:0], 1'b0} ;
                        Count <= Count + 1 ;
                        State <= 4 ;
                    end
            4 :
                    begin
                        if (C == 1'b1)
                        begin
                            V <= 1'b1 ;
                            State <= 0 ;
                        end
                        else
                        begin
                            Dividend <= {Dividend[30:0], 1'b0} ;
                            Count <= Count + 1 ;
                            State <= 5 ;
                        end
                    end
            5 :
                    begin
                        if (C == 1'b1)
                        begin
                            `Acc <= Sum ;
                            Dividend[0] <= 1'b1 ;
                        end
                        else
                        begin
                            Dividend <= {Dividend[30:0], 1'b0} ;
                            if (Count == 15)
                            begin
                                State <= 6 ;
                            end
                            Count <= Count + 1 ;
                        end
                    end
            6 :
                    begin
                        State <= 0 ;
                        if (C == 1'b1)
                        begin
                            `Acc <= Sum ;
                            Dividend[0] <= 1'b1 ;
                            State <= 6 ;
                        end
                        else if ((Sign ^ Divisor[15]) == 1'b1)
                        begin
                            Dividend[15:0] <= ~Dividend[15:0] + 1 ;
                            if(Sign == 1)
                            begin
```

```
                            Dividend [31:16] <= ~Dividend[31:16] + 1;
                         end
                    end
                    else begin
                         if(Sign && Divisor[15])
                         begin
                            Dividend [31:16] <= ~Dividend[31:16] + 1;
                         end
                    end
               end
          endcase
     end

endmodule
```

We are now ready to test the divider design by using the Verilog simulator. We will need a comprehensive set of test examples that will test all the different special cases that can arise in the division process. To start with, we need to test the basic operation of the divider for all the different combinations of signs for the divisor and dividend ($+\ +, +\ -, -\ +,$ and $-\ -$). We also need to test the overflow detection for these four cases. Limiting cases must also be tested, including largest quotient, zero quotient, and so forth. Use of a Verilog test bench is convenient, because the test data must be supplied in sequence at certain times and the length of time to complete the division is dependent on the test data. Figure 4-55 shows a test bench for the divisor. The test bench contains a dividend array and a divisor array for the test data. The notation 32'h07FF00BB is the hexadecimal representation of a bit string. The process in testsdiv first puts the upper dividend on *Dbus* and supplies a start signal. After waiting for the clock, it puts the lower dividend on *Dbus*. After the next clock, it puts the divisor on *Dbus*. It then waits until the *Rdy* signal indicates that division is complete before continuing. *Count* is set equal to the loop-index, so that the change in *Count* can be used to trigger the listing output.

FIGURE 4-55: Test Bench for Signed Divider

```
module testsdiv;
  parameter N = 12;
  reg[31:0] dividendarr[1:N];
  reg[15:0] divisorarr[1:N];

  //inputs to sdiv module should be reg types
  reg CLK;
  reg St;
  reg[15:0] Dbus;

  //outputs from sdiv module should be wire types
  wire[15:0] Quotient;
  wire[15:0] Remainder;
  wire V;
  wire Rdy;
```

```verilog
    reg[15:0] Divisor;
    reg[31:0] Dividend;
    reg[3:0] Count;
    reg[31:0] dividendarr_tmp;
    reg [15:0] quotientarr[1:N];
    reg [15:0] remainderarr[1:N];

    integer i;

    always
    begin
      #10 CLK <= ~CLK;
    end

    initial
    begin
      //initialization of dividend array
      dividendarr[1]  = 32'h0000006F;
      dividendarr[2]  = 32'h07FF00BB;
      dividendarr[3]  = 32'hFFFFFE08;
      dividendarr[4]  = 32'hFF80030A;
      dividendarr[5]  = 32'h3FFF8000;
      dividendarr[6]  = 32'h3FFF7FFF;
      dividendarr[7]  = 32'hC0008000;
      dividendarr[8]  = 32'hC0008000;
      dividendarr[9]  = 32'hC0008001;
      dividendarr[10] = 32'h00000000;
      dividendarr[11] = 32'hFFFFFFFF;
      dividendarr[12] = 32'hFFFFFFFF;

      //initialization of divisor array
      divisorarr[1]  = 16'h0007;
      divisorarr[2]  = 16'hE005;
      divisorarr[3]  = 16'h001E;
      divisorarr[4]  = 16'hEFFA;
      divisorarr[5]  = 16'h7FFF;
      divisorarr[6]  = 16'h7FFF;
      divisorarr[7]  = 16'h7FFF;
      divisorarr[8]  = 16'h8000;
      divisorarr[9]  = 16'h7FFF;
      divisorarr[10] = 16'h0001;
      divisorarr[11] = 16'h7FFF;
      divisorarr[12] = 16'h0000;

      //initialization of quotient array
      quotientarr[1] = 16'h000F;
      quotientarr[2] = 16'hBFFE;
      quotientarr[3] = 16'hFFF0;
      quotientarr[4] = 16'h07FC;
      quotientarr[5] = 16'h0000;
      quotientarr[6] = 16'h7FFF;
      quotientarr[7] = 16'h0000;
```

```verilog
      quotientarr[8]  = 16'h7FFF;
      quotientarr[9]  = 16'h8001;
      quotientarr[10] = 16'h0000;
      quotientarr[11] = 16'h0000;
      quotientarr[12] = 16'h0002;

      //initialization of remainder array
      remainderarr[1]  = 16'h0006;
      remainderarr[2]  = 16'h00C5;
      remainderarr[3]  = 16'hFFE8;
      remainderarr[4]  = 16'hF2F2;
      remainderarr[5]  = 16'h7FFF;
      remainderarr[6]  = 16'h7FFE;
      remainderarr[7]  = 16'h7FFF;
      remainderarr[8]  = 16'h0000;
      remainderarr[9]  = 16'h8002;
      remainderarr[10] = 16'h0000;
      remainderarr[11] = 16'hFFFF;
      remainderarr[12] = 16'h0000;

   CLK = 0;
   Count = 0;
@(posedge CLK);
@(negedge CLK);
   for(i = 1 ; i <= N ; i = i + 1)
   begin
      St = 1'b1;
      dividendarr_tmp = dividendarr[i];
      Dbus = dividendarr_tmp[31:16];

      @(posedge CLK);
      Dbus = dividendarr_tmp[15:0];

      @(posedge CLK);
      Dbus = divisorarr[i];
      St = 1'b0;
      Dividend = dividendarr_tmp[31:0];
      Divisor = divisorarr[i];

      @(posedge Rdy);
      Count = i;
       if(quotientarr[i] == Quotient)
         begin
         $display("quotient[%d] is correct",i);
         end
        else
         begin
         $display("quotient[%d] is wrong",i);
         end
       if(remainderarr[i] == Remainder)
         begin
         $display("remainder[%d] is correct",i);
         end
```

```
        else
        begin
            $display("remainder[%d] is wrong",i);
        end
    end
    $display("TESTS DONE");
end

    sdiv sdiv1(CLK, St, Dbus, Quotient, Remainder, V, Rdy);
endmodule
```

Figure 4-56 shows the simulator command file and output. The -notrigger, together with the -trigger count in the list statement, causes the output to be displayed only when the *count* signal changes. Examination of the simulator output shows that the divider operation is correct for all of the test cases, except for the following case:

$$C0008000h \div 7FFFh = -3FFF8000 \div 7FFFh = -8000h = 8000h$$

In this case, the overflow is turned on and division never occurs. In general, the divider will indicate an overflow whenever the quotient should be 8000h (the most negative value). This occurs because the divider basically divides positive numbers and the largest positive quotient is 7FFFh. If it is important to be able to generate the quotient 8000h, the overflow detection can be modified so it does not generate an overflow in this special case.

FIGURE 4-56: Simulation Test Results for Signed Divider

```
// Command file to test results of signed divider
add list -hex -notrigger Dividend Divisor Quotient Remainder V -trigger Count
run 5300

    ns   delta   dividend   divisor   quotient   remainder   v   count
    0     +0    xxxxxxxx    xxxx       xxxx        xxxx      x     x
  490     +3    0000006F    0007       000F        0006      0     1
  930     +3    07FF00BB    E005       BFFE        00C5      0     2
 1350     +3    FFFFFE08    001E       FFF0        FFE8      0     3
 1930     +3    FF80030A    EFFA       07FC        F2F2      0     4
 2030     +3    3FFF8000    7FFF       0000        7FFF      1     5
 2730     +3    3FFF7FFF    7FFF       7FFF        7FFE      0     6
 2830     +3    C0008000    7FFF       0000        7FFF      1     7
 3530     +3    C0008000    8000       7FFF        0000      0     8
 4230     +3    C0008001    7FFF       8001        8002      0     9
 4630     +3    00000000    0001       0000        0000      0     A
 5030     +3    FFFFFFFF    7FFF       0000        FFFF      0     B
 5130     +3    FFFFFFFF    0000       0002        0000      1     C
```

In this chapter, we have presented several design examples. The examples included several arithmetic and non-arithmetic circuits. A 7-segment display, a BCD adder, a traffic light controller, a scoreboard, and a keypad scanner are examples of non-arithmetic circuits presented in the chapter. We also described algorithms for addition, multiplication, and division of unsigned and signed binary numbers. Specific designs such as the carry look-ahead adder and the array multiplier were presented. We designed digital systems to implement these algorithms. After developing a block diagram for such a system and defining the required control signals, we used state graphs to define a sequential machine that generates control signals in the proper sequence. We used Verilog to describe the systems at several different levels so that we can simulate and test for correct operation of the systems we have designed.

• • • • • • • • • •

Problems

4.1 This problem involves the design of a parallel adder-subtracter for 8-bit numbers expressed in sign and magnitude notation. The inputs X and Y are in sign and magnitude, and the output Z must be in sign and magnitude. Internal computation may be done in either 2's complement or 1's complement (specify which you use), but no credit will be given if you assume the inputs X and Y are in 1's or 2's complement. If the input signal $Sub = 1$, then $Z = X - Y$, else $Z = X + Y$. Your circuit must work for all combinations of positive and negative inputs for both add and subtract. You may use only the following components: an 8-bit adder, a 1's complementer (for the input Y), a second complementer (which may be either 1's complement or 2's complement—specify which you use), and a combinational logic circuit to generate control signals. *Hint*: $-X + Y = -(X - Y)$. Also generate an overflow signal that is 1 if the result cannot be represented in 8-bit sign and magnitude.

  (a) Draw the block diagram. No registers, multiplexers, or tristate busses are allowed.

  (b) Give a truth table for the logic circuit that generates the necessary control signals. Inputs for the table should be Sub, Xs, and Ys in that order, where Xs is the sign of X and Ys is the sign of Y.

  (c) Explain how you would determine the overflow and give an appropriate equation.

4.2 A block diagram for a divider that divides an 8-bit unsigned number by a 4-bit unsigned number to give a 4-bit quotient is shown subsequently. Note that the X_i inputs to the subtractors are shifted over one position to the left. This means that the shift-and-subtract operation can be completed in one clock time instead of two. Depending on the borrow from the subtractor, a shift or shift-and-subtract operation occurs at each clock time, and the division can always be completed in four clock times after the registers are loaded. Ignore overflow. When the start signal (St) is 1, the X and Y registers are loaded. Assume that the start signal (St) is 1 for only one clock time. Sh causes X to shift left with 0 fill. $SubSh$ causes the subtractor output to be loaded into the left part of X and at the same time the rest of X is shifted left.

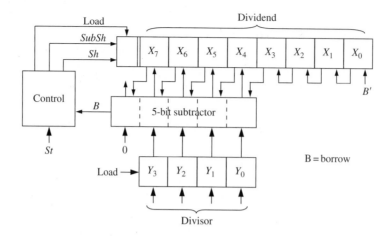

(a) Draw a state graph for the controller (five states).
(b) Complete the Verilog code that follows. Registers and signals should be of type unsigned so that overloaded operators may be used. Write behavioral code that uses a single always block.

```
module divu(dividend, divisor, St, clk, quotient);
input[7:0] dividend;
input[3:0] divisor;
input St,clk;
output[3:0] quotient;
 .
 .
 .

endmodule
```

4.3 Four pushbuttons (B_0, B_1, B_2, and B_3) are used as inputs to a logic circuit. Whenever a button is pushed, it is debounced, after which the circuit loads the button number in binary into a 2-bit register (N). For example, if B_2 is pushed, the register output becomes $N = 10_2$. The register holds this value until another button is pushed. Use a total of two flip-flops for debouncing. Use a 10-bit counter as a clock divider to provide a slow clock for debouncing. Kd is a signal that is 1 when any button has been pushed and debounced.

(a) Draw a state graph (two states) to generate the signal that loads the register when $Kd = 1$.
(b) Draw a logic circuit diagram showing the 10-bit counter, the 2-bit register N, and all necessary gates and flip-flops.

4.4 An older model Thunderbird car has three left (LA, LB, LC) and three right (RA, RB, RC) tail lights, which flash in unique patterns to indicate left and right turns.

LEFT turn pattern RIGHT turn pattern

```
    LC  LB  LA  RA  RB  RC        LC  LB  LA  RA  RB  RC
    ○   ○   ○   ○   ○   ○         ○   ○   ○   ○   ○   ○
    ○   ○   ●   ○   ○   ○         ○   ○   ○   ●   ○   ○
    ○   ●   ●   ○   ○   ○         ○   ○   ○   ●   ●   ○
    ●   ●   ●   ○   ○   ○         ○   ○   ○   ●   ●   ●
```

Design a Moore sequential circuit to control these lights. The circuit has three inputs: *LEFT*, *RIGHT*, and *HAZ*. *LEFT* and *RIGHT* come from the driver's turn signal switch and cannot be 1 at the same time. As indicated in the diagram, when *LEFT* = 1 the lights flash in a pattern *LA* on, *LA* and *LB* on, *LA*, *LB*, and *LC* on, all off; then the sequence repeats. When *RIGHT* = 1, a similar sequence appears on lights RA, RB, and RC, as indicated on the right side of the diagram. If a switch from *LEFT* to *RIGHT* (or vice versa) occurs in the middle of a flashing sequence, the circuit should immediately go to the IDLE (lights off) state and then start the new sequence. *HAZ* comes from the hazard switch, and when *HAZ* = 1, all six lights flash on and off in unison. *HAZ* takes precedence if *LEFT* or *RIGHT* is also on.

Assume that a clock signal is available with a frequency equal to the desired flashing rate.

(a) Draw the state graph (eight states).
(b) Realize the circuit using six D flip-flops, and make a one-hot state assignment such that each flip-flop output drives one of the six lights directly. (You may use *LogicAid*.)
(c) Realize the circuit using three D flip-flops, using the guidelines from Section 1.7 to determine a suitable encoded state assignment. Note the tradeoff between more flip-flops and more gates in (b) and (c).

4.5 Design a 4 × 4 keypad scanner for the following keypad layout.

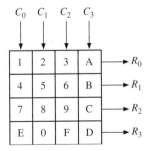

(a) Assuming only one key can be pressed at a time, find the equations for a number decoder given $R_{3\text{-}0}$ and $C_{3\text{-}0}$, whose output corresponds to the binary value of the key. For example, the F key will return $N_{3\text{-}0} = 1111$ in binary, or 15.
(b) Design a debouncing circuit that detects when a key has been pressed or depressed. Assume switch bounce will die out in one or two clock cycles. When a key has been pressed, $K = 1$ and Kd is the debounced signal.

(c) Design and draw a state graph that performs the keyscan and issues a valid pulse when a valid key has been pressed using inputs from part **(b)**.

(d) Write a Verilog description of your keypad scanner and include the decoder, the debouncing circuit, and the scanner.

4.6 Design a sequential circuit to control the motor of a tape player. The logic circuit will have five inputs and three outputs. Four of the inputs are the control buttons on the tape player. The input PL is 1 if the play button is pressed, the input RE is 1 if the rewind button is pressed, the input FF is 1 if the fast-forward button is pressed, and the input ST is 1 if the stop button is pressed. The fifth input to the control circuit is M, which is 1 if the special "music sensor" detects music at the current tape position. The three outputs of the control circuit are P, R, and F, which make the tape play, rewind, and fast forward, respectively, when 1. No more than one output should ever be on at a time; all outputs off causes the motor to stop. The buttons control the tape as follows: If the play button is pressed, the tape player will start playing the tape (output $P = 1$). If the play button is held down and the rewind button is pressed and released, the tape player will rewind to the beginning of the current song (output $R = 1$ until $M = 0$) and then start playing. If the play button is held down and the fast forward button is pressed and released, the tape player will fast forward to the end of the current song (output $F = 1$ until $M = 0$) and then start playing. If rewind or fast forward is pressed while play is released, the tape player will rewind or fast forward the tape. Pressing the stop button at any time should stop the tape player motor.

(a) Construct a state graph chart for the tape player controller. You may assume that only one of the four buttons can be pressed at any given time.

(b) Write Verilog code for the controller.

4.7 This problem concerns the design of a divider for unsigned binary numbers that will divide a 16-bit dividend by an 8-bit divisor to give an 8-bit quotient. Assume that the start signal ($ST = 1$) is 1 for exactly one clock time. If the quotient would require more than 8 bits, the divider should stop immediately and output $V = 1$ to indicate an overflow. Use a 17-bit dividend register and store the quotient in the lower 8 bits of this register. Use a 4-bit counter to count the number of shifts, together with a subtract-shift controller.

(a) Draw a block diagram of the divider.

(b) Draw a state graph for the subtract-shift controller (three states).

(c) Write a Verilog description of the divider. Use two always blocks, as in Figure 4-40.

(d) Write a test bench for your divider (similar to Figure 4-55).

4.8 A block diagram for a 16-bit 2's complement serial subtracter is given here. When $St = 1$, the registers are loaded and then subtraction occurs. The shift counter, C, produces a signal $C15 = 1$ after 15 shifts. V should be set to 1 if an overflow occurs. Set the carry flip-flop to 1 during load in order to form the 2's complement. Assume that St remains 1 for one clock time.

(a) Draw a state diagram for the control (two states).
(b) Write Verilog code for the system. Use two always blocks. The first always block should determine the next state and control signals; the second always block should update the registers on the rising edge of the clock.

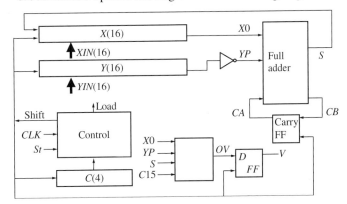

4.9 A block diagram and state graph for a divider for unsigned binary numbers is shown subsequently. This divider divides a 16-bit dividend by a 16-bit divisor to give a 16-bit quotient. The divisor can be any number in the range 1 to $2^{16} - 1$. The only case where an overflow can occur is when the divisor is 0. Control signals are defined as follows: $Ld1$—load the divisor from the input bus; $Ld2$—load the dividend from the input bus and clear ACC; Sh—left shift ACC and Dividend; Su—load the subtracter output into ACC and set the lower quotient bit to 1; $K = 1$ when 15 shifts have been made. Write complete Verilog code for the divider. Use always blocks.

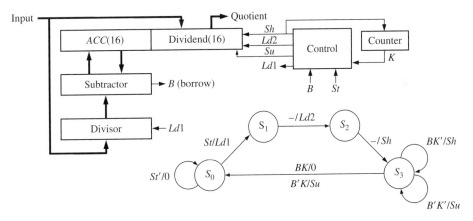

4.10 This problem involves the design of a BCD-to-binary converter. Initially a 3-digit BCD number is placed in the A register. When an St signal is received, conversion to binary takes place and the resulting binary number is stored in the B register. At each step of the conversion, the entire BCD number (along with the binary number) is shifted one place to the right. If the result in a given decade is greater than or equal to 1000, the correction circuit subtracts 0011 from that decade. (If the result is less than 1000, the correction circuit leaves the contents of the decade unchanged.) A shift counter is provided to count the number of shifts. When conversion is complete, the maximum value of B will be 999 (in binary). *Note*: B is 10 bits.

(a) Illustrate the algorithm starting with the BCD number 857, showing A and B at each step.
(b) Draw the block diagram of the BCD-to-binary converter.
(c) Draw a state diagram of the control circuit (three states). Use the following control signals: *St*: start conversion; *Sh*: shift right; *Co*: subtract correction if necessary; and *C9*: counter is in state 9, or *C10*: counter is in state 10. (Use either *C9* or *C10* but not both.)
(d) Write a Verilog description of the system.

4.11 An $n \times n$ array multiplier, as in Figure 4-29, takes $3n - 4$ adder delays $+ 1$ gate delay to calculate a product. Design an array multiplier that is faster than this for $n > 4$. (*Hint*: Instead of passing carry output to the left adder, pass it to the diagonally lower one, speeding up the critical path. This topology is called "multiplier using carry-save adder.")

4.12 This problem involves the design of a circuit that finds the square root of an 8-bit unsigned binary number N using the method of subtracting out odd integers. To find the square root of N, we subtract 1, then 3, then 5, and so on, until we can no longer subtract without the result going negative. The number of times we subtract is equal to the square root of N. For example, to find $\sqrt{27}$: $27 - 1 = 26; 26 - 3 = 23; 23 - 5 = 18; 18 - 7 = 11; 11 - 9 = 2; 2 - 11$ (can't subtract). Since we subtracted five times, $\sqrt{27} = 5$. Note that the final odd integer is $11_{10} = 1011_2$, and this consists of the square root ($101_2 = 5_{10}$) followed by a 1.

(a) Draw a block diagram of the square rooter that includes a register to hold N, a subtracter, a register to hold the odd integers, and a control circuit. Indicate where to read the final square root. Define the control signals used on the diagram.
(b) Draw a state graph for the control circuit using a minimum number of states. The N register should be loaded when $St = 1$. When the square root is complete, the control circuit should output a done signal and wait until $St = 0$ before resetting.

4.13 The block diagram for a multiplier for signed (2's complement) binary numbers is shown in Figure 4-33. Give the contents of the A and B registers after each clock pulse when multiplicand $= -1/8$ and multiplier $= -3/8$.

4.14 This problem concerns the design of a multiplier for unsigned binary numbers that multiplies a 4-bit number by a 16-bit number to give a 20-bit product. To speed up the multiplication, a 4-by-4 array multiplier is used so that we can multiply by 4 bits in one clock time instead of by only 1 bit at each clock time. The hardware includes a 24-bit accumulator register that can be shifted right 4 bits at a time using a control signal *Sh4*. The array multiplier multiplies 4 bits by 4 bits to give an 8-bit product. This product is added to the accumulator using an *Ad* control signal. When an *St* signal occurs, the 16-bit multiplier is loaded into the lower part of the A register. A done signal should be turned on when the multiplication is complete. Since both the array multiplier and adder are combinational circuits, the 4-bit multiply and the 8-bit add can both be completed in the same clock cycle. Do NOT include the array multiplier logic in your code; just use the overloaded "*" operator. If D and E are 4-bit unsigned numbers, $D * E$ will compute an 8-bit product.

(a) Draw a state graph for the controller (10 states)
(b) Write Verilog code for the multiplier. Use two always blocks (a combinational always block and a clocked always block).

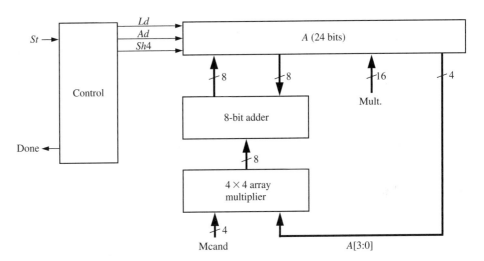

4.15 In Section 4.10 we developed an algorithm for multiplying signed binary fractions, with negative fractions represented in 2's complement.

(a) Illustrate this algorithm by multiplying 1.0111 by 1.101.
(b) Draw a block diagram of the hardware necessary to implement this algorithm for the case where the multiplier is 4 bits, including sign, and the multiplicand is 5 bits, including sign.

4.16 (a) Draw the organization of an 8×8 array multiplier and calculate how many full adders, half adders, and AND gates are required.
(b) Highlight the critical path in your answer to (a). (If there are many equivalent ones, highlight any one of them).
(c) What is the longest delay in an 8×8 array multiplier, assuming an AND gate delay is $t_g = 1$ ns, and an adder delay (full adder and half adder) is $t_{ad} = 2$ ns?
(d) For an 8-bit $\times$ 8-bit add-and-shift multiplier (similar to Figure 4-25), how fast must the clock be in order to complete the multiplication in the same time as in part (c)?

4.17 The objective of this problem is to use Verilog to describe and simulate a multiplier for signed binary numbers using Booth's algorithm. Negative numbers should be represented by their 2's complement. Booth's algorithm works as follows, assuming each number is n bits including sign: Use an $(n + 1)$-bit register for the accumulator

(A) so the sign bit will not be lost if an overflow occurs. Also, use an $(n + 1)$-bit register (B) to hold the multiplier and an n-bit register (C) to hold the multiplicand.

1. Clear A (the accumulator), load the multiplier into the upper n bits of B, clear B_0, and load the multiplicand into C.
2. Test the lower two bits of B ($B_1 B_0$).

    If $B_1 B_0 = 01$, add C to A (C should be sign-extended to $n + 1$ bits and added to A using an $(n + 1)$-bit adder).

    If $B_1 B_0 = 10$, add the 2's complement of C to A.

    If $B_1 B_0 = 00$ or 11, skip this step.
3. Shift A and B together right one place with sign extended.
4. Repeat steps 2 and 3, $n - 1$ more times.
5. The product will be in A and B, except ignore B_0.

Example for $n = 5$: Multiply -9 by -13.

		A	B	$B_1 B_0$	
1.	Load registers.	000000	100110	10	$C = 10111$
2.	Add 2's comp. of C to A.	001001			
		001001	100110		
3.	Shift $A \& B$.	000100	110011	11	
3.	Shift $A \& B$.	000010	011001	01	
2.	Add C to A.	110111			
		111001	011001		
3.	Shift $A \& B$.	111100	101100	00	
3.	Shift $A \& B$.	111110	010110	10	
2.	Add 2's comp. of C to A.	001001			
		000111	010110		
3.	Shift $A \& B$.	000011	101011		

Final result: $0001110101 = +117$

(a) Draw a block diagram of the system for $n = 8$. Use 9-bit registers for A and B, a 9-bit full adder, an 8-bit complementer, a 3-bit counter, and a control circuit. Use the counter to count the number of shifts.

(b) Draw a state graph for the control circuit. When the counter is in state 111, return to the start state at the time the last shift occurs (three states should be sufficient).

(c) Write behavioral Verilog code for the multiplier.

(d) Simulate your Verilog design using the following test cases (in each pair, the second number is the multiplier):

$$01100110 \times 00110011$$
$$10100110 \times 01100110$$
$$01101011 \times 10001110$$
$$11001100 \times 10011001$$

Verify that your results are correct.

4.18 **(a)** Estimate how many AND gates and adders will be required for a 16-bit × 16-bit array multiplier.
(b) What is the longest delay in a 16 × 16 array multiplier, assuming an AND gate delay is t_g, and an adder delay (full adder and half adder) is t_{ad}?

4.19 Design a multiplier that will multiply two 16-bit signed binary integers to give a 32-bit product. Negative numbers should be represented in 2's complement form. Use the following method: First complement the multiplier and multiplicand if they are negative, multiply the positive numbers, and then complement the product if necessary. Design the multiplier so that after the registers have been loaded, the multiplication can be completed in 16 clocks.

(a) Draw a block diagram of the multiplier. Use a 4-bit counter to count the number of shifts. (The counter will output a signal $K = 1$ when it is in state 15.) Define all condition and control signals used on your diagram.
(b) Draw a state diagram for the multiplier control using a minimum number of states (five states). When the multiplication is complete, the control circuit should output a done signal and then wait for $ST = 0$ before returning to state S_0.
(c) Write a Verilog behavioral description of the multiplier without using control signals (e.g,. see Figure 4-35) and test it.
(d) Write a Verilog behavioral description using control signals (e.g., see Figure 4-40) and test it.

4.20 Design the correction circuit for a BCD adder that computes Zdigit 0 and C for S_0 (see Figures 4-5 and 4-6). This correction circuit adds 0110 to S_0 if $S_0 > 9$. This is the same as adding 0AA0 to S_0, where $A = 1$ if $S_0 > 9$. Draw a block diagram for the correction circuit using one full adder, three half adders, and a logic circuit to compute A. Design a circuit for A using a minimum number of gates. Note that the maximum possible value of S_0 is 10010.

4.21 **(a)** If gate delays are 5 ns, what is the delay of the fastest 4-bit ripple carry adder? Explain your calculation.
(b) If gate delays are 5 ns, what is the delay of the fastest 4-bit adder? What kind of adder will it be? Explain your calculation.

4.22 Develop a Verilog model for a 16-bit carry look-ahead adder utilizing the 4-bit adder from Figure 4-10 as a module.

4.23 Derive generates, propagates, group generates, group-propagates, and the final sum and carry out for the 16-bit carry look-ahead adder of Figure 4-9, while adding 0101 1010 1111 1000 and 0011 1100 1100 0011.

4.24 **(a)** Write a Verilog module that describes one bit of a full adder with accumulator. The module should have two control inputs, Ad and L. If $Ad = 1$, the Y input (and carry input) are added to the accumulator. If $L = 1$, the Y input is loaded into the accumulator.
(b) Using the module defined in **(a)**, write a Verilog description of a 4-bit subtracter with accumulator. Assume negative numbers are represented in 1's complement. The subtracter should have control inputs Su (subtract) and Ld (load).

4.25 (a) Implement the traffic-light controller of Figure 4-14 using a module 13 counter with added logic. The counter should increment every clock, with two exceptions. Use a ROM to generate the outputs.
(b) Write a Verilog description of your answer to (a).
(c) Write a test bench for part (b) and verify that your controller works correctly. Use concurrent statements to generate test inputs for Sa and Sb.

4.26 Make the necessary additions to the following state graph so that it is a proper, completely specified state graph. Demonstrate that your answer is correct. Convert the graph to a state table using 0's and 1's for inputs and outputs.

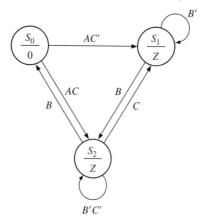

4.27 Write synthesizable Verilog code that will generate the given waveform (W). Use a single always block. Assume that a clock with a 1 µs period is available as an input.

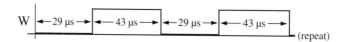

4.28 A BCD adder adds two BCD numbers (each of range 0 to 9) and produces the sum in BCD form. For example, if it adds 9 (1001) and 8 (1000) the result would be 17 (1 0111). Implement such a BCD adder using a 4-bit binary adder and appropriate control circuitry. Assume that the two BCD numbers are already loaded into two 4-bit registers (A and B) and there is a 5-bit sum register (S) available. You need some kind of correction to get the sum in the BCD form, because the binary adder produces results in the range 0000 to 1111 (plus a carry in some cases). If any addition is required for this correction, use the same adder (i.e., you can use only one adder). Use multiplexers at the adder inputs to steer the appropriate numbers to the adder in each cycle. Assume a start signal to initiate the addition and a done signal to indicate completion.

(a) Draw a block diagram of the system. Label each component appropriately to indicate its functionality and size.
(b) Describe step by step the algorithm that you would use to perform the addition. Explain and illustrate the correction step.
(c) Draw a state graph for the controller.

4.29 Write Verilog code for a shift register module that includes a 16-bit shift register, a controller, and a 4-bit down counter. The shifter can shift a variable number of bits depending on a count provided to the shifter module. Inputs to the module are a number *N* (indicating shift count) in the range 1 to 15, a 16-bit vector *par_in*, a clock, and a start signal, *St*. When $St = 1$, *N* is loaded into the down counter, and *par_in* is loaded into the shift register. Then the shift register does a cycle left shift *N* times, and the controller returns to the start state. Assume that *St* is only 1 for one clock time. All operations are synchronous on the falling edge of the clock.

   (a) Draw a block diagram of the system and define any necessary control signals.
   (b) Draw a state graph for the controller (two states).
   (c) Write Verilog code for the shift-register module. Use two always blocks (one for the combinational part of the circuit and one for updating the registers).

4.30 **(a)** Figure 4-12 shows the block diagram for a 32-bit serial adder with accumulator. The control circuit uses a 5-bit counter, which outputs a signal $K = 1$ when it is in state 11111. When a start signal (*St*) is received, the registers should be loaded. Assume that *St* will remain 1 until the addition is complete. When the addition is complete, the control circuit should go to a stop state and remain there until *St* is changed back to 0. Draw a state diagram for the control circuit (excluding the counter).
   (b) Write the Verilog for the complete system and verify its correct operation.

SM Charts and Microprogramming

A state machine is often used to control a digital system that carries out a step-by-step procedure or algorithm. State diagrams or state graphs with circles representing states and arcs representing transitions have traditionally been used to specify the operation of the controller state machine. As an alternative to using state graphs, a special type of flowchart, called a *state machine chart*, or **SM chart**, may be use d to describe the behavior of a state machine. These charts are also called *algorithmic state machine charts*, or **ASM charts**. SM charts are often used to design control units for digital systems.

In this chapter, we first describe the properties of SM charts and how they are used in the design of state machines. Then we show examples of SM charts for a multiplier and a dice game controller. We construct Verilog descriptions of these systems from the SM charts, and we simulate the Verilog code to verify correct operation. We then proceed with the design and show how the SM chart can be realized with hardware. Finally, we introduce **microprogramming** as a technique to implement the SM chart.

5.1 State Machine Charts

SM charts resemble software flowcharts. Flowcharts have been very useful in software design for decades, and in a similar fashion, SM charts have been useful in hardware design. This is especially true in behavioral-level design entry.

SM charts offer several advantages over state graphs. It is often easier to understand the operation of a digital system by inspection of the SM chart rather than the equivalent state graph. A proper state graph has to obey some conditions: (i) one and exactly one transition from a state must be true at any time and (ii) the next state must be uniquely defined for every input combination. These conditions are automatically satisfied for an SM chart. An SM chart also directly leads to a hardware realization. A given SM chart can be converted into several equivalent forms, and different forms might naturally result in different implementations. Hence, a designer may optimize and transform SM charts to suit the implementation style/technology that he/she is looking for.

An SM chart differs from an ordinary flowchart in that certain specific rules must be followed in constructing the SM chart. When these rules are followed, the SM chart is equivalent to a state graph, and it leads directly to a hardware realization.

Figure 5-1 shows the three principal components of an SM chart. The state of the system is represented by a *state box*. The state box contains a *state name*, followed by a slash (/) and an optional *output list*. After a state assignment has been made, a *state code* may be placed outside the box at the top. A *decision box* is represented by a diamond-shaped symbol with true and false branches. The *condition* placed in the box is a Boolean expression that is evaluated to determine which branch to take. The *conditional output box,* which has curved ends, contains a *conditional output list.* The conditional outputs depend on both the state of the system and the inputs.

FIGURE 5-1: Principal Components of an SM Chart

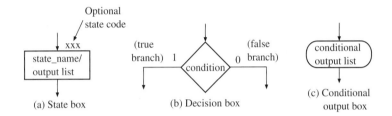

(a) State box (b) Decision box (c) Conditional output box

An SM chart is constructed from *SM blocks.* Each SM block (Figure 5-2) contains exactly one state box, together with the decision boxes and conditional output boxes associated with that state. An SM block has one *entrance path* and one or more *exit paths*. Each SM block describes the machine operation during the time that the machine is in one state. When a digital system enters the state associated with a given SM block, the outputs on the output list in the state box become true.

FIGURE 5-2: Example of an SM Block

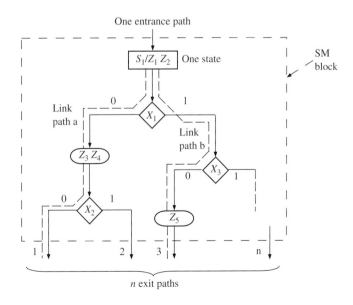

The conditions in the decision boxes are evaluated to determine which paths are followed through the SM block. When a conditional output box is encountered along such a path, the corresponding conditional outputs become true. If an output is not encountered along a path, that output is false by default. A path through an SM block from entrance to exit is referred to as a *link path*.

For the example of Figure 5-2, when state S_1 is entered, outputs Z_1 and Z_2 become 1. If input $X_1 = 0$, Z_3 and Z_4 also become 1. If $X_1 = X_2 = 0$, at the end of the state time the machine goes to the next state via exit path 1. On the other hand, if $X_1 = 1$ and $X_3 = 0$, the output Z_5 is 1 and exiting to the next state will occur via exit path 3. Since Z_3 and Z_4 are not encountered along this link path, $Z_3 = Z_4 = 0$ by default.

A given SM block can generally be drawn in several different forms. Figure 5-3 shows two equivalent SM blocks. In both (a) and (b), the output $Z_2 = 1$ if $X_1 = 0$; the next state is S_2 if $X_2 = 0$ and S_3 if $X_2 = 1$. As illustrated in this example, the order in which the inputs are tested may affect the complexity of the SM chart.

FIGURE 5-3: Equivalent SM Blocks

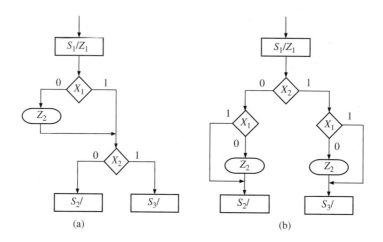

The SM charts of Figure 5-4(a) and (b) each represent a combinational circuit, since there is only one state and no state change occurs. The output is $Z_1 = 1$

FIGURE 5-4: Equivalent SM Charts for a Combinational Circuit

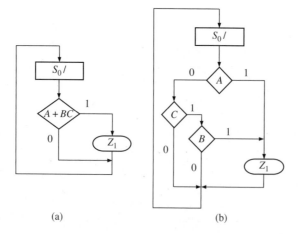

if $A + BC = 1$; otherwise $Z_1 = 0$. Figure 5-4(b) shows an equivalent SM chart in which the input variables are tested individually. The output is $Z_1 = 1$ if $A = 1$ or if $A = 0$, $B = 1$, and $C = 1$. Hence,

$$Z_1 = A + A'BC = A + BC$$

which is the same output function realized by the SM chart of Figure 5-4(a).

Certain rules must be followed when constructing an SM block. First, for every valid combination of input variables, there must be exactly one exit path defined. This is necessary since each allowable input combination must lead to a single next state. Second, no internal feedback within an SM block is allowed. Figure 5-5 shows incorrect and correct ways of drawing an SM block with feedback.

FIGURE 5-5: SM Block with Feedback

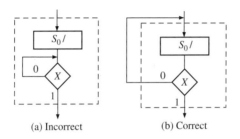

As shown in Figure 5-6(a), an SM block can have several parallel paths that lead to the same exit path, and more than one of these paths can be active at the same time. For example, if $X_1 = X_2 = 1$ and $X_3 = 0$, the link paths marked with dashed lines are active, and the outputs Z_1, Z_2, and Z_3 are 1. Although Figure 5-6(a) would not be a valid flowchart for a program for a serial computer, it presents no problems

FIGURE 5-6: Equivalent SM Blocks

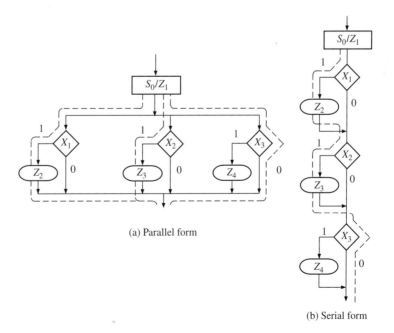

for a state machine implementation. The state machine can have a multiple-output circuit that generates Z_1, Z_2, and Z_3 at the same time. Figure 5-6(b) shows a serial SM block that is equivalent to Figure 5-6(a). In the serial block, only one active link path between entrance and exit is possible. For any combination of input values, the outputs will be the same as in the equivalent parallel form. The link path for $X_1 = X_2 = 1$ and $X_3 = 0$ is shown with a dashed line, and the outputs encountered on this path are Z_1, Z_2, and Z_3. Regardless of whether the SM block is drawn in serial or parallel form, all the tests take place within one clock time. In the remainder of this text, we use only the serial form for SM charts.

It is easy to convert a state graph for a sequential machine to an equivalent SM chart. The state graph of Figure 5-7(a) has both Moore and Mealy outputs. The equivalent SM chart has three blocks—one for each state. The Moore outputs (Z_a, Z_b, Z_c) are placed in the state boxes, since they do not depend on the input. The Mealy outputs (Z_1, Z_2) appear in conditional output boxes, since they depend on both the state and input. In this example, each SM block has only one decision box, since only one input variable must be tested. For both the state graph and SM chart, Z_c is always 1 in state S_2. If $X = 0$ in state S_2, $Z_1 = 1$ and the next state is S_0. If $X = 1$, $Z_2 = 1$ and the next state is S_2. We have added a state assignment ($S_0 = 00$, $S_1 = 01$, $S_2 = 11$) next to the state boxes.

FIGURE 5-7: Conversion of a State Graph to an SM Chart

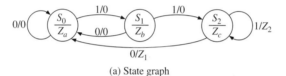

(a) State graph

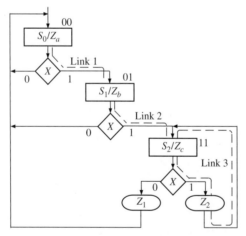

(b) Equivalent SM chart

Figure 5-8 shows a timing chart for the SM chart of Figure 5-7 with an input sequence $X = 1, 1, 1, 0, 0, 0$. In this example, all state changes occur immediately after the rising edge of the clock. Since the Moore outputs (Z_a, Z_b, Z_c) depend on the state, they can change only immediately following a state change. The Mealy outputs (Z_1, Z_2) can change immediately after a state change or an input change. In any case, all outputs will have their correct values at the time of the active clock edge.

FIGURE 5-8: Timing Chart for Figure 5-7

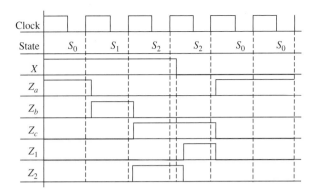

5.2 Derivation of SM Charts

The method used to derive an SM chart for a sequential control circuit is similar to that used to derive the state graph. First, we should draw a block diagram of the system we are controlling. Next, we should define the required input and output signals to the control circuit. Then we can construct an SM chart that tests the input signals and generates the proper sequence of output signals. In this section, we give two examples of derivation of SM charts.

5.2.1 Binary Multiplier

The first example is an SM chart for control of the binary multiplier shown in Figures 4-26 and 4-28(a). The add-shift control generates the required sequence of add and shift signals. The counter counts the number of shifts and outputs $K = 1$ just before the last shift occurs. The SM chart for the multiplier control (Figure 5-9) corresponds closely to the state graph of Figure 4-28(c). In state S_0, when the start signal St is 1, the registers are loaded. In S_1, the multiplier bit M is tested. If $M = 1$, an add signal is generated and the next state is S_2. If $M = 0$, a shift signal is generated and K is tested. If $K = 1$, this will be the last shift and the next state is S_3. In S_2, a shift signal is generated, since a shift must always follow an add. If $K = 1$, the circuit goes to S_3 at the time of the last shift; otherwise, the next state is S_1. In S_3, the done signal is turned on.

Conversion of an SM chart to a Verilog process is straightforward. A **case** statement can be used to specify what happens in each state. Each condition box corresponds directly to an **if** statement (or an **else**). Figure 5-10 shows the Verilog code for the SM chart in Figure 5-9. Two always statements are used. The first always statement represents the combinational part of the circuit, and the second always statement updates the state register on the rising edge of the clock. The signals *Load*, *Sh*, and *Ad* are turned on in the appropriate states, and they must be turned off when the state changes. A convenient way to do this is to set them all to 0 at the start of the process. This Verilog code only models the controller. It assumes the presence of adders and shifters (shift registers) in the architecture and generates the appropriate signals to load the registers, add and/or shift.

FIGURE 5-9: SM Chart for Binary Multiplier

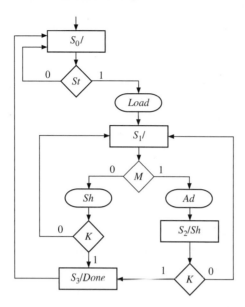

FIGURE 5-10: Behavioral Verilog for Multiplier Controller (SM Chart of Figure 5-9)

```verilog
module Mult (CLK, St, K, M, Load, Sh, Ad, Done);
   input CLK;
   input St;
   input K;
   input M;
   output Load;
   output Sh;
   output Ad;
   output Done;

   reg Ad;
   reg Sh;
   reg Load;
   reg Done;

   reg[1:0] State;
   reg[1:0] Nextstate;

   initial
   begin
       State = 0;
       Nextstate = 0;
   end

   always @(St or K or M or State)
   begin
       Load = 1'b0;
       Sh = 1'b0;
       Ad = 1'b0;
       Done = 1'b0;
```

```verilog
        case (State)
            0 :
                    begin
                        if (St == 1'b1) begin
                            Load = 1'b1;
                            Nextstate = 1;
                        end
                        else begin
                            Nextstate = 0;
                        end
                    end
            1 :
                    begin
                        if (M == 1'b1) begin
                            Ad = 1'b1;
                            Nextstate = 2;
                        end
                        else begin
                            Sh = 1'b1;
                            if (K == 1'b1) begin
                                Nextstate = 3;
                            end
                            else begin
                                Nextstate = 1;
                            end
                        end
                    end
            2 :
                    begin
                        Sh = 1'b1;
                        if (K == 1'b1) begin
                            Nextstate = 3;
                        end
                        else begin
                            Nextstate = 1;
                        end
                    end
            3 :
                    begin
                        Done = 1'b1;
                        Nextstate = 0;
                    end
        endcase
    end

    always @(posedge CLK)
    begin
        State <= Nextstate;
    end
endmodule
```

5.2.2 A Dice Game

As a second example of SM chart construction, we will design an electronic dice game. This game is popularly known as "craps" in the United States. The game involves two dice, each of which can have a value between 1 and 6. Two counters are used to simulate the roll of the dice. Each counter counts in the sequence 1, 2, 3, 4, 5, 6, 1, 2, ... Thus, after the "roll" of the dice, the sum of the values in the two counters will be in the range 2 through 12. The rules of the game are as follows:

1. After the first roll of the dice, the player wins if the sum is 7 or 11. The player loses if the sum is 2, 3, or 12. Otherwise, the sum the player obtained on the first roll is referred to as a point, and he or she must roll the dice again.
2. On the second or subsequent roll of the dice, the player wins if the sum equals the point, and he or she loses if the sum is 7. Otherwise, the player must roll again until he or she finally wins or loses.

Figure 5-11 shows the block diagram for the dice game. The inputs to the dice game come from two push buttons, *Rb* (roll button) and *Reset*. *Reset* is used to initiate a new game. When the roll button is pushed, the dice counters count at a high speed, so the values cannot be read on the display. When the roll button is released, the values in the two counters are displayed.

FIGURE 5-11: Block Diagram for Dice Game

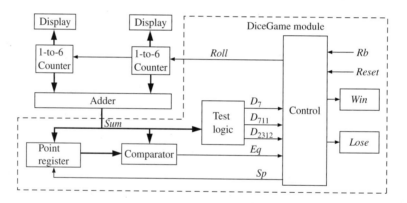

Figure 5-12 shows a flowchart for the dice game. After the dice is rolled, the sum is tested. If it is 7 or 11, the player wins; if it is 2, 3, or 12, he or she loses. Otherwise the sum is saved in the point register, and the player rolls again. If the new sum equals the point, the player wins; if it is 7, he or she loses. Otherwise, the player rolls again. If the *Win* light or *Lose* light is not on, the player must push the roll button again. After winning or losing, he or she must push *Reset* to begin a new game. We will assume at this point that the push buttons are properly debounced and that changes in *Rb* are properly synchronized with the clock. A method for debouncing and synchronization was discussed in Chapter 4.

The components for the dice game shown in the block diagram (Figure 5-11) include an adder, which adds the two counter outputs, a register to store the point, test logic to determine conditions for win or lose, and a control circuit. Input signals to the control circuit are defined as follows:

FIGURE 5-12: Flowchart for Dice Game

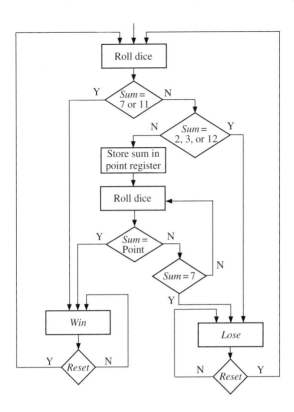

D_7	= 1 if the sum of the dice is 7.
D_{711}	= 1 if the sum of the dice is 7 or 11.
D_{2312}	= 1 if the sum of the dice is 2, 3, or 12.
Eq	= 1 if the sum of the dice equals the number stored in the point register.
Rb	= 1 when the roll button is pressed.
$Reset$	= 1 when the reset button is pressed.

Outputs from the control circuit are defined as follows:

$Roll$	= 1 enables the dice counters.
Sp	= 1 causes the sum to be stored in the point register.
Win	= 1 turns on the win light.
$Lose$	= 1 turns on the lose light.

The Rb and $Roll$ signals may look synonymous; however, they are different. We are using electronic dice counters and $Roll$ is the signal to let the counters continue to count. Rb is a push-button signal requesting that the dice be rolled. Thus, Rb is an input to the control circuit, while $Roll$ is an output from the control circuit. When the control circuit is in a state looking for a new roll of the dice, whenever the pushbutton is pressed (i.e., Rb is activated), the control circuit will generate the $Roll$ signal to the electronic dice.

We now convert the flowchart for the dice game to an SM chart for the control circuit using the control signals defined previously. Figure 5-13 shows the resulting SM chart.

FIGURE 5-13: SM Chart for Dice Game

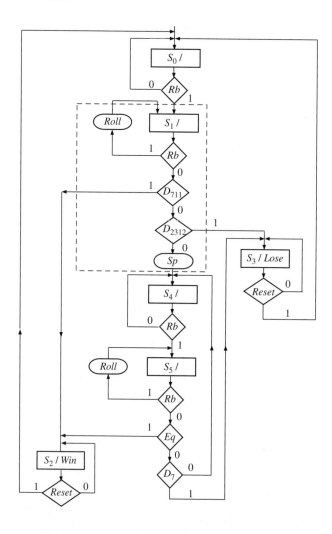

The control circuit waits in state S_0 until the roll button is pressed ($Rb = 1$). Then, it goes to state S_1, and the roll counters are enabled as long as $Rb = 1$. As soon as the roll button is released ($Rb = 0$), D_{711} is tested. If the sum is 7 or 11, the circuit goes to state S_2 and turns on the *Win* light; otherwise, D_{2312} is tested. If the sum is 2, 3, or 12, the circuit goes to state S_3 and turns on the *Lose* light; otherwise, the signal *Sp* becomes 1 and the sum is stored in the point register. It then enters S_4 and waits for the player to "roll the dice" again. In S_5, after the roll button is released, if $Eq = 1$, the sum equals the point and state S_2 is entered to indicate a win. If $D_7 = 1$, the sum is 7 and S_3 is entered to indicate a loss. Otherwise, control returns to S_4 so that the player can roll again. When in S_2 or S_3, the game is reset to S_0 when the *Reset* button is pressed.

Instead of using an SM chart, we could construct an equivalent state graph from the flowchart. Figure 5-14 shows a state graph for the dice game controller. The state graph has the same states, inputs, and outputs as the SM chart. The arcs have been labeled consistently with the rules for proper state graphs given in Section 4.2. Thus, the arcs leaving state S_1 are labeled Rb, $Rb'D_{711}$, $Rb'D'_{711}D_{2312}$, and $Rb'D'_{711}D'_{2312}$.

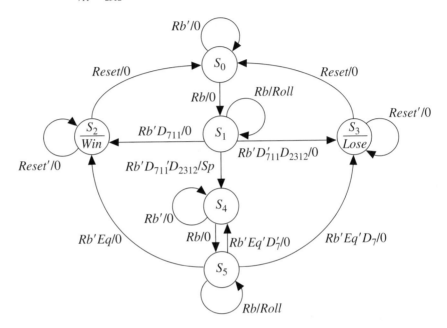

FIGURE 5-14: State Graph for Dice-Game Controller

Before proceeding with the design, it is important to verify that the SM chart (or state graph) is correct. We will write a behavioral Verilog description based on the SM chart and then write a test bench to simulate the roll of the dice. Initially, we will write a dice game module that contains the control circuit, point register, and comparator (see Figure 5-11). Later, we will add the counters and adder so that we can simulate the complete dice game.

The Verilog code for the dice game in Figure 5-15 corresponds directly to the SM chart of Figure 5-13. The **case** statement in the first **always** statement tests the state, and in each state nested **if-else** (or **else if**) statements are used to implement the conditional tests. In State 1 the *Roll* signal is turned on when Rb is 1. If all conditions test false, Sp is set to 1 and the next state is 4. In the second **always** statement, the state is updated after the rising edge of the clock, and if Sp is 1, the sum is stored in the point register.

FIGURE 5-15: Behavioral Model for Dice-Game Controller

```verilog
module DiceGame (Rb, Reset, CLK, Sum, Roll, Win, Lose);
    input Rb;
    input Reset;
    input CLK;
```

```verilog
input[3:0] Sum;
output Roll;
output Win;
output Lose;

reg Roll;
reg Win;
reg Lose;

reg[2:0] State;
reg[2:0] Nextstate;
reg[3:0] Point;
reg Sp;

initial
begin
    State = 0;
    Nextstate = 0;
    Point = 2;
end

always @(Rb or Reset or Sum or State)
begin
    Sp = 1'b0;
    Roll = 1'b0;
    Win = 1'b0;
    Lose = 1'b0;
    Nextstate = 0;
    case (State)
      0 :
                begin
                    if (Rb == 1'b1) begin
                        Nextstate = 1;
                    end
                end
      1 :
                begin
                    if (Rb == 1'b1) begin
                        Roll = 1'b1;
                    end
                    else if (Sum == 7 | Sum == 11) begin
                        Nextstate = 2;
                    end
                    else if (Sum == 2 | Sum == 3 | Sum == 12) begin
                        Nextstate = 3;
                    end
                    else begin
                        Sp = 1'b1;
                        Nextstate = 4;
                    end
                end
```

```verilog
            2 :
                        begin
                            Win = 1'b1;
                            if (Reset == 1'b1) begin
                                Nextstate = 0;
                            end
                        end
            3 :
                        begin
                            Lose = 1'b1;
                            if (Reset == 1'b1) begin
                                Nextstate = 0;
                            end
                        end
            4 :
                        begin
                            if (Rb == 1'b1) begin
                                Nextstate = 5;
                            end
                        end
            5 :
                        begin
                            if (Rb == 1'b1) begin
                                Roll = 1'b1;
                            end
                            else if (Sum == Point) begin
                                Nextstate = 2;
                            end
                            else if (Sum == 7) begin
                                Nextstate = 3;
                            end
                            else begin
                                Nextstate = 4;
                            end
                        end
        default :
                        begin
                            Nextstate = 0;
                        end
        endcase
    end

    always @(posedge CLK)
    begin
            State <= Nextstate;
            if (Sp == 1'b1) begin
                Point <= Sum;
            end
    end
endmodule
```

We are now ready to test the behavioral model of the dice game. It is not convenient to include the counters that generate random numbers in the initial test, since we want to specify a sequence of dice rolls that will test all paths on the SM chart. We could prepare a simulator command file that would generate a sequence of data for *Rb*, *Sum*, and *Reset*. This would require careful analysis of the timing to make sure that the input signals change at the proper time. A better approach for testing the dice game is to design a Verilog test bench module to monitor the output signals from the dice-game module and supply a sequence of inputs in response.

Figure 5-16 shows the `DiceGame` connected to a module called `GameTest`. `GameTest` needs to perform the following functions:

1. Initially supply the *Rb* signal.
2. When the `DiceGame` responds with a *Roll* signal, supply a *Sum* signal, which represents the sum of the two dice.
3. If no *Win* or *Lose* signal is generated by the `DiceGame`, repeat steps 1 and 2 to roll again.
4. When a *Win* or *Lose* signal is detected, generate a *Reset* signal and start again.

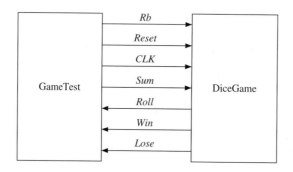

FIGURE 5-16: Dice Game with Test Bench

Figure 5-17 shows an SM chart for the `GameTest` module. *Rb* is generated in state T_0. When `DiceGame` detects *Rb*, it goes to S_1 and generates *Roll*. When `GameTest` detects *Roll*, the *Sum* that represents the next roll of the dice is read from *Sumarray(i)* and *i* is incremented. When the state goes to T_1, *Rb* goes to 0. The `DiceGame` goes to S_2, S_3, or S_4 and `GameTest` goes to T_2. The *Win* and *Lose* outputs are tested in state T_2. If *Win* or *Lose* is detected, a *Reset* signal is generated before the next roll of the dice. After *N* rolls of the dice, `GameTest` goes to state T_3, and no further action occurs.

`GameTest` (Figure 5-18) implements the SM chart for the `GameTest` module. It contains an array of test data, a concurrent statement that generates the clock, and two processes. The first always statement generates *Rb*, *Reset*, and *Tnext* (the next state) whenever *Roll*, *Win*, *Lose*, or *Tstate* changes. The second always statement updates *Tstate* (the state of `GameTest`). When running the simulator, we want to display only one line of output for each roll of the dice. To facilitate this, we have added a signal *Trig1*, which changes every time state T_2 is entered.

5.2 Derivation of SM Charts

FIGURE 5-17: SM Chart for Dice-Game Test

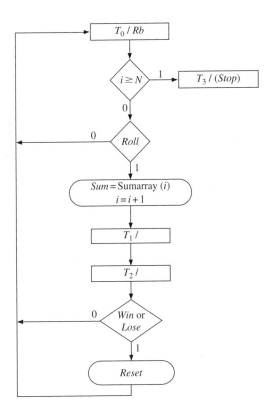

FIGURE 5-18: Dice-Game Test Module

```
module GameTest (Rb, Reset, Sum, CLK, Roll, Win, Lose);

    output Rb;
    output Reset;
    output[3:0] Sum;
    input CLK;
    input Roll;
    input Win;
    input Lose;

    reg[3:0] Sum;
    reg Reset;
    reg Rb;
    reg[1:0] Tstate;
    reg[1:0] Tnext;
    reg Trig1;

    integer Sumarray[0:11];
    integer i;

    initial
    begin
        Sumarray[0] = 7;
        Sumarray[1] = 11;
```

```verilog
            Sumarray[2] = 2;
            Sumarray[3] = 4;
            Sumarray[4] = 7;
            Sumarray[5] = 5;
            Sumarray[6] = 6;
            Sumarray[7] = 7;
            Sumarray[8] = 6;
            Sumarray[9] = 8;
            Sumarray[10] = 9;
            Sumarray[11] = 6;
            i = 0;
            Tstate = 0;
            Tnext = 0;
            Trig1 = 0;
        end
        always @(Roll or Win or Lose or Tstate)
        begin
            case (Tstate)
                0 :
                            begin
                                Rb = 1'b1;
                                Reset = 1'b0;
                                if (i >= 12) begin
                                    Tnext = 3;
                                end
                                else if (Roll == 1'b1) begin
                                    Sum = Sumarray[i];
                                    i = i + 1;
                                    Tnext = 1;
                                end
                            end
                1 :
                            begin
                                Rb = 1'b0;
                                Tnext = 2;
                            end
                2 :
                            begin
                                Tnext = 0;
                                Trig1 = ~Trig1;
                                if ((Win || Lose) == 1'b1) begin
                                    Reset = 1'b1;
                                end
                            end
                3 :
                            begin
                            end
            endcase
        end
        always @(posedge CLK)
```

```verilog
    begin
        Tstate <= Tnext;
    end
endmodule
```

Tester (Figure 5-19) connects the `DiceGame` and `GameTest` components so that the game can be tested. Figure 5-20 shows the simulator command file and output. The listing is triggered by *Trig1* once for every roll of the dice. The run 2000 command runs for more than enough time to process all the test data.

FIGURE 5-19: Tester for Dice Game

```verilog
module tester ();
    wire rb1;
    wire reset1;
    reg clk1;
    wire roll1;
    wire win1;
    wire lose1;
    wire[3:0] sum1;

    initial
      begin
          clk1 = 0;
      end

    always #20 clk1 <= ~clk1;

    DiceGame Dice (rb1, reset1, clk1, sum1, roll1, win1, lose1);
    GameTest Dicetest (rb1, reset1, sum1, clk1, roll1, win1, lose1);
endmodule
```

FIGURE 5-20: Simulation and Command File for Dice Game Tester

add list /Dicetest/Trig1 -notrigger sum1 win1 lose1 /Dice/Point run 2000

ns	delta	trig1	sum1	win1	lose1	point
0	+1	0	x	x	x	2
100	+3	0	7	1	0	2
260	+3	0	11	1	0	2
420	+3	0	2	0	1	2
580	+2	1	4	0	0	4
740	+3	1	7	0	1	4
900	+2	0	5	0	0	5
1060	+2	1	6	0	0	5
1220	+3	1	7	0	1	5
1380	+2	0	6	0	0	6
1540	+2	1	8	0	0	6
1700	+2	0	9	0	0	6
1860	+3	0	6	1	0	6

5.3 Realization of SM Charts

Methods used to realize SM charts are similar to the methods used to realize state graphs. As with any sequential circuit, the realization will consist of a combinational subcircuit, together with flip-flops for storing the state of the circuit. In some cases, it may be possible to identify equivalent states in an SM chart and eliminate redundant states using the same method as was used for reducing state tables. However, an SM chart is usually incompletely specified in the sense that all inputs are not tested in every state, which makes the reduction procedure more difficult. Even if the number of states in an SM chart can be reduced, it is not always desirable to do so, since combining states may make the SM chart more difficult to interpret.

Before deriving next-state and output equations from an SM chart, a state assignment must be made. The best way of making the assignment depends on how the SM chart is realized. If gates and flip-flops (or the equivalent PLD realization) are used, the guidelines for state assignment given in Section 1.7 may be useful. If programmable gate arrays are used, a one-hot assignment may be best, as explained in Section 6.9.

As an example of realizing an SM chart, consider the SM chart in Figure 5-21.

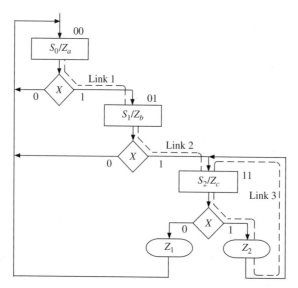

FIGURE 5-21: Example SM Chart for Implementation

We have made the state assignment $AB = 00$ for S_0, $AB = 01$ for S_1, and $AB = 11$ for S_2. After a state assignment has been made, output and next-state equations can be read directly from the SM chart. Since the Moore output Z_a is 1 only in state 00, $Z_a = A'B'$. Similarly, $Z_b = A'B$ and $Z_c = AB$. The conditional output $Z_1 = ABX'$, since the only link path through Z_1 starts with $AB = 11$ and takes the $X = 0$ branch. Similarly, $Z_2 = ABX$. There are three link paths (labeled link 1, link 2, and link 3 in Figure 5-7(b)), which terminate in a state that has $B = 1$. Link 1 starts with a present state $AB = 00$, takes the $X = 1$ branch, and terminates on a state in which $B = 1$. Therefore, the next state of B (B^+) equals 1 when $A'B'X = 1$. Link 2 starts in state

01, takes the $X = 1$ branch, and ends in state 11, so B^+ has a term $A'BX$. Similarly, B^+ has a term ABX from link 3. The next state equation for B thus has three terms corresponding to the three link paths:

$$B^+ = \underset{\text{link 1}}{A'B'X} + \underset{\text{link 2}}{A'BX} + \underset{\text{link 3}}{ABX}$$

Similarly, two link paths terminate in a state with $A = 1$, so

$$A^+ = A'BX + ABX$$

These output and next state equations can be simplified with Karnaugh maps using the unused state assignment ($AB = 10$) as a "don't care" condition.

As illustrated previously for flip-flops A and B, the procedure for deriving the next state equation for a flip-flop Q from the SM chart is as follows:

1. Identify all of the states in which $Q = 1$.
2. For each of these states, find all the link paths that lead *into* the state.
3. For each of these link paths, find a term that is 1 when the link path is followed. That is, for a link path from S_i to S_j, the term will be 1 if the machine is in state S_i and the conditions for exiting to S_j are satisfied.
4. The expression for Q^+ (the next state of Q) is formed by ORing together the terms found in step 3.

Implementation of Binary Multiplier Controller

Next, consider the SM chart for the multiplier control repeated here, shown in Figure 5-22.

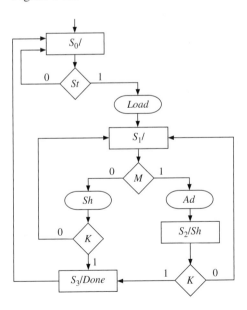

FIGURE 5-22: SM Chart for Multiplier Controller

We can realize this SM chart with two D flip-flops and a combinational circuit. Let us assume that the state assignments are $AB = 00$ for S_0, $AB = 01$ for S_1, $AB = 10$ for S_2, and $AB = 11$ for S_3.

The logic equations for the multiplier control and the next state equations can be derived by tracing link paths on the SM chart and then simplifying the resulting equations. First, let us consider the control signals. *Load* is true only in S_0 and only if *St* is true. Hence, $Load = S_0 \cdot St = A'B'St$. Similarly, *Ad* is true only in S_1 and only if *M* is true. Hence, $Ad = A'BM$. *Done* is a Moore output in S_3, and hence, $Done = S_3 = AB$. In summary, the logic equations for the multiplier control are

$$Load = A'B'St$$
$$Sh = A'BM'(K' + K) + AB'(K' + K) = A'BM' + AB'$$
$$Ad = A'BM$$
$$Done = AB$$

The next state equations can be derived by inspection of the SM chart and considering the state assignments. *A* is true in states S_2 and S_3. State S_2 is the next state when current state is S_1 and *M* is true ($A'BM$). State S_3 is the next state when current state is S_1, *M* is false, and *K* is true ($A'BM'K$) and when current state is S_2 and *K* is true ($AB'K$). Hence, we can write that

$$A^+ = A'BM'K + A'BM + AB'K = A'B(M + K) + AB'K$$

Similarly one can derive the next-state equation for *B* by inspection of the ASM diagram:

$$B^+ = A'B'St + A'BM'(K' + K) + AB'(K' + K) = A'B'St + A'BM' + AB'$$

The multiplier controller can be implemented in a hardwired fashion by two flip-flops and a few logic gates. The logic gates implement the next-state equations and the control signal equations. The circuit can be implemented with discrete gates or in a PLA, CPLD, or FPGA.

Table 5-1 illustrates a state transition table for the multiplier control. Each row in the table corresponds to one of the link paths in the SM chart. Since S_0 has two exit paths, the table has two rows for present state S_0. The first row corresponds to the $St = 0$ exit path, so the next state and outputs are 0. In the second row, $St = 1$, so the next state is 01 and the other PLA outputs are 1000. Since *St* is not tested in states S_1, S_2, and S_3, *St* is a "don't care" in the corresponding rows. The outputs for each row can be filled in by tracing the corresponding link paths on the SM chart. For example, the link path from S_1 to S_2 passes through conditional output *Ad*, so $Ad = 1$ in this row. Since S_2 has a Moore output *Sh*, $Sh = 1$ in both of the rows for which $AB = 10$.

TABLE 5-1: State Transition Table for Multiplier Control

	A	B	St	M	K	A^+	B^+	Load	Sh	Ad	Done
S_0	0	0	0	–	–	0	0	0	0	0	0
	0	0	1	–	–	0	1	1	0	0	0
S_1	0	1	–	0	0	0	1	0	1	0	0
	0	1	–	0	1	1	1	0	1	0	0
	0	1	–	1	–	1	0	0	0	1	0
S_2	1	0	–	–	0	0	1	0	1	0	0
	1	0	–	–	1	1	1	0	1	0	0
S_3	1	1	–	–	–	0	0	0	0	0	1

The design may also be implemented with ROM. If it has to be implemented using the ROM method, one can calculate the size of the ROM as follows. There are five different inputs to the combinational circuit here (A, B, St, M, and K). Hence, the ROM will have 32 entries. The combinational circuit should generate six signals (four control signals plus two next states). Hence, each entry has to be 6 bits wide. Thus, this design can be implemented using a 32 × 6 ROM and two D flip-flops.

If a ROM is used, the table must be expanded to $2^5 = 32$ rows since there are five inputs. To expand the table, the dashes in each row must be replaced with all possible combinations of 0s and 1s. If a row has n dashes, it must be replaced with 2^n rows. For example, the fifth row in Table 5-1 would be replaced with the following four rows:

```
0  1  0  1  0 | 1  0  0  0  1  0
0  1  0  1  1 | 1  0  0  0  1  0
0  1  1  1  0 | 1  0  0  0  1  0
0  1  1  1  1 | 1  0  0  0  1  0
```

The added entries are printed in boldface.

5.4 Implementation of the Dice Game

We can realize the SM chart for the dice game (Figure 5-13) using combinational circuitry and three D flip-flops, as shown in Figure 5-23. We use a straight binary state assignment. The combinational circuit has 9 inputs and 7 outputs. Three of the inputs correspond to current state, and three of the outputs provide the

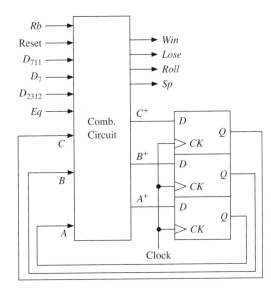

FIGURE 5-23: Realization of Dice-Game Controller

next-state information. All inputs and outputs are listed at the top of Table 5-2. The state transition table has one row for each link path on the SM chart. In state $ABC = 000$, the next state is $A^+B^+C^+ = 000$ or 001, depending on the value of Rb. Since state 001 has four exit paths, the table has four corresponding rows. When Rb is 1, *Roll* is 1 and there is no state change. When $Rb = 0$ and D_{711} is 1, the next state is 010. When $Rb = 0$ and $D_{2312} = 1$, the next state is 011. For the link path from state 001 to 100, Rb, D_{711}, and D_{2312} are all 0, and Sp is a conditional output. This path corresponds to row 4 of the state transition table, which has $Sp = 1$ and $A^+B^+C^+ = 100$. In state 010, the *Win* signal is always on, and the next state is 010 or 000, depending on the value of *Reset*. Similarly, *Lose* is always on in state 011. In state 101, $A^+B^+C^+ = 010$ if $Eq = 1$; otherwise, $A^+B^+C^+ = 011$ or 100, depending on the value of D_7. Since states 110 and 111 are not used, the next states and outputs are don't cares when $ABC = 110$ or 111.

One can use Table 5-2 and derive equations for the control signals and the next state equations. The required equations can be derived from Table 5-2 using the method of map-entered variables (see Chapter 1) or using a CAD program such as *LogicAid*. These equations can also be derived by tracing link paths on the SM chart and then simplifying the resulting equations using the "don't care" next states.

Figure 5-24 shows K-maps for A^+, B^+, and *Win*, which were plotted directly from the table. Since A, B, C, and Rb have assigned values in most of the rows of the table, these four variables are used on the map edges, and the remaining variables are entered within the map. (Chapter 1 described the K-map technique that uses map-entered variables.) E_1, E_2, E_3, and E_4 on the maps represent the expressions given below the maps. From the A^+ column in the table, A^+ is 1 in row 4, so we should enter $D'_{711}D'_{2312}$ in the $ABCRb = 0010$ square of the map. To save space, we define $E_1 = D'_{711}D'_{2312}$ and place E_1 in the square. Since A^+ is 1 in rows 11, 12,

TABLE 5-2: State Transition Table for Dice Game

	ABC	Rb	Reset	D_7	D_{711}	D_{2312}	Eq	A^+	B^+	C^+	Win	Lose	Roll	Sp
1	000	0	—	—	—	—	—	0	0	0	0	0	0	0
2	000	1	—	—	—	—	—	0	0	1	0	0	0	0
3	001	1	—	—	—	—	—	0	0	1	0	0	1	0
4	001	0	—	—	0	0	—	1	0	0	0	0	0	1
5	001	0	—	—	0	1	—	0	1	1	0	0	0	0
6	001	0	—	—	1	—	—	0	1	0	0	0	0	0
7	010	—	0	—	—	—	—	0	1	0	1	0	0	0
8	010	—	1	—	—	—	—	0	0	0	1	0	0	0
9	011	—	1	—	—	—	—	0	0	0	0	1	0	0
10	011	—	0	—	—	—	—	0	1	1	0	1	0	0
11	100	0	—	—	—	—	—	1	0	0	0	0	0	0
12	100	1	—	—	—	—	—	1	0	1	0	0	0	0
13	101	0	—	0	—	—	0	1	0	0	0	0	0	0
14	101	0	—	1	—	—	0	0	1	1	0	0	0	0
15	101	0	—	—	—	—	1	0	1	0	0	0	0	0
16	101	1	—	—	—	—	—	1	0	1	0	0	1	0
17	110	—	—	—	—	—	—	—	—	—	—	—	—	—
18	111	—	—	—	—	—	—	—	—	—	—	—	—	—

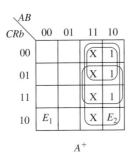

FIGURE 5-24: Maps Derived from Table 5-2

and 16, 1s are placed on the map squares $ABCRb = 1000, 1001$, and 1011. From row 13, we place $E_2 = D'_7 Eq'$ in the 1010 square. In rows 7 and 8, *Win* is always 1 when $ABC = 010$, so 1s are plotted in the corresponding squares of the *Win* map.

$$E_1 = D'_{711} D'_{2312} \qquad R = Reset$$
$$E_2 = D'_7 Eq' \qquad E_3 = D_{711} + D'_{711} D_{2312} = D_{711} + D_{2312}$$
$$E_4 = Eq + Eq' D_7 = Eq + D_7$$

The resulting equations are

$$A^+ = A'B'C\,Rb'D'_{711}D'_{2312} + AC' + ARb + AD'_7 Eq' \qquad (5\text{-}1)$$
$$B^+ = A'B'C\,Rb'(D_{711} + D_{2312}) + BReset' + AC\,Rb'(Eq + D_7)$$
$$C^+ = B'Rb + A'B'C\,D'_{711}D_{2312} + BC\,Reset' + AC\,D_7 Eq'$$
$$Win = BC'$$
$$Lose = BC$$
$$Roll = B'CRb$$
$$Sp = A'B'C\,Rb'D'_{711}D'_{2312}$$

These equations can be implemented in any standard technology (using discrete gates, PALs, GALs, CPLDs, or FPGAs).

The dice-game controller can also be realized using a ROM. A ROM (LUT) implementation of the game controller will need 512 entries (since there are nine inputs). Each entry must be 7 bits wide (3 bits for next states and 4 bits for outputs). The ROM is very large because of the large number of inputs involved. The ROM method is, hence, not very desirable for state machines with a large number of inputs.

We now write a data flow Verilog model for the dice game controller based on the block diagram of Figure 5-11 and Equations 5-1. The corresponding Verilog code is shown in Figure 5-25. The always statement updates the flip-flop states and the point register when the rising edge of the clock occurs. Generation of the control signals and D flip-flop input equations is done using concurrent statements. In particular, D_7, D_{711}, D_{2312}, and Eq are implemented using conditional signal assignments. As an alternative, all the signals and D input equations could have been implemented in an always statement with a sensitivity list containing A, B, C,

Sum, *Point*, *Rb*, D_7, D_{711}, D_{2312}, *Eq*, and *Reset*. If the Verilog module of Figure 5-25 is used with the test bench of Figure 5-19, the results are identical to those obtained with the behavioral architecture in Figure 5-15.

To complete the Verilog implementation of the dice game, we add two modulo-six counters as shown in Figures 5-26 and 5-27. The counters are initialized to 1, so the sum of the two dice will always be in the range 2 through 12. When *Cnt1* is in state 6, the next clock sets it to state 1, and *Cnt2* is incremented (or *Cnt2* is set to 1 if it is in state 6).

FIGURE 5-25: Data Flow Model for Dice Game (Based on Equations 5-1)

```verilog
module DiceGame (Rb, Reset, Clk, Sum, Roll, Win, Lose);
    input Rb;
    input Reset;
    input Clk;
    input[3:0] Sum;
    output Roll;
    output Win;
    output Lose;

    wire Sp;
    wire Eq;
    wire D7;
    wire D711;
    wire D2312;
    wire DA;
    wire DB;
    wire DC;
    reg A;
    reg B;
    reg C;
    reg[3:0] Point;

    initial
    begin
        A = 0;
        B = 0;
        C = 0;
    end

    always @(posedge Clk)
    begin
        A <= DA ;
        B <= DB ;
        C <= DC ;
        if (Sp == 1'b1)
            Point <= Sum ;
    end

    assign Win = B & ~C ;
    assign Lose = B & C ;
    assign Roll = ~B & C & Rb ;
```

```
    assign Sp = ~A & ~B & C & ~Rb & ~D711 & ~D2312 ;
    assign D7 = (Sum == 7) ? 1'b1 : 1'b0 ;
    assign D711 = ((Sum == 11) | (Sum == 7)) ? 1'b1 : 1'b0 ;
    assign D2312 = ((Sum == 2) | (Sum == 3) | (Sum == 12)) ? 1'b1 : 1'b0 ;
    assign Eq = (Point == Sum) ? 1'b1 : 1'b0 ;
    assign DA = (~A & ~B & C & ~Rb & ~D711 & ~D2312) | (A & ~C) | (A & Rb) | (A & ~D7
            & ~Eq) ;
    assign DB = ((~A & ~B & C & ~Rb) & (D711 | D2312)) | (B & ~Reset) | ((A & C & ~Rb)
            & (Eq | D7)) ;
    assign DC = (~B & Rb) | (~A & ~B & C & ~D711 & D2312) | (B & C & ~Reset) | (A & C
            & D7 & ~Eq) ;
endmodule
```

FIGURE 5-26: Counter for Dice Game

```
module Counter (Clk, Roll, Sum);

    input Clk;
    input Roll;
    output [3:0] Sum;

    reg [2:0] Cnt1;
    reg [2:0] Cnt2;

    initial
    begin
        Cnt1 = 1;
        Cnt2 = 1;
    end

    always @(posedge Clk)
    begin
        if (Roll == 1'b1) begin
            if (Cnt1 == 6) begin
                Cnt1 <= 1;
            end
            else begin
                Cnt1 <= Cnt1 + 1 ;
            end
            if (Cnt1 == 6) begin
                if (Cnt2 == 6) begin
                    Cnt2 <= 1;
                end
                else begin
                    Cnt2 <= Cnt2 + 1;
                end
            end
        end
    end

    assign Sum = Cnt1 + Cnt2;
endmodule
```

FIGURE 5-27: Complete Dice Game

```
module Game (Rb, Reset, Clk, Win, Lose);
   input Rb;
   input Reset;
   input Clk;
   output Win;
   output Lose;

   wire roll1;
   wire[3:0] sum1;

   DiceGame Dice (Rb, Reset, Clk, sum1, roll1, Win, Lose);
   Counter Count (Clk, roll1, sum1);
endmodule
```

This section has illustrated one way of realizing an SM chart. The implementation can use discrete gates, a PLA, a ROM, or a PAL. Alternative procedures are available that make it possible to reduce the size of the PLA or ROM by adding some components to the circuit. These methods are generally based on transformation of the SM chart to different forms and techniques, such as microprogramming.

5.5 Microprogramming

Microprogramming is a technique to implement the control unit of a digital system. In order to realize a control unit, one can inspect the state diagram or SM chart, write the logic equations for the control outputs and the next states, and implement the state machine using gates and flip-flops. Sections 5.3 and 5.4 demonstrated this process for the binary multiplier and the dice game, respectively. This method of implementation is called **hardwiring**, to indicate that the control signals are generated using fixed (hardwired) logic circuitry.

In contrast, an alternative approach called **microprogramming** has been developed for designing control units for complex digital systems. Proposed by Maurice Wilkes in 1951, microprogramming is building a special computer for executing the algorithmic flow chart describing the controller of a system. This development stemmed out of the separation of architecture and controller, which was described at the beginning of Chapter 4. Once the architecture and controller are clearly delineated, the controller flow chart systematically specifies all the controller signals that should be generated at each time during the flow of control from the reset state through each of the other states. By inspection of the SM chart for the shift and add binary multiplier in Figure 5-28(a), one can write pseudocode for the multiplier controller operation, as illustrated in Figure 5-28(b). This multiplier was presented in detail in Chapter 4.

Such a description of the controller easily makes one see the correspondence of the controller activity to a normal computer program. Microprogramming developed from exactly this realization.

FIGURE 5-28 SM Chart and Operation Flow of the Multiplier

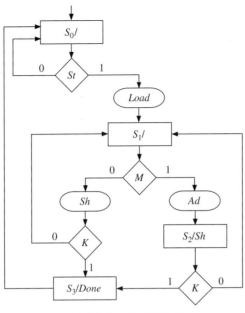

(a) SM chart for Multiplier

S_0: if St is true, produce Load Signal and go to S_1,
  else return to S_0
S_1: if M is true, produce Ad and go to S_2,
  else produce Sh, check whether K is 1;
  if K is 1 go to S_3;
  if K is 0, go to S_1;
S_2: produce Sh;
  if $K = 0$, go to S_1;
  else go to S_3;
S_3: produce Done and go to S_0

(b) Pseudocode representing the operation of the multiplier controller

If a memory can store all control signals and the next state information corresponding to each state for each input condition, one should be able to realize the controller by just "**sequencing**" through the memory. For this reason, microprogrammed controllers are also often called sequencers. The memory that stores the control words is called the **control store** or **microprogram memory**.

Microprogramming seemed extremely attractive in an era where the complexity of digital systems was growing prohibitively. Since debugging was done manually in those days, it was very hard to identify and correct errors. The systematic nature of microprogramming made debugging systems easier. Changes to systems can be implemented relatively easily. Errors can be identified and corrected easily. This made microprogramming very popular.

The disadvantage of microprogramming is that it is slow. A memory access is required to access the control word from the control store. Hardwiring results in faster systems, because hardwired control signals are generated by logic gates and they are typically faster than memory.

Early microprocessors such as Intel 8086 and Motorola 68000 were microprogrammed. These microprocessors supported a variety of memory-addressing modes with base registers and index registers. They allowed operands to be accessed directly from memory and results to be written directly to memory. Many complex instructions that performed a series of fundamental operations were available on these processors. Microprogramming was convenient when the control signals for the several operations needed for a complex instruction could be systematically specified in the microprogram word. It would have been extremely hard to implement these microprocessors with hardwiring.

Many things have changed since then. In the late 1970s, it was observed that in many microprocessors, more than half of the chip area was spent in the controller—that is, the data path of the processor occupied less than half the chip area. The complexity of the microprocessors led researchers and designers to the reduced instruction set computing (RISC) era. RISC microprocessors are simpler, have fewer memory-addressing modes, and need simpler control units. Computer-aided design (CAD) tools have improved, and the designers' capability to debug has improved. Today microprogramming may be used only for microprocessors with complex instruction set architectures (ISAs), however, it is a powerful concept and a very elegant one.

Microprogramming can be implemented in a variety of ways. The general idea is to store a control word corresponding to each state. The control word is also called a **microinstruction**. The microinstruction specifies the outputs to be generated. It also specifies where the next microinstruction can be found. This corresponds to the state transitions in the state diagram or SM chart.

Figure 5-29 illustrates a suitable hardware arrangement for a typical microprogram implementation. Each ROM location stores a control word or microinstruction. The only inputs to the ROM come from the state register. A multiplexer with each of the inputs can be used to selectively test at most one variable in each state. This multiplexer is used to indicate whether the selected control signal (as indicated by TEST) is true or false. Another multiplexer is used to select which next state should control branch to.

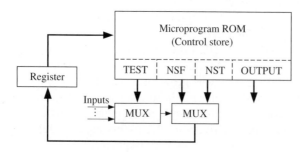

FIGURE 5-29: Typical Hardware Arrangement for Microprogramming

The ROM output has four fields: TEST, NSF, NST, and OUTPUT. TEST controls the input MUX, which selects one of the inputs to be tested in each state. If this input is 0 (false), the second MUX selects the NSF field as the next state. If the input is 1 (true), it selects the NST field as the next state. The OUTPUT bits correspond to the control signals. Note that in order to use this hardware arrangement, the SM chart must have only Moore outputs, since the outputs can be a function only of the state.

SM Chart Transformations for Microprogramming

Transformations are performed on the SM chart to facilitate easy and efficient microprogramming. One does not want a naïve look-up table method where all combinations of inputs and present states are directly specified. We transform the SM chart in such a way that only one entry is required per state. Some of the transformations do increase the number of states; however, the achieved microprogram size is still significantly smaller than the ROM size in a naïve LUT method.

Eliminate Conditional Outputs

It is desirable to construct the controller as a Moore machine so that there will be no conditional control signals. If control signals are conditional on some inputs, one should store control signals corresponding to different combinations of inputs. Hence, the first step in transforming a state diagram or SM chart for easy microprogramming is to convert it into a Moore state machine. Any Mealy machine can be converted into a Moore machine by adding an appropriate number of additional states.

Allow Only One Qualifier Per State

The inputs that are tested in each state of the state machine are called **qualifiers** in the microprogram literature. For example, in Figure 5-28, St, M, and K are qualifiers. States S_0 and S_2 contain only one qualifier, but state S_1 tests qualifiers M and K. The multiple qualifiers in S_1 led to nested **if** statements in the pseudocode in Figure 5-28. Although microprogramming can be done with multiple qualifiers per state, it is simpler to implement microprogramming when only one variable is tested in each state. Thus, microprogramming becomes easy if the following two transformations are done on SM charts:

   I. Eliminate all conditional outputs by transforming to a Moore machine.
   II. Test only one input (qualifier) in each state.

Let us transform the SM chart of the multiplier for microprogramming. First, we will convert it to a Moore machine by adding a state for each conditional output (i.e., each oval in the SM chart). That results in additional states S_{01} in state S_0 for the conditional output $Load$, S_{11} in the original state S_1 for the conditional output Ad, and S_{12} in S_1 for the conditional output Sh. Fortunately, no more than one qualifier is tested in any state. The modified SM chart is shown in Figure 5-30.

FIGURE 5-30: Multiplier SM Chart with No Conditional Outputs (Derived from Figure 5-28)

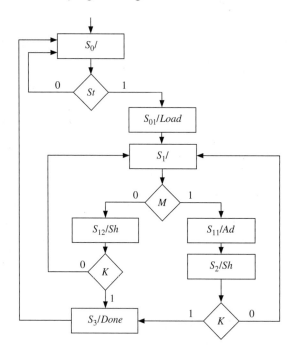

The corresponding actions can be described by the following pseudocode:

S_0:   if St is true, go to S_{01},
             else go to S_0;
S_{01}: produce Load; Go to S_1;
S_1:   if M is true, go to S_{11}, else go to S_{12};
S_{11}: produce Ad; go to S_2;
S_{12}: produce Sh; if K = 0, go to S_1; else go to S_3;
S_2:   produce Sh;
             if K=0, go to S_1;
             else go to S_3;
S_3:   produce Done; go to S_0;

At this stage, the transformed SM chart can be inspected for eliminating redundant states. Can states S_{11} and S_2 be combined? Since the add operation has to be performed before shift, the Ad control signal should appear ahead of the Sh control signal. Hence, S_{11} and S_2 cannot be combined.

Now, let us inspect states S_{12} and S_2. States S_{12} and S_2 perform exactly the same tasks and have the same next states. Hence, they can be combined. This is an example of potential state minimizations after the transformation. Let us denote the new combined state as S_2. The improved SM chart is shown in Figure 5-31.

The microprogram will look as in Table 5-3, assuming a straight binary state assignment in the sequence S_0, S_{01}, S_1, S_{11}, S_2, and S_3. Since there are three inputs, St, M, and K, a 4-to-1 MUX will be sufficient to select the appropriate qualifier. The multiplexer connections are assumed to be as in Figure 5-32.

5.5 Microprogramming

FIGURE 5-31: Modified Multiplier SM Chart after State Minimization Is Applied to Figure 5-30

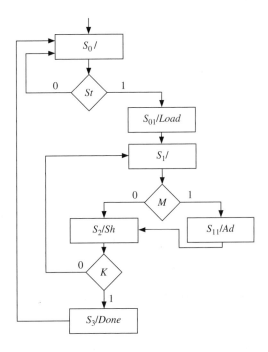

TABLE 5-3: Two-Address Microprogram for Multiplier—Both NST and NSF Specified (Corresponds to Figure 5-29b)

State	ABC	TEST	NSF	NST	Load	Ad	Sh	Done
S_0	000	00	000	001	0	0	0	0
S_{01}	001	11	010	010	1	0	0	0
S_1	010	01	100	011	0	0	0	0
S_{11}	011	11	100	100	0	1	0	0
S_2	100	10	010	101	0	0	1	0
S_3	101	11	000	000	0	0	0	1

Let us look at the first row in Table 5-3. It corresponds to state S_0, which is encoded as 000. The input tested is St. Since St is connected to input 0 of the multiplexer, the TEST field for this row is 00. If St is false, the next state is S_0, leading to 000 in the NSF field. If St is true, the next state is S_{01}, leading to the 001 bits in the NST field. The control signals $Load$, Ad, Sh, and $Done$ are 0 in state S_0.

The microcode for state S_{01} is shown in the second row. State S_{01} generates the $Load$ signal and the controller transitions to state S_1. No input signals are tested. In the multiplexer in Figure 5-32, we provide a value of 1 to the last unused muliplexer input. So, we can mark the TEST field as 11, corresponding to the last input of the

FIGURE 5-32: 4-1 MUX for Microprogramming the Multiplier (Two-Address Microcode)

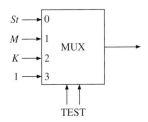

multiplexer. In state S_1, input signal M is tested. Since M is connected to input 1 of the multiplexer, the TEST field for the third row is 01. In a similar fashion, all rows of Table 5-3 are filled.

Since there are six states, three flip-flops will be required. The ROM that stores this microprogram will need six entries, one for each state. Each entry will need 12 bits, including 2 bits for TEST, 3 bits for NSF, 3 bits for NST, and 4 bits for control signals *Load, Ad, Sh,* and *Done. ABC* represents the address at which the microinstruction is stored.

The hardware arrangement in Figure 5-29 is for microprogramming with two next-state addresses and a single qualifier per state. Single-qualifier microprogramming means that only one input can be tested in a state. Two-address microcoding means that next states for both possible input values—next state if the input is true (NST) and next state if the input is false (NSF)—are explicitly specified in the control word. (Figure 5-29 could be modified to allow Mealy outputs by replacing the OUTPUT field with OUTPUTF and OUTPUTT and adding a MUX to select one of the two output fields.)

Single-Qualifier Single-Address Microcode

In the preceding microprogram, each microinstruction can specify two potential next states: the next state if the input is true and the next state if the input is false. The microcode for the different states can be located in any sequence, because the next microinstruction for each state is specified without assuming any default flow of control.

The preceding microprogram has resemblance to software, but in conventional programs, control flows in sequence except when branch and jump instructions alter the control flow. If a branch is not taken, control simply flows to the next instruction. If one could take advantage of a similar structure, each microprogram entry will need to specify only one next-state address.

Let us consider what we should do in order to make the default next state be the state located in the next row. In that case, the state assignments should be such that, if the qualifier (input) is false, the next state should be the current state incremented by 1. The next state when the qualifier is true will be the only next state explicitly specified in the microcode. If the qualifier is false, control simply goes to the next row to obtain the succeeding microinstruction.

This type of microprogram can be implemented using the hardware arrangement shown in Figure 5-33. Since control normally just advances to the next location, a counter can be effectively used. This counter is analogous to a program counter (PC)

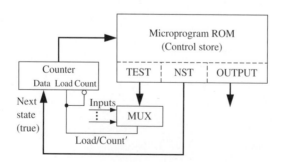

FIGURE 5-33: Microprogrammed System with Single-Address Microcode

in a microprocessor. The counter points to the current state of the controller, analogous to a PC pointing to the next instruction to be fetched. Each ROM location stores a control word or microinstruction. The OUTPUT bits correspond to the control signals. The TEST bits specify the qualifier being tested, and the NST bits indicate the target microinstruction if the qualifier is true. A multiplexer is used to indicate whether the selected control signal (as indicated by TEST) is true or false. If the qualifier is false, the counter increments to point to the next microinstruction. This corresponds to the default next state. If the qualifier is true, the counter should load the NST bits as the location of the next microinstruction. This is the explicitly specified next state. A counter with parallel load capability is the ideal building block for this module. The multiplexer selects the relevant qualifier, and its output is used to decide whether the counter should count sequentially or load the next state indicated by NST.

The state assignment for the single-address microcoding has to be done carefully. (In contrast, in the two-address microcoding that was discussed earlier, any state assignment was acceptable.) In the current technique, the assignments should meet the condition that for every state, one of the next states should be current state's assignment incremented by one (the default next state). For each condition box, for the false branch, the next state must be assigned in sequence, if possible. If this is not possible, extra states (called X-states) must be added. The required number of X-states can be reduced by assigning long strings of states in sequence. To facilitate this, it may be necessary to complement some of the variables that are tested.

Figure 5-34 illustrates the modified SM chart for a binary multiplier with a serial state assignment for single microcoding. For state S_0, input St is complemented, so that S_{01} can be the default next state, as shown in Figure 5-34(a). If input St is not complemented, an extra state will be required as shown in Figure 5-34(b). State S_2 is the default successor for state S_1. In state S_2, we use K' so that S_3 can be the default successor to S_2. Thus, in Figure 5-34(a), states $S_0, S_{01}, S_1, S_2,$ and S_3 can be assigned sequential values from 0 to 4. The explicit next state, corresponding to the

FIGURE 5-34: Modified SM Chart for Binary Multiplier with Serial State Assignment for Single-Address Microcoding

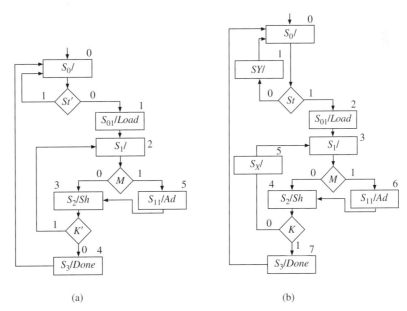

(a) (b)

qualifier being true, can have any assignment. We assign 5 to state S_{11}. If variable K was used instead of K', an extra state would be required on the path from S_2 to S_1, when K equals 0. As Figure 5-34(b) illustrates, two extra states will be required if input variables cannot be complemented.

Table 5-4 illustrates the single address microprogram for the multiplier. The modified SM chart, with the minimum number of states (Figure 5-34(a)), is used. Since there are three inputs, St', M, and K', a 4-to-1 MUX will be sufficient to select the appropriate qualifier. The multiplexer connections are assumed to be as in Figure 5-35.

TABLE 5-4: Single-Address Microprogram for Multiplier (Only NST Specified)

State	ABC	TEST	NST	Load	Ad	Sh	Done
S_0	000	00	000	0	0	0	0
S_{01}	001	11	010	1	0	0	0
S_1	010	01	101	0	0	0	0
S_2	011	10	010	0	0	1	0
S_3	100	11	000	0	0	0	1
S_{11}	101	11	011	0	1	0	0

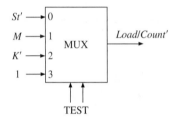

FIGURE 5-35: Multiplexer for Microprogramming the Multiplier (Single-Address Microcode)

The single address microprogram in Table 5-4 consists of six entries of 9 bits each in contrast to the two-address microprogram in Table 5-3, which needs six entries of 12 bits each.

If the multiplier controller is implemented by a standard ROM (LUT) method, the ROM size must be 32 × 6. There are four states, necessitating two flip-flops and two next-state equations. There are 3 inputs—St, M, and K. Hence, the state table for this state machine will have 32 rows. There will be two next-state equations and 4 outputs, necessitating 6 bits in each entry. A comparison of the ROM (LUT) method with the microcoded implementations is shown in Table 5-5. If the state machine had a large number of inputs, the size of the ROM in naïve LUT method will be prohibitively large.

TABLE 5-5: Comparison of Different Implementations of the Multiplier Control

Method	Size of ROM	
	#entries × width	# bits
ROM method with original SM chart	32 × 6	192 bits
Two-address microcode	6 × 12	72 bits
Single-address microcode	6 × 9	54 bits

5.5.1 Microprogramming the Dice Controller

Let us realize the dice controller that we described previously by microprogramming. It can be microprogrammed using two-address microcoding or single-address microcoding.

Two-Address Microcode Implementation for the Dice Controller

We first discuss the two-address microcoding of the dice controller using the hardware arrangement in Figure 5-29. In order to perform microcoding, we need to modify the SM chart. First, all the outputs must be converted to Moore outputs. Second, only one input variable must be tested in each state. This corresponds directly to the block diagram of Figure 5-29, since the TEST field can select only one input to test in each state and the output depends only on the state. Figure 5-36 shows a modified version of the dice game SM chart.

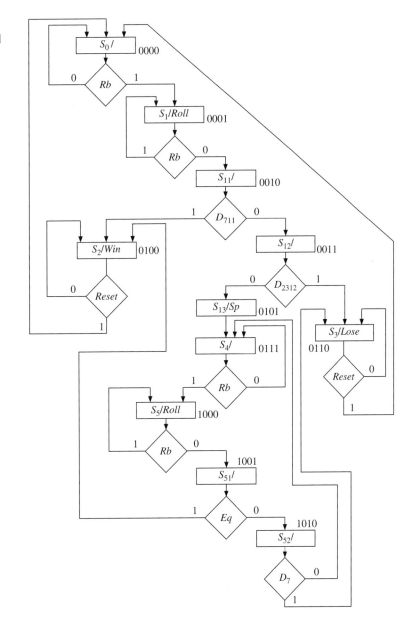

FIGURE 5-36: SM Chart with Moore Outputs and One Qualifier per State

Next, we derive the microprogram (Table 5-6) using a straight binary state assignment. The variables Rb, D_{711}, D_{2312}, Eq, D_7, and $Reset$ must be tested, so we will use an 8-to-1 MUX (Figure 5-37). When $TEST = 001$, Rb is selected, and so forth. In state S_{13} the next state is always 0111, so $NSF = NST = 0111$ and the TEST field is a "don't care." Each row in the ROM table corresponds to a link path on the SM chart. For example, in S_2, the test field 110 selects $Reset$. If $Reset = 0$, $NSF = 0100$ is selected, and if $Reset = 1$, $NST = 0000$ is selected. In S_2, the output $Win = 1$ and the other outputs are 0.

TABLE 5-6: Two-Address Microprogram for Dice Game

State	ABCD	TEST	NSF	NST	ROLL	Sp	Win	Lose
S_0	0000	001	0000	0001	0	0	0	0
S_1	0001	001	0010	0001	1	0	0	0
S_{11}	0010	010	0011	0100	0	0	0	0
S_{12}	0011	011	0101	0110	0	0	0	0
S_2	0100	110	0100	0000	0	0	1	0
S_{13}	0101	xxx	0111	0111	0	1	0	0
S_3	0110	110	0110	0000	0	0	0	1
S_4	0111	001	0111	1000	0	0	0	0
S_5	1000	001	1001	1000	1	0	0	0
S_{51}	1001	100	1010	0100	0	0	0	0
S_{52}	1010	101	0111	0110	0	0	0	0

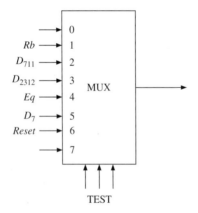

FIGURE 5-37: MUX for Two-Address Microcoding of Dice Game

Single-Address Microcode for the Dice Controller

Single-address microcode will use the hardware as in the block diagram of Figure 5-33. This circuit uses a counter instead of the state register. Only one target, the NST field, is specified. The TEST field selects one of the inputs to be tested in each state. If the selected input is 1 (true), the NST field is loaded into the counter. If the selected input is 0, the counter is incremented.

This method requires that the SM chart be modified, as shown in Figure 5-38, and that the state assignment be made in a serial fashion. If serial state assignment is not possible, extra states are added. The required number of X-states can be reduced by assigning long strings of states in sequence. To facilitate this, it may be necessary to complement some of the variables that are tested. In Figure 5-38, Rb and $Reset$ have each been complemented in two places, and the 0 and 1 branches

FIGURE 5-38: SM Chart with Serial State Assignment and Added X-State

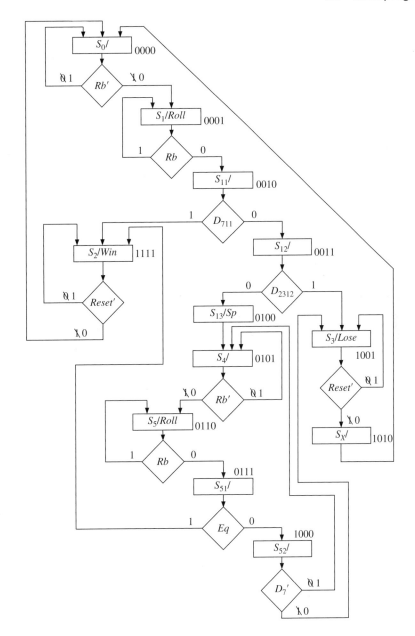

have been interchanged accordingly. With this change, states 0000, 0001, ... , 1000 are in sequence. S_3 has been assigned 1001, and before adding an X-state, *NSF* was 0000 and *NST* was 1001, so neither next state was in sequence. Therefore, X-state S_X was added with a sequential assignment 1010; the next state of S_X is always 0000. If we assign 1011 to S_2, the next states would be 1011 and 0000, and neither next state would be in sequence. We could solve the problem by adding an X-state. A better approach is to assign 1111 to S_2, as shown. Since incrementing 1111 goes to 0000, one of the next states is in sequence, and no X-state is required.

The inputs tested by the MUX in Figure 5-39 are similar to Figure 5-37, except that D_7 and Reset have been complemented and both Rb and Rb' are needed. Since NST is always 0101 in state S_{13}, a 1 input to multiplexer is needed. The corresponding microprogram ROM table is given in Table 5-7.

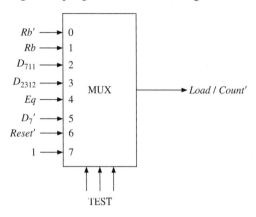

FIGURE 5-39: MUX for Single-Address Microcoding of Dice Game

TABLE 5-7: Microprogram for Dice Game with Single-Address Microcoding

State	ABCD	TEST	NST	ROLL	Sp	Win	Lose
S_0	0000	000	0000	0	0	0	0
S_1	0001	001	0001	1	0	0	0
S_{11}	0010	010	1111	0	0	0	0
S_{12}	0011	011	1001	0	0	0	0
S_{13}	0100	111	0101	0	1	0	0
S_4	0101	000	0101	0	0	0	0
S_5	0110	001	0110	1	0	0	0
S_{51}	0111	100	1111	0	0	0	0
S_{52}	1000	101	0101	0	0	0	0
S_3	1001	110	1001	0	0	0	1
S_x	1010	111	0000	0	0	0	0
S_2	1111	110	1111	0	0	1	0

A comparison of the naïve LUT (ROM) method implementation with the microprogrammed implementations is given in Table 5-8. The ROM method with original SM chart (Figure 5-13) needs 2^9 entries because it needs three state variables and six inputs. The three next-state variables and four outputs necessitate 7 bits in each entry. The two-address microcode entry is based on Table 5-7, and the single-address microcode entry is based on Table 5-6.

TABLE 5-8: Comparison of Different Implementations of Dice Controller

Method	Size of ROM	
	#entries × width	#bits
ROM method with original SM chart	512 × 7	3584 bits
Two-address microcode	11 × 15	165 bits
Single-address microcode	12 × 11	132 bits

The methods we have just studied for implementing SM charts are examples of microprogramming. The counter in Figure 5-33 is analogous to the program counter in a computer, which provides the address of the next instruction to be executed.

The ROM output is a **microinstruction**, which is executed by the remaining hardware. Each microinstruction is like a conditional branch instruction that tests an input and branches to a different address if the test is true; otherwise, the next instruction in sequence is executed. The output field in the microinstruction has bits that control the operation of the hardware.

5.6 Linked State Machines

When a sequential machine becomes large and complex, it is desirable to divide the machine into several smaller machines that are linked together. Each of the smaller machines is easier to design and implement. Also, one of the submachines may be "called" in several different places by the main machine. This is analogous to dividing a large software program into procedures that are called by the main program.

Figure 5-40 shows the SM charts for two serially linked state machines. The main machine (machine A) executes a sequence of "some states" until it is ready to call the submachine (machine B). When state S_A is reached, the output signal Z_A activates machine B. Machine B then leaves its idle state and executes a sequence of "other states." When it is finished, it outputs Z_B before returning to the idle state. When machine A receives Z_B, it continues to execute "other states." Figure 5-40 assumes that the two machines have a common clock.

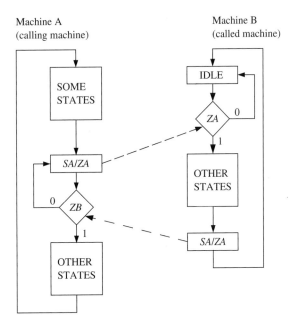

FIGURE 5-40: SM Charts for Serially Linked State Machines

As an example of using linked state machines, we split the SM chart of Figure 5-13 into two linked SM charts. In Figure 5-13, Rb is used to control the roll of the dice in states S_0 and S_1 and in an identical way in states S_4 and S_5. Since this function is repeated in two places, it is logical to use a separate machine for the roll control (Figure 5-41(b)). Use of the separate roll control allows the main dice control (Figure 5-41(a)) to be reduced from six states to four states. The main control generates an *En_roll* (enable roll) signal in T_0 and then waits for a *Dn_roll* (done rolling) signal before continuing. Similar action occurs in T_1. The roll-control machine waits in state S_0 until it receives an *En_roll* signal from the main dice-game control. Then, when the roll button is pressed ($Rb = 1$), the machine goes to S_1 and generates a *Roll* signal. It remains in S_1 until $Rb = 0$, in which case, the *Dn_roll* signal is generated and the machine goes back to state S_0.

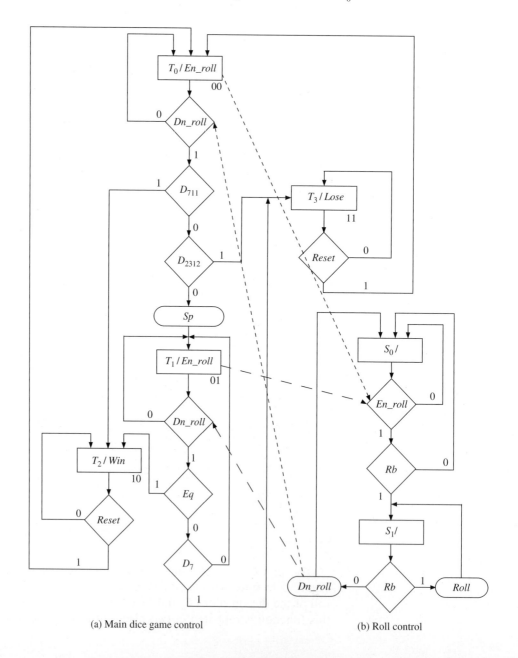

FIGURE 5-41: Linked SM Charts for Dice Game

(a) Main dice game control

(b) Roll control

In this chapter we described a procedure for digital system design based on SM charts. An SM chart is equivalent to a state graph, but it is usually easier to understand the system operation by inspection of the SM chart. After we have drawn a block diagram for a digital system, we can represent the control unit by an SM chart. Next we can write a behavioral Verilog description of the system based on this chart. Using a test bench written in Verilog, we can simulate the Verilog code to verify that the system functions according to specifications. After making any necessary corrections to the Verilog code and the SM chart, we can proceed with the detailed logic design of the system. Rewriting the Verilog architecture to describe the system operation in terms of control signals and logic equations allows us to verify that our design is correct.

We also presented techniques for implementing control units: hardwiring and microprogramming. We showed how logic equations can easily be derived by tracing link paths on an SM chart. Hardwired control units can easily be implemented from these equations. Then we presented microprogramming. In this technique, control words are stored in the microprogram memory. The size of the microprogram is reduced by transforming the SM chart into a form in which only one input is tested in each state. For complex systems, we can split the control unit into several sections by using linked state machines.

Problems

5.1 Draw an SM chart for the BCD-to-binary converter of Problem 4.10.

5.2 **(a)** Construct an SM chart equivalent to the following state table. Test only one variable in each decision box. Try to minimize the number of decision boxes.
(b) Write a Verilog description of the state machine based on the SM chart.

Present State	Next State $X_1X_2 =$ 00	01	10	11	Output (Z_1Z_2) $X_1X_2 =$ 00	01	10	11
S_0	S_3	S_2	S_1	S_0	00	10	11	01
S_1	S_0	S_1	S_2	S_3	10	10	11	11
S_2	S_3	S_0	S_1	S_1	00	10	11	01
S_3	S_2	S_2	S_1	S_0	00	00	01	01

5.3 Draw an SM chart for the square root circuit of Problem 4.12.

5.4 Construct an SM chart that is equivalent to the following state table. Test only one variable in each decision box. Try to minimize the number of decision boxes. Show Mealy and Moore outputs on the SM chart.

Present State	Next State $X_1X_2 =$ 00	01	10	11	Output $(Z_1Z_2Z_3)$ $X_1X_2 =$ 00	01	10	11
S_0	S_1	S_1	S_1	S_1	000	100	110	010
S_1	S_1	S_1	S_0	S_0	001	001	001	001

5.5 Draw an SM chart for the binary multiplier of Problem 4.19.

5.6 An association has 15 voting members. Executive meetings of this association can be held only if more than half (i.e., at least eight) of the members are present (i.e., eight is the minimum quorum required to hold meetings). Classified matters can be discussed and voted on only if two-thirds of the members are present. The chairman can cast two votes if the quorum is met and an even number of members (including the chairman) are present. Above the room door there are three lights—GREEN, BLUE, and RED—to indicate the quorum status. Derive an SM chart for a system that will indicate whether a minimum quorum has been met (GREEN), classified matters can be discussed (BLUE), or a quorum has been met and an even number of members are present (RED). GREEN and RED lights may be lit at the same time or GREEN, BLUE, and RED lights may be lit simultaneously.

Assume that there is a single door to the meeting room and that it is fitted with two photocells. One photocell (PHOTO1) is on the inner side of the door, and the other (PHOTO2) is on the outer side. Light beams shine on each photocell, producing a false output from the cell; a true output from a photocell arises when the light beam is interrupted. Assume that once a person starts through a door, the process is completed before another one can enter or leave (i.e., only one person enters or leaves at a time). If PHOTO1 is followed by PHOTO2, a sequencer generates a *LEAVE* signal and if PHOTO2 is followed by PHOTO1, the sequencer generates an *ENTER* signal. At most one *ENTER* or *LEAVE* will be true at any time. Assume that these signals will be true until you read them. Basically, you read the signal and provide a signal to the door controller indicating that the door is *READY* to let the next person in or out.

(a) Draw a block diagram for the data section of this circuit. Assume that *ENTER* and *LEAVE* signals are available for you (i.e., you do not need to generate them for this part of the question).
(b) Draw an SM chart for the controller. Write the steps required to accomplish the design. Define all control signals used.
(c) Draw an SM chart for a circuit that generates *ENTER* and *LEAVE*.

5.7 Design a binary-to-BCD converter that converts a 10-bit binary number to a 3-digit BCD number. Assume that the binary number is ≤ 999. Initially the binary number is placed in register B. When an *St* signal is received, conversion to BCD takes place and the resulting BCD number is stored in the A register (12 bits). Initially A contains 0000 0000 0000. The conversion algorithm is as follows: If the digit in any decade of A is ≥ 0101, add 0011 to that decade. Then shift the A register together with the B register one place to the left. Repeat until 10 shifts have occurred. At each step, as the left shift occurs, this effectively multiplies the BCD number by 2 and adds in the next bit of the binary number.

(a) Illustrate the algorithm by converting 100011101 to BCD.
(b) Draw the block diagram of the binary-to-BCD converter. Use a counter to count the number of shifts. The counter should output a signal C_{10} after 10 shifts have occurred.

(c) Draw an SM chart for the converter (three states).
(d) Write a Verilog description of the converter.

5.8 (a) Draw the block diagram for a divider that divides an 8-bit dividend by a 5-bit divisor to give a 3-bit quotient. The dividend register should be loaded when $St = 1$.
(b) Draw an SM chart for the control circuit.
(c) Write a Verilog description of the divider based on your SM chart. Your Verilog code should explicitly generate the control signals.
(d) Give a sequence of simulator commands that would test the divider for the case 93 divided by 17.

5.9 The block diagram for an elevator controller for a building with two floors is shown in the following diagram. The inputs FB_1 and FB_2 are floor buttons in the elevator. The inputs $CALL_1$ and $CALL_2$ are call buttons in the hall. The inputs FS_1 and FS_2 are floor switches that output a 1 when the elevator is at the first or second floor landing. Outputs UP and $DOWN$ control the motor, and the elevator is stopped when $UP = DOWN = 0$. N_1 and N_2 are flip-flops that indicate when the elevator is needed on the first or second floor. R_1 and R_2 are signals that reset these flip-flops. $DO = 1$ causes the door to open, and $DC = 1$ indicates that the door is closed. Draw an SM chart for the elevator controller (four states).

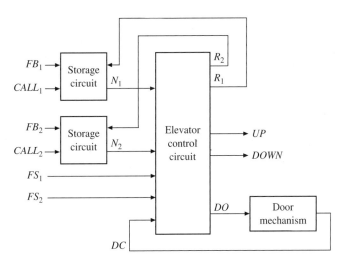

5.10 Realize the following SM chart using a ROM with a minimum number of inputs, a multiplexer, and a loadable counter (like the 74163). The ROM should generate NST. The multiplexer inputs are selected as shown in the table beside the SM chart.

(a) Draw the block diagram.
(b) Convert the SM chart to the proper format. Add a minimum number of extra states.
(c) Make a suitable state assignment and give the first five rows of the ROM table.
(d) Write a Verilog description of the system using a ROM.

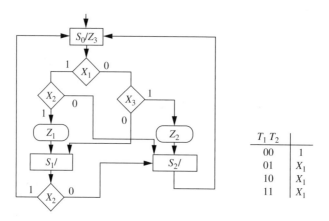

5.11 Design a multiplier for 16-bit binary integers. Use a design similar to Figures 4-33 and 4-34.

   (a) Draw the block diagram. Add a counter to the control circuit to count the number of shifts.
   (b) Draw the SM chart for the controller (three states). Assume that the counter outputs $K = 1$ after 15 shifts have occurred.
   (c) Write Verilog code for your design.

5.12 The SM charts for three linked machines are given here. All state changes occur during the falling edge of a common clock. Complete a timing chart including ST, Wa, A, B, C, and D. All state machines start in the state with an asterisk (*).

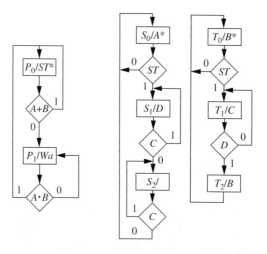

5.13 The following SM chart is to be realized using the two-address microprogramming structure shown in Figure 5-29.

   (a) Convert the SM chart to the proper form by adding a minimum number of states to the given diagram. Make a suitable state assignment.
   (b) Write the microprogram required to implement this SM chart.

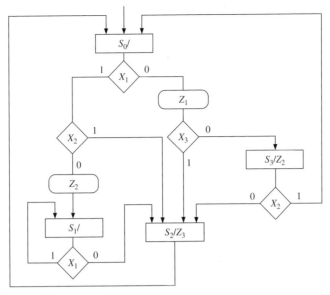

(c) Draw a block diagram showing how the SM chart can be realized using a ROM, multiplexers, and flip-flops.

$Q_c\ Q_b\ Q_a$	$T_1\ T_0\ C_F\ B_F\ A_F\ C_T\ B_T\ A_T\ Z_1\ Z_2\ Z_3$
0 0 0	

5.14 (a) Write Verilog code that describes the following SM chart. Assume that state changes occur on the falling edge of the clock. Use two processes.

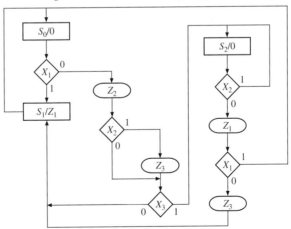

(b) The SM chart is to be implemented using a PLA and two flip-flops (A and B). Complete the state transition table (PLA table) by tracing link paths. Find the equation for A^+ by inspection of the PLA table.

$A\ B\ X_1\ X_2\ X_3$	$A^+\ B^+\ Z_1\ Z_2\ Z_3$

(c) Complete the following timing diagram.

[Timing diagram: Clock, X_1, X_2, X_3, State S_0, Z_1, Z_2, Z_3]

5.15 (a) What are the conditions an SM chart must satisfy in order to realize it using single-address microprogramming with a counter, a ROM, and a multiplexer as in Fig 5-33?

(b) Give the modified SM chart and the required state assignment if the SM chart of Problem 5.13 is realized with this kind of microprogramming.

5.16 SM charts for two linked state machines are shown here. Machine T starts in state T_0, and machine S starts in state S_0. Draw a timing chart that shows CLK, the states of T and S, and signals P, R, and D for 10 clocks. All state changes occur on the rising edge of the clock.

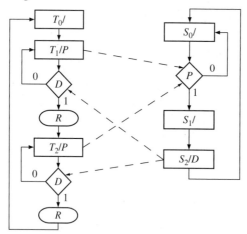

5.17 The SM charts for two linked state machines are given in the following diagram.

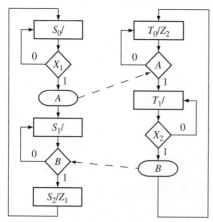

(a) Complete the timing diagram shown here.
(b) For the SM chart on the left, make a one-hot state assignment and derive D flip-flop input equations and output equations by inspection.

5.18 The SM chart for a simplified vending machine is shown here. The vending machine accepts only dimes and nickels. One soda costs 15 cents at this vending machine.

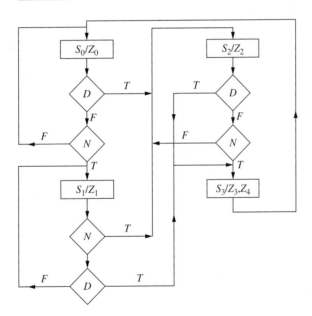

D: Dime **N**: Nickel Z_0: 0 cent Z_1: 5 cents Z_2: 10 cents Z_3: 15 cents Z_4: soda

(a) Modify this figure to add one more functionality on resetting the vending machine. This reset could cancel the order.
(b) Derive the next-state equation and draw the state table.
(c) If the given SM chart (with reset function) is implemented using a traditional ROM method, how big a ROM is needed? Explain your calculation.
(d) Convert the SM chart (with reset function) to the proper form by adding a minimum number of extra states.
(e) Write the microcode for implementing the proper SM chart using single address microprogramming.

5.19 The following SM chart is to be realized using single-address microprogramming.

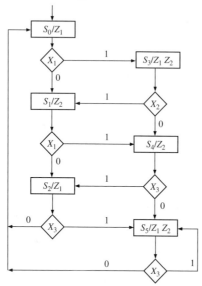

(a) Show the new SM chart and show the state assignments. The MUX inputs are 1, X_1, X_2, and X_3. Do not invert inputs. Add extra states if necessary.
(b) Write the microcode for implementing this state machine using single-address microprogramming.
(c) If the given (original) SM chart is implemented using a traditional ROM method, how big a ROM is needed? Explain your calculation.

5.20 Given the following ASM chart,

(a) Derive the next state and output equations, assuming the following state assignment: $S_0 = 00$, $S_1 = 01$, $S_2 = 10$.
(b) Convert the ASM chart to a form where it can be implemented by single-address microprogramming, with only next state true (NST) specified in the microprogram. Show the new SM chart and show the new state assignments.
(c) Write the single-address microprogram required to implement this circuit.
(d) What is the size of the microprogram ROM for single-address microprogramming of the modified SM chart?

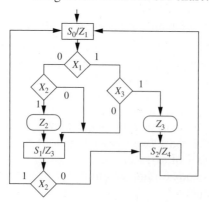

5.21 Realize the SM chart of Problem 5.10 using the two-address microprogramming structure shown in Figure 5-29.

   (a) Convert the SM chart to the proper form by adding a minimum number of states to the given chart.
   (b) Write the microprogram required to implement the circuit.
   (c) What is the size of the ROM required for microprogramming?
   (d) What is the size of the ROM if no microprogram is used but the traditional ROM method is used to implement the original SM chart?

5.22 Write a test bench for the elevator controller of Problem 5.9. The test bench has two functions: to simulate the operation of the elevator (including the door operation) and to provide a sequence of button pushes to test the operation of the controller.

To simulate the elevator: if the elevator is on the first floor ($FS_1 = 1$) and an UP signal is received, wait 1 second and turn off FS_1; then wait 10 seconds and turn on FS_2; this simulates the elevator moving from the first floor to the second. Similar action should occur if the elevator is on the second floor ($FS_2 = 1$) and a $DOWN$ signal is received. When a door open signal is received ($DO = 1$), set door closed (DC) to 0, wait 5 seconds, and then set $DC = 1$.

Test sequence: CALL1, 2, FB2, 4, FB1, 1, CALL2, 10, FB2

Assume each button is held down for 1 second and then released. The numbers between buttons are the delays in seconds between button pushes; this delay is in addition to the 1 second the button is held down.

Complete the following test bench:

```
module test_el;
.
.
.
elev_control eltest(CALL1, CALL2, FB1, FB2, FS1, FS2, DC, CLK,
UP, DOWN, DO);
.
.
.
endmodule
```

5.23 **(a)** What are the conditions an SM chart must satisfy in order to realize it using single-address microprogramming with a counter, a ROM, and a multiplexer as in Fig 5-33?
   (b) Give the modified SM chart and the required state assignment if the SM chart of Problem 5.10 is realized with this kind of microprogramming.

5.24 For the following SM chart,

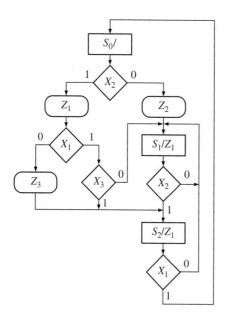

(a) Draw a timing chart that shows the clock, the state (S_0, S_1, or S_2), the inputs (X_1, X_2, and X_3), and the outputs. The input sequence is $X_1 \ X_2 \ X_3 = 011$, 101,111,010,110,101,001. Assume that all state changes occur on the rising edge of the clock and the inputs change between clock pulses.

(b) Use the state assignment S_0: $AB = 00$; S_1: $AB = 01$; S_2: $AB = 10$. Derive the next-state and output equations by tracing link paths. Simplify these equations using the don't care state ($AB = 11$).

(c) Realize the chart using a PLA and D flip-flops. Give the PLA table (state transition table).

(d) If a ROM is used instead of a PLA, what size ROM is required? Give the first five rows of the ROM table. Assume a naïve ROM method is used (i.e., a full look-up table).

5.25 Realize the SM chart given here using a ROM, a counter, and a 4-to-1 multiplexer.

(a) Draw a block diagram. Show the MUX inputs.

(b) Change the SM chart to the proper form. Mark required changes on the following chart.

(c) Make a suitable state assignment. Give the first six rows of the ROM table.

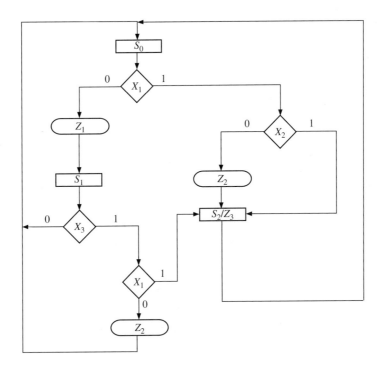

5.26 For the given SM chart,

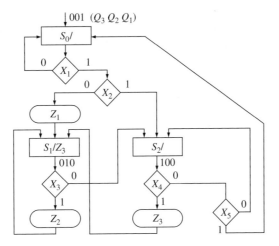

(a) Complete the following timing diagram (assume that $X_1 = 1$, $X_2 = 0$, $X_3 = 0$, $X_5 = 1$, and X_4 is as shown). Flip-flops change state on falling edge of clock.

[Waveform diagram with signals: Clock, X_4, Q_2, Q_3, Z_3]

(b) Using the given one-hot state assignment, derive the minimum next state and output equations by inspection of the SM chart.

(c) Write a Verilog description of the digital system.

5.27 Realize the SM chart of Problem 5.15 using the two-address microprogramming hardware structure shown in Figure 5-29.

(a) Convert the SM chart to the proper form by adding a minimum number of states to the given diagram. What are the changes needed?

(b) Write the microcode for implementing this state machine using the indicated hardware. You may indicate states in the microcode using the state names S_0, S_1, and so forth instead of using a bit assignment. Indicate the MUX connections (inputs) necessary to understand your microcode.

(c) What is the size of the microcode ROM? Explain your calculation.

(d) If the given (original) SM chart is implemented using a traditional ROM method, how big a ROM is needed? Explain your calculation.

5.28 (a) Draw an SM chart that is equivalent to the state graph of Figure 4-46.

(b) If the SM chart is implemented using a PLA and three flip-flops (A, B, C), give the PLA-table (state transition table). Use a straight binary state assignment.

(c) Give the equation for A^+ determined by inspection of the PLA table.

(d) If a one-hot state assignment is used, give the next-state and output equations.

CHAPTER 6

Designing with Field Programmable Gate Arrays

This chapter describes various issues related to implementing designs in field programmable gate arrays (FPGAs) A few simple designs are hand-mapped into FPGA building blocks to illustrate tradeoffs arising from the structure of the basic FPGA building block. Shannon's expansion for decomposition of large functions into smaller functions is presented. Issues of the one-hot method of state assignment, which is particularly suitable for FPGA-like technology, are discussed. The design flow is described, and synthesis, mapping, and placement issues are discussed briefly. Features of several commercial FPGAs appear in discussions and examples, but we avoid presenting the entire architecture of any commercial FPGA family. Instead, the basic principles are presented in a general fashion. Once you understand the fundamentals, you will be able to refer to manufacturer's data books and web pages for more detailed descriptions of the particular devices you want to use or understand in more detail.

6.1 Implementing Functions in FPGAs

Typically behavioral, RTL, or structural models of designs are created in a language such as VHDL or Verilog, and automatic CAD software is used to synthesize, map, partition, place, and route the design into an FPGA. To understand issues associated with partitioning a design into an FPGA, let us design some small components using FPGAs.

Let us assume that we want to design a 4-to-1 multiplexer using an FPGA whose logic block is represented by Figure 6-1(a). This building block contains two 4-variable function generators, X and Y, and two flip-flops. The X function generator can generate any functions of X_1, X_2, X_3, and X_4. Similarly the Y function generator can create any function of Y_1, Y_2, Y_3, and Y_4. Latched or unlatched forms of the generated functions can be brought to the output of the logic block. The latched outputs are QX and QY; the combinational outputs are X and Y. Assuming that the

341

FIGURE 6-1: Example Building Blocks for an FPGA (a) With Look up Tables and flip-flops (b) With Multiplexers

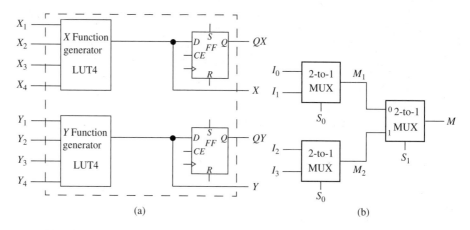

multiplexer inputs are I_0, I_1, I_2, and I_3 and that the multiplexer selects are S_1 and S_0, the output equation for the multiplexer can be written as follows:

$$M = S_1'S_0'I_0 + S_1'S_0I_1 + S_1S_0'I_2 + S_1S_0I_3 \qquad (6\text{-}1)$$

A 4-to-1 multiplexer can be decomposed into three 2-to-1 multiplexers as illustrated in Figure 6-1(b):

$$M_1 = S_0'I_0 + S_0I_1$$
$$M_2 = S_0'I_2 + S_0I_3$$

A third 2-to-1 multiplexer must now be used to create the output of the 4-to-1 multiplexer:

$$M = S_1'M_1 + S_1M_2$$

The output is the same as the expected output of the 4-to-1 multiplexer (M). Two of the 2-to-1 multiplexers (M_1 and M_2) can be implemented in one logic block, and a second logic block can be used to implement the third multiplexer (M). Thus, two logic blocks will be required to implement a 4-to-1 multiplexer using this type of logic block. The functions generated by the first logic block are

$$X = M_1 = S_0'I_0 + S_0I_1$$
$$Y = M_2 = S_0'I_2 + S_0I_3$$

Only half of the second logic block is used. The X function generator creates the function

$$M = S_1'M_1 + S_1M_2$$

The path used by M_1 and M_2 are highlighted in Figure 6-2. The flip-flops are unused in this design.

Many modern FPGAs use a 4-input look-up table (**LUT**) as a basic building block. Many designers refer to this building block as **LUT4**. It can implement a function (1-bit) of any four variables. It takes 16 bits of SRAM to realize the 4-input LUT using the SRAM technology.

6.1 Implementing Functions in FPGAs 343

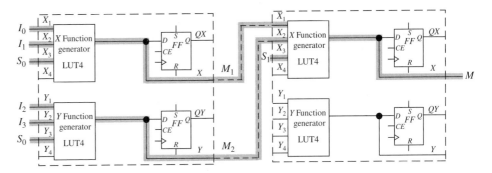

FIGURE 6-2: Highlighting Paths for a 4-to-1 MUX

Example

What are the contents of the look-up tables implementing the multiplexer in Figure 6-2?

Answer: As illustrated in the figure, three look up tables are used to implement functions M_1, M_2, and M. Each of them are essentially 2-to-1 multiplexers. Assuming X_1 and Y_1 are the LSBs and X_4 and Y_4 are the MSBs of the LUT addresses, one can create the truth tables for each LUT as shown. When S_0 is 0, the output (X) equals I_0, and when S_0 is 1, the output equals I_1. Let us denote the three LUTs as LUT-M1, LUT-M2, and LUT-M.

	Inputs			Output
X_4	X_3 (S_0)	X_2 (I_1)	X_1 (I_0)	X
x	0	0	0	0
x	0	0	1	1
x	0	1	0	0
x	0	1	1	1
x	1	0	0	0
x	1	0	1	0
x	1	1	0	1
x	1	1	1	1

The MSB of each LUT is unused. The contents of the first eight locations of the LUT should be duplicated for the next eight locations, since irrespective of the value of X_4, we expect it to behave like a 2-to-1 multiplexer. Hence, the contents of LUT-M1 are the following:

LUT-M1 – 0, 1, 0, 1, 0, 0, 1, 1, 0, 1, 0, 1, 0, 0, 1, 1

Since all three LUTs in Figure 6-2 are implementing 2-to-1 multiplexers, they have identical contents for the input connections shown. The contents of the second and third LUTs are the following:

LUT-M2 - 0, 1, 0, 1, 0, 0, 1, 1, 0, 1, 0, 1, 0, 0, 1, 1
LUT-M3 - 0, 1, 0, 1, 0, 0, 1, 1, 0, 1, 0, 1, 0, 0, 1, 1

Some FPGAs provide two 4-variable function generators and a method to combine the output of the two function generators. Consider the logic block in Figure 6-3. This programmable logic block has nine logic inputs (X_1, X_2, X_3, X_4, Y_1, Y_2, Y_3, Y_4, and C). It can generate two independent functions of four variables:

$$f_1(X_1, X_2, X_3, X_4) \text{ and } f_2(Y_1, Y_2, Y_3, Y_4)$$

FIGURE 6-3: Example Programmable Logic Block with Three Look-Up Tables. Based on Xilinx.

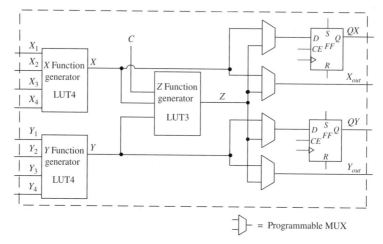

It can also generate a function Z, which depends on f_1, f_2, and C. Several programmable multiplexers are used to select what is brought out at the combinational outputs (X, Y) and the sequential outputs (QX, QY). It can generate any function of five variables in the form $Z = f_1(F_1, F_2, F_3, F_4) \cdot C' + f_2(F_1, F_2, F_3, F_4) \cdot C$. It can also generate some functions of six, seven, eight, and nine variables. A Xilinx FPGA from the past, the XC4000, uses a similar structure for its logic blocks.

Now, consider the implementation of a 4-to-1 multiplexer using this FPGA building block. A 4-to-1 multiplexer can be implemented using a single logic block of this FPGA, as highlighted in Figure 6-4. The X function generator (LUT4) implements the function $M_1 = S_0'I_0 + S_0I_1$, the Y function generator (LUT4) implements the function $M_2 = S_0'I_2 + S_0I_3$, and the Z function generator implements the function $M = S_1'M_1 + S_1M_2$. The input C is used to feed in select signal S_1 for use in the Z function generator. This design needs no flip-flops or latches.

FIGURE 6-4: A 4-to-1 Multiplexer in a Programmable Logic Block with Three Function Generators

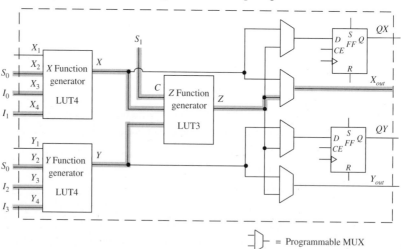

Often, there are many ways to map the same design. The 4-to-1 multiplexer (shown in Figure 6-4) was generated using the C input of the block. The multiplexer can be created even without using the C input. The first two terms of the

multiplexer's equation (Equation 6-1) have four variables S_0, S_1, I_0, and I_1. The third and fourth terms of the equation have four variables S_0, S_1, I_2, and I_3. Thus, a 4-variable function generator can implement the first two terms, and another 4-variable function generator can implement the third and fourth terms. However, now the outputs of the two 4-variable function generators must be combined. The Z function generator can be used for this purpose. In this case, the X function generator (LUT4) generates the function:

$$F_1 = S_1'S_0'I_0 + S_1'S_0I_1 \tag{6-1.a}$$

which is the first half of the function in Equation 6-1.

The Y function generator (LUT4) generates the function

$$F_2 = S_1S_0'I_2 + S_1S_0I_3 \tag{6-1.b}$$

which is the second half of the function in Equation 6-1. The Z function generator (LUT3) performs an OR function of the F_1 and F_2 functions.

$$Z = F_1 + F_2 \tag{6-2}$$

In this case, the C input is not required. This is an example of how mapping software has choices in the mapping of circuitry into resources available in the target technology.

The example on the previous pages illustrated that it is very expensive to create multiplexers using LUTs. Three 4-input function generators (LUTs) are required to create a 4-to-1 multiplexer. Since 16 SRAM cells are required to create a 4-variable function generator, 48 memory cells are required to create a 4-to-1 multiplexer using the FPGA building block in Figure 6-2.

Eight memory cells are required to create a 3-variable function generator (LUT3). Hence, the multiplexer in Figure 6-3 needs 40 memory cells (16 cells for X, 16 cells for Y, and 8 cells for Z). The contents of these memory cells are part of what one needs to download into the FPGA in order to program it.

When the programmable logic block of an FPGA is a large unit with ability to realize a fairly complex multivariable function, it is possible that a large part of each logic block may go unused. Let us consider an example. Assume that one has to design a 4-bit circular shift register in an FPGA, whose building block is similar to the one in Figure 6-1(a). In a circular shift register, the output of the rightmost flip-flop is fed back to the input of the leftmost flip-flop. Such a shift register is also called a **ring counter**. Since four flip-flops are required for a 4-bit shift register, two such basic building blocks will be required to realize this circuit. The four next-state equations are $D_1 = Q_4$, $D_2 = Q_1$, $D_3 = Q_2$, and $D_4 = Q_3$. Two next-state equations can be realized using the combinatorial function generators in one CLB. Figure 6-5(b) highlights the active paths for the shift register. The X function generator is used to generate $D_1 = Q_4$ and the Y function generator is used to generate $D_2 = Q_1$.

Notice that the 4-variable function generators are largely unused in this example, because the next-state equations for the flip-flops are rather simple; they depend only on the current state of the preceding flip-flop (i.e., a single variable

FIGURE 6-5: Shift Register Implementation in a LUT-based FPGA (a) 4-bit Shift Register (b) Highlighted paths for the implementation FPGA

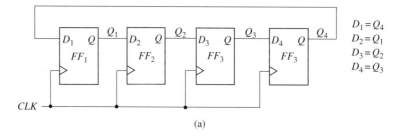

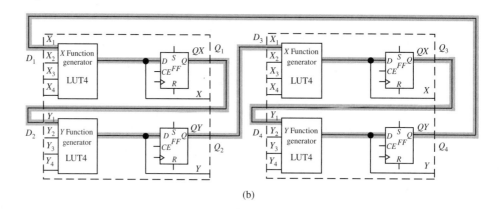

function). However, even if a function generator is used for a single variable function, the rest of the function generator cannot be used for anything else.

Example

How many programmable logic blocks similar to the one in Figure 6-1(a) will be required to create a 3-to-8 decoder?

Answer: Four. A 3-to-8 decoder has three inputs and eight outputs. Each output will need a 3-variable function generator. Since what is available in the logic block in Figure 6-1(a) is a 4-variable function generator, one will have to use one such function generator to create one output. Thus, eight function generators (i.e., eight 4-input LUTs) will be required to create a 3-to-8 decoder. One logic block shown in Figure 6-1(a) can generate two outputs. Consequently, four such programmable logic blocks will be required to create a 3-to-8 decoder.

If the LUTs are SRAM based, 128 SRAM cells are required to implement the 3-to-8 decoder using the LUT-based FPGA. This decoder will need eight only 3-input AND gates and three inverters if it is implemented using logic gates. Thus, LUTs are very expensive for implementing certain functions.

Some FPGAs use multiplexers and gates as basic building blocks. Some FPGAs (e.g., the Xilinx Spartan) provide LUTs and multiplexers. The mapping software looks at the resources available in the target technology (i.e., the specific FPGA that is used) and translates the design into the available building blocks.

6.2 Implementing Functions Using Shannon's Decomposition

Shannon's expansion theorem can be used to decompose functions of large numbers of variables into functions of fewer variables. In the previous section, we decomposed a 4-to-1 multiplexer into 2-to-1 multiplexers in order to implement it in a logic block with 4-variable function generators. Shannon's expansion offers a general decomposition technique for any function.

Let us illustrate Shannon's decomposition for realizing any 6-variable function $Z(a, b, c, d, e, f)$. First, expand the function as follows:

$$Z(a, b, c, d, e, f) = a' \cdot Z(0, b, c, d, e, f) + a \cdot Z(1, b, c, d, e, f) = a'Z_0 + aZ_1 \quad (6\text{-}3)$$

One can verify that Equation 6-3 is correct by first setting a to 0 on both sides and then setting a to 1 on both sides. Since the equation is true for both $a = 0$ and $a = 1$, it is always true. Equation 6-3 leads directly to the circuit of Figure 6-6(a), which uses two cells to realize Z_0 and Z_1. Half of a third cell is used to realize the 3-variable function, $Z = a'Z_0 + aZ_1$.

FIGURE 6-6: Realization of Functions by Decomposition (a) 6-Variable Function Using 5-Variable Function Generators (b) 6-Variable Function Using 4-Variable Function Generators

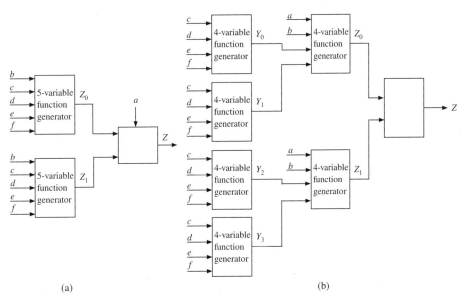

As an example, consider the following function:

$$Z = abcd'ef' + a'b'c'def' + b'cde'f$$

Setting $a = 0$ gives

$$Z_0 = 0 \cdot bcd'ef' + 1 \cdot b'c'def' + b'cde'f = b'c'def' + b'cde'f$$

and setting $a = 1$ gives

$$Z_1 = 1 \cdot bcd'ef' + 0 \cdot b'c'def' + b'cde'f = bcd'ef' + b'cde'f$$

Since Z_0 and Z_1 are 5-variable functions, each of them needs a 5-input LUT. Irrespective of the number of terms in a function, as long as there are only five variables, it can be realized by one 5-input LUT. Then, a 2-to-1 multiplexer or another LUT5 will be required to generate Z from Z_0 and Z_1.

If only 4-input LUTs are available, the 5-variable functions should be further decomposed into 4-variable functions. This can be done by applying Shannon's expansion theorem twice, first expanding about a and then expanding about b. Or, it can be done in one step by decomposing into four component functions as follows:

$$Z(a, b, c, d, e, f) = a'b' \cdot Z(0, 0, c, d, e, f) + a'b \cdot Z(0, 1, c, d, e, f)$$
$$+ ab' \cdot Z(1, 0, c, d, e, f) + ab \cdot Z(1, 1, c, d, e, f)$$
$$= a'b' \cdot Y_0 + a'b \cdot Y_1 + ab' \cdot Y_2 + ab \cdot Y_3 \qquad (6\text{-}4)$$

Figure 6-6(b) illustrates the realization of a general 6-variable function using 4-variable functions.

Now, let us consider the decomposition of the function

$$Z = abcd'ef' + a'b'c'def' + b'cde'f$$

into 4-variable functions. Let us apply Shannon's expansion around a and b.

- Substituting $a = b = 0$ gives $Y_0 = c'def' + cde'f$
- Substituting $a = 0, b = 1$ gives $Y_1 = 0$
- Substituting $a = 1, b = 0$ gives $Y_2 = cde'f$,
- Substituting $a = b = 1$ gives $Y_3 = cd'ef'$

In a general implementation, seven 4-variable function generators will be required to implement a 6-variable function as in Figure 6-6(b). However, in this particular example, one of the 4-variable functions obtained by decomposing is the null function, which results in a simpler function:

$$Z = a'b' \cdot Y_0 + ab' \cdot Y_2 + ab \cdot Y_3$$

Five 4-variable function generators will be sufficient to implement this function, one each for Y_0, Y_2, and Y_3, one for generating $Z_1 = ab' \cdot Y_2 + ab \cdot Y_3$, and another one for generating $a'b' \cdot Y_0 + Z_1$. Figure 6-7 illustrates the implementation of

FIGURE 6-7: Example Function Implementation Using 4-Variable Function Generators

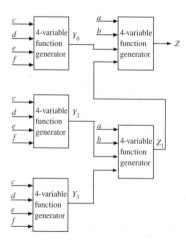

the function $Z = abcd'ef' + a'b'c'def' + b'cde'f$, using only 4-variable function generators.

Any 7-variable function can be realized with 6 or fewer LUT5s. The expansion for a general 7-variable function is

$$Z(a, b, c, d, e, f, g) = a'b' \cdot Z(0, 0, c, d, e, f, g) + a'b \cdot Z(0, 1, c, d, e, f, g)$$
$$+ ab' \cdot Z(1, 0, c, d, e, f, g) + ab \cdot Z(1, 1, c, d, e, f, g)$$
$$= a'b' \cdot Y_0 + a'b \cdot Y_1 + ab' \cdot Y_2 + ab \cdot Y_3 \qquad (6\text{-}5)$$

Here Y_0, Y_1, Y_2, and Y_3 are 5-variable functions of c, d, e, f, and g. Equation 6-5 can be obtained by applying the expansion theorem twice, first expanding about a and then expanding about b. As an example, consider the 7-variable function

$$Z = c'de'fg + bcd'e'fg' + a'c'defg + a'b'd'ef'g' + ab'defg'$$

- Substituting $a = b = 0$ gives $Y_0 = c'de'fg + c'def'g + d\cdot ef'g'$
- Substituting $a = 0, b = 1$ gives $Y_1 = c'de'fg + cd\cdot e'fg' + c'def'g$
- Substituting $a = 1, b = 0$ gives $Y_2 = c'de'fg + defg'$
- Substituting $a = b = 1$ gives $Y_3 = c'de'fg + cd'e'fg'$

This function can be implemented using six 5-variable function generators. Four of the function generators will implement the functions, Y_0, Y_1, Y_2, and Y_3. A fifth function generator implements the 4-variable function, $Z_0 = a'b' \cdot Y_0 + a'b \cdot Y_1$, and the remaining function generator implements a 5-variable function, $Z = Z_0 + ab' \cdot Y_2 + ab \cdot Y_3$.

Shannon's decomposition allows one to decompose an n-variable function into two n-1 variable functions and a 2-to-1 multiplexer. As we saw in the earlier part of this chapter, it is very inefficient to realize multiplexers using LUTs. As the number of variables (n) increases, the number of look-up tables required to realize an n-variable function increases rapidly. The availability of multiplexers can greatly reduce the number of LUTs needed. For this reason, some FPGAs provide multiplexers in addition to LUT4s.

Example

Implement a 7-variable function using 4-input LUTs and 2-to-1 multiplexers.

Answer: Shannon's expansion can be used to obtain the following decompositions:

7-variable function generator = two 6-variable function generators + a 2-to-1 mux ... (i)

6-variable function generator = two 5-variable function generators + a 2-to-1 mux ... (ii)

5-variable function generator = two 4-variable function generators + a 2-to-1 mux ... (iii)

Substituting (iii) into (ii), we obtain

6-variable function generator = Four 4-variable function generators + three 2-to-1 muxes ... (iv)

Substituting (iv) into (i), we obtain

7-variable function generator = Eight 4-variable function generators + seven 2-to-1 muxes

Thus a 7-variable function can be implemented as in Figure 6-8.

FIGURE 6-8: A 7-Variable Function Using 4-Input LUTs and 2-to-1 MUXes

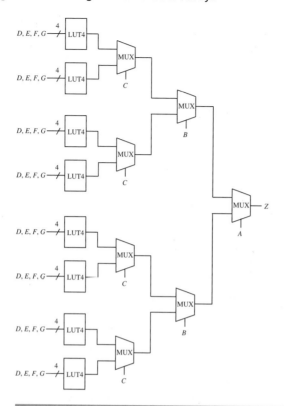

If only 4-variable LUTs are available, a 7-variable function needs 15 4-variable LUTs. A 2-to-1 multiplexer is cheaper than a 4-input LUT; hence, it is implemented using eight 4-input LUTs and seven 2-to-1 multiplexers in Figure 6-8.

The Xilinx Spartan FPGA is an example of an FPGA that provides multiplexers in addition to the general 4-variable LUTs. A logic unit in these FPGAs is called a **slice** and a slice may be represented in a simple fashion as in Figure 6-9. It contains

FIGURE 6-9: Simplified View of Resources in a Xilinx Spartan Slice. Based on Xilinx.

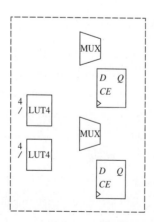

two 4-input LUTs and two 2-to-1 multiplexers (plus other logic not shown here). A 7-variable function can be realized using four such slices, as in Figure 6-10. Dotted lines are used to indicate each slice.

FIGURE 6-10:
Implementing a 7-Variable Function Using Four Xilinx Spartan Slices. Based on Xilinx.

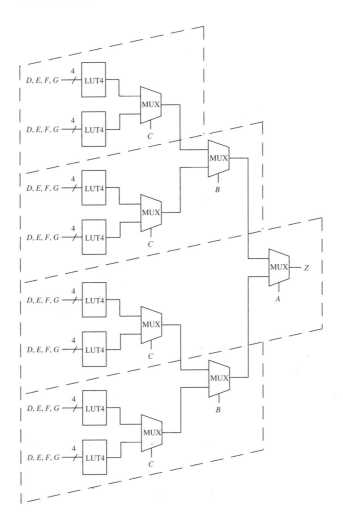

As another example, let us generate a parity function using 4-variable function generators. The parity function is defined as

$$F = A \oplus B \oplus C \oplus D \oplus E$$

which has 16 terms when expanded to a sum-of-products, but it is a 5-variable function. Any 5-variable function can be decomposed into two 4-variable functions using Shannon's expansion and can be realized using two 4-input LUTs and a 2-to-1 multiplexer. If only 4-variable function generators are available, three 4-variable function generators will be required.

6.3 Carry Chains in FPGAs

The most naïve method to create an adder with FPGAs would be to use FPGA logic blocks to generate the sum and carry for each bit. A 4-variable look-up table (which is currently the standard building block) can generate the sum, and another LUT4 will typically be required to realize the carry equation. The carry output from each bit has to be forwarded to the next bit using interconnect resources. But since addition is a fundamental and commonplace operation, many FPGAs provide dedicated circuitry for generating and propagating carry bits to subsequent higher bits. Typically, a dedicated carry chain is implemented. As an example, consider

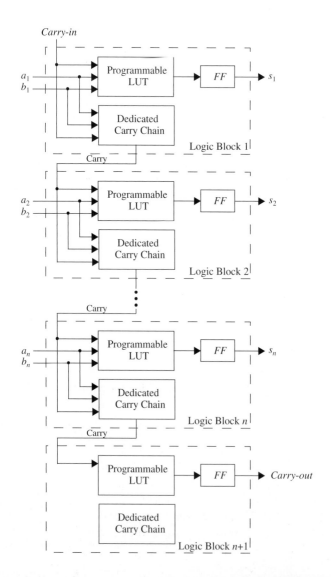

FIGURE 6-11: Carry Chains for Fast Addition. Based on Xilinx.

the carry chain illustrated in Figure 6-11. Each LUT generates the sum bit of the corresponding input bits (*a*, *b*, and *Carry-in*). The carry chain generates the carry in parallel and feeds it using the dedicated interconnect to the LUT, implementing the sum of the next bit.

Without such a carry chain, an n-bit adder typically will take $2n$ logic blocks (if a logic block is a LUT4), whereas with the carry chain, n logic blocks (albeit with additional dedicated circuitry) are sufficient. Dedicated circuitry generates the carry and routes it directly to the next LUT4. The hardware for the carry generation will be unused in many circuits, but because addition is a common operation, it is generally worthwhile to include such circuitry in the FPGA logic block.

6.4 Cascade Chains in FPGAs

Some FPGAs contain support for cascading outputs from FPGA blocks in series. The common types of cascading are the AND configuration and the OR configuration. These are extremely useful while creating wide-AND and wide-OR gates. Instead of using separate function generators to perform AND or OR functions of logic block outputs, the output from one logic block can be directly fed to the cascade circuitry to create AND or OR functions of the logic block outputs. Figure 6-12 illustrates the cascade chains in an example FPGA that uses 4-input LUTs for function generation. If an OR operation of 32 variables is desired, one can accomplish it using eight logic blocks. Each logic block will generate a 4-variable OR and the cascading OR gate can be used to OR the output from the previous logic block. Cascading AND and exclusive OR gates are also provided in some FPGAs. In look-up-table–based FPGAs, these types of cascade chains may be called LUT chains.

Example

How many logic blocks with LUT4s will be needed to create a 32-variable AND function with and without the AND cascade chain?

Answer: Without the AND cascade chain, 11 logic blocks will be needed. Eight logic blocks can be used to make AND of 4-variables each, resulting in eight sub-functions. These eight functions can be fed to two logic blocks, and then another logic block will be needed to generate the overall AND. With the AND cascade chain, only eight logic blocks will be needed. The AND cascade chain will be used to AND the outputs of the eight logic blocks.

Register Chains in FPGAs

In many FPGAs, the only input to the flip flops in the logic blocks is through the LUTs or logic elements. That is why in Figure 6-5, the shift register has to use the LUT to simply act as a wire passing the input via the LUT to the flip flop. Additionally the logic block in Figure 6-5 cannot implement any other circuitry.

FIGURE 6-12: Example Cascade Chains (a) AND Cascade Chain (b) OR Cascade Chain (c) Register Cascade Chain. (b) …Chain: Based on Xilinx. After (c) …Chain: Based on Altera.

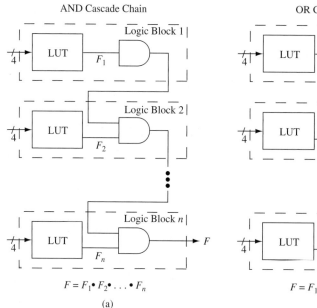
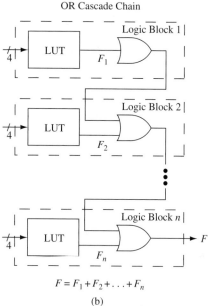

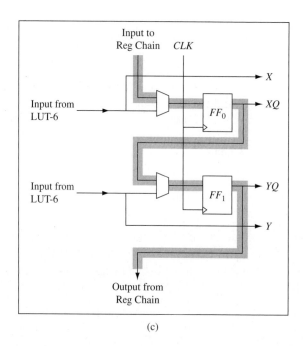

However, some FPGAs such as the Altera Stratix IV contain support for register chains without using the LUTs, as illustrated in Figure 6-12(c). There is a separate register chain input to the flip-flop, and it is possible to use the LUT for other logic functions and route their outputs using the combinational outputs X and Y, potentially increasing the utilization of the FPGA blocks.

6.5 Examples of Logic Blocks in Commercial FPGAs

We provide three examples of commercial FPGA logic blocks. They are from Xilinx, Altera, and Microsemi. The Xilinx and Altera architectures both use 6-variable look-up tables as their basic building block. Microsemi has an architecture that uses a 4-variable look-up table and another that uses multiplexers and gates.

The Xilinx Kintex Configurable Logic Block

The Xilinx Kintex FPGA uses four copies of a basic block called a **slice**, illustrated in Figure 6-13, to form a Configurable Logic Block (CLB). CLB is the Xilinx terminology for the programmable logic block in its FPGAs. The Kintex uses 6-variable look-up tables.

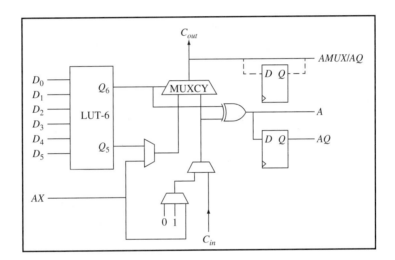

FIGURE 6-13: Simplified View of the Xilinx Kintex "Slice" (1/4 of a CLB). Based on Xilinx.

In contrast, the Xilinx Virtex and Spartan FPGAs use LUT4s. Each slice contains two function generators—the G function generator and the F function generator. Additionally, there are two multiplexers, F_5 and FX, for function implementation. In order to implement a 4-variable LUT, 16 SRAM bits are required; therefore, a slice contains 32 bits of SRAM in order to generate the combinational function. The F_5 multiplexer can be used to combine the outputs of two 4-variable function generators to form a 5-variable function generator. The select input of the multiplexer is available to feed in the fifth input variable. All inputs of the FX multiplexer are accessible allowing the creation of any general 2-variable function. This multiplexer can be used to combine the F_5 outputs from two slices to form a 6-input function. Each slice also contains two flip-flops that can be configured as edge-sensitive D flip-flops or as level-sensitive latches. There is support for fast carry generation for addition. There is also additional logic to generate a few specific logic functions in addition to the general 4-variable LUT.

The Altera Stratix IV Logic Module

Altera's name for its basic logic block is the logic module (LM). Figure 6-14 illustrates a simplified view of the logic block of the Altera Stratix IV FPGA. Each LM contains two 6-variable look-up tables (LUTs) and two flip-flops. Each LUT6 has two independent inputs and four shared inputs. Basically a pair of LUTs share four of the inputs, as illustrated in Figure 6-14. It can implement two functions of six variables. The output can come out directly from the combinational logic or from the flip-flop. There are two 1-bit built-in adders with carry-chaining. Another special feature of the Stratix IV logic module is the register chaining which allows to create shift registers or other types of flip-flop arrays and to use the flip-flops separately from the LUTs. Since the figure presents a simplified view, many details are left out. The Stratix V logic module is similar, except that four flip flops exist per logic module instead of the two in Stratix IV.

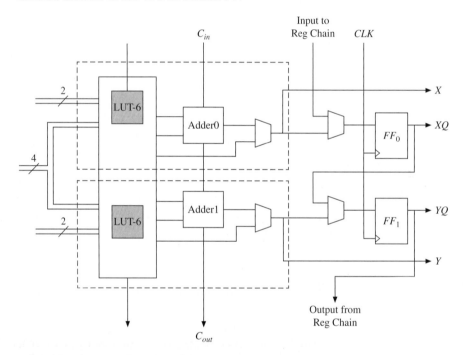

FIGURE 6-14: Simplified View of the Altera Stratix IV Logic Module. Based on Altera.

The Microsemi Fusion VersaTile

Microsemi also makes FPGAs which are LUT based, but in the interest of showing a building block that is very different from the typical LUT-based architecture; the building block in the Microsemi Fusion architecture, consisting of multiplexers and gates, is illustrated in Figure 6-15. Microsemi calls its basic block the **VersaTile**. The VersaTile block has four inputs—X_1, X_2, X_3, and X_c—as illustrated in Figure 6-15. Each VersaTile can be configured to be any of the following:

- a 3-input logic function
- a latch with a clear or set
- a D-flip-flop with clear or set
- a D flip-flop with enable, clear, or set

FIGURE 6-15: Simplified View of the Microsemi Fusion Logic Block. Based on Microsemi.

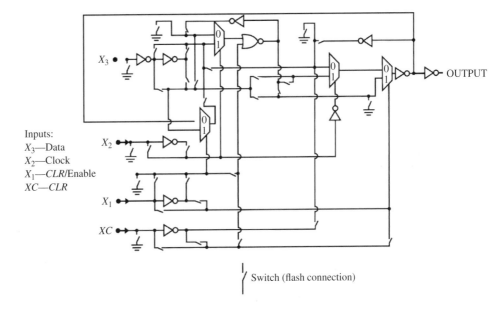

Inputs:
X_3—Data
X_2—Clock
X_1—CLR/Enable
XC—CLR

When used as a 3-input logic function, the inputs are X_1, X_2, and X_3. When used for the latch/flip-flop, input X_2 is typically used for the clock. Inputs X_1 and X_c are used for flip-flop enable, and clear signals. The logic block provides duplicate outputs tailored for fast local connections or efficient long-line connections, but for simplicity we show only one output in Figure 6-15. The VersaTile is of significantly finer grain than the 4-input LUTs in many other FPGAs. The granularity of this building block is comparable to that of standard gate arrays (i.e., traditional gate arrays that are mask programmable).

6.6 Dedicated Memory in FPGAs

Many applications need memory. It could be for storing a table of constants to be used as coefficients during processing, or it could be for implementing instruction and data memories for an embedded processor that is being designed using the FPGA. Early FPGAs did not contain any dedicated memory. Designers typically interfaced the FPGAs to external memory chips when memory was desired. As chip densities have increased, FPGA designers started to incorporate dedicated memory on FPGA chips, eliminating the need to interface them with external memory chips.

Modern FPGAs include 16K to 10M bits of dedicated memory. Table 6-1 presents the amount of dedicated RAM in some FPGAs. As an example, the Xilinx Virtex-5 contains 1 to 10M bits of dedicated memory. Similarly, the Altera Stratix II contains 409K to 9M bits of memory. The Microsemi Fusion contains 27 to 270K bits of memory. The dedicated memory is typically implemented using a few (4–1000) large blocks of dedicated SRAM located in the FPGA. Figure 6-16

TABLE 6-1: Size of Dedicated RAM in Example FPGAs

FPGA Family	Dedicated RAM Size (Kbits)	Organization
Xilinx Kintex 7	4680–34380	270–1910 18Kb blocks 135–955 36Kb blocks
Xilinx Artix 7	1800–13140	100–730 18Kb blocks 50–365 36Kb blocks
Xilinx Virtex 6	5616–38304	312–2128 18Kb blocks 156–1064 36Kb blocks
Xilinx Virtex 5	1152–18576	64–1032 18Kb blocks 32–516 36Kb blocks
Xilinx Spartan 3E	72–648	4–36 18Kb blocks
Xilinx Spartan 6	216–4824	12–268 18Kb blocks
Altera Stratix V	13760–53200	688–2660 20Kb blocks
Altera Stratix IV	6462–20736	462–1280 9Kb blocks 16–64 144Kb blocks
Altera Cyclone IV	540–6480	60–720 9Kb blocks
Altera Arria V GT	10510–24140	1051–2414 10Kb blocks
Altera Arria II GZ	11366–16794	1235–1248 9Kb blocks 24–36 144 Kb blocks
Lattice SC	1054–7987	56–424 18Kb blocks
Microsemi Fusion	27–270	6–60 4Kb blocks

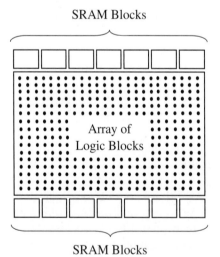

FIGURE 6-16: Embedded RAMs in FPGAs

shows a typical organization for the dedicated RAM blocks. In many FPGAs, they are situated outside the region of the logic block arrays (e.g., Xilinx Virtex/Spartan and Microsemi Fusion). In some FPGAs (e.g., Altera Stratix), there are columns of memory in a few different locations in the FPGA. In many FPGAs, the SRAM blocks are of one size (e.g., 18Kb in Xilinx Virtex). In some FPGAs, there are blocks of different sizes. For example, the Altera FPGAs have embedded memory built from sizes such as 4Kb, 9Kb, 20Kb, 144Kb, and 512Kb blocks. Many FPGAs have multiple types of memory building blocks in the same chip; for instance Altera Stratix IV contains 9Kb and 144Kb memory blocks. The dedicated memory on the Xilinx FPGAs is called **block RAM**. Some FPGAs provide parity bits in the embedded RAM. The parity bits are included when calculating the dedicated RAM size in the literature from some vendors; other vendors exclude the parity bits and count only the usable dedicated RAM.

A key feature of the dedicated RAM on modern FPGAs is the ability to adjust the width of the RAM. As shown in Table 6-1, there are several tiles or blocks of memory. They can be placed in various ways to achieve different aspect ratios. Let us assume that there are 32K bits of SRAM provided as blocks of RAM. This RAM can be used as $32K \times 1$, $16K \times 2$, $8K \times 4$, or $4K \times 8$. Thus, the width of the RAM can be adjusted depending on the needs of the application. One application may need byte-wide memories; another application may need 64-bit-wide memories.

LUT-based FPGAs offer another alternative for memory. If only small amounts of memory are required, it is possible to create that memory using the bits in the LUTs (i.e., without using the dedicated memory). As you know, a 4-variable LUT contains 16 bits of storage. One can create small amounts of memory by combining the storage cells from the LUTs. Two 4-input LUTs (as shown in Figure 6-17) can be used to create a 32×1 memory or a 16×2 memory. When used as a 32×1 memory, there must be five address lines and one data line (i.e., D_1 and D_2 must be connected). The top LUT must be enabled when the MSB of the address is 0, and the bottom LUT must be enabled when the MSB of the address is 1. This can be done using the highest bit of the address and an inverter. When used as a 16×2 memory, both LUTs are enabled, and data lines D_1 and D_2 are brought out in parallel.

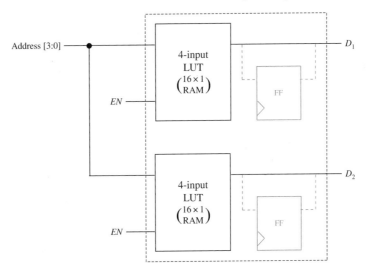

FIGURE 6-17: Creating Memory from LUTs. Based on Xilinx.

Memory created from LUT cells is called **Distributed Memory** (in Xilinx terminology). As the term indicates, this memory is distributed throughout the chip inside the logic blocks. A disadvantage of distributed memory is that once the LUT memory is used in this fashion, the logic block is generally unusable. The LUT memory can be used as asynchronous memory; it can also be combined with the logic block flip-flops to create synchronous memory. Table 6-2 presents the amount of LUT-based memory available in some FPGAs.

TABLE 6-2: LUT-Based RAM in Some FPGAs

FPGA Family	LUT-Based RAM (Kb)	No. of LUTs
Xilinx Kintex 7	838–6788	41000–298600
Xilinx Artix 7	400–2888	63400–134600
Xilinx Virtex 6	1045–8280	46560–474240
Xilinx Virtex 5	320–3420	19200–207360
Xilinx Virtex 4	96–987	12288–126336
Xilinx Virtex-II	8–1456	512–93184
Xilinx Spartan 3E	15–231*	1920–29504
Altera Stratix V	2781–11225	44496–179600
Altera Stratix IV	195–2242**	12480–143520
Altera Cyclone II	72–1069**	4608–68416
Altera Arria V	1200–8064	76800–516096
Altera Arria II	705–4007	45125–256500
Lattice SC	245–1884	15200–115200
Lattice ECP2	12–136	6000–68000

* does not use all of the LUTs as distributed RAM.
**calculated from LUT counts.

Verilog Models for Inferring Memory in FPGAs

Embedded memory on FPGAs can be instantiated using behavioral Verilog models. Memories can be synchronous or asynchronous. An asynchronous read operation means that the data from the addressed location is available on the output bus after the access time, irrespective of the clock. In contrast, in synchronous memory read and write control lines will have an impact only if the clock is active. In some memories, write is synchronous and read is asynchronous.

Modern synthesis tools provided by FPGA vendors are capable of inferring embedded memory from high-level constructs. Figure 6-18 illustrates Verilog code that creates a synchronous-write, asynchronous-read memory. The memory array is represented by an array of registers. The memory array, DATAMEM, is not initialized here; however, it may be initialized to any desired values. The write operation is performed inside the always statement and only at the positive edge of the clock. The read operation is outside the always statement; hence, it occurs irrespective of the clock. Synthesis using current Xilinx tools with the "auto" option results in distributed memory for this code. Distributed memory is ideal for asynchronous memory, since the LUT generates its output asynchronously. In contrast, the code in Figure 6-19 infers block RAM. In this code sequence, the read statement appears inside the always statement, and read also happens only at the clock edge.

FIGURE 6-18: Behavioral Verilog Code that Typically Infers LUT-Based Memory

```verilog
module Memory (Address, CLK, MemWrite, Data_In, Data_Out);
   input[6:0] Address;
   input CLK;
   input MemWrite;
   input[31:0] Data_In;
   output[31:0] Data_Out;
   wire[31:0] Data_Out;

   reg[31:0] DataMEM[0:127];

   always @(posedge CLK)
   begin
     if (MemWrite == 1'b1)
     begin
        DataMEM[Address] <= Data_In;   // Synchronous Write
     end
   end
   assign Data_Out = DataMEM[Address];      // Asynchronous Read
endmodule
```

FIGURE 6-19: Behavioral Verilog Code that Typically Infers Dedicated Memory but Can Be Forced to Yield LUT RAM Using the Tool Menu

```verilog
module Memory (Address, CLK, MemWrite, Data_In, Data_Out);
   input[6:0] Address;
   input CLK;
   input MemWrite;
   input[31:0] Data_In;
   output[31:0] Data_Out;
   reg[31:0] Data_Out;

   reg[31:0] DataMEM[0:127];

   always @(posedge CLK)
   begin
     if (MemWrite == 1'b1)
     begin
        DataMEM[Address] <= Data_In;   // Synchronous Write
     end
     Data_Out <= DataMEM[Address];     // Synchronous Read
   end
endmodule
```

In modern FPGA synthesis tools, it is possible to indicate whether LUT-based RAM or Block RAM is desired. The menu options in the tools allow this. Hence, although the code in Figure 6-19 will by default synthesize to block RAM, it is possible to force LUT-based RAM by using the menu options. However, for the code in Figure 6-18, which synthesizes by default to LUT-based RAM, if block RAM is forced, the tool will give a warning that asynchronous read cannot be obtained from the block RAM and synthesize it into LUT RAM. In summary, synchronous reads/writes can be obtained with either block RAM or LUT RAM, but asynchronous reads can be obtained only with LUT RAM.

If the ROM method is used for implementing circuits, the synthesis tools may infer RAM in order to implement the look-up tables. As an example, consider the creation of a 4 × 4 multiplier using a look-up table method, as illustrated by the Verilog code in Figure 6-20. Because it uses the look-up table method, the product values for each of the input combinations are stored in a look-up table. Since the multiplicand and multiplier are four bits each, there are 256 possible combinations of inputs. A constant array is used to store the product array. The multiplicand is 0000 for the first 16 entries; hence, the product is 0 for the first 16 entries. The multiplicand is 0001 for the next 16 entries; hence, the product ranges from 0 to 15 (decimal) as the multiplier changes from 0 to 15.

Verilog code for this multiplier is presented in Figure 6-20. If this code is synthesized, current Xilinx tools infer distributed RAM to store the product values. Distributed RAM is inferred to implement asynchronous reads since the LUTs in the logic blocks can continuously update the outputs as the inputs change. No clock is required. If one tries to force the creation of block RAM using menu options in the synthesis tool, current Xilinx tools give no warning or error, however, only LUT RAM is created. If one desires to store the arrays in the dedicated block RAM, especially if one does not want to waste LUTs for realizing memory, the statement

```
assign         PRODuct = PROD_ROM[{Mplier, Mcand}];
```

must be made synchronous, as in the following code:

```
always @(posedge CLK)
begin
  PRODuct = PROD_ROM[{Mplier, Mcand}];
  // read Product Synchronously
end
```

With this modification in the code, the synthesis tools from Xilinx will infer dedicated block RAM to store the 256 product values. If reads are synchronous, menu options from the synthesis tool can be used to force either LUT RAM or block RAM. One should always check the synthesis reports and verify that the desired type of memory was actually created, rather than relying on menu options and warning messages.

FIGURE 6-20: Look-Up Table-Based 4 × 4 Multiplier

```
module LUTmult (Mplier, Mcand, PRODuct);
   input    [3:0] Mplier;
   input    [3:0] Mcand;
   output   [7:0] PRODuct;

   reg      [7:0] PROD_ROM [0:255];

   initial
   begin
```

```
PROD_ROM[0]  = 8'h00;
PROD_ROM[1]  = 8'h00;
PROD_ROM[2]  = 8'h00;
PROD_ROM[3]  = 8'h00;
PROD_ROM[4]  = 8'h00;
PROD_ROM[5]  = 8'h00;
PROD_ROM[6]  = 8'h00;
PROD_ROM[7]  = 8'h00;
PROD_ROM[8]  = 8'h00;
PROD_ROM[9]  = 8'h00;
PROD_ROM[10] = 8'h00;
PROD_ROM[11] = 8'h00;
PROD_ROM[12] = 8'h00;
PROD_ROM[13] = 8'h00;
PROD_ROM[14] = 8'h00;
PROD_ROM[15] = 8'h00;
PROD_ROM[16] = 8'h00;
PROD_ROM[17] = 8'h01;
PROD_ROM[18] = 8'h02;
PROD_ROM[19] = 8'h03;
PROD_ROM[20] = 8'h04;
PROD_ROM[21] = 8'h05;
PROD_ROM[22] = 8'h06;
PROD_ROM[23] = 8'h07;
PROD_ROM[24] = 8'h08;
PROD_ROM[25] = 8'h09;
PROD_ROM[26] = 8'h0A;
PROD_ROM[27] = 8'h0B;
PROD_ROM[28] = 8'h0C;
PROD_ROM[29] = 8'h0D;
PROD_ROM[30] = 8'h0E;
PROD_ROM[31] = 8'h0F;
PROD_ROM[32] = 8'h00;
PROD_ROM[33] = 8'h02;
PROD_ROM[34] = 8'h04;
PROD_ROM[35] = 8'h06;
PROD_ROM[36] = 8'h08;
PROD_ROM[37] = 8'h0A;
PROD_ROM[38] = 8'h0C;
PROD_ROM[39] = 8'h0E;
PROD_ROM[40] = 8'h10;
PROD_ROM[41] = 8'h12;
PROD_ROM[42] = 8'h14;
PROD_ROM[43] = 8'h16;
PROD_ROM[44] = 8'h18;
PROD_ROM[45] = 8'h1A;
PROD_ROM[46] = 8'h1C;
PROD_ROM[47] = 8'h1E;
PROD_ROM[48] = 8'h00;
PROD_ROM[49] = 8'h03;
PROD_ROM[50] = 8'h06;
PROD_ROM[51] = 8'h09;
```

```verilog
PROD_ROM[52]  = 8'h0C;
PROD_ROM[53]  = 8'h0F;
PROD_ROM[54]  = 8'h12;
PROD_ROM[55]  = 8'h15;
PROD_ROM[56]  = 8'h18;
PROD_ROM[57]  = 8'h1B;
PROD_ROM[58]  = 8'h1E;
PROD_ROM[59]  = 8'h21;
PROD_ROM[60]  = 8'h24;
PROD_ROM[61]  = 8'h27;
PROD_ROM[62]  = 8'h2A;
PROD_ROM[63]  = 8'h2D;
PROD_ROM[64]  = 8'h00;
PROD_ROM[65]  = 8'h04;
PROD_ROM[66]  = 8'h08;
PROD_ROM[67]  = 8'h0C;
PROD_ROM[68]  = 8'h10;
PROD_ROM[69]  = 8'h14;
PROD_ROM[70]  = 8'h18;
PROD_ROM[71]  = 8'h1C;
PROD_ROM[72]  = 8'h20;
PROD_ROM[73]  = 8'h24;
PROD_ROM[74]  = 8'h28;
PROD_ROM[75]  = 8'h2C;
PROD_ROM[76]  = 8'h30;
PROD_ROM[77]  = 8'h34;
PROD_ROM[78]  = 8'h38;
PROD_ROM[79]  = 8'h3C;
PROD_ROM[80]  = 8'h00;
PROD_ROM[81]  = 8'h05;
PROD_ROM[82]  = 8'h0A;
PROD_ROM[83]  = 8'h0F;
PROD_ROM[84]  = 8'h14;
PROD_ROM[85]  = 8'h19;
PROD_ROM[86]  = 8'h1E;
PROD_ROM[87]  = 8'h23;
PROD_ROM[88]  = 8'h28;
PROD_ROM[89]  = 8'h2D;
PROD_ROM[90]  = 8'h32;
PROD_ROM[91]  = 8'h37;
PROD_ROM[92]  = 8'h3C;
PROD_ROM[93]  = 8'h41;
PROD_ROM[94]  = 8'h46;
PROD_ROM[95]  = 8'h4B;
PROD_ROM[96]  = 8'h00;
PROD_ROM[97]  = 8'h06;
PROD_ROM[98]  = 8'h0C;
PROD_ROM[99]  = 8'h12;
PROD_ROM[100] = 8'h18;
PROD_ROM[101] = 8'h1E;
PROD_ROM[102] = 8'h24;
```

```verilog
PROD_ROM[103] = 8'h2A;
PROD_ROM[104] = 8'h30;
PROD_ROM[105] = 8'h36;
PROD_ROM[106] = 8'h3C;
PROD_ROM[107] = 8'h42;
PROD_ROM[108] = 8'h48;
PROD_ROM[109] = 8'h4E;
PROD_ROM[110] = 8'h54;
PROD_ROM[111] = 8'h5A;
PROD_ROM[112] = 8'h00;
PROD_ROM[113] = 8'h07;
PROD_ROM[114] = 8'h0E;
PROD_ROM[115] = 8'h15;
PROD_ROM[116] = 8'h1C;
PROD_ROM[117] = 8'h23;
PROD_ROM[118] = 8'h2A;
PROD_ROM[119] = 8'h31;
PROD_ROM[120] = 8'h38;
PROD_ROM[121] = 8'h3F;
PROD_ROM[122] = 8'h46;
PROD_ROM[123] = 8'h4D;
PROD_ROM[124] = 8'h54;
PROD_ROM[125] = 8'h5B;
PROD_ROM[126] = 8'h62;
PROD_ROM[127] = 8'h69;
PROD_ROM[128] = 8'h00;
PROD_ROM[129] = 8'h08;
PROD_ROM[130] = 8'h10;
PROD_ROM[131] = 8'h18;
PROD_ROM[132] = 8'h20;
PROD_ROM[133] = 8'h28;
PROD_ROM[134] = 8'h30;
PROD_ROM[135] = 8'h38;
PROD_ROM[136] = 8'h40;
PROD_ROM[137] = 8'h48;
PROD_ROM[138] = 8'h50;
PROD_ROM[139] = 8'h58;
PROD_ROM[140] = 8'h60;
PROD_ROM[141] = 8'h68;
PROD_ROM[142] = 8'h70;
PROD_ROM[143] = 8'h78;
PROD_ROM[144] = 8'h00;
PROD_ROM[145] = 8'h09;
PROD_ROM[146] = 8'h12;
PROD_ROM[147] = 8'h1B;
PROD_ROM[148] = 8'h24;
PROD_ROM[149] = 8'h2D;
PROD_ROM[150] = 8'h36;
PROD_ROM[151] = 8'h3F;
PROD_ROM[152] = 8'h48;
PROD_ROM[153] = 8'h51;
```

```verilog
PROD_ROM[154] = 8'h5A;
PROD_ROM[155] = 8'h63;
PROD_ROM[156] = 8'h6C;
PROD_ROM[157] = 8'h75;
PROD_ROM[158] = 8'h7E;
PROD_ROM[159] = 8'h87;
PROD_ROM[160] = 8'h00;
PROD_ROM[161] = 8'h0A;
PROD_ROM[162] = 8'h14;
PROD_ROM[163] = 8'h1E;
PROD_ROM[164] = 8'h28;
PROD_ROM[165] = 8'h32;
PROD_ROM[166] = 8'h3C;
PROD_ROM[167] = 8'h46;
PROD_ROM[168] = 8'h50;
PROD_ROM[169] = 8'h5A;
PROD_ROM[170] = 8'h64;
PROD_ROM[171] = 8'h6E;
PROD_ROM[172] = 8'h78;
PROD_ROM[173] = 8'h82;
PROD_ROM[174] = 8'h8C;
PROD_ROM[175] = 8'h96;
PROD_ROM[176] = 8'h00;
PROD_ROM[177] = 8'h0B;
PROD_ROM[178] = 8'h16;
PROD_ROM[179] = 8'h21;
PROD_ROM[180] = 8'h2C;
PROD_ROM[181] = 8'h37;
PROD_ROM[182] = 8'h42;
PROD_ROM[183] = 8'h4D;
PROD_ROM[184] = 8'h58;
PROD_ROM[185] = 8'h63;
PROD_ROM[186] = 8'h6E;
PROD_ROM[187] = 8'h79;
PROD_ROM[188] = 8'h84;
PROD_ROM[189] = 8'h8F;
PROD_ROM[190] = 8'h9A;
PROD_ROM[191] = 8'hA5;
PROD_ROM[192] = 8'h00;
PROD_ROM[193] = 8'h0C;
PROD_ROM[194] = 8'h18;
PROD_ROM[195] = 8'h24;
PROD_ROM[196] = 8'h30;
PROD_ROM[197] = 8'h3C;
PROD_ROM[198] = 8'h48;
PROD_ROM[199] = 8'h54;
PROD_ROM[200] = 8'h60;
PROD_ROM[201] = 8'h6C;
PROD_ROM[202] = 8'h78;
PROD_ROM[203] = 8'h84;
PROD_ROM[204] = 8'h90;
```

```
PROD_ROM[205] = 8'h9C;
PROD_ROM[206] = 8'hA8;
PROD_ROM[207] = 8'hB4;
PROD_ROM[208] = 8'h00;
PROD_ROM[209] = 8'h0D;
PROD_ROM[210] = 8'h1A;
PROD_ROM[211] = 8'h27;
PROD_ROM[212] = 8'h34;
PROD_ROM[213] = 8'h41;
PROD_ROM[214] = 8'h4E;
PROD_ROM[215] = 8'h5B;
PROD_ROM[216] = 8'h68;
PROD_ROM[217] = 8'h75;
PROD_ROM[218] = 8'h82;
PROD_ROM[219] = 8'h8F;
PROD_ROM[220] = 8'h9C;
PROD_ROM[221] = 8'hA9;
PROD_ROM[222] = 8'hB6;
PROD_ROM[223] = 8'hC3;
PROD_ROM[224] = 8'h00;
PROD_ROM[225] = 8'h0E;
PROD_ROM[226] = 8'h1C;
PROD_ROM[227] = 8'h2A;
PROD_ROM[228] = 8'h38;
PROD_ROM[229] = 8'h46;
PROD_ROM[230] = 8'h54;
PROD_ROM[231] = 8'h62;
PROD_ROM[232] = 8'h70;
PROD_ROM[233] = 8'h7E;
PROD_ROM[234] = 8'h8C;
PROD_ROM[235] = 8'h9A;
PROD_ROM[236] = 8'hA8;
PROD_ROM[237] = 8'hB6;
PROD_ROM[238] = 8'hC4;
PROD_ROM[239] = 8'hD2;
PROD_ROM[240] = 8'h00;
PROD_ROM[241] = 8'h0F;
PROD_ROM[242] = 8'h1E;
PROD_ROM[243] = 8'h2D;
PROD_ROM[244] = 8'h3C;
PROD_ROM[245] = 8'h4B;
PROD_ROM[246] = 8'h5A;
PROD_ROM[247] = 8'h69;
PROD_ROM[248] = 8'h78;
PROD_ROM[249] = 8'h87;
PROD_ROM[250] = 8'h96;
PROD_ROM[251] = 8'hA5;
PROD_ROM[252] = 8'hB4;
PROD_ROM[253] = 8'hC3;
PROD_ROM[254] = 8'hD2;
PROD_ROM[255] = 8'hE1;
```

```
        end
    assign         PRODuct = PROD_ROM[{Mplier, Mcand}];
endmodule
```

6.7 Dedicated Multipliers in FPGAs

Many modern FPGAs provide dedicated multipliers. Suppose that a designer wants a 16 × 16 multiplier. If dedicated multipliers are not provided, several programmable logic blocks will be used to create the 16 × 16 multiplier. Such a multiplier will be expensive in terms of the number of blocks and interconnect resources used; it will also be slow because of the switches involved in interconnecting the parts of the multiplier. Dedicated multipliers will be more area efficient and will be faster than multipliers realized using logic blocks. Since multiplication is an important operation in many applications involving FPGAs, many commercial FPGAs provide dedicated multipliers. For instance, Xilinx Virtex-4/Spartan-3 and Altera Stratix/Cyclone FPGAs contain 18 × 18 multipliers. These multipliers take two 18-bit operands and produce a 36-bit product as illustrated in Figure 6-21. It is possible to load the multiplicand and multiplier into optional registers and load the product into an optional product register. The inputs to the multipliers can come from external pins or from other logic in the FPGA.

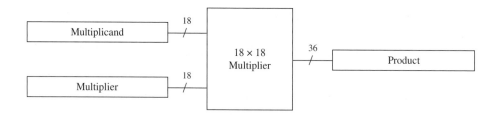

FIGURE 6-21: Dedicated Multipliers

When multiplication of numbers larger than 18 bits is required, several of the dedicated built-in multipliers can be put together. If A and B are 32 bits, and C, D, E, and F are the 16 bit components of A and B such that

$$A = C \times 2^{16} + D$$
$$B = E \times 2^{16} + F$$

then $AB = CE \times 2^{32} + (DE + CF) \times 2^{16} + DF$. This means that four multipliers are required to generate the partial products CE, DE, CF, and DF, and several adders are required to add the partial products.

Synthesis tools are capable of inferring dedicated multipliers on FPGAs that provide them. For instance, if the Verilog code in Figure 6-22 is synthesized for Xilinx Spartan devices using Xilinx ISE tools, the synthesis tool infers 4 dedicated 18 × 18 multipliers. When the code in Figure 6-22 is synthesized, several logic blocks in the FPGA are used in addition to the four multipliers. The logic blocks are used

FIGURE 6-22: Verilog Code that Infers Dedicated Multipliers

```
module multiplier (A, B, C);
    input[31:0] A;
    input[31:0] B;
    output[63:0] C;
    wire[63:0] C;

    assign C = A * B ;
endmodule
```

to realize the adders for the partial products. Sixty-four I/O pins are used to provide the multiplicand and multiplier, and 64 I/O pins are used for the output. Here external pins are used to provide inputs to the multipliers, but the inputs to the multipliers may also come from the embedded memory in the FPGAs or the optional registers.

6.8 Cost of Programmability

The programmability in an FPGA comes with a significant amount of hardware cost. In an SRAM based FPGA, such as the Xilinx Virtex and Spartan families, SRAM is used for creating the logic blocks, the programmable interconnects, and the programmable I/O blocks. The logic blocks in many modern FPGAs contain 4-variable function generators. A 4-variable function generator takes 16 bits of SRAM. Logic functions are realized by loading appropriate bits into the LUTs. Additionally, several multiplexers are used to select between various generated functions, to choose between latched and unlatched outputs, or to generate functions of more variables. One bit of SRAM is required to implement the select input of the 2-to-1 multiplexers, and two bits of SRAM is required for select lines of the programmable 4-to-1 multiplexers. Consider the logic block shown in Figure 6-23. The small boxes with "M" marked in them indicate memory cells required to program the multiplexers. A memory cell is used to select an external clock-enable signal. Another memory cell is used to invert the clock. A total of 46 memory cells are required to configure this logic block. The 40 memory cells in the three function generators (LUTs) might be implementing a simple one-variable function or a complex 5-variable function.

We will use one more example to illustrate the overhead of programmability. Figure 6-15 illustrated a logic block of the Microsemi Fusion FPGA. Each switch shown in the figure needs a flash memory cell. The various flash memory cells required to program this logic block constitute the overhead of programmability of this logic block.

The I/O blocks also contain several programmable points. Consider the I/O block in Figure 6-24. Memory bits for controlling the configuration are indicated by the boxes marked with "M." They are used to enable tristate output, to invert outputs, to enable the latching of output, to control the slew rate of the signal, to enable pull-up resistors, and so forth.

370 Chapter 6 Designing with Field Programmable Gate Arrays

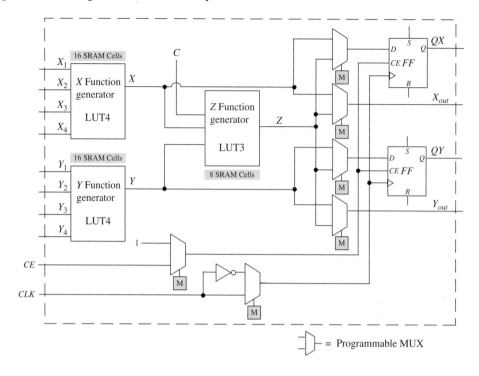

FIGURE 6-23: Logic Block with Several Programmable SRAM Cells

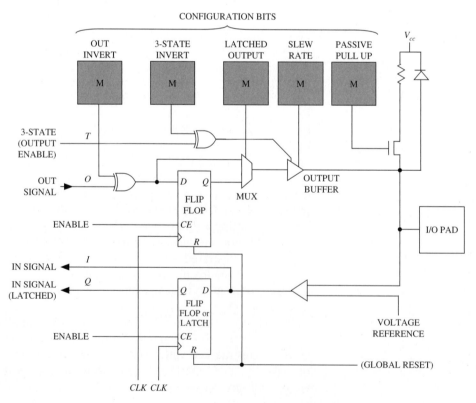

FIGURE 6-24: Programmable Points in FPGA I/O Block (Indicated by Boxes with "M"). Based on Xilinx.

Each SRAM cell typically takes six transistors. A flash memory cell consumes approximately 25% of an SRAM cell's area. The various programmable points add flexibility to the FPGA; however, the flexibility comes with the cost associated with the SRAM/flash memory cells. Table 6-3 shows the number of configuration bits in a few Xilinx Spartan and Virtex FPGAs. A Virtex-II FPGA, the XC2V40, which has 512 4-variable LUTs, needs 338,976 configuration bits. Another Virtex-II FPGA, the XC2V8000, has 93,184 4-variable LUTs and needs more than 26 million configuration bits. Thus, it is clear that the flexibility and programmability of the FPGA comes at a high cost.

TABLE 6-3: Number of Configuration Bits in Example FPGAs

Vendor	Device Family	Device	No. of Configuration Bits	No. of Logic Blocks	No. of LUTs	No. of Usable I/O Pins
Xilinx	Kintex 7	7K70T	23 M	10,250	41,000	300
		7K40T	143 M	74,650	298,600	400
Xilinx	Artix 7	7A100T	29 M	15,850	63,400	300
		7A200T	74 M	33,650	134,600	500
Xilinx	Virtex-6	XC6VLX75T	26 M	11,640	46,560	360
		XC6VLX760	177 M	118,560	474,240	120
Xilinx	Virtex-5	XC5VLX30	8.4 M	4,800	19,200	400
		XC5VLX330	79.7 M	51,840	207,360	1200
Xilinx	Virtex-II	XC2V40	0.3 M	256	512	88
		XC2V8000	26.2 M	46,592	93,184	1108
Xilinx	Spartan 3E	XC3S100E	0.6 M	960	1,920	108
		XC3S1600E	6.0 M	14,752	29,504	376
Xilinx	Spartan 6	XC6SLX4	2.7 M	600	2400	132
		XC6SLX150	33.9 M	23,038	92,152	576
Altera	Stratix II	EP2S15	4.7 M	6,240	12,480	366
		EP2S180	49.8 M	71,760	143,520	1170
Altera	Stratix	EP1S10	3.5 M	10,570	10,570	426
		EP1S80	23.8 M	79,040	79,040	1238
Altera	Arria V	5AGXA1	71 M	28,302	76,800	416
		5AGXB7	185.9 M	190,240	516,096	704
Altera	Arria II	EP2AGX45	29.6 M	18,050	45,125	364
		EP2AGX260	86.9 M	102,600	256,500	612
Altera	Cyclone II	EP2C5	1.3 M	4,608	4,608	158
		EP2C70	14.3 M	68,416	68,416	622

6.9 FPGAs and One-Hot State Assignment

When designing with FPGAs, it may not be important to minimize the number of flip-flops used in the design. Instead, we should try to reduce the total number of logic cells used and try to reduce the interconnections between cells. In order to design faster logic, we should try to reduce the number of cells required to realize each equation. Using a *one-hot state assignment* will often help to accomplish this. One-hot assignment takes more flip-flops than encoded assignment, but this is generally not a problem for FPGA-based designs because most commercial FPGA logic blocks contain two flip-flops. The next-state equations for flip-flops are often simpler in the one-hot method than the equations in the encoded method, which helps to reduce the number of necessary logic blocks.

The one-hot assignment uses one flip-flop for each state, so a state machine with N states requires N flip-flops. Exactly one flip-flop is set to 1 in each state. For example, a system with four states (T_0, T_1, T_2, and T_3) could use four flip-flops (Q_0, Q_1, Q_2, and Q_3) with the following state assignment:

$$T_0: Q_0Q_1Q_2Q_3 = 1000, \qquad T_1: 0100, \qquad T_2: 0010, \qquad T_3: 0001 \qquad (6\text{-}6)$$

The other 12 combinations are not used.

We can write next-state and output equations by inspection of the state graph or by tracing link paths on an SM chart. Consider the partial state graph given in Figure 6-25. The next state equation for flip-flop Q_3 could be written as

$$Q_3^+ = X_1 Q_0 Q_1' Q_2' Q_3' + X_2 Q_0' Q_1 Q_2' Q_3'$$
$$+ X_3 Q_0' Q_1' Q_2 Q_3' + X_4 Q_0' Q_1' Q_2' Q_3$$

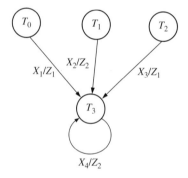

FIGURE 6-25: Partial State Graph

However, since $Q_0 = 1$ implies $Q_1 = Q_2 = Q_3 = 0$, the $Q_1' Q_2' Q_3'$ term is redundant and can be eliminated. Similarly, all the primed state variables can be eliminated from the other terms, so the next state equation reduces to

$$Q_3^+ = X_1 Q_0 + X_2 Q_1 + X_3 Q_2 + X_4 Q_3$$

Note that each term contains exactly one state variable. Similarly, each term in each output equation contains exactly one state variable:

$$Z_1 = X_1 Q_0 + X_3 Q_2, \qquad Z_2 = X_2 Q_1 + X_4 Q_3$$

When a one-hot assignment is used, the next state equation for each flip-flop will contain one term for each arc leading into the corresponding state (or for each link path leading into the state). In general, each term in every next-state equation and in every output equation will contain exactly one state variable.

When a one-hot assignment is used, resetting the system requires that one flip-flop be set to 1 instead of resetting all flip-flops to 0. If the flip-flops used do not have a preset input (as is the case for the Xilinx 3000 series), then we can modify the one-hot assignment by replacing Q_0 with Q_0' throughout. For the preceding assignment, the modification is

$$T_0: Q_0Q_1Q_2Q_3 = 0000, \qquad T_1: 1100, \qquad T_2: 1010, \qquad T_3: 1001 \tag{6-7}$$

and the modified equations are

$$Q_3^+ = X_1Q_0' + X_2Q_1 + X_3Q_2 + X_4Q_3$$
$$Z_1 = X_1Q_0' + X_3Q_2, \qquad Z_2 = X_2Q_1 + X_4Q_3$$

Another way to solve the reset problem without modifying the one-hot assignment is to add an extra term to the equation for the flip-flop, which should be 1 in the starting state. If the system is reset to state 0000 after power-up, we can add the term $Q_0'Q_1'Q_2'Q_3'$ to the equation for Q_0^+. Then, after the first clock the state will change from 0000 to 1000 (T_0), which is the correct starting state.

In general, both an assignment with a minimum number of state variables and a one-hot assignment should be tried to see which one leads to a design with the smallest number of logic cells. Alternatively, if speed of operation is important, the design that leads to the fastest logic should be chosen. When a one-hot assignment is used, more next-state equations are required, but in general both the next-state and output equations will contain fewer variables. An equation with fewer variables generally requires fewer logic cells to realize. The more cells are cascaded, the longer the propagation delay and the slower the operation.

6.10 FPGA Capacity: Maximum Gates versus Usable Gates

Designers like to know whether a design that typically consumes X number of gates in the ASIC world will fit in a particular FPGA. In order to help designers to answer this question, FPGA vendors often provide some capacity metrics, either as equivalent gate counts or number of logic blocks. As you know by now, most FPGAs are not structured as arrays of gates. Some are simply arrays of look-up tables rather than arrays of gates. So what does the gate count of an FPGA mean?

The number of raw gates that have gone into building an FPGA is not an interesting or useful metric to an FPGA user. What is useful to the user is a count of the circuitry that can fit into a particular FPGA. This is called the **equivalent gate count**. But as one might guess, this type of achievable gate count will depend on the type of circuitry, the type of interconnections between different parts of the circuitry, the routing resources available in the FPGA, and so forth. This type of gate count is extremely difficult to compute.

Gate counts are estimated in many different ways. An approximate equivalent gate count can be established for a logic block by considering typical circuits that can be implemented in a logic block, For instance, a 2-to-1 multiplexer is considered to be four gates, and a 3-input XOR is considered to be six gates. A 4-input XOR is 9 gates and a flip-flop with clear is considered to be 6-7 gates. An equivalent gate count can be obtained for a programmable logic block in an FPGA in this fashion, and the total gate count can be estimated by multiplying it with the number of logic blocks in the FPGA. This type of gate count is likely to be higher than the gate count of practical circuitry that can be realized in the FPGA.

A better gate count estimate can be derived using benchmark circuits. The **Programmable Electronics Performance Company (PREP)** benchmark suite was an early attempt to facilitate standard benchmark circuits for ASIC and FPGA benchmarking. Assume that a particular circuitry typically takes 2000 gates in ASIC, and if an FPGA device can fit 20 copies of that circuitry, an FPGA vendor may estimate the maximum gate count of their FPGA as 40K. Since the circuit is simply replicated and no actual interconnection exists between the copies, this count is also likely to be higher than the gate count of practical circuitry that can be realized in the FPGA. Some FPGA vendors provide a typical gate count by adjusting the maximum gate count with some weighting schemes.

It is very difficult to estimate gate counts of FPGAs in which logic is implemented with LUTs. A 4-input LUT may be used to implement a 4-variable logic function with one or more product terms, or it can be used to store 16 bits of information. When the LUTs are used as RAM, higher gate counts may be obtained. Hence, depending on the portion of LUTs used as RAM, one can estimate different gate counts for the same FPGA. Vendors often compute their "system gates" count by considering a fraction of CLBs (say 20–30%) as RAM.

Altera provides two types of gate counts for their APEX family: maximum gates and usable gates. The APEX II devices range from 1.9 million to 5.25 million maximum gates, but the typical gate count is published as 600K to 3 million. Due to the difficulty in estimating equivalent gate counts, many FPGA vendors provide their chip capacities with a count of the logic blocks (logic elements) rather than a gate count.

PREP Benchmarks

The **Programmable Electronics Performance Company (PREP)** was a non-profit organization that gathered and distributed a series of benchmarks for programmable logic chips in the early days of FPGAs. The nine PREP benchmark circuits in the PREP 1.3 suite were as follows:

1. An 8-bit data path consisting of a 4:1 MUX, a register, and a shift-register
2. An 8-bit timer-counter consisting of two registers, a 4:1 MUX, a counter, and a comparator
3. A small state machine (8 states, 8 inputs, and 8 outputs)
4. A larger state machine (16 states, 8 inputs, and 8 outputs)

> 5. An ALU consisting of a 4 × 4 multiplier, an 8-bit adder, and an 8-bit register
> 6. A 16-bit accumulator
> 7. A 16-bit counter with synchronous load and enable
> 8. A 16-bit prescaled counter with load and enable
> 9. A 16-bit address decoder
>
> PREP also made additional synthesis benchmarks available, including a bit-slice processor, a multiplier, and an R4000 MIPS RISC microprocessor. The PREP circuits are very small in comparison to the size of modern FPGAs. No recent activity has been observed from the PREP organization.

6.11 Design Translation (Synthesis)

In the preceding sections of this chapter, we hand-mapped some designs into FPGA logic blocks. This process is analogous to writing assembly language programs for microprocessors. It is tedious. The productivity of designers will be very low if they can enter designs only at that level. Just as the majority of the programs in the modern-day world are written in high-level languages such as C and translated by a compiler, modern-day digital designs are done at behavioral or RTL level and translated to target devices. This applies not only for FPGAs but also for ASIC design.

A number of CAD tools are now available that take a Verilog/VHDL description of a digital system and automatically generate a circuit description that implements the digital system. The term **synthesis** refers to the translation of an abstract high-level design to a circuit description, typically in the form of a logic schematic. The input to the CAD tool is a behavioral or structural VHDL/Verilog model. The output from the synthesis tools may be a logic schematic together with an associated wirelist, which implements the digital system as an interconnection of gates, flip-flops, registers, counters, multiplexers, adders, and other basic logic blocks. This representation is called a **netlist**. The circuit can now be targeted for an FPGA, a CPLD, or an ASIC.

Typical computer-aided design flow involves the following steps:

- Design translation (synthesis) and optimization
- Mapping
- Placement
- Routing

These steps are illustrated in Figure 6-26. In this section, we describe design translation and optimization techniques. The mapping, placement, and routing of designs are described in the following section.

FIGURE 6-26: CAD Design Flow

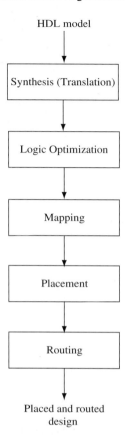

Even if Verilog/VHDL code compiles and simulates correctly, it may not necessarily synthesize correctly. And even if the Verilog/VHDL code does synthesize correctly, the resulting implementation may not be very efficient. In general, synthesis tools will accept only a subset of Verilog as input. Other changes must be made in the Verilog/VHDL code so the synthesis tool "understands" the intent of the designer. Further changes in the Verilog/VHDL code may be required in order to produce an efficient implementation.

In Verilog, a signal may represent the output of a flip-flop or register, or it may represent the output of a combinational logic block. The synthesis tool will attempt to determine what is intended from the context. For example, the concurrent statement

```
assign A = B & C;
```

implies that *A* should be implemented using combinational logic. On the other hand, if the sequential statements

```
always @(posedge CLK)
begin
   A = B & C;
end
```

appear in an always statement, this implies that *A* represents a register (or flip-flop) that changes state on the rising edge of the clock.

Most Verilog synthesizers do a line-by-line translation of Verilog into gates, registers, multiplexers, and other general components with very little optimization up front. Then the resulting design is optimized. Synthesizers associate particular Verilog constructs with particular hardware structures. For instance, case statements typically result in multiplexers. Use of "+," "−," and comparison results in the use of an adder; use of shift operators results in the use of a shift register, and so on.

During the initial translation of the Verilog code and during the optimization phase, the synthesis tool will select components from those available in its library. Several different component libraries may be provided to allow implementation with different technologies. In the ASIC world, the libraries can include parts that are specifically targeted for low power, low area, or high speed. Depending on the specific goals of the design, the synthesis process can be instructed to meet specific requirements of area, power, or speed. Of course, there are tradeoffs between these and designs will have to prioritize between the requirements.

Synthesis of a Case Statement

The example of Figure 6-27 shows how the Synopsys Design Compiler implements a case statement using multiplexers and gates. Figure 6-27(a) shows the code. The inputs *a* and *b* are each implemented with 2-bit binary numbers. Two 4-to-1 multiplexers are required. The two bits of *a* are used as control inputs to the multiplexer. The multiplexer inputs are hardwired to a logic 1 or a logic 0. Figure 6-27(b) shows the hardware that will be generated by a typical synthesizer.

Most modern synthesizers will also perform optimizations to reduce the logic that is generated. Because the MUX inputs are constants, elimination of the MUX and several gates is possible by inspection of the truth table in Figure 6-27(c). The optimized output equations are $b_1 = a_1'a_0 = (a_1 + a_0')'$ and $b_0 = (a_1 a_0')'$. An optimized circuit for the code in Figure 6-27(a) consists only of a NOR, a NAND, and a NOT gate. Figure 6-27(d) shows the resulting circuit after optimization.

FIGURE 6-27: Synthesis of a Case Statement

```
module case_example (a, b);

    input[1:0] a;
    output[1:0] b;
    reg[1:0] b;

    always @(a)
    begin
        case (a)
            0 :
                    begin
                        b <= 1 ;
                    end
            1 :
                    begin
                        b <= 3 ;
                    end
            2 :
```

```
                begin
                    b <= 0 ;
                end
        3 :
                begin
                    b <= 1 ;
                end
    endcase
end

endmodule
```

(a) Verilog code for case example

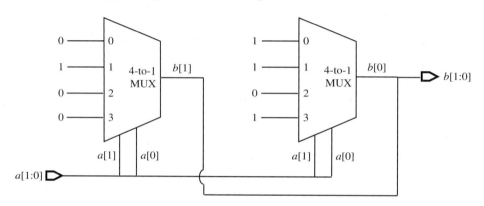

(b) Synthesized circuit before optimization

a_1	a_0	b_1	b_0
0	0	0	1
0	1	1	1
1	0	0	0
1	1	0	1

$b_1 = a_1' \cdot a_0$

$ = (a_1 + a_0')'$

$b_0 = (a_1 \cdot a_0')'$

(c) Logic optimization

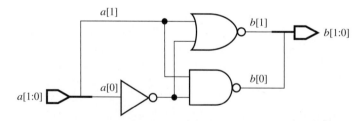

(d) Synthesized circuit after optimization

Unintentional Latch Creation

In general, when a Verilog signal is assigned a value, it will hold that value until it is assigned a new value. Because of this property, some Verilog synthesizers will infer a latch when none is intended by the designer. Figure 6-28(a) shows an example of a case statement that creates an unintended latch. The case statement results in a 4-to-1 multiplexer whose data inputs are set to the values in each case. The select lines are controlled by the value of *a*. Since the value of *b* is not specified if *a* is not equal to 0, 1, or 2, the synthesizer assumes that the value of *b* should be held in a latch if *a* = 3.

When *a* = 3, the previous value of *b* should be used as the output. This necessitates a latch whose **D** input = a_0. In order to hold the value in the latch, the latch gate control signal *G* should be 0 when **a** = **3**. Thus $G = (a_1 a_0)'$. A naïve synthesizer might generate a 4-to-1 multiplexer and a latch as in Figure 6-28(c). The latch can be eliminated by adding the Verilog code b <= 1'b0 for a = 3 as in Figure 6-28(b). If this change is made, most synthesizers will generate only a multiplexer and no latch.

Most modern synthesizers also perform optimizations to reduce the logic that is generated. For example, a 4-to-1 multiplexer is not required for this circuit. As easy way to derive the optimized circuit is by inspection of the truth table in Figure 6-28(d). One may easily observe that when *a* equals 0, 1, or 2, $b = a_0'$. An optimizing synthesizer might generate a single NOT gate for the code, as shown in Figure 6-28(e). If the *a* = 3 case was not added, this optimizing synthesizer will generate a latch also as shown in Figure 6-28(d)—that is, with the unintended latch.

FIGURE 6-28: Example of Unintentional Latch Creation

```
module latch_example (a, b);

   input[1:0] a;
   output b;
   reg b;

   always @(a)
   begin
      case (a)
         0 :
                     b <= 1'b1 ;
         1 :
                     b <= 1'b0 ;
         2 :
                     b <= 1'b1 ;
      endcase
   end
endmodule
```

(a) Verilog code that infers a latch

```
module latch_example (a, b);

   input[1:0] a;
   output b;
   reg b;

   always @(a)
```

```
    begin
        case (a)
            0 :
                        begin
                            b <= 1'b1 ;
                        end
            1 :
                        begin
                            b <= 1'b0 ;
                        end
            2 :
                        begin
                            b <= 1'b1 ;
                        end
            3 :
                        begin
                            b <= 1'b0 ;
                        end
        endcase
    end
endmodule
```

(b) Modified code not resulting in a latch

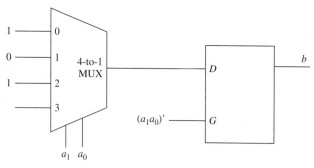

(c) Synthesized circuit for code in (a)

a_1	a_0	b
0	0	1
0	1	0
1	0	1
1	1	previous b

a_1	a_0	b
0	0	1
0	1	0
1	0	1
1	1	0

$a_0 \rightarrow\!\!\triangleright\!\!\circ\, b_1$

$b_1 = a_0'$

(d) Optimized circuit for code in (a)

(e) Optimized circuit for code in (b)

Synthesis of `if` Statements

When `if` statements are used, care should be taken to specify a value for each branch. For example, if a designer writes

```
if (A == 1'b1)
begin
    Nextstate <= 3;
    Z <= 1;
end
```

he or she may intend for *Nextstate* to retain its previous value if $A \neq 1$ and the code will simulate correctly. However, the synthesizer might interpret this code to mean if $A \neq 1$, then *Nextstate* is unknown ('*X*'), and the result of the synthesis may be incorrect. Also, it will result in latches for *Z*. For this reason, it is always best to include an `else` clause in every `if` statement. For example,

```
if (A == 1'b1)
begin
    Nextstate <= 3;
    Z <= 1;
end
else
begin
    Nextstate <= 2;
    Z <= 0;
end
```

is unambiguous.

The example of Figure 6-29 shows how a typical synthesizer implements an `if-then-else` `if-else` statement using a multiplexer and gates. Figure 6-29(b) represents the truth table corresponding to the various input combinations. *C* is selected if $A = 1$; *D* is selected if $A = 0$ and $B = 0$; and *E* is selected if $A = 0$ and $B = 1$. Figure 6-29(c) indicates the synthesized hardware. *A* and *B* are used as select signals of the multiplexer.

FIGURE 6-29: Synthesis of an `if` Statement

```verilog
module if_example (A, B, C, D, E, Z);

    input A;
    input B;
    input[2:0] C;
    input[2:0] D;
    input[2:0] E;
    output[2:0] Z;

    reg[2:0] Z;

    always @(A or B)
    begin
```

```
        if (A == 1'b1)
        begin
            Z <= C ;
        end
        else if (B == 1'b0)
        begin
            Z <= D ;
        end
        else
        begin
            Z <= E ;
        end
    end
endmodule
```

(a) Verilog code for if example

A	B	Z
0	0	D
0	1	E
1	0	C
1	1	C

(b) Equivalent truth table

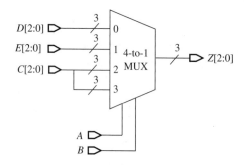

(c) Synthesized hardware for code in (a)

Example

What hardware does the statement

    `assign LE = (A <= B);`

result in? Assume that A and B are 4-bit vectors.

Answer: A 4-bit comparator. The <= symbol between A and B is a relational operator. The right side of the assignment symbol returns a TRUE or 1 if A is less than B. Hence, if A is less than B, LE is set to 1. Otherwise, LE will be 0.

One could have a statement

    `LE <= (A <= B);`

inside an always block to yield the same hardware. This statement can look confusing because the two <= symbols have different meanings. The first one is an assignment whereas the second one is a relational operator.

Most standard comparators come with EQUAL_TO (*EQ*), GREATER_THAN (*GT*), and LESS_THAN (*LT*) outputs. In this case, *LE* should be 1 if EQUAL_TO or LESS_THAN is true. Figure 6-30 illustrates the hardware.

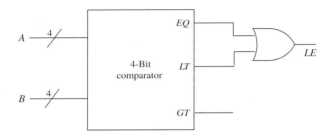

FIGURE 6-30: Hardware for Less Than or Equal To Checker

Synthesis of Arithmetic Components

CAD tools for synthesis have design libraries that include components to implement the operations defined in the numeric packages. When this code is synthesized, the result includes library components that implement a 4-bit comparator, a 4-bit binary adder with a 4-bit accumulator register, and a 4-bit counter. Some synthesis tools will implement the counter with a 4-bit adder with a 0001 input and then optimize the result to eliminate unneeded gates. The resulting hardware is shown in Figure 6-31(b).

FIGURE 6-31: Verilog Code Example for Synthesis and Corresponding Hardware

```verilog
module examples (clock, A, B, ge, acc, count);
   input clock;
   input[3:0] A;
   input[3:0] B;
   output ge;
   inout[3:0] acc;
   inout[3:0] count;

   reg[3:0] acc_temp;
   reg[3:0] count_temp;

   assign acc = acc_temp;
   assign count = count_temp;

   assign ge = (A >= B) ;

   always @(posedge clock)
   begin
      acc_temp <= acc + B ;
      count_temp <= count + 1 ;
   end
endmodule
```

(a) Verilog code

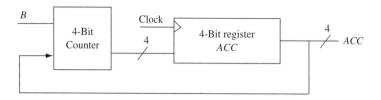
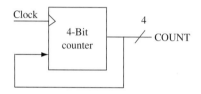
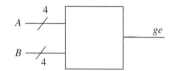

(b) Synthesized hardware for the VHDL code in (a)

Example

Generate optimized hardware for the following statement, assuming A is a 4-bit vector:

```
assign EQ3 = (A == 3);
```

Answer: A 4-bit comparator can be used to realize this statement. One input to the comparator will be A, and the other input will be the number 3—that is, 0011 (binary). But since we know that one input is constantly 3 we could optimize it further to result in an AND gate and two inverters as shown in Figure 6-32.

FIGURE 6-32: Optimized Hardware for Equality Checker

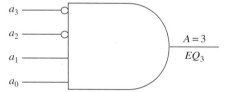

Some synthesizers may not automatically provide this optimized hardware. Under such circumstances, one can alter the Verilog source code to

```
assign EQ3 = ~A[3] & ~A[2] & A[1] & A[0];
```

This statement will result in the 4-input AND gate of Figure 6-32.

Different kinds of optimizations are required for different target technologies. For instance, reduction in absolute number of gates is important for a gate-based target technology, but if an FPGA with LUTs is the target technology, the optimization does not need to consider absolute number of gates in the design. Instead, it needs to optimize the number of LUTs.

Many FPGAs include specific arithmetic components—for example, built-in multipliers. The synthesis tools for these FPGAs recognize functions that can be directly mapped into these hardware components. Section 6.7 has already illustrated synthesis into multipliers. Use of carry-chains is another FPGA-specific optimization that synthesis tools for FPGAs can handle.

Area, Power, and Delay Optimizations

Most Verilog synthesizers allow the design to be optimized for maximum speed or for minimum chip area. Power consumption has also recently become a major design constraint along with area and delay. Typically, optimizing for one constraint will worsen the performance on another. For example, if speed is improved, area might worsen. Improving speed often means that an operation that is being performed serially, reusing some gates, may have to be performed in parallel. Hence, improving the speed often results in increasing the number of components. Consider a serial adder that is used to perform 4-bit addition compared with a fully parallel combinational 4-bit adder that uses a lot more hardware to achieve much better speed. When optimizing for area, an effort is made to decrease the number of components, which in turn often increases the critical path. **Critical path** means the longest delay in the circuit.

CAD tools incorporate gate libraries. The libraries provide various options for achieving requirements regarding area, speed, and power. Gates and building blocks that are optimized individually for area, speed, or power or collectively for two or more of these can be obtained, and depending on the designer's specifications, appropriate elements from the libraries can be used.

Area and delay of a circuit are often inversely related to each other. Energy and delay are also inversely related. The **Area-Time** (AT) product and **Energy-Delay** (ED) product are popularly-used metrics to describe the quality of a circuit. **Area-Time^2** (AT^2) and **Energy-Delay^2** (ED^2) are also used as metrics to measure the quality of circuits and systems.

In spite of the inverse relationships between area and delay or energy and delay, there are optimizations that simultaneously improve area, delay, and power. For example, consider the optimizations in Figure 6-27(b) to (d) and the optimization in Figure 6-28(c) to (e). These optimizations at the logic level perform the required task in an effective way, resulting in less hardware, less area, less power, and surprisingly smaller critical path as well.

When designing with FPGAs, we should keep in mind that optimizations for discrete gates are not necessarily the best optimizations for FPGAs. As an example, consider function minimization—reducing the number of terms in an expression, which is extremely important when implementing the design using gates. In an SRAM FPGA, the important issue is to minimize the number of variables in an expression. Minimizing the number of terms in an equation is not required, because the entire truth table is stored in LUT form.

Major Vendors of CAD Tools

Cadence
Synopsys
Mentor Graphics

Major Vendors of FPGA CAD Tools

XIlinx
Altera
Microsemi
Lattice

6.12 Mapping, Placement, and Routing

Once the design is translated by synthesis and the netlist is generated, the resulting design has to be mapped into a specific implementation technology. Implementation technologies include gate arrays, FPGAs, CPLDs, and ASIC standard cell designs. Mapping, placement, and routing are the three major steps that happen in order to transform the design in netlist form to the appropriate target technology.

Mapping

Mapping is the process of binding technology-dependent circuits of the target technology to the technology-independent circuits in the design. As is generally known, a design can be implemented in many ways: using multiplexers, using ROM or lookup tables (LUT), using NAND gates, using NOR gates, or using AND-OR gates. Designs can also be implemented as a combination of several of these technologies.

If one is using a gate-array based on standard cells, the netlist needs to be "mapped" into the standard cells. If one is using a field-programmable gate array with LUTs, the design needs to be transferred or "mapped" into the LUTs. If one is using a field-programmable gate array with only 4-to-1 multiplexers, the design must be mapped into a structure that requires only multiplexers. If a target technology contains only 2-input NAND gates, the design must be mapped to a form that uses only 2-input NAND gates. We did this process manually for a shift register and multiplexer at the beginning of this chapter. CAD tools use mapping software to accomplish this task.

> **Standard Cell Approach:** Standard cell design is a common technique for integrated circuit design. The design is mapped into a library of standard logic gates. Typically NOT, AND, NAND, OR, NOR, XOR, XNOR, among others, are available. CAD tools that support standard cell-design methodology will also usually contain a library of complex functions and standard building blocks such as multiplexers, decoders, encoders, comparators, counters, and the like. The design is mapped into a form that contains only cells available in the library. The cells are placed in rows that are separated by routing channels as shown in Figure 6-33. Some cells may be used only for routing between rows of cells. Such cells are called feedthrough cells. For the standard cell methodology to be effective, the height of cells should be the same. But it is possible to include memory modules, specialized arithmetic modules, and so on.

Place and Route

Placement is the process of taking defined logic and I/O blocks (modules) from the technology mapper and assigning them to physical locations of the target implementation. It involves determining the positions of the sub-blocks in the design area. Placement choices matter because they affect subsequent routing. A good placement algorithm will try to reduce area and delay. Area and delay are partly determined by wiring. Algorithms typically estimate wire length and decide on appropriate placement choices. Complicated placement algorithms are not desirable because they consume too much run time.

Routing is the process of interconnecting the sub-blocks in a design. The choices for routing are greatly dependent on placement; hence, place and route are often done in tandem. Routing may be done in multiple steps. Global routing decisions can be made to minimize routing wire length, and then detailed routing of sections can be done. When only a part of a circuit is changed, incremental routing is useful.

Usually heuristics are used to perform placement. Most placement techniques start with an initial solution and then try to improve it with alternative placements.

FIGURE 6-33: Example Standard Cell Layout

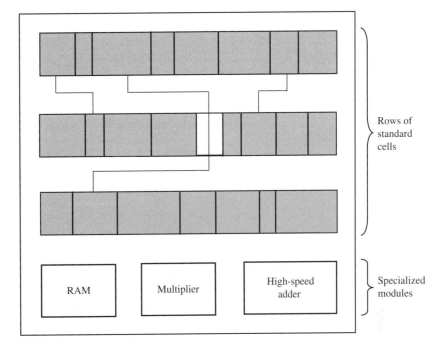

For instance, two blocks in one placement can be swapped to arrive at an alternative placement, and wire length is evaluated for both the choices. The process is repeated until no further improvements are possible.

Simulated annealing techniques are used in the place and route process. Annealing is a term from metallurgy. Simulated annealing algorithms quickly and effectively optimize solutions over large state spaces. It does not guarantee the optimal solution, but it can produce a solution close to the global minimum in much less time than an exhaustive search. The simulated annealing process starts with a feasible solution (i.e., legal but not necessarily optimal) and searches for better solutions by making random modifications (permutations). An **iterative improvement** algorithm accepts only better solutions in each step. Algorithms that accept only better moves are considered **greedy algorithms**. But if one accepts only better placements, one could be caught in a local minimum. It has been shown that it is beneficial to occasionally accept "bad moves." Often, these "bad moves" will let the algorithm reach a global minimum.

Accepting a bad move is certainly a risk. One can take more risks in the beginning of the simulated annealing process, but one needs to be more conservative during the later stages because there might not be sufficient time left to refine the solution to an acceptable level. In simulated annealing algorithms, the algorithms have a concept of a temperature, as in physical annealing in metallurgy. The temperature is high in the beginning and keeps reducing. Simulated annealing algorithms allow risky moves depending on the temperature. As the temperature is reduced, the probability of accepting bad moves decreases. Eventually, the algorithm defaults to a greedy algorithm that accepts only positive moves. Figure 6-34 illustrates the difference between simulated annealing and iterative improvement algorithms. The y-axis is the cost (or figure of merit) of the solution. The x-axis indicates the steps during the process.

FIGURE 6-34: Simulated Annealing versus Iterative Improvement Algorithms

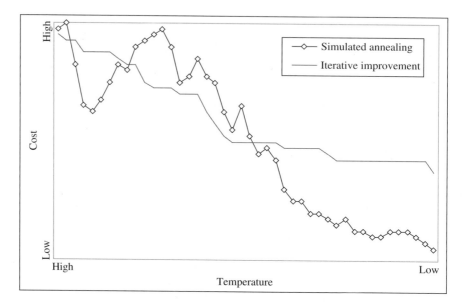

In simulated annealing place-and-route algorithms, an initial placement is assumed and cost of alternate placement is estimated. Typically, the cost of a placement indicates the amount of routing that is needed. A move is considered better if it produces a better cost figure (for example, wire length).

The ability of the tools to map and route designs depends on the algorithms in the tools and the granularity of the resources. Figure 6-35 shows a routed FPGA implementing an example design. (It is actually the dice game of Chapter 5, implemented in an early Xilinx FPGA, the XC3000.) The boxes on the periphery are the I/O blocks. Obviously, only a few of them on the top-left corner and on the bottom side are used. The logic blocks in the middle are utilized, while several logic blocks are unused. Synthesis tools will provide a synthesis report giving the number and percentage of logic blocks used, number and percentage of flip-flops used, and so on.

The utilization of an FPGA depends on the nature of the logic blocks, the efficiency of the mapping tools, the routing resources, and the efficiency of the routing tools, among other things. If logic blocks are of large granularity, it is very likely that parts of logic blocks will be unused. For instance, we saw that the shift register design in Figure 6-5 did not utilize a large part of the function generator. Similarly, the multiplexer designs in Figures 6-2 and 6-4 did not utilize the flip-flops on the logic block. If the logic blocks are of fine granularity, the utilization of logic blocks can be higher but more routing resources will be needed for interconnection, often resulting in slower circuits.

In this chapter we described several types of FPGAs and procedures for designing with these devices. Currently, sophisticated CAD tools are available to assist with the design of systems using programmable gate arrays. However, in this chapter, several hand designs were presented first to illustrate the underlying steps in CAD tools. Techniques to decompose functions of several variables into functions with fewer variables were illustrated. Features of modern FPGAs such as embedded memory, embedded multipliers, and carry and cascade chains were described. A brief overview of the synthesis, mapping, placement, and routing process was presented.

6.12 Mapping, Placement, and Routing

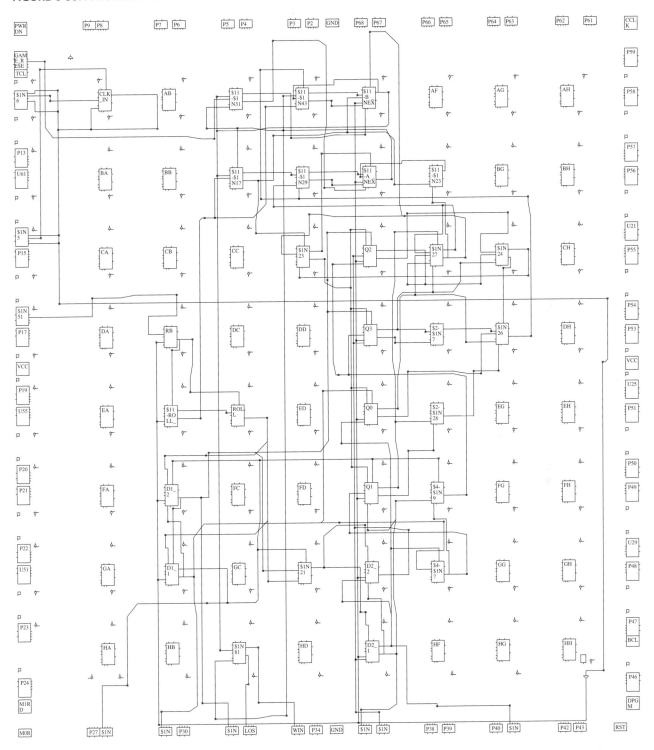

FIGURE 6-35: A Routed FPGA

Problems

6.1 An 8-bit right-shift register with parallel load is to be implemented using an FPGA with logic blocks as shown in Figure 6-1(a). The flip-flops are labeled $X_7X_6X_5X_4X_3X_2X_1X_0$. The control signals N and S operate as follows: $N = 0$, do nothing; $NS = 11$, right shift; $NS = 10$, load. The serial input for right shift is SI.

(a) How many logic blocks are required?
(b) Show the required connections for the rightmost block on a copy of Figure 6-1(a). Connect N to CE.
(c) Give the function generator outputs for this block.

6.2 Implement a 2-bit binary counter using one logic block as shown in Figure 6-1(a). A_0 is the least significant bit, and A_1 is the most significant bit of the counter. The counter has a synchronous load (Ld). The counter operates as follows:

$En = 0$	No change.	
$En = 1$,	$Ld = 1$	Load A_0 and A_1 with external inputs U and V on rising edge of clock.
$En = 1$,	$Ld = 0$	Increment counter on rising edge of clock.

(a) Give the next-state equations for A_0 and A_1.
(b) Show all required inputs and connections on a copy of Figure 6-1(a). Show the connection paths with heavy lines. Use the CE input. Give the function realized by each 4-input LUT.

6.3 Design a 4-bit right-shift register using an FPGA with logic blocks as shown in Figure 6-1(a). When the register is clocked, the register loads if $Ld = 1$ and $En = 1$; it shifts right when $Ld = 0$ and $En = 1$; and nothing happens when $En = 0$. S_i and S_o are the shift input and output of the register. $D_{3\text{-}0}$ and $Q_{3\text{-}0}$ are the parallel inputs and outputs, respectively. The next-state equation for the leftmost flip-flop is $Q_3^+ = En'Q_3 + En\,(Ld\,D_3 + Ld'\,S_i)$.

(a) Give the next-state equations for the other three flip-flops.
(b) Determine the minimum number of Figure 6-1(a) logic blocks required to implement the shift register.
(c) For the left block, give the input connections and the internal paths on a copy of Figure 6-1(a). In addition, give the X and Y functions.

6.4 The next-state equations for a sequential circuit with two flip-flops (Q_1 and Q_2), input signals R, S, T, and an output P are

$$D_1 = Q_1^+ = Q_2R + Q_1S$$
$$D_2 = Q_2^+ = Q_1 + Q_2'T$$

The output equation is $P = Q_2RT + Q_1ST$

(a) Explain how this sequential circuit can be implemented using a single Figure 6-3 logic block. Write the equation that each function generator in the block will implement.

(b) Mark (highlight) the input signals, the state and output variables, and the activated paths on a copy of Figure 6-3.

6.5 (a) Implement an 8-to-1 multiplexer using a minimum number of logic blocks of the type shown in Figure 6-1(a). Give the *X* and *Y* functions for each block and show the connections between blocks.

(b) Repeat (a) using the logic blocks of Figure 6-3. Give *X*, *Y*, and *Z* for each block.

(c) What are the LUT contents for the design in part (a)?

(d) What are the LUT contents for the design in part (b)?

6.6 (a) Write Verilog code that describes the logic block shown in Figure 6-1(a). Use the following module:

```
module Figure6_1a(X_in,Y_in,clk,CE,Qx,Qy,X,Y,XLUT,YLUT);
input[1:4]X_in,Y_in;
input CE,clk;
input[0:15] XLUT,YLUT;
inout X,Y;
output Qx,Qy;
.
.
.
endmodule
```

(b) Write structural Verilog code that instantiates two Figure 6-1(a) block components to implement the 4-to-1 MUX of Figure 6-2. When you instantiate a block, use the actual bit patterns stored in *XLUT* and *YLUT* to specify the function generated by each of the LUTs.

6.7 (a) Write Verilog code that describes the logic block shown in Figure 6-3. Use a module similar to that used in Problem 6.6(a), except add *ZLUT* and *SA*, *SB*, *SC*, and *SD*. *SA*, *SB*, *SC*, and *SD* represent the programmable select bits that control the four MUXes. These bits should be assigned values of 0 or 1 when the block component is instantiated.

(b) Write structural Verilog code that instantiates two Figure 6-3 block components to implement the code converter shown in Figure 1-26. When you instantiate a block component, use the actual bit patterns stored in *XLUT*, *YLUT*, and *ZLUT* to specify the function generated by each of the LUTs.

6.8 (a) How many logic blocks as shown in Figure 6-1(a) are required to create a 4-to-16 decoder?

(b) Give the contents of the LUTs in the first logic block.

6.9 (a) How many logic blocks as shown in Figure 6-3 are required to create an 8-to-3 priority encoder?

(b) Give the contents of the LUTs in the first logic block.

6.10 Show how to realize the following combinational function using two Figure 6-1(a) logic blocks. Show the connections on a copy of Figure 6-1(a) and give the functions X and Y for both blocks.

$$F = X_1'X_2X_3'X_6 + X_2'X_3'X_4X_6' + X_2X_3'X_4' + X_2X_3X_4'X_6 + X_3'X_4X_5X_6' + X_7$$

6.11 Realize the following next-state equation using a minimum number of Figure 6-1(a) logic blocks. Draw a diagram that shows the connections to the logic blocks and give the functions X and Y for each cell. (The equation is already in minimum form.)

$$Q^+ = UQV'W + U'Q'VX'Y' + UQX'Y + U'Q'V'Y + U'Q'XY + UQVW' + U'Q'V'X$$

6.12 Show how to realize the following next-state equations using a minimum number of the Kintex logic slice (Figure 6-13). Draw a diagram to show the connections between the logic elements and indicate how many CLBs you have used.

$$Q_1^+ = X_1'X_2'Q_1 + Q_2 X_3 X_4$$
$$Q_2^+ = X_3 X_2 Q_1 + X_1' Q_3$$
$$Q_3^+ = X_4 Q_2$$
$$C = X_1' Q_2 Q_3 Q_1'$$

6.13 What is the minimum number of Figure 6-3 logic blocks required to realize the following function?

$$X = X_1'X_2'X_3'X_4'X_5 + X_1X_2X_3X_4X_5 + X_5'X_6X_7'X_8'X_9 + X_5'X_6'X_7X_8X_9'$$

If your answer is 1, show the required input connections on a copy of Figure 6-3 and mark the internal connection paths with heavy lines. If your answer is greater than 1, draw a block diagram showing the cell inputs and interconnections between cells. In any case, give the functions to be realized by each X, Y, and Z function generator.

6.14 Illustrate how to realize the following equations using a minimum number of the Stratix IV logic module. Display the input/output by drawing a diagram and highlight the data path on the diagram.

$$X^+ = A'BY + E'FX + B'E + A'F';$$
$$Y^+ = A'D'X' + BC + AB'C'G;$$
$$Z^+ = X'Y'ZA + B'G + YBA';$$

6.15 Given $Z(T, U, V, W, X, Y) = VW'X + U'V'WY + TV'WY'$,

(a) Show how Z can be realized using a single Figure 6-3 logic block. Show the cell inputs on a copy of Figure 6-3; indicate the internal connections in the cell; and specify the functions X, Y, and Z.

(b) Show how Z can be realized using two Figure 6-1(a) logic blocks. Draw a diagram showing the inputs to each cell, the interconnections between cells, and the X and Y functions for each cell.

6.16 Decompose the following function using Shannon's decomposition around the variable X_6. Do not simplify the function.

$$F = X_1'X_2X_3'X_4X_6 + X_2'X_3'X_4X_6' + X_2'X_4' + X_3X_4X_5X_6 + X_3'X_4X_6' + X_1X_3$$

Write an expression for F in terms of the decomposed functions and X_6.

6.17 Use Shannon's expansion theorem around a and b for the function

$$Y = abcde + cde'f + a'b'c'def + bcdef' + ab'cd'ef' + a'bc'de'f + abcd'e'f$$

so that it can be implemented using only 4-variable function generators. Draw a block diagram to indicate how Y can be implemented using only 4-variable function generators. Indicate the function realized by each 4-variable function generator.

6.18 Use Shannon's expansion theorem around e and f for the function

$$Y = ab'cdef + a'bc'd'e + b'c'ef' + abcde'f$$

so that it can be implemented using a minimum number of 4-variable functions. Rewrite Y to indicate how it will be implemented using 4-variable function generators and draw a block diagram. Indicate the function generated by each function generator.

6.19 **(a)** Use Shannon's expansion theorem around a for the function

$$Y = ab'cd'e + a'bc'd'e + b'c'e + abcde$$

so that it can be implemented using 4-variable functions.
(b) Use the expanded function to show how Y can be implemented using one Figure 6-3 logic block. Mark (highlight) the input signals and the activated paths on a copy of Figure 6-3.
(c) Give the contents of the three LUTs.

6.20 **(a)** If logic blocks of Figure 6-1(a) are used, how many LUTs are required to build a 4-bit adder with accumulator?
(b) If an FPGA with built-in carry-chain logic as shown in Figure 6-11 is used, how many 4-input LUTs are required?
(c) Design a 4-bit adder-subtracter with accumulator using an FPGA with carry-chain logic and 4-input LUTs. Assume a control signal Su which is 0 for addition and 1 for subtraction. Show the required connections on a diagram similar to that shown in Figure 6-11 and give the function realized by each LUT.

6.21 A 4×4 array multiplier (Figure 4-29) is to be implemented using an FPGA.

(a) Partition the logic so that it fits in a minimum number of Figure 6-1(a) logic blocks. Draw loops around each set of components that will fit in a single logic block. Determine the total number of 4-input LUTs required.
(b) Repeat part **(a)**, except assume that carry-chain logic is available.

6.22 **(a)** Use Shannon's expansion theorem to expand the following function around A and then expand each sub-function around D:

$$Z = AB'CD'E'F + A'BC'D'EF' + B'C'E'F + A'BC'E'F' + ABCDE$$

(b) Explain how the expanded function could be implemented using two Xilinx Kintex FPGA slices (Figure 6-13). On the slice diagrams, label the inputs to the LUTs (function generators) and draw the connection paths within the slice. Give the function implemented by each LUT.

6.23 **(a)** Indicate the connections of the switches in Figure 6-15 to realize the function

$$Z = AB'C + A'BC' + BC$$

(b) Indicate the connections of the switches in Figure 6-15 to realize the function

$$F = AB + A'C$$

(c) Indicate the connections of the switches in Figure 6-15 to realize a latch as shown in Figure 2-18.

(d) Indicate the connections of the switches in Figure 6-15 to realize a D-flip-flop.

6.24 The logic equations for a sequential network with five inputs, two flip-flops, and two outputs are

$$Q_1^+ = Q_1(Q_2ABC) + Q_1'(Q_2'CDE)$$
$$Q_2^+ = Q_1'$$
$$Z_1 = Q_1'Q_2'AB + Q_1'Q_2'A'B' + Q_1Q_2'AB' + Q_1Q_2(A' + B + C)$$
$$Z_2 = Q_1A' + Q_1B + Q_2'$$

How many Kintex slices (Figure 6-13) are required to implement the logic equations, including the flip-flops? Specify the inputs to each slice and the functions realized by each LUT.

6.25 Indicate whether the following structure created using four slices from Xilinx SPARTAN FPGAs can implement each of the following ("All" means any of all possible functions; "Some" means at least one):

(i) All 32-variable functions
(ii) Some 32-variable functions
(iii) All 8-variable functions
(iv) Some 8-variable functions
(v) All 7-variable functions
(vi) Some 7-variable functions
(vii) All 6-variable functions
(viii) Some 6-variable functions
(ix) All 36-variable functions
(x) Some 36-variable functions

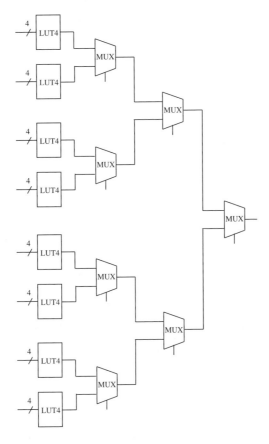

- **(xi)** All 39-variable functions
- **(xii)** Some 39-variable functions
- **(xiii)** Some 40-variable functions

6.26 Perform a survey of FPGA chips now on the market.

- **(a)** Generate a table like Table 6-1 for current FPGAs.
- **(b)** Generate a table like Table 6-2 for current FPGAs.

6.27 Stratix IV and Stratix V are the popular Altera FPGAs on the market. Indicate the most significant differences between these chips' logic modules.

6.28 Show how 32 × 32-bit unsigned multiplication can be accomplished using four 16 × 16-bit multipliers and several adders. Draw a block diagram showing the required connections.

6.29 Fast shifting can be accomplished by using dedicated multipliers. Shifting left N places is equivalent to multiplying by 2^N.

- **(a)** Given that A is a 16-bit unsigned number and $0 \leq N \leq 15$, show how to construct a left shifter using a multiplier and a decoder.

(b) Write Verilog code that infers this type of shifter.
(c) Repeat (a) and (b) for a right shifter. (*Hint*: multiply by 2^{15-N} and select the appropriate 16-bits of the 32-bit product.)

6.30 Make a one-hot state assignment for Figure 4-28(c). Derive the next-state and output equations by inspection.

6.31 Make a one-hot state assignment for Figure 4-53 and write the next-state and output equations by inspection. Then change the state assignment so that S_0 is assigned 0000000, S_1 is assigned 1100000, S_2 is assigned 1010000, and so forth and rewrite the equations for this assignment.

6.32 Assume that a sequential system with four states is to be implemented using a one-hot state assignment, but the flip-flops do not have a preset input. The flip-flops do have a reset input; hence, it is beneficial to have 0000 as the starting state. What should be the state assignments for the other states to take advantage of the one-hot assignment scheme? Explain.

6.33 For the given state graph:

(a) Derive the simplified next-state and output equations by inspection. Use the following one-hot state assignment for flip-flops $Q_0Q_1Q_2Q_3$; S_0, 1000; S_1, 0100; S_2, 0010; S_3, 0001.
(b) How many Kintex slices (Figure 6-13) are required to implement these equations?

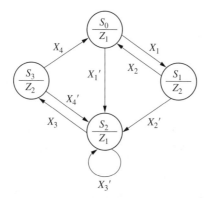

6.34 Make any necessary changes in the Verilog code for the traffic light controller (Figure 4-15) so that it can be synthesized without latches using whatever synthesis tool you have available. Synthesize the code using a suitable FPGA or CPLD as a target.

6.35 Synthesize the behavioral model of the 2's complement multiplier (Figure 4-35) using whatever synthesis tool you have available. Then synthesize the model with control signals (Figure 4-40) and compare the results (number of flip-flops, number of LUTs, number of slices, etc.). Try different synthesis options such as optimizing for area or speed, and different finite-state machine encoding algorithms such as one-hot, compact, and so forth and compare the results. Which combination of options uses the least resources?

6.36 Consider the Verilog code

```
module example(a,b);
input[1:0] a;
output[1:0] b;
reg[1:0] b;

always @(a)
begin
   case(a)
      0: b = 2'd3;
      1: b = 2'd2;
      2: b = 2'd1;
      3: b = 2'd1;
   endcase
end

endmodule
```

(a) Show the hardware you would obtain if you synthesize the foregoing Verilog code without any optimizations. Explain your reasoning.

(b) Show optimized hardware emphasizing minimum area. Show the steps and the reasoning by which you obtained the optimized hardware.

6.37 Draw the hardware structures that will be inferred by typical synthesizers from the code excerpts that follow. A, B, and E are 4-bit vectors, and C and D are 2-bit numbers; *clock* is a 1-bit signal. Draw the structure and mark the inputs and outputs.

(a)
```
always @(clock)
begin
      A <= {A[3], A[3:1]};
      B <= {A[0], B[3:1]};
end
```

(b)
```
always @(C)
begin
      case(C)
         0: D = 2'b11;
         1: D = 2'b10;
         2: D = 2'b00;
         default: begin
         end
      endcase
end
```

(c)
```
always @(C)
begin
      case(C)
         0: E = A + B;
```

```
            1: E = A >>> 2;
            2: E = A - B;
            3: E = A;
            endcase
    end
```

6.38 (a) Draw a logic diagram (use gates, adders, MUXes, D flip-flops, etc.) that shows the result of synthesizing the following Verilog code. *A*, *B*, and *C* are 3-bit unsigned vectors.

```
always @(negedge CLK)
begin
        if(CO == 1)
                C <= ~A;
        if(Ad == 1)
                C <= A + B;
        if(Sh == 1)
                C <= C >>> 1;
end
```

(b) Describe in one or two sentences what this circuit does.

6.39 Draw the hardware structures that will be inferred by typical synthesizers from the code excerpts that follow. If any ambiguities exist in the code, mention what you are assuming. Show optimized and unoptimized hardware.

(a)
```
always @(a)
begin
        case(a)
        0: b = 2;
        1: b = 0;
        2: b = 3;
        3: b = 1;
        endcase
end
```

(b)
```
if((arg1 > arg2)&&(arg1 > arg3))
        result <= arg1;
else
        result <= 0;
```

6.40 What hardware does the statement

$$F = (A >= B);$$

result in? Assume that *A* and *B* are 8-bit vectors.

6.41 Generate optimized hardware for the following statement, assuming *A* is a 4-bit vector:

$$F = (A = 9);$$

Floating-Point Arithmetic

Floating-point numbers are frequently used for numerical calculations in computing systems. Arithmetic units for floating-point numbers are considerably more complex than those for fixed-point numbers. Floating-point numbers allow very large or very small numbers to be specified. This chapter first describes a simple representation for floating-point numbers. Then it describes the IEEE floating-point standard. Next, an algorithm for floating-point multiplication is developed and tested using Verilog. Then the design of the floating-point multiplier is completed and implemented using an FPGA. Floating-point addition, subtraction, and division are also briefly described.

7.1 Representation of Floating-Point Numbers

A simple representation of a floating-point (or real) number (N) uses a fraction (F), base (B), and exponent (E), where $N = F \times B^E$. The base can be 2, 10, 16, or any other integer larger than 1. The fraction and the exponent can be represented in many formats. For example, they can be represented by 2's complement formats, sign-magnitude form, or another negative number representation. There are a variety of floating-point formats depending on how many bits are available for F and E, what the base is, and how negative numbers are represented for F and E. The base can be implied or explicit. Depending on all these choices, a wide variety of floating-point formats have existed in the past.

7.1.1 A Simple Floating-Point Format Using 2's Complement

In this section, we look at a floating-point format where negative exponents and fractions are represented using the 2's complement form. The base for the exponent is 2. Hence, the value of the number is $N = F \times 2^E$. In a typical floating-point number system, F is 16 to 64 bits long and E is 8 to 15 bits long. In order to keep the examples in this section simple and easy to follow, we will use a 4-bit fraction and a 4-bit exponent, but the concepts presented here can easily be extended to more bits.

The fraction and the exponent in this system will use 2's complement. (Refer to Section 4.10 for a discussion of 2's complement fractions.) We will use 4 bits for the fraction and 4 bits for the exponent. The fractional part will have a leading sign bit and 3 actual fraction bits. The implied binary point is after the first bit. The sign bit is 0 for positive numbers and 1 for negative numbers.

As an example, let us represent decimal 2.5 in this 8-bit 2's complement floating-point format.

$$2.5 = 0010.1000$$
$$= 1.010 \times 2^1 \quad \text{(standardized normal representation)}$$
$$= 0.101 \times 2^2 \quad \text{(4-bit 2's complement fraction)}$$

Therefore,

$$F = 0.101 \quad E = 0010 \quad N = 5/8 \times 2^2$$

If the number is -2.5, the same exponent can be used, but the fraction must have a negative sign. The 2's complement representation for the fraction is 1.011.

Therefore,

$$F = 1.011 \quad E = 0010 \quad N = -5/8 \times 2^2$$

Other examples of floating-point numbers using a 4-bit fraction and a 4-bit exponent are

$$F = 0.101 \quad E = 0101 \quad N = 5/8 \times 2^5$$
$$F = 1.011 \quad E = 1011 \quad N = -5/8 \times 2^{-5}$$
$$F = 1.000 \quad E = 1000 \quad N = -1 \times 2^{-8}$$

In order to utilize all the bits in F and have the maximum number of significant figures, F should be normalized so that its magnitude is as large as possible. If F is not normalized, we can normalize F by shifting it left until the sign bit and the next bit are different. Shifting F left is equivalent to multiplying by 2, so every time we shift we must decrement E by 1 to keep N the same. After normalization, the magnitude of F will be as large as possible, since any further shifting would change the sign bit. In the following examples, F is unnormalized to start with and then it is normalized by shifting left.

Unnormalized: $\quad F = 0.0101 \quad E = 0011 \quad N = 5/16 \times 2^3 = 5/2$
Normalized: $\quad F = 0.101 \quad E = 0010 \quad N = 5/8 \times 2^2 = 5/2$
Unnormalized: $\quad F = 1.11011 \quad E = 1100 \quad N = -5/32 \times 2^{-4} = -5 \times 2^{-9}$
(shift F left) $\quad F = 1.1011 \quad E = 1011 \quad N = -5/16 \times 2^{-5} = -5 \times 2^{-9}$
Normalized: $\quad F = 1.011 \quad E = 1010 \quad N = -5/8 \times 2^{-6} = -5 \times 2^{-9}$

The exponent can be any number between -8 and $+7$. The fraction can be any number between -1 and $+0.875$.

Zero cannot be normalized, so $F = 0.000$ when $N = 0$. Any exponent could then be used; however, it is best to have a uniform representation of 0. In this format, we will associate the negative exponent with the largest magnitude with the fraction 0. In a 4-bit 2's complement integer number system, the most negative number is 1000, which represents −8. Thus when F and E are 4 bits, 0 is represented by

$$F = 0.000 \qquad E = 1000 \qquad N = 0.000 \times 2^{-8}$$

Some floating-point systems use a biased exponent whereby $E = 0$ is associated with $F = 0$.

7.1.2 The IEEE 754 Floating-Point Formats

The IEEE 754 is a floating-point standard established by the IEEE in 1985. It contains two representations for floating-point numbers—the IEEE single precision format and the IEEE double precision format. The IEEE 754 single precision representation uses 32 bits, and the double precision system uses 64 bits.

Although 2's complement representations are very common for negative numbers, the IEEE floating-point representations do not use 2's complement for either the fraction or the exponent. The designers of IEEE 754 desired a format that was easy to sort and hence adopted **a sign-magnitude system** for the **fractional part** and a **biased notation** for the **exponent**.

The IEEE 754 floating-point formats need three sub-fields: sign, fraction, and exponent. The fractional part of the number is represented using a sign-magnitude representation in the IEEE floating-point formats—that is, there is an explicit sign bit (S) for the fraction. The sign is 0 for positive numbers and 1 for negative numbers. In a binary normalized scientific notation, the leading bit before the binary point is always 1. Therefore, the designers of the IEEE format decided to make it implied, representing only the bits after the binary point. In general, the number is of the form

$$N = (-1)^S \times (1 + F) \times 2^E$$

where S is the sign bit, F is the fractional part, and E is the exponent. The base of the exponent is 2. The base is implied—that is, it is not stored anywhere in the representation. The magnitude of the number is $1 + F$ because of the omitted leading 1. The term significand means the magnitude of the fraction and is $1 + F$ in the IEEE format. But often the terms significand and fraction are used interchangeably by many and are so used in this book.

The exponent in the IEEE floating-point formats uses what is known as a biased notation. A biased representation is one in which every number is represented by the number plus a certain bias. In the IEEE single precision format, the bias is 127. Hence, if the exponent is +1, it will be represented by $+1 + 127 = 128$. If the exponent is −2, it will be represented by $-2 + 127 = 125$. Thus, exponents less than 127 indicate actual negative exponents, and exponents greater than 127 indicate actual positive exponents. The bias is 1023 in the double precision format.

If a positive exponent becomes too large to fit in the exponent field, the situation is called **overflow**, and if a negative exponent is too large to fit in the exponent field, that situation is called **underflow**.

The IEEE Single Precision Format

The IEEE single precision format uses 32 bits for representing a floating-point number, divided into three subfields, as illustrated in Figure 7-1. The first field is the sign bit for the fractional part. The next field consists of 8 bits, which are used for the exponent. The third field consists of the remaining 23 bits and is used for the fractional part.

FIGURE 7-1: IEEE Single Precision Floating-Point Format

S	Exponent	Fraction
1 bit	8 bits	23 bits

The sign bit reflects the sign of the fraction. It is 0 for positive numbers and 1 for negative numbers. In order to represent a number in the IEEE single precision format, first it should be converted to a normalized scientific notation with exactly one bit before the binary point, simultaneously adjusting the exponent value.

The exponent representation that goes into the second field of the IEEE 754 representation is obtained by adding 127 to the actual exponent of the number when it is represented in the normalized form. Exponents in the range 1–254 are used for representing normalized floating-point numbers. Exponent values 0 and 255 are reserved for special cases, which will be discussed subsequently.

The representation for the 23-bit fraction is obtained from the normalized scientific notation by dropping the leading 1. Zero cannot be represented in this fashion; hence, it is treated as a special case (explained later). Since every number in the normalized scientific notation will have a leading 1, this leading 1 can be dropped so that one more bit can be packed into the significand (fraction). Thus, a 24-bit fraction can be represented using the 23 bits in the representation. The designers of the IEEE formats wanted to make the highest use of all the bits in the exponent and fraction fields.

In order to understand the IEEE format, let us represent 13.45 in the IEEE floating-point format. One can see that 0.45 is a recurring binary fraction and hence,

13.45 = 1101.01 1100 1100 1100 … … …   with the bits 1100 continuing to recur.

Normalized scientific representation yields

$$13.45 = 1.10101\ 1100\ 1100\ \ldots \times 2^3$$

Since the number is positive, the sign bit for the IEEE 754 representation is 0.

The exponent in the biased notation will be 127 + 3 = 130, which in binary format is 10000010. The fraction is 1.10101 1100 1100 … … … (with 1100 recurring). Omitting the leading 1, the 23 bits for the fractional part are

10101 1100 1100 1100 1100 11

Thus, the 32 bits are

0 100 0001 0 101 01 11 00 11 00 11 00 11 00 11

as illustrated in Figure 7-2.

FIGURE 7-2: IEEE Single Precision Floating-Point Representation for 13.45

S	Exponent	Fraction
0	10000010	1010111001100110011

The 32 bits can be expressed more conveniently in a hexadecimal (hex) format as

$$4157\ 3333$$

The number -13.45 can be represented by changing only the sign bit—that is, the first bit must be 1 instead of 0. Hence, the hex number C157 3333 represents -13.45 in IEEE 754 single precision format.

The IEEE Double Precision Format

The IEEE double precision format uses 64 bits for representing a floating-point number, as illustrated in Figure 7-3. The first bit is the sign bit for the fractional part. The next 11 bits are used for the exponent, and the remaining 52 bits are used for the fractional part.

FIGURE 7-3: IEEE Double Precision Floating-Point Format

S	Exponent	Fraction
1 bit	11 bits	52 bits

As in the single precision format, the sign bit is 0 for positive numbers and 1 for negative numbers.

The exponent representation used in the second field is obtained by adding the bias value of 1023 to the actual exponent of the number in the normalized form. Exponents in the range 1–2046 are used for representing normalized floating-point numbers. Exponent values 0 and 2047 are reserved for special cases.

The representation for the 52-bit fraction is obtained from the normalized scientific notation by dropping the leading 1 and considering only the next 52 bits. As an example, let us represent 13.45 in IEEE double precision floating-point format. Converting 13.45 to a binary representation,

$$13.45 = 1101.01\ 1100\ 1100\ 1100\ \ldots\ldots\ldots$$ with the bits 1100 continuing to recur.

In normalized scientific representation,

$$13.45 = 1.10101\ 1100\ 1100\ \ldots \times 2^3$$

The exponent in biased notation will be $1023 + 3 = 1026$, which in binary representation is

$$10000000010$$

The fraction is $1.10101\ 1100\ 1100\ \ldots\ldots\ldots$ (with 1100 recurring). Omitting the leading 1, the 52 bits of the fractional part are

$$10101\ 1100\ 1100\ 1100\ 1100\ 1100\ 1100\ 1100\ 1100\ 1100\ 1100\ 1100\ 110$$

Thus, the 64 bits are

$$0\ 10000000010\ 10101\ 1100\ 1100\ 1100\ 1100\ 1100\ 1100\ 1100\ 1100\ 1100\ 1100\ 110$$

as illustrated in Figure 7-4.

FIGURE 7-4: IEEE Double Precision Floating-Point Representation for 13.45

S	Exponent	Fraction
0	1 0 0 0 0 0 0 0 0 1 0	1 0 1 0 1 1 1 0 0 1 1 0 0 1 1 0 0 1 1 0

Fraction (cont'd)
0 1 1 0 0 1 1 0 0 1 1 0 0 1 1 0 0 1 1 0 0 1 1 0 0 1 1 0 0 1 1 0

The 64 bits can be expressed more conveniently in a hexadecimal format as

402A E666 6666 6666

The number −13.45 can be represented by changing only the sign bit—that is, the first bit must be 1 instead of 0. Hence, the hex number C02A E666 6666 6666 represents −13.45 in IEEE 754 double precision format.

Special Cases in the IEEE 754 Standard

The IEEE 754 standard has several special cases, which are illustrated in Figure 7-5. These include 0, infinity, denormalized numbers, and NaN (Not a Number) representations. The smallest and the highest exponents are used to denote these special cases.

FIGURE 7-5: Special Cases in the IEEE 754 Floating-Point Formats

Single Precision		Double Precision		Object Represented
Exponent	Fraction	Exponent	Fraction	
0	0	0	0	0
0	nonzero	0	nonzero	± denormalized number
255	0	2047	0	± infinity
255	nonzero	2047	nonzero	NaN (Not a number)

Zero

The IEEE format specifies 0 to be the representation with 0s in all bits—that is, all exponent and fraction bits are 0. Zero is specified as a special case in the format due to the difficulty in representing 0 in a normalized format. When using the usual convention for IEEE format normalized numbers, one would add a leading 1 to the fractional part, but that would make it impossible to represent 0.

Denormalized Numbers

The smallest normalized number that the single precision format can represent is

$$1.0 \times 2^{-126}$$

Numbers between this number and 0 cannot be expressed in the normalized format. If normalization is not made a requirement of the format, one could represent numbers smaller than 1.0×2^{-126}. Hence, the IEEE floating-point format allows denormalized numbers as a special case. If the exponent is 0, and the fraction is non-zero, the number is considered denormalized. Now, the smallest number that can be represented is

$$0.00000000000000000000001 \times 2^{-126}, \text{ which is } 1.0 \times 2^{-149}.$$

Thus, denormalization allows numbers between 1.0×2^{-126} and 1.0×2^{-149} to be represented. For double precision, the denormalized range allows numbers between 1.0×2^{-1022} and 1.0×2^{-1074}.

Infinity

Infinity is represented by the highest exponent value together with a fraction of 0. In the case of single precision representation, the exponent is 255, and for double precision, it is 2047.

Not a Number (NaN)

The IEEE 754 standard has a special representation to represent the result of invalid operations, such as 0/0. This special representation is called NaN or Not a Number. If the exponent is 255 and the fraction is any non-zero number, it is considered to be NaN or Not a Number.

Rounding

When the number of bits available is smaller than the number of bits required to represent a number, rounding is employed. It is desirable to round to the nearest value. One can round up if the number is higher than halfway and round down if the number is less than halfway. Another option is to truncate, ignoring the bits beyond the allowable number of bits. One has to keep more bits in intermediate representations to achieve higher accuracy. The IEEE standard requires two extra bits in intermediate representations in order to facilitate better rounding. The two bits are called **guard and round**. Sometimes, a third intermediate bit is used in rounding in addition to the guard and round bits. It is called the **sticky bit**. The sticky bit is set whenever there are non-zero bits to the right of the round bit.

The biggest challenge comes when the number falls halfway in between. The IEEE standard has four different rounding modes:

- **Round up:** Round towards positive infinity; round up to the next higher number.
- **Round down:** Round towards negative infinity; round down to the nearest smaller number.
- **Truncate:** Round towards zero. Ignore bits beyond the allowable number of bits. Same as truncation in sign magnitude.
- **Unbiased:** Round to nearest. If the number falls halfway, round up half the time and round down half the time. In order to achieve rounding up half the time, add 1 if the lowest bit retained is 1, and truncate if it is 0. This is based on the assumption that a 0 or 1 appears in the lowest retained bit with an equal probability. One consequence of this rounding scheme is that the rounded number always has a 0 in the lowest place.

7.2 Floating-Point Multiplication

Given two floating-point numbers, $F_1 \times 2^{E_1}$ and $F_2 \times 2^{E_2}$, the product is

$$(F_1 \times 2^{E_1}) \times (F_2 \times 2^{E_2}) = (F_1 \times F_2) \times 2^{(E_1 + E_2)} = F \times 2^E$$

The fraction part of the product is the product of the fractions, and the exponent part of the product is the sum of the exponents. Hence, a floating-point multiplier consists of two major components:

1. A fraction multiplier
2. An exponent adder.

The details of floating-point multiplication depend on the precise formats in which the fraction multiplication and exponent addition are performed.

Fraction multiplication can be done in many ways. If the IEEE format is used, multiplication of the magnitude can be done and then the signs can be adjusted. If 2's complement fractions are used, one can use a fraction multiplier that handles signed 2's complement numbers directly. Such a fraction multiplier was discussed in Chapter 4.

Addition of the exponents can be done with a binary adder. If the IEEE formats are directly used, the representations have to be carefully adjusted in order to obtain the correct result. For instance, if exponents of two floating-point numbers in the biased format are added, the sum contains twice the bias value. To get the correct exponent, the bias value must be subtracted from the sum.

The 2's complement system has several interesting properties for performing arithmetic. Hence, many floating-point arithmetic units convert the IEEE notation to 2's complement and then use the 2's complement internally for carrying out the floating-point operations. Then the final result is converted back to IEEE standard notation.

The general procedure for performing floating-point multiplication is the following:

1. Add the two exponents
2. Multiply the two fractions (significands).
3. If the product is 0, adjust the representation to the proper representation for 0.
4. a. If the product fraction is too big, normalize by shifting it right and incrementing the exponent.
   b. If the product fraction is too small, normalize by shifting left and decrementing the exponent.
5. If an exponent underflow or overflow occurs, generate an exception or error indicator.
6. Round to the appropriate number of bits. If rounding resulted in loss of normalization, go to step 4 again.

One may note that, in addition to adding the exponents and multiplying the fractions, several steps such as normalizing the product, handling overflow and underflow, rounding to the appropriate number of bits, and so on need to be done. We assume that the two numbers are properly normalized to start with, and we want the final result to be normalized.

Now, we discuss the design of a floating-point multiplier. We use 4-bit fractions and 4-bit exponents, with negative numbers represented in 2's complement. The fundamental steps are to add the exponents (step 1) and multiply the fractions (step 2). However, several special cases must be considered. If F is 0, we must set the exponent E to the largest negative value (1000) (step 3). A special situation occurs if we multiply -1 by -1 (1.000 × 1.000). The result should be $+1$. Since we cannot represent $+1$ as a 2's complement fraction with a 4-bit fraction, this special case necessitates right shifting as in step 4. To correct this situation, we right shift the significand (fraction) and increment the exponent. Essentially, we set $F = 1/2$ (0.100) and add 1 to E. This results in the correct answer, since $1 \times 2^E = 1/2 \times 2^{E+1}$.

When we multiply the fractions, the result could be unnormalized. For example,

$$(0.1 \times 2^{E_1}) \times (0.1 \times 2^{E_2}) = 0.01 \times 2^{E_1 + E_2} = 0.1 \times 2^{E_1 + E_2 - 1}$$

This is situation 4.b. In this case, we normalize the result by shifting the fraction left one place and subtracting 1 from the exponent to compensate. Finally, if the resulting exponent is too large in magnitude to represent in our number system, we have an exponent overflow. (An overflow in the negative direction is referred to as an *underflow*.) Since we are using 4-bit exponents, if the exponent is not in the range 1000 to 0111 (-8 to $+7$), an overflow has occurred. Since an exponent overflow cannot be corrected, an overflow indicator should be turned on (step 5).

A flowchart for this floating-point multiplier is shown in Figure 7-6. After the fraction is multiplied, all the special cases are tested for. Since F_1 and F_2 are normalized, the smallest possible magnitude for the product is 0.01, as indicated in the preceding example. Therefore, only one left shift is required to normalize F.

The hardware required to implement the multiplier (Figure 7-7) consists of an exponent adder, a fraction multiplier, and a control unit that provides the signals to perform the appropriate operations of right shifting, left shifting, exponent incrementing/decrementing, and so forth.

Exponent Adder: Since 2's complement addition results with the sum in the proper format, the design of the exponent adder is straightforward. A 5-bit full adder is used as the exponent adder as demonstrated in Figure 7-7. When the fraction is normalized, the exponent will have to be correspondingly incremented or decremented. Also, in the special case when product is 0, the register should be set to the value 1000. The register has control signals for incrementing, decrementing, and setting to the most negative value (*SM8*).

The register that holds the sum is made into a 5-bit register to handle special situations. When the exponents are added, an overflow can occur. If E_1 and E_2 are positive and the sum (E) is negative, or if E_1 and E_2 are negative and the sum is positive, the result is a 2's complement overflow. However, this overflow might be corrected when 1 is added to or subtracted from E during normalization or correction of fraction overflow. To allow for this case, we have made the X register

FIGURE 7-6: Flowchart for Floating-Point Multiplication with 2's Complement Fractions/Exponents

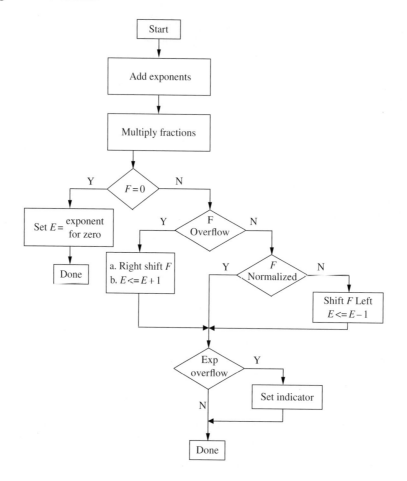

5 bits long. When E_1 is loaded into X, the sign bit must be extended so that we have a correct 2's complement representation. Since there are two sign bits, if the addition of E_1 and E_2 produces an overflow, the lower sign bit will be changed, but the high-order sign bit will remain unchanged. Each of the following examples has an overflow, since the lower sign bit has the wrong value:

$$7 + 6 = 00111 + 00110 = 01101 = 13 \quad \text{(maximum allowable value is 7)}$$

$$-7 + (-6) = 11001 + 11010 = 10011 = -13 \quad \text{(maximum allowable negative value is } -8\text{)}$$

The following example illustrates the special case where an initial fraction overflow and exponent overflow occur, but the exponent overflow is corrected when the fraction overflow is corrected:

$$(1.000 \times 2^{-3}) \times (1.000 \times 2^{-6}) = 01.000000 \times 2^{-9} = 00.100000 \times 2^{-8}$$

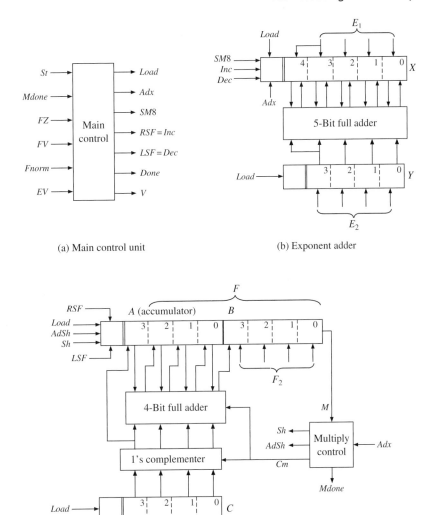

FIGURE 7-7: Major Components of a Floating-Point Multiplier

(a) Main control unit

(b) Exponent adder

(c) Fraction multiplier

Fraction Multiplier: The fraction multiplier that we designed in Chapter 4 handles 2's complement fractions in a straightforward manner. Hence, we adapt that design for the floating-point multiplier. It implements a shift and add multiplier algorithm. Since we are multiplying 3 bits plus sign by 3 bits plus sign, the result will be 6 bits plus sign. After the fraction multiplication, the 7-bit result (F) will be the lower 3 bits of A concatenated with B. The multiplier has its own control unit that generates appropriate shift and add signals depending on the multiplier bits.

Main Control Unit: The SM chart for the main controller (Figure 7-8) of the floating-point multiplier is based on the flowchart. This controller is called the main

FIGURE 7-8: SM Chart for Floating-Point Multiplication

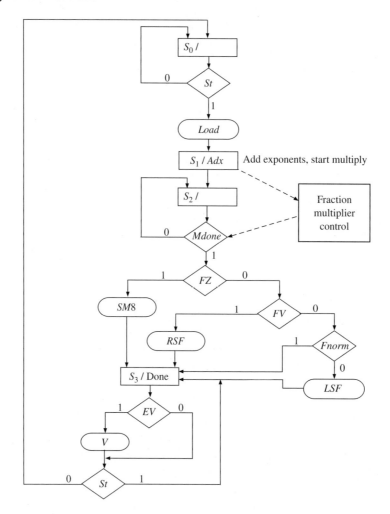

controller to distinguish it from the controller for the multiplier, which is a separate state machine that is linked into the main controller.

The SM chart uses the following inputs and control signals:

St:	Start the floating-point multiplication.
Mdone:	Fraction multiply is done.
FZ:	Fraction is zero.
FV:	Fraction overflow (fraction is too big).
Fnorm:	F is normalized.
EV:	Exponent overflow.
Load:	Load F_1, E_1, F_2, E_2 into the appropriate registers (also clear A in preparation for multiplication).
Adx:	Add exponents; this signal also starts the fraction multiplier.
SM8:	Set exponent to minus 8 (to handle special case of 0).

RSF: Shift fraction right; also increment *E*.
LSF: Shift fraction left; also decrement *E*.
V: Overflow indicator.
Done: Floating-point multiplication is complete.

The SM chart for the main controller has four states. In S_0, the registers are loaded when the start signal is 1. In S_1, the exponents are added, and the fraction multiply is started. In S_2, we wait until the fraction multiply is done, and then test for special cases and take appropriate action. It may seem surprising that the tests on *FZ*, *FV*, and *Fnorm* can all be done in the same state, since they are done in sequence on the flowchart. However, *FZ*, *FV*, and *Fnorm* are generated by combinational circuits that operate in parallel and hence can be tested in the same state. However, we must wait until the exponent has been incremented or decremented at the next clock before we can check for exponent overflow in S_3. In S_3, the *Done* signal is turned on, and the controller waits for $St = 0$ before returning to S_0.

The state graph for the multiplier control (Figure 7-9) is similar to that shown in Figure 4-34, except that the load state is not needed because the registers are loaded by the main controller. Add and shift operations are performed in one state because, as seen in Figure 7-7 (c), the sum wires from the adder are shifted by 1 before loading into the accumulator register. When $Adx = 1$, the multiplier is started, and *Mdone* is turned on when the multiplication is completed.

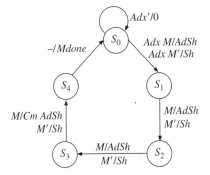

FIGURE 7-9: State Graph for Multiplier Control

The Verilog behavioral description (Figure 7-10) uses three always statements. The main always statements generates control signals based on the SM chart. A second always statement generates the control signals for the fraction multiplier. The third always statement tests the control signals and updates the appropriate registers on the rising edge of the clock. In state S_2 of the main always statements, $A = 0000$ implies that $F = 0$ ($FZ = 1$ on the SM chart). If we multiply 1.000×1.000, the result is $A \,\&\, B = 01000000$, and a fraction overflow has occurred ($FV = 1$). If $A(2) = A(1)$, the sign bit of *F* and the following bit are the same and *F* is unnormalized (*Fnorm* = 0). In state S_3, if the two high-order bits of *X* are different, an exponent overflow has occurred ($EV = 1$).

The registers are updated in the third always statements. The variable *addout* represents the output of the 4-bit full adder, which is part of the fraction multiplier.

FIGURE 7-10: Verilog Code for Floating-Point Multiplier

```verilog
module FMUL (CLK, St, F1, E1, F2, E2, F, V, done);
    input CLK;
    input St;
    input[3:0] F1;
    input[3:0] E1;
    input[3:0] F2;
    input[3:0] E2;
    output[6:0] F;
    output V;
    output done;

    reg[6:0] F;
    reg done;
    reg V;

    reg[3:0] A;
    reg[3:0] B;
    reg[3:0] C;
    reg[4:0] X;
    reg[4:0] Y;
    reg Load;
    reg Adx;
    reg SM8;
    reg RSF;
    reg LSF;
    reg AdSh;
    reg Sh;
    reg Cm;
    reg Mdone;
    reg[1:0] PS1;
    reg[1:0] NS1;
    reg[2:0] State;
    reg[2:0] Nextstate;

    initial
    begin
        State = 0;
        PS1 = 0;
        NS1 = 0;
        Nextstate =0;
    end

    always @(PS1 or St or Mdone or X or A or B)
    begin : main_control
        Load = 1'b0 ;
        Adx = 1'b0 ;
        NS1 = 0 ;
        SM8 = 1'b0 ;
        RSF = 1'b0 ;
        LSF = 1'b0 ;
        V = 1'b0 ;
```

```verilog
            F = 7'b0000000 ;
            done = 1'b0 ;
            case (PS1)
               0 :
                           begin
                               F = 7'b0000000 ;
                               done = 1'b0 ;
                               V = 1'b0 ;
                               if (St == 1'b1)
                               begin
                                   Load = 1'b1 ;
                                   NS1 = 1 ;
                               end
                           end
               1 :
                           begin
                               Adx = 1'b1 ;
                               NS1 = 2 ;
                           end
               2 :
                           begin
                               if (Mdone == 1'b1)
                               begin
                                   if (A == 0)
                                   begin
                                       SM8 = 1'b1 ;
                                   end
                                   else if (A == 4 & B == 0)
                                   begin
                                       RSF = 1'b1 ;
                                   end
                                   else if (A[2] == A[1])
                                   begin
                                       LSF = 1'b1 ;
                                   end
                                   NS1 = 3 ;
                               end
                               else
                               begin
                                   NS1 = 2 ;
                               end
                           end
               3 :
                           begin
                               if (X[4] != X[3])
                               begin
                                   V = 1'b1 ;
                               end
                               else
                               begin
                                  V = 1'b0 ;
                               end
```

```verilog
                        done = 1'b1 ;
                        F = {A[2:0], B} ;
                        if (St == 1'b0)
                        begin
                            NS1 = 0 ;
                        end
                    end
            endcase
    end
    always @(State or Adx or B)
    begin : mul2c
        AdSh = 1'b0 ;
        Sh = 1'b0 ;
        Cm = 1'b0 ;
        Mdone = 1'b0 ;
        Nextstate = 0 ;
        case (State)
            0 :
                    begin
                        if (Adx == 1'b1)
                        begin
                            if ((B[0]) == 1'b1)
                            begin
                                AdSh = 1'b1 ;
                            end
                            else
                            begin
                                Sh = 1'b1 ;
                            end
                            Nextstate = 1 ;
                        end
                    end
            1, 2 :
                    begin
                        if ((B[0]) == 1'b1)
                        begin
                            AdSh = 1'b1 ;
                        end
                        else
                        begin
                            Sh = 1'b1 ;
                        end
                        Nextstate = State + 1 ;
                    end
            3 :
                    begin
                        if ((B[0]) == 1'b1)
                        begin
                            Cm = 1'b1 ;
                            AdSh = 1'b1 ;
                        end
```

```verilog
                        else
                        begin
                            Sh = 1'b1 ;
                        end
                        Nextstate = 4 ;
                    end
        4 :
                    begin
                        Mdone = 1'b1 ;
                        Nextstate = 0 ;
                    end
    endcase
end

wire[3:0] addout;
assign addout = (Cm == 1'b0)? (A + C) : (A - C);

always @(posedge CLK)
begin : update
    PS1 <= NS1 ;
    State <= Nextstate ;
    if (Load == 1'b1)
    begin
        X <= {E1[3], E1} ;
        Y <= {E2[3], E2} ;
        A <= 4'b0000 ;
        B <= F1 ;
        C <= F2 ;
    end
    if (Adx == 1'b1)
    begin
        X <= X + Y ;
    end
    if (SM8 == 1'b1)
    begin
        X <= 5'b11000 ;
    end
    if (RSF == 1'b1)
    begin
        A <= {1'b0, A[3:1]} ;
        B <= {A[0], B[3:1]} ;
        X <= X + 1 ;
    end
    if (LSF == 1'b1)
    begin
        A <= {A[2:0], B[3]} ;
        B <= {B[2:0], 1'b0} ;
        X <= X + 31 ;
    end
    if (AdSh == 1'b1)
    begin
```

```
                A <= {(C[3] ^ Cm), addout[3:1]} ;
                B <= {addout[0], B[3:1]} ;
            end
            if (Sh == 1'b1)
            begin
                A <= {A[3], A[3:1]} ;
                B <= {A[0], B[3:1]} ;
            end
        end
endmodule
```

This adder adds the 2's complement of C to A when $Cm = 1$. When $Load = 1$, the sign-extended exponents are loaded into X and Y. When $Adx = 1$, vectors X and Y are added. When $SM8 = 1$, -8 is loaded into X. When $AdSh = 1$, A is loaded with the sign bit of C (or the complement of the sign bit if $Cm = 1$), concatenated with bits 3 to 1 of the adder output, and the remaining bit of *addout* is shifted into the B register.

Testing the Verilog code for the floating-point multiplier must be done carefully to account for all the special cases in combination with positive and negative fractions, as well as positive and negative exponents. Figure 7-11 shows a command file and some test results. This is not a complete test.

FIGURE 7-11: Test Data and Simulation Results for Floating-Point Multiplier

```
add list  F X F1  E1 F2  E2  V done
force F1  7 0, 9 200, 8 400, 0 600, 7 800
force E1  1 0, 9 200, 7 400, 8 600, 7 800
force F2  7 0, 9 200, 8 400, 0 600, 9 800
force E2  8 0, 1 200, 9 400, 8 600, 1 800
force St  1 0, 0 20,  1 200, 0 220, 1 400, 0 420, 1 600, 0 620, 1 800, 0 820
force CLK 0 0, 1 10 -repeat 20
run 1000

ns  delta     F       X     F1   E1   F2   E2  V  done
 0    +0  xxxxxxx  xxxxx  0111 0001 0111 1000  x   x
 0    +2  0000000  xxxxx  0111 0001 0111 1000  0   0      $(0.111 \times 2^1) \times (0.111 \times 2^{-8})$
10    +0  0000000  00001  0111 0001 0111 1000  0   0
30    +0  0000000  11001  0111 0001 0111 1000  0   0
110   +1  0110001  11001  0111 0001 0111 1000  0   1      $= 0.110001 \times 2^{-7}$
130   +1  0000000  11001  0111 0001 0111 1000  0   0
200   +0  0000000  11001  1001 1001 1001 0001  0   0      $(1.001 \times 2^{-7}) \times (1.001 \times 2^1)$
230   +0  0000000  11010  1001 1001 1001 0001  0   0
310   +1  0110001  11010  1001 1001 1001 0001  0   1      $= 0.110001 \times 2^{-6}$
330   +1  0000000  11010  1001 1001 1001 0001  0   0
400   +0  0000000  11010  1000 0111 1000 1001  0   0      $(1.000 \times 2^7) \times (1.000 \times 2^{-7})$
410   +0  0000000  00111  1000 0111 1000 1001  0   0
430   +0  0000000  00000  1000 0111 1000 1001  0   0
510   +0  0000000  00001  1000 0111 1000 1001  0   0
```

ns	delta	F	X	F1	E1	F2	E2	V	done	
510	+1	0100000	00001	1000	0111	1000	1001	0	1	$= 0.100000 \times 2^1$
530	+1	0000000	00001	1000	0111	1000	1001	0	0	
600	+0	0000000	00001	0000	1000	0000	1000	0	0	$(0.000 \times 2^{-8}) \times (0.000 \times 2^{-8})$
610	+0	0000000	11000	0000	1000	0000	1000	0	0	
630	+0	0000000	10000	0000	1000	0000	1000	0	0	
710	+0	0000000	11000	0000	1000	0000	1000	0	0	
710	+1	0000000	11000	0000	1000	0000	1000	0	1	$= 0.0000000 \times 2^{-8}$
730	+1	0000000	11000	0000	1000	0000	1000	0	0	
800	+0	0000000	11000	0111	0111	1001	0001	0	0	$(0.111 \times 2^7) \times (1.001 \times 2^1)$
810	+0	0000000	00111	0111	0111	1001	0001	0	0	
830	+0	0000000	01000	0111	0111	1001	0001	0	0	
910	+1	1001111	01000	0111	0111	1001	0001	1	1	$= 1.001111 \times 2^8$ (overflow)
930	+1	0000000	01000	0111	0111	1001	0001	0	0	

When the Verilog code was synthesized for the Xilinx Spartan-3/Virtex-4 architectures using the Xilinx ISE tools, the result was 9.5 CLBs, 38 slices, 29 flip-flops, 72 4-input LUTs, 27 I/O blocks and one global clock circuitry. The output signals V, $Done$, and F were set to zero at the beginning of the process to eliminate unwanted latches. An RTL-level design was also attempted, but the RTL design was not superior to the synthesized behavioral design.

Now that the basic design has been completed, we need to determine how fast the floating-point multiplier will operate and to determine the maximum clock frequency. Most CAD tools provide a way of simulating the final circuit taking into account both the delays within the logic blocks and the interconnection delays. If this timing analysis indicates that the design does not operate fast enough to meet specifications, several options are possible. Most FPGAs come in several different speed grades, so one option is to select a faster part. Another approach is to determine the longest delay path in the circuit and to attempt to reroute the connections or redesign that part of the circuit to reduce the delays.

7.3 Floating-Point Addition

Next, we consider the design of an adder for floating-point numbers. Two floating-point numbers will be added to form a floating-point sum:

$$(F_1 \times 2^{E_1}) + (F_2 \times 2^{E_2}) = F \times 2^E$$

Again, we will assume that the numbers to be added are properly normalized and that the answer should be put in normalized form. In order to add two fractions, the associated exponents must be equal. Thus, if the exponents E_1 and E_2 are different, we must unnormalize one of the fractions and adjust the exponent accordingly. The smaller number is the one that should be adjusted so that if significant digits are lost, the effect is not significant. To illustrate the process, we add

$$F_1 \times 2^{E_1} = 0.111 \times 2^5 \text{ and } F_2 \times 2^{E_2} = 0.101 \times 2^3$$

Since $E_2 \neq E_1$, we unnormalize the smaller number F_2 by shifting right two times, and adding 2 to the exponent

$$0.101 \times 2^3 = 0.0101 \times 2^4 = 0.00101 \times 2^5$$

Note that shifting right one place is equivalent to dividing by 2, so each time we shift we must add 1 to the exponent to compensate. When the exponents are equal, we add the fractions

$$(0.111 \times 2^5) + (0.00101 \times 2^5) = 01.00001 \times 2^5$$

This addition caused an overflow into the sign bit position, so we shift right and add 1 to the exponent to correct the fraction overflow. The final result is

$$F \times 2^E = 0.100001 \times 2^6$$

When one of the fractions is negative, the result of adding fractions may be unnormalized, as illustrated in the following example:

$(1.100 \times 2^{-2}) + (0.100 \times 2^{-1})$
$= (1.110 \times 2^{-1}) + (0.100 \times 2^{-1})$ (after shifting F_1)
$= 0.010 \times 2^{-1}$ (result of adding fractions is unnormalized)
$= 0.100 \times 2^{-2}$ (normalized by shifting left and subtracting 1 from exponent)

In summary, the steps required to carry out floating-point addition are as follows:

1. Compare exponents. If the exponents are not equal, shift the fraction with the smaller exponent right and add 1 to its exponent; repeat until the exponents are equal.
2. Add the fractions (significands).
3. If the result is 0, set the exponent to the appropriate representation for 0 and exit.
4. If fraction overflow occurs, shift right and add 1 to the exponent to correct the overflow.
5. If the fraction is unnormalized, shift left and subtract 1 from the exponent until the fraction is normalized.
6. Check for exponent overflow. Set overflow indicator, if necessary.
7. Round to the appropriate number of bits. Is it still normalized? If not, go back to step 4.

Figure 7-12 illustrates this procedure graphically. An optimization can be added to step 1. We can identify cases where the two numbers are vastly different. If $E_1 \gg E_2$ and F_2 is positive, F_2 will become all 0s as we right shift F_2 to equalize the exponents. In this case, the result is $F = F_1$ and $E = E_1$, so it is a waste of time to do the shifting. If $E_1 \gg E_2$ and F_2 is negative, F_2 will become all 1s (instead of all 0s) as we right shift F_2 to equalize the exponents. When we add the fractions, we will get the wrong answer. To avoid this problem, we can skip the shifting when $E_1 \gg E_2$ and set $F = F_1$ and $E = E_1$. Similarly, if $E_2 \gg E_1$, we can skip the shifting and set $F = F_2$ and $E = E_2$.

7.3 Floating-Point Addition **419**

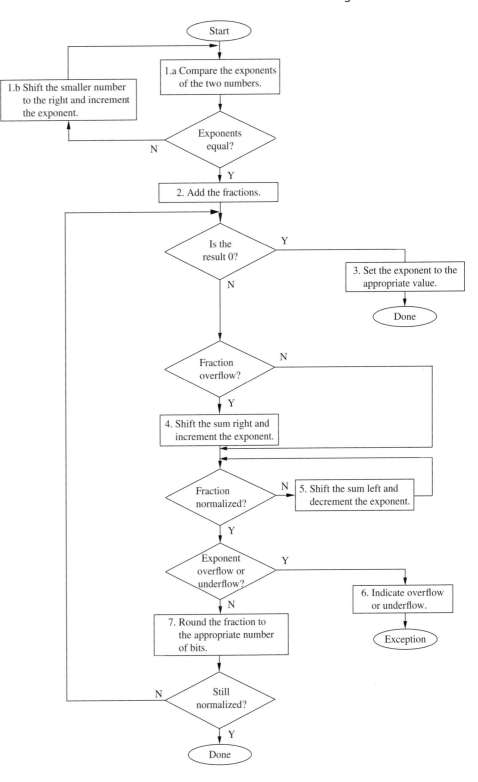

FIGURE 7-12: Flowchart for Floating-Point Addition

For the 4-bit fractions in our example, if $|E_1 - E_2| > 3$, we can skip the shifting. For IEEE single precision numbers, there are 23 bits after the binary point; hence, if the exponent difference is greater than 23, the smaller number will become 0 before the exponents are equal. In general, if the exponent difference is greater than the number of available fractional bits, the sum should be set to the larger number. If $E_1 \gg E_2$, set $F = F_1$ and $E = E_2$. If $E_2 \gg E_1$, set $F = F_2$ and $E = E_2$.

Inspection of this procedure illustrates that the following hardware units are required to implement a floating-point adder:

- Adder (subtractor) to compare exponents (step 1a)
- Shift register to shift the smaller number to the right (step 1b)
- ALU (adder) to add fractions (step 2)
- Bidirectional shifter, incrementer/decrementer (steps 4, 5)
- Overflow detector (step 6)
- Rounding hardware (step 7)

Many of these components can be combined. For instance, the register that stores the fractions can be made a shift register in order to perform the shifts. The register that stores the exponent can be a counter with increment/decrement capability. Figure 7-13 shows a hardware arrangement for the floating-point adder. The major components are the exponent comparator and the fraction adder. Fraction addition can be done using 2's complement addition. It is assumed that the operands are delivered on an I/O bus. If the numbers are in a signed-magnitude form as in the IEEE format, they can be converted to 2's complement numbers and then added. Special cases should be handled according to the requirements of the format. The sum is written back into the addend register in Figure 7-13.

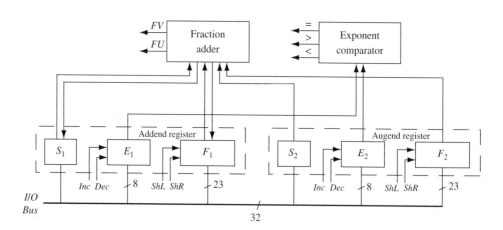

FIGURE 7-13: Overview of a Floating-Point Adder

Figure 7-14 shows Verilog code for a floating-point adder based on the IEEE single precision floating-point format. This code is not a complete implementation of the standard. It handles the special case of 0, but it does not deal with infinity, unnormalized, and Not-a-Number formats. The final result is truncated instead of rounded. Sign and magnitude format and biased exponents are used throughout, except 2's complement is used for the fraction addition.

FIGURE 7-14: Verilog Code for a Floating-Point Adder

```verilog
module FPADD (CLK, St, done, ovf, unf, FPinput, FPsum);
   input CLK;
   input St;
   output done;
   output ovf;
   output unf;
   input[31:0] FPinput;
   output[31:0] FPsum;

   reg done;
   reg ovf;
   reg unf;

   reg[25:0] F1;
   reg[25:0] F2;
   reg[7:0] E1;
   reg[7:0] E2;
   reg S1;
   reg S2;
   wire FV;
   wire FU;
   wire[27:0] F1comp;
   wire[27:0] F2comp;
   wire[27:0] Addout;
   wire[27:0] Fsum;
   reg[2:0] State;

   initial
   begin
     State = 0;
     done = 0;
     ovf = 0;
     unf = 0;
     S1 = 0;
     S2 = 0;
     F1 = 0;
     F2 = 0;
     E1 = 0;
     E2 = 0;
   end

   assign F1comp = (S1 == 1'b1) ? ~({2'b00, F1}) + 1 : {2'b00, F1} ;
   assign F2comp = (S2 == 1'b1) ? ~({2'b00, F2}) + 1 : {2'b00, F2} ;
   assign Addout = F1comp + F2comp ;
   assign Fsum = ((Addout[27]) == 1'b0) ? Addout : ~Addout + 1 ;
   assign FV = Fsum[27] ^ Fsum[26] ;
   assign FU = ~F1[25] ;
   assign FPsum = {S1, E1, F1[24:2]} ;

   always @(posedge CLK)
   begin
```

```verilog
case (State)
    0 :
        begin
            if (St == 1'b1)
            begin
                E1 <= FPinput[30:23] ;
                S1 <= FPinput[31] ;
                F1[24:0] <= {FPinput[22:0], 2'b00} ;
                if (FPinput == 0)
                begin
                    F1[25] <= 1'b0 ;
                end
                else
                begin
                    F1[25] <= 1'b1 ;
                end
                done <= 1'b0 ;
                ovf <= 1'b0 ;
                unf <= 1'b0 ;
                State <= 1 ;
            end
        end
    1 :
        begin
            E2 <= FPinput[30:23] ;
            S2 <= FPinput[31] ;
            F2[24:0] <= {FPinput[22:0], 2'b00} ;
            if (FPinput == 0)
            begin
                F2[25] <= 1'b0 ;
            end
            else
            begin
                F2[25] <= 1'b1 ;
            end
            State <= 2 ;
        end
    2 :
        begin
            if (F1 == 0 | F2 == 0)
            begin
                State <= 3 ;
            end
            else
            begin
                if (E1 == E2)
                begin
                    State <= 3 ;
                end
                else if (E1 < E2)
                begin
```

```verilog
                        F1 <= {1'b0, F1[25:1]} ;
                        E1 <= E1 + 1 ;
                    end
                else
                    begin
                        F2 <= {1'b0, F2[25:1]} ;
                        E2 <= E2 + 1 ;
                    end
            end
        end
3 :
        begin
            S1 <= Addout[27] ;
            if (FV == 1'b0)
                begin
                    F1 <= Fsum[25:0] ;
                end
            else
                begin
                    F1 <= Fsum[26:1] ;
                    E1 <= E1 + 1 ;
                end
            State <= 4 ;
        end
4 :
        begin
            if (F1 == 0)
                begin
                    E1 <= 8'b00000000 ;
                    State <= 6 ;
                end
            else
                begin
                    State <= 5 ;
                end
        end
5 :
        begin
            if (E1 == 0)
                begin
                    unf <= 1'b1 ;
                    State <= 6 ;
                end
            else if (FU == 1'b0)
                begin
                    State <= 6 ;
                end
            else
                begin
                    F1 <= {F1[24:0], 1'b0} ;
                    E1 <= E1 - 1 ;
```

```
                        end
                    end
        6 :
                    begin
                        if (E1 == 255)
                        begin
                            ovf <= 1'b1 ;
                        end
                        done <= 1'b1 ;
                        State <= 0 ;
                    end
            endcase
    end
endmodule
```

FPinput is an input bus, and we assume that the input numbers represent normalized floating-point numbers in IEEE standard format. In state 0, the first number is split and loaded into S_1, F_1, and E_1. These represent the sign of the fraction, the magnitude of the fraction, and the biased exponent. When F_1 is loaded, the 23-bit fraction is prefixed by a 1 except in the special case of 0, in which case the leading bit is a 0. Two 0s are appended at the end of the fraction to conform to the IEEE standard requirements (guard and round bits). In state 1, the second number to be added is loaded into S_2, F_2, and E_2. In state 2, the fraction with the smallest exponent is unnormalized by shifting right and incrementing the exponent. When this operation is complete, the exponents are equal, except in the special case when F_1 or F_2 equals 0.

The fractions are added using 2's complement arithmetic, which is performed by concurrent statements. The input numbers are first converted to 2's complement representation. Two sign bits (00) are prefixed to F_1, and the 2's complement is formed if S_1 is 1 (negative). Two sign bits are used so that the sign is not lost if the fraction addition overflows into the first sign bit. F_2 is processed in a similar way. The resulting numbers, F1comp and F2comp, are added, and the sum is assigned to Addout. The adder output is read in state 3. Fsum represents the magnitude of the fraction, so Addout must be complemented if it is negative. Normally, the two sign bits of Fsum are 00, so they are discarded, and the result is stored back in F_1, which serves as a floating-point accumulator. The sign bit is extracted from the MSB of Addout. Fraction overflow and underflow are indicated by FV and FU respectively. Fraction overflow can be detected by exclusive-OR of the highest two bits of Addout. This is done as a concurrent statement. In case of fraction overflow, the sign bits of Fsum are 01, so FV = 1, Fsum is right shifted before it is stored in F_1, and E_1 is incremented. If the result of addition $F_1 = 0$, E1 is set to 0 in state 4 and the floating-point addition is complete. If F_1 is unnormalized, it is normalized in state 5 by shifting F_1 left and decrementing E_1. Exponent overflow and underflow are represented by ovf and unf, respectively. Since the normal range of biased exponents is 1 to 254, an underflow occurs if E_1 is decremented to 0, and unf is set to 1 before exiting state 5. In state 6, if $E_1 = 255$, this indicates an exponent overflow, and ovf is set to 1. The done signal is turned on before exiting state 6. The numbers in S_1, E_1, and F_1 are merged by a concurrent statement to give the final sum, FPsum, in IEEE format.

The *FP* adder was tested for the following test cases:

Addend		Augend		Sum	
Number (binary)	IEEE Single Precision	Number (binary)	IEEE Single Precision	Number (binary)	IEEE Single Precision
0	x00000000	0	x00000000	0	x00000000
1×2^0	x3F800000	1×2^0	x3F800000	1×2^1	x40000000
-1×2^0	xBF800000	-1×2^0	xBF800000	-1×2^1	xC0000000
1×2^0	x3F800000	-1×2^0	xBF800000	0	x00000000
$1.111... \times 2^{127}$	x7F7FFFFF	1×2^0	x3F800000	$1.111... \times 2^{127}$	x7F7FFFFF
$-1.111... \times 2^{127}$	xFF7FFFFF	-1×2^0	xBF800000	$-1.111... \times 2^{127}$	xFF7FFFFF
$1.111... \times 2^{127}$	x7F7FFFFF	$1.111... \times 2^{127}$	x7F7FFFFF	overflow	
$-1.111... \times 2^{127}$	xFF7FFFFF	$-1.111... \times 2^{127}$	xFF7FFFFF	overflow	
1.11×2^8	x43E00000	-1.11×2^6	xC2E00000	1.0101×2^8	x43A80000
-1.11×2^8	xC3E00000	1.11×2^6	x42E00000	-1.0101×2^8	xC3A80000
$1.111... \times 2^{127}$	x7F7FFFFF	$0.0...01 \times 2^{127}$	x73800000	overflow	
$-1.111... \times 2^{127}$	xFF7FFFFF	$-0.0...01 \times 2^{127}$	xF3800000	overflow	
$1.1...10 \times 2^{127}$	x7F7FFFFE	$0.0...01 \times 2^{127}$	x73800000	$1.111... \times 2^{127}$	x7F7FFFFF
$-1.1...10 \times 2^{127}$	xFF7FFFFE	$-0.0...01 \times 2^{127}$	xF3800000	$-1.111... \times 2^{127}$	xFF7FFFFF
1.1×2^{-126}	X00C00000	-1.0×2^{-126}	x80800000	underflow	

7.4 Other Floating-Point Operations

Subtraction

Floating-point subtraction is the same as floating-point addition, except that we must subtract the fractions instead of adding them. The rest of the steps remain the same.

Division

The quotient of two floating-point numbers is

$$(F_1 \times 2^{E_1}) \div (F_2 \times 2^{E_2}) = (F_1 / F_2) \times 2^{(E_1 - E_2)} = F \times 2^E$$

Thus, the basic rule for floating-point division is to divide the fractions and subtract the exponents. In addition to considering the same special cases as for multiplication, we must test for divide by 0 before dividing. If F_1 and F_2 are normalized, then the largest positive quotient (F) will be

$$0.1111\ldots / 0.1000\ldots = 01.111\ldots$$

which is less than 10_2, so the fraction overflow is easily corrected. For example,

$$(0.110101 \times 2^2) \div (0.101 \times 2^{-3}) = 01.010 \times 2^5 = 0.101 \times 2^6$$

Alternatively, if $F_1 \geq F_2$, we can shift F_1 right before dividing and avoid fraction overflow in the first place. In the IEEE format, when divide by 0 is involved, the result can be set to NaN (Not a Number).

In this chapter, we presented different representations of floating-point numbers. IEEE floating-point single precision and double precision formats were discussed. A floating-point format with 2's complement numbers was also presented. Then, we presented a floating-point multiplier. We also presented a procedure to perform addition of floating-point numbers. In the process of designing the multiplier we used the following steps:

1. Develop an algorithm for floating-point multiplication, taking all of the special cases into account.
2. Draw a block diagram of the system and define the necessary control signals.
3. Construct an SM chart (or state graph) for the control state machine using a separate linked state machine for controlling the fraction multiplier.
4. Write behavioral Verilog code.
5. Test the Verilog code to verify that the high-level design of the multiplier is correct.
6. Use the CAD software to synthesize the multiplier. Then implement the multiplier in the desired target technology (e.g., ASIC, FPGA, etc.).

Problems

7.1 **(a)** What is the biggest number that can be represented in the 8-bit 2's complement floating-point format with 4 bits for exponent and 4 for fraction?
(b) What is the smallest number that can be represented in the 8-bit 2's complement format with 4 bits for exponent and 4 for fraction?
(c) What is the biggest normalized number that can be represented in the IEEE single precision floating-point format?
(d) What is the smallest normalized number that can be represented in the IEEE single precision floating-point format?
(e) What is the biggest normalized number that can be represented in the IEEE double precision floating-point format?
(f) What is the smallest normalized number that can be represented in the IEEE double precision floating-point format?

Problems 427

7.2 Convert the following decimal numbers in the IEEE single precision format:
(i) 25.25, (ii) 2000.22, (iii) 1, (iv) 0, (v) 1000, (vi) 8000, (vii) 10^6, (viii) -5.4, (ix) 1.0×2^{-140}, (x) 1.5×10^9

7.3 Convert the following decimal numbers to IEEE double precision format:
(i) 25.25, (ii) 2000.22, (iii) 1, (iv) 0, (v) 1000, (vi) 8000, (vii) 10^6, (viii) -5.4, (ix) 1.0×2^{-140}, (x) 1.5×10^9

7.4 What do the following hex representations mean if they are in IEEE single precision format?
(i) ABABABAB, (ii) 45454545, (iii) FFFFFFFF, (iv) 00000000, (v) 11111111, (vi) 01010101

7.5 What do the following hex representations mean if they are in IEEE double precision format?
(i) ABABABAB 00000000, (ii) 45454545 00000001, (iii) FFFFFFFF 10001000, (iv) 00000000 00000000, (v) 11111111 10001000, (vi) 01010101 01010101

7.6 (a) Represent -35.25 in IEEE single precision floating-point format.
(b) What does the hex number ABCD0000 represent if it is in an IEEE single precision floating-point format?

7.7 (a) Represent 25.625 in IEEE single precision floating-point format.
(b) Represent -15.6 in IEEE single precision floating-point format.

7.8 This problem concerns the design of a digital system that converts an 8-bit signed integer (negative numbers are represented in 2's complement) to a floating-point number. Use a floating-point format similar to the ones used in Section 7.1.1 except that the fraction should be 8 bits and the exponent 4 bits. The fraction should be properly normalized.

7.9 (a) Multiply the following two floating-point numbers to give a properly normalized result. Assume 4-bit 2's complement format.

$$F_1 = 1.011, E_1 = 0101, F_2 = 1.001, E_2 = 0011$$

(b) Repeat (a) for

$$F_1 = 1.011, E_1 = 1011, F_2 = 0.110, E_2 = 1101$$

7.10 A floating-point number system uses a 4-bit fraction and a 4-bit exponent with negative numbers expressed in 2's complement. Design an *efficient* system that will multiply the number by -4 (minus four). Take all special cases into account and give a properly normalized result. Assume that the initial fraction is properly normalized or zero. *Note*: This system multiplies *only* by -4.

(a) Give examples of the normal and special cases that can occur (for multiplication by -4).
(b) Draw a block diagram of the system.
(c) Draw an SM chart for the control unit. Define all signals used.

7.11 Redesign the floating-point multiplier in Figure 7.7 using a common 5-bit full adder connected to a bus instead of two separate adders for the exponents and fractions.

  (a) Redraw the block diagram, being sure to include the connections to the bus, and include all control signals.

  (b) Draw a new SM chart for the new control.

  (c) Write the Verilog description for the multiplier or specify the changes that need to be made to an existing description.

7.12 This problem concerns the design of a circuit to find the square of a floating-point number, $F \times 2^E$. F is a normalized 5-bit fraction, and E is a 5-bit integer; negative numbers are represented in 2's complement. The result should be properly normalized. Take advantage of the fact that $(-F)^2 = F^2$.

  (a) Draw a block diagram of the circuit. (Use only one adder and one complementer.)

  (b) State your procedure, taking all special cases into account. Illustrate your procedure for

$$F = 1.0110 \qquad E = 00100$$

  (c) Draw an SM chart for the main controller. You may assume that multiplication is carried out using a separate control circuit, which outputs $Mdone = 1$ when multiplication is complete.

  (d) Write a Verilog description of the system.

7.13 Write a behavioral Verilog code for a floating-point multiplier using the IEEE single precision floating-point format. Use an overloaded multiplication operator instead of using an add-shift multiplier. Ignore special cases such as infinity, denormalized, and not-a-number formats. Truncate the final result instead of rounding.

7.14 Write a test bench for the floating-point adder of Figure 7-14.

7.15 Add the following floating-point numbers (show each step). Assume that each fraction is 5 bits (including sign) and each exponent is 5 bits (including sign) with negative numbers in 2's complement.

$$F_1 = 0.1011 \qquad E_1 = 11111$$
$$F_2 = 1.0100 \qquad E_2 = 11101$$

7.16 Two floating-point numbers are added to form a floating-point sum:

$$(F_1 \times 2^{E_1}) + (F_2 \times 2^{E_2}) = F \times 2^E$$

Assume that F_1 and F_2 are normalized and the result should be normalized.

  (a) List the steps required to carry out floating-point addition, including all special cases.

  (b) Illustrate these steps for $F_1 = 1.0101$, $E_1 = 1001$, $F_2 = 0.1010$, and $E_2 = 1000$. Note that the fractions are 5 bits, including sign, and the exponents are 4 bits, including sign.

  (c) Write a Verilog description of the system.

7.17 For the floating-point adder of Figure 7-14, modify the Verilog code so that
  (a) It handles IEEE standard single precision denormalized numbers both as input and output.
  (b) In state 2, it speeds up the processing when the exponents differ by more than 23.
  (c) It rounds up instead of truncating the resulting fraction.

7.18 (a) Perform the floating point addition.
  (b) Draw an SM chart for a floating-point adder that adds $F_1 \times 2^{E_1}$ and $F_2 \times 2^{E_2}$. Assume that the fractions are initially normalized (or zero) and the final result should be normalized (or zero). A zero fraction should have an exponent of -8. Set an exponent overflow flag (EV) if the final answer has an exponent overflow. Each number to be added consists of a 4-bit fraction and a 4-bit exponent, with negative numbers represented in 2's complement. Assume that all registers ($F_1, E_1, F_2,$ and E_2) can be loaded in one clock time when a start signal (St) is received. If $E_1 > E_2$, the control signal $GT = 1$, and if $E_1 < E_2$, the control signal $LT = 1$. Define all other control signals used. Include the special case where $|E_1 - E_2| > 3$.

7.19 (a) Draw a block diagram for a floating-point subtracter. Assume that the inputs to the subtracter are properly normalized, and the answer should be properly normalized. The fractions are 8 bits including sign, and the exponents are 5 bits including sign. Negative numbers are represented in 2's complement.
  (b) Draw an SM chart for the control circuit for the floating-point subtracter. Define the control signals used, and give an equation for each control signal used as an input to the control circuit.
  (c) Write the Verilog description of the floating-point subtracter.

7.20 (a) State the steps necessary to carry out floating-point subtraction, including special cases. Assume that the numbers are initially in normalized form and the final result should be in normalized form.
  (b) Subtract the following (fractions are in 2's complement):

$$(1.0111 \times 2^{-3}) - (1.0101 \times 2^{-5})$$

  (c) Write a Verilog description of the system. Fractions are 5 bits including sign, and exponents are 4 bits including sign.

7.21 This problem concerns the design of a divider for floating-point numbers:

$$(F_1 \times 2^{E_1})/(F_2 \times 2^{E_2}) = F \times 2^E$$

Assume that F_1 and F_2 are properly normalized fractions (or 0), with negative fractions expressed in 2's complement. The exponents are integers with negative numbers expressed in 2's complement. The result should be properly normalized if it is not zero. Fractions are 8 bits including sign, and exponents are 5 bits including sign.

(a) Draw a flowchart for the floating-point divider. Assume that a divider is available that will divide two binary fractions to give a fraction as a result. Do not show the individual steps in the division of the fractions on your flowchart; just say "divide." The divider requires that $|F_2| > |F_1|$ before division is carried out.

(b) Illustrate your procedure by computing

$$0.111 \times 2^3 / 1.011 \times 2^{-2}$$

When you divide F_1 by F_2, you don't need to show the individual steps, just the result of the division.

(c) Write a Verilog description for the system.

7.22 Assume that A, B, and C are floating-point numbers expressed in IEEE single precision floating-point format and that floating-point addition is performed. If $A = 2^{40}$, $B = -2^{40}$, $C = 1$, then

(a) What is A + (B + C)? (B + C is done first, and then A is added to it.)
(b) What is (A + B) + C? (A + B is done first, and then C is added to it.)

7.23 Assume that A, B, and C are floating-point numbers expressed in IEEE double precision floating-point format and that floating-point addition is performed. If $A = 2^{40}$, $B = -2^{40}$, $C = 1$, then

(a) What is A + (B + C)? (B + C is done first, and then A is added to it.)
(b) What is (A + B) + C? (A + B is done first, and then C is added to it.)

7.24 Assume that A, B and C are floating-point numbers expressed in IEEE single precision floating-point format and that floating-point addition is performed. If $A = 2^{65}$, $B = -2^{65}$, $C = 1$, then

(a) What is A + (B + C)? (B + C is done first, and then A is added to it.)
(b) What is (A + B) + C? (A + B is done first, and then C is added to it.)

7.25 Assume that A, B, and C are floating-point numbers expressed in IEEE double precision floating-point format and that floating-point addition is performed. If $A = 2^{65}$, $B = -2^{65}$, $C = 1$, then

(a) What is A + (B + C)? (B + C is done first, and then A is added to it.)
(b) What is (A + B) + C? (A + B is done first, and then C is added to it.)

Additional Topics in Verilog

Up to this point, we have described the basic features of Verilog and how they can be used in the digital system design process. In this chapter, we describe additional features of Verilog that illustrate its power and flexibility. Verilog functions and tasks are presented. Several additional features such as user-defined primitives, generate statements, compiler directives, built-in primitives, file I/O, and others are presented. A simple memory model is presented to illustrate the use of tristate signals.

8.1 Verilog Functions

A key feature of VLSI circuits is the repeated use of similar structures. Verilog provides functions and tasks to easily express repeated invocation of the same functionality or the repeated use of structures. These functions are described in this section. A Verilog function is similar to a Verilog task, which will be described later. They have small differences: a function cannot drive more than one output, nor can a function contain delays.

The functions can be defined in the module that they will be used in. The functions can also be defined in separate files, and the compile directive `include will be used to include the function in the file (compiler directives are explained in Section 8.12). The functions should be executed in zero time delay, which means that the functions cannot include timing delay information. The functions can have any number of inputs but can return only a single output.

A function returns a value by assigning the value to the function name. The function's return value can be a single bit or multiple bits. The variables declared within the function are local variables, but the functions can also use global variables when no local variables are used. When local variables are used, basically output is assigned only at the end of function execution. The functions can call other functions but cannot call tasks. Functions are frequently used to do type conversions.

A function executes a sequential algorithm and returns a single value to the calling program. When the following function is called, it returns a value equal to

the input rotated one position to the right:
```verilog
function [7:0] rotate_right;
  input [7:0] reg_address;

  begin
    rotate_right = reg_address >> 1;
  end
endfunction
```

A function call can be used anywhere that an expression can be used. For example, if $A = 10010101$, the statement

```verilog
B <= rotate_right(A);
```

would set B equal to 11001010 and leave A unchanged.

The general form of a function declaration is

```verilog
function <range or type> function-name
  input [declarations]
  <declarations>      // reg, parameter, integer, etc.

  begin
    sequential statements
  end
endfunction
```

The general form of a function call is

```verilog
function_name(input-argument-list)
```

The number and type of parameters on the `input-argument-list` must match the input [declaration] in the function declaration. The parameters are treated as input values and cannot be changed during the execution of the function.

Example

Write a Verilog function for generating an even parity bit for a 4-bit number. The input is a 4-bit number and the output is a code word that contains the data and the parity bit.

The answer is shown in Figure 8-1. The function name and the final output to be returned from the function must be the same.

FIGURE 8-1: Parity Generation Using a Function

```verilog
// Function example code without a loop
// This function takes a 4-bit vector
// It returns a 5-bit code with even parity

function [4:0] parity;
  input [3:0] A;
  reg temp_parity;

  begin
    temp_parity = A[0] ^ A[1] ^ A[2] ^ A[3];
```

```
      parity = {A, temp_parity};
  end
endfunction
```

If parity circuits are used in several parts in a system, one could call the function each time it is desired. The function can be called as follows:

```
module function_test(Z);
  output reg [4:0] Z;
  reg [3:0] INP;
  initial
  begin
    INP = 4'b0101;
    Z = parity(INP);
  end

endmodule
```

Figure 8-2 illustrates a function using a **for** loop. In Figure 8-2, the loop index *i* will be initialized to 0 when the **for** loop is entered, and the sequential statements will be executed. Execution will be repeated for $i = 1$, $i = 2$, and $i = 3$; then the loop will terminate.

FIGURE 8-2: Add Function

```
// This function adds two 4-bit vectors and a carry.
// Illustrates function creation and use of loop.
// It returns a 5-bit sum.

function [4:0] add4;
  input [3:0] A;
  input [3:0] B;
  input       cin;

  reg [4:0] sum;
  reg       cout;

  begin
    integer i;
    for (i=0; i<=3; i=i+1)
    begin
      cout = (A[i] & B[i]) | (A[i] & cin) | (B[i] & cin);
      sum[i] = A[i] ^ B[i] ^ cin;
      cin = cout;
    end

    sum[4] = cout;
    add4 = sum;
  end
endfunction
```

The function given in Figure 8-2 adds two 4-bit vectors plus a carry and returns a 5-bit vector as the sum. The function name is *add4*; the input arguments are *A*, *B*, and *carry*; and the add4 will be returned. The local variables C_{out} and *sum* are defined to hold intermediate values during the calculation. When the function is called, C_{in} will be initialized to the value of the carry. The **for** loop adds the bits of *A* and *B* serially in the same manner as a serial adder.

The first time through the loop, C_{out} and *sum[0]* are computed using *A[0]*, *B[0]*, and C_{in}. Then the C_{in} value is updated to the new C_{out} value, and execution of the loop is repeated. During the second time through the loop, C_{out} and *sum[1]* are computed using *A[1]*, *B[1]*, and the new C_{in}. After four times through the loop, all values of *sum[i]* have been computed and *sum* is returned. The function call is of the form

```
add4(A, B, carry)
```

A and *B* may be replaced with any expressions that evaluate to a return value with dimensions 3 **downto** 0, and *carry* may be replaced with any expression that evaluates to a bit. For example, the statement

```
Z <= add4(X, ~Y, 1);
```

calls the function *add4*. Parameters *A*, *B*, and *carry* are set equal to the values of *X*, ~*Y*, and 1, respectively. *X* and *Y* must be multiple bits dimensioned 3:0. The function computes

$$Sum = A + B + carry = X + \sim Y + 1$$

and returns this value. Since ~*Y* + 1 equals the 2's complement of *Y*, the computation is equivalent to subtracting by adding the 2's complement. If we ignore the carry stored in *Z[4]*, the result is *Z[3:0]* = *X* − *Y*.

Figure 8-3 illustrates the function to compute square of numbers. The function as well as the function call is illustrated. In the illustrated call to the function, the number is 4 bits wide.

FIGURE 8-3: A Function to Compute Squares and Its Call

```
module test_squares (CLK);

   input CLK;
   reg[3:0] FN;
   reg[7:0] answer;

   function [7:0] squares;
      input[3:0] Number;

      begin
         squares = Number * Number;
      end
   endfunction

   initial
   begin
```

```
      FN = 4'b0011;
   end

   always @(posedge CLK)
   begin
      answer = squares(FN);
   end
endmodule
```

A function executes in zero simulation time. Functions must not contain any timing control statements or delay events. Functions must have at least one input argument. Functions cannot have **output** or **inout** arguments.

Functions can be recursive. Recursive functions must be declared as **automatic**, which causes distinct memory to be allocated each time a function is called. Otherwise, if a function is called recursively, each call will overwrite the memory allocated in the previous call.

8.2 Verilog Tasks

Tasks facilitate decomposition of Verilog code into modules. Unlike functions, which return only a single value implicitly with the function name, tasks can return any number of values. The form of a task declaration is

```
task task_name
  input [declarations]
  output [declarations]

  <declarations>      // reg, parameter, integer, etc.

begin
  sequential statements
end task_name;
```

The `formal-parameter-list` specifies the inputs and outputs to the task and their types. A task call is a sequential or concurrent statement of the form

```
task_name(actual-parameter-list);
```

Unlike functions, a task can be executed in non-zero simulation time. Tasks may contain delay, event, or timing control statements. Tasks can have zero or more arguments of type input, output, or inout. Tasks do not return with a value, but they can pass multiple values through **output** and **inout** arguments.

As an example, we will write a task *Addvec*, which will add two *4*-bit data and a carry and return a *4*-bit sum and a carry. We will use a task call of the form

```
Addvec(A, B, Cin, Sum, Cout);
```

where *A*, *B*, and *Sum* are *4*-bit data and C_{in} and C_{out} are 1-bit data.

Figure 8-4 gives the task definition. *Add1*, *Add2*, and C_{in} are input parameters, and *sum* and C_{out} are output parameters. The addition algorithm is essentially the

FIGURE 8-4: Task for Adding Multiple Bits

```verilog
// This task adds two 4-bit data and a carry and
// returns an n-bit sum and a carry. Add1 and Add2 are assumed
// to be of the same length and dimensioned 3 downto 0.

    task Addvec;
        input [3:0] Add1;
        input [3:0] Add2;
        input Cin;
        output [3:0] sum;
        output cout;

        reg C;

        begin
            C = Cin;
            integer i;
            for(i = 0; i <= 4; i = i + 1)
                begin
                    sum[i] = Add1[i] ^ Add2[i] ^ C ;
                    C = (Add1[i] & Add2[i]) | (Add1[i] & C) | (Add2[i] & C);
                end
            cout = C ;
        end
    endtask
```

same as the one used in the *add4* function. *C* must be a variable, since the new value of *C* is needed each time through the loop; hence it is declared as **reg**. After *4* times through the loop, all 4 bits of *sum* have been computed. It is desirable not to mix blocking and non-blocking statements in tasks.

The tasks can be defined in the module that the functions will be used in. The tasks can also be defined in separate files, and the compile directive '**include should be used** (') (described in Section 8.12) will be used to include the task in the file. The variables declared within the task are local to the task. When no local variables are used, global variables will be used for input and output. When only local variables are used within the task, the variables from the last execution within the task will be passed to the caller.

> **Functions:**
> At least one input arguments, but no output or inout arguments
> Returns a single value by assigning the value to the function name
> Can call other functions, but cannot call tasks
> Cannot embed delays, wait statements or any time-controlled statement
> Executes in zero time
> Can be recursive
> Cannot contain non-blocking assignment or procedural continuous assignments

> **Tasks:**
> Any number of input, output or inout arguments
> Outputs need not use task name
> Can call other functions or tasks
> May contain time-controlled statements
> Task can recursively call itself

8.3 Multivalued Logic and Signal Resolution

In previous chapters, we have mostly used two-valued bit logic in our Verilog code. In order to represent tristate buffers and buses, it is necessary to be able to represent a third value, Z, which represents the high-impedance state. It is also at times necessary to have a fourth value, X, to represent an unknown state. This unknown state may occur if the initial value of a signal is unknown, or if a signal is simultaneously driven to two conflicting values, such as 0 and 1. If the input to a gate is Z, the gate output may assume an unknown value, X.

8.3.1 A 4-Valued Logic System

Signals in a 4-valued logic can assume the four values: X, 0, 1, and Z, where each of the symbols represent the following:

- X Unknown
- 0   0
- 1   1
- Z High impedance

The high impedance state is used for modeling tristate buffers and buses. This unknown state can be used if the initial value of a signal is unknown, or if a signal is simultaneously driven to two conflicting values, such as 0 and 1. Verilog uses the 4-valued logic system by default.

Let us model tristate buffers using the 4-valued logic. Figure 8-5 shows two tristate buffers with their outputs tied together, and Figure 8-6 shows the

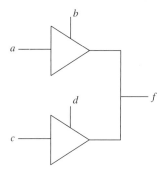

FIGURE 8-5: Tristate Buffers with Active-High Output Enable

corresponding Verilog representation in two different ways. Data type X01Z, which can have the four values X, 0, 1, and Z, is assumed. The tristate buffers have an active-high output enable, so that when $b = 1$ and $d = 0$, $f = a$; when $b = 0$ and $d = 1$, $f = c$; and when $b = d = 0$, the f output assumes the high-Z state. If $b = d = 1$, an output conflict can occur. Tristate buffers are used for bus management whereby only one active-high output should be enabled to avoid the output conflict.

FIGURE 8-6: Verilog Code for Tristated Buffers

```
module t_buff_exmpl (a, b, c, d, f);

   input a;
   input b;
   input c;
   input d;
   output f;
   reg f;

   always @(a or b)
   begin : buff1
      if (b == 1'b1)
         f = a ;
      else
         f = 1'bz ; //"drive" the output high Z when not enabled
   end

   always @(c or d)
   begin : buff2
      if (d == 1'b1)
         f = c ;
      else
         f = 1'bz ; //"drive" the output high Z when not enabled
   end

endmodule
```

(a) Tristate module with always statements

```
module t_buff_exmpl2 (a, b, c, d, f);
   input a;
   input b;
   input c;
   input d;
   output f;

   assign f = b ? a: 1'bz ;
   assign f = d ? c: 1'bz ;

endmodule
```

(b) Tristate module with assign statements

The operation of a tristate bus with the 4-valued logic, is specified by the following table:

	X	0	1	Z
X	X	X	X	X
0	X	0	X	0
1	X	X	1	1
Z	X	0	1	Z

This table gives the resolved value of a signal for each pair of input values: Z resolved with any value returns that value, X resolved with any value returns X, and 0 resolved with 1 returns X. If individual wires s(0), s(1), s(2) were to change as indicated in the following table at the times indicated, signal R in the last column shows the result of the resolution

Time	s(0)	s(1)	s(2)	R
0	Z	Z	Z	Z
2	0	Z	Z	0
4	0	1	Z	X
6	Z	1	Z	1
8	Z	1	1	1
10	Z	1	0	X

AND and OR functions for the 4-valued logic may be defined using the following tables:

AND	X	0	1	Z
X	X	0	X	X
0	0	0	0	0
1	X	0	1	X
Z	X	0	X	X

OR	X	0	1	Z
X	X	X	1	X
0	X	0	1	X
1	1	1	1	1
Z	X	X	1	X

The first table corresponds to the way an AND gate with 4-valued inputs would work. If one of the AND gate inputs is 0, the output is always 0. If both inputs are 1, the output is 1. In all other cases, the output is unknown (X), since a high-Z gate input may act like either a 0 or a 1. For an OR gate, if one of the inputs is 1, the output is always 1. If both inputs are 0, the output is 0. In all other cases, the output is X.

8.4 Built-in Primitives

The focus of this book is on behavioral-level modeling, and hence we focused on Verilog constructs that allow that. However, Verilog allows modeling at switch-level details and has several predefined primitives for that. It also has predefined gate-level primitives with drive strength among other things. There are 14 predefined logic gate primitives and 12 predefined switch primitives to provide the gate- and switch-level modeling facility. Modeling at the gate and switch level has several advantages:

(i) Synthesis tools can easily map it to the desired circuitry since gates provide a very close one-to-one mapping between the intended circuit and the model.

(ii) There are primitives such as the bidirectional transfer gate that cannot be otherwise modeled using continuous assignments.

The Verilog module in Figure 2-7 is defined in Figure 8-7 using built-in primitives. The **and** and **or** are built-in primitives with one output and multiple inputs. The output terminal must be listed first followed by inputs as in

and (out, in_1, in_2, ..., in_n);

FIGURE 8-7: Verilog Gate Module Using Built-In Primitives

```
module two_gates (A, B, D, E);      // Figure 2-7 shows the same module using
   output E;                         // concurrent statements instead of
   input A, B, D;                    // built-in primitives

   wire C;

   or    (E,C,D);                   // output port first followed by inputs
   and   (C,A,B);                   // output port first

endmodule
```

Verilog also provides built-in primitives for tristate gates. The **bufif0** is a non-inverting buffer primitive with active-low control input while **bufif1** has active-high control input. The **notif0** and **notif1** are inverting buffers with active-low and active-high controls, respectively. These primitives can support multiple outputs. The outputs must be listed first followed by input and finally the tristate control input. An array of four inverting tristate buffers can be created as shown in Figure 8-8.

FIGURE 8-8: Verilog Array of Tristate Buffers Using Built-In Primitives

```
module tri_driver (in, out, tri_en);
   input [3:0] in;
   output [3:0] out;
   input tri_en;

   bufif0 buf_array[3:0] (out, in, tri_en); // array of three-state buffers
endmodule
```

The statement

bufif0 buf_array[3:0] (out, in, tri_en);   // instance name is
                                              // indexed here

in Figure 8-8 uses buf_array[3:0] as instance name. Notice that the instance name is indexed here. This statement could be replaced using

bufif0 buf_array3 (out[3], in[3], tri_en); // instance name not
                                               // indexed
bufif0 buf_array2 (out[2], in[2], tri_en);
bufif0 buf_array1 (out[1], in[1], tri_en);
bufif0 buf_array0 (out[0], in[0], tri_en);

where the instance names are not indexed. The instance names are not signal names.

The circuit in Figure 8-5, presented using the behavioral model in Figure 8-6, can be written using Verilog built-in primitives as in Figure 8-9.

FIGURE 8-9: Verilog Tristate Buffer Module Using Built-In Primitives

```
module tri_buffer (a,b,c,d,f);
   input a,b,c,d;
   output f;

   bufif1 buf_one (f,a,b); //output first followed by inputs, control last.
   bufif1 buf_two (f,c,d);

endmodule
```

Table 8.1 lists the built-in primitives in Verilog. The first 12 are gate-level primitives, and the rest are switch-level primitives. Switch-level models are not discussed in this book, but they are useful if low-level models of designs are to be built.

TABLE 8-1: Built-In Gate and Switch Primitives in Verilog

Built-in Primitive Type	Primitives
n-input gates	and, nand, nor, or, xnor, xor
n-output gates	buf, not
Three-state gates	bufif0, bufif1, notif0, notif1
Pull gates	pulldown, pullup
MOS switches	cmos, nmos, pmos, rcmos, rnmos, rpmos
Bidirectional Switches	rtran, rtranif0, rtranif1, tran, transif0, tranif1

The built-in primitives can have an optional delay specification. A delay specification can contain up to three delay values, depending on the gate type. For a three-delay specification,

(i) the first delay refers to the transition to the rise delay (i.e., transition to 1),

(ii) the second delay refers to the transition to the fall delay (i.e., transition to 0), and

(iii) the third delay refers to the transition to the high-impedance value (i.e., turn-off).

If only one delay is specified, it is rise delay. If there are only two delays specified, they are rise delay and fall delay. If turn-off delay must be specified, all three delays must be specified. The **pullup** and **pulldown** instance declarations must not include delay specifications. The following are examples of built-in primitives with one, two, and three delays:

```
and    #(10)       a1 (out, in1, in2);       // only one delay
                                              // (so rise delay=10)
and    #(10,12)    a2 (out, in1, in2);       // rise delay=10 and
                                              // fall delay=12
bufif0 #(10,12,11) b3 (out, in, ctrl);       // rise, fall, and
                                              // turn-off delays
```

8.5 User-Defined Primitives

The predefined gate primitives in Verilog can be augmented with new primitive elements called user-defined primitives (UDPs). UDPs define the functionality of the primitive in truth table or state table form. Once these primitives are specified by the user, instances of these new UDPs can be created in exactly the same manner as built-in gate primitives are instantiated. While the built-in primitives are only combinational, UDPs can be combinational or sequential.

A *combinational UDP* uses the value of its inputs to determine the next value of its output. A *sequential UDP* uses the value of its inputs and the current value of its output to determine the value of its output. Sequential UDPs provide a way to model sequential circuits such as flip-flops and latches. A sequential UDP can model both level-sensitive and edge-sensitive behavior.

A UDP can have multiple input ports but has exactly one output port. Bidirectional inout ports are not permitted on UDPs. All ports of a UDP must be scalar—that is, vector ports are not permitted. Each UDP port can be in one of three states: 0, 1, or X. The tristate or high-impedance state value Z is not supported. If Z values are passed to UDP inputs, they shall be treated the same as X values. In sequential UDPs, the output always has the same value as the internal state.

A UDP begins with the keyword **primitive**, followed by the name of the UDP. The functionality of the primitive is defined with a truth table or state table, starting with the keyword **table** and ending with the keyword **endtable**. The UDP definition then ends with the keyword **endprimitive**. The truth table for a UDP consists of a section of columns, one for each input followed by a colon and finally the output column. The multiplexer we defined in Figure 2-36 using continuous assign statements is redefined as a UDP in Figure 8-10.

FIGURE 8-10: User-Defined Primitive (UDP) for a 2-to-1 Multiplexer

```
primitive mux1 (F, A, I0, I1);
  output F;
  input A, I0, I1; //A is the select input

table
  //  A   I0  I1       F
      0   1   0    :   1   ;
```

```
            0   1   1   :   1   ;
            0   1   x   :   1   ;
            0   0   0   :   0   ;
            0   0   1   :   0   ;
            0   0   x   :   0   ;
            1   0   1   :   1   ;
            1   1   1   :   1   ;
            1   x   1   :   1   ;
            1   0   0   :   0   ;
            1   1   0   :   0   ;
            1   x   0   :   0   ;
            x   0   0   :   0   ;
            x   1   1   :   1   ;
    endtable

endprimitive
```

The first entry in the truth table in Figure 8-10 can be explained as follows: when A equals 0, I0 equals 1, and I1 equals 0, then output F equals 1. The input combination 0xx (A=0, I0=x, I1=x) is not specified. If this combination occurs during simulation, the value of output port F will become x. Each row of the table in the UDP is terminated by a semicolon.

The multiplexer model can also be specified more concisely in a UDP by using ?. The ? means that the signal listed with ? can take the values of 0,1, or x. Using ? the multiplexer can be defined as in Figure 8-11.

FIGURE 8-11: User-Defined Primitive (UDP) for a 2-to-1 Multiplexer Using?

```
primitive mux2 (F, A, I0, I1);
  output F;
  input A, I0, I1;

table
//  A   I0  I1      F
    0   1   ?   :   1   ;   // ? can equal 0, 1, or x
    0   0   ?   :   0   ;
    1   ?   1   :   1   ;
    1   ?   0   :   0   ;
    x   0   0   :   0   ;
    x   1   1   :   1   ;
endtable

endprimitive
```

UDPs are instantiated just as are built-in primitives. For instance, the preceding multiplexer can be instantiated by using the statement

    mux1 (outF, Sel, in1, in2)

where outF, Sel, in1, and in2 are the names of the output, select, data input1, and data input2 signals.

As an example of a sequential UDP, we present the description of a D flip-flop in Figure 8-12. In a sequential UDP, the output must be defined as a **reg**. Edge-sensitive behavior can be represented in tabular form by listing the value before and after the edge. For instance, (01) means a positive (rising) edge and (10) means a negative (falling) edge. In the sequential UDP, there is an additional colon separating the inputs to the present state in the state table.

FIGURE 8-12: User-Defined Primitive (UDP) for a D Flip-Flop

```
primitive DFF (Q, CLK, D);
  output Q;
  input CLK, D;

  reg Q;

table
  // CLK,  D,    Q,   Q⁺
     (01)  0  : ? : 0  ; //rising edge with input 0
     (01)  1  : ? : 1  ; //rising edge with input 1
     (0?)  1  : 1 : 1  ; //Present state 1, either rising edge or steady clock
     (?0)  ?  : ? : -  ; //Falling edge or steady clock, no change in output
      ?   (??) : ? : -  ; //Steady clock, ignore inputs, no change in output
endtable

endprimitive
```

The (01) in the first line of the table indicates a rising edge clock. The '-' in the output column means that the output should not change for any of the circumstances covered by that line. For instance, the line

(?0) ? ? : - ;

means that if there is a falling edge or steady clock, whether the input and present state are 0,1, or x, the output must not change. This line actually represents 27 possibilities, because each ? represents three possibilities. If this line was omitted in the code, the simulator would yield an x output in these situations. It is important to make the truth table as unambiguous as possible by specifying all possible cases. The last line of code clarifies further that under steady clock (i.e., no clock edges), the flip-flop must ignore all inputs.

? (??) : ? : - ;

The rows in the primitive truth table do not need to be in order. Hence the row order

0 1 0 : 1 ;
0 0 0 : 0 ;

is acceptable. A simulator scans the truth table from top to bottom. If level-sensitive behavior such as asynchronous set and reset are in a table along with edge-sensitive behavior for data, the level-sensitive behavior should be listed before the edge-sensitive behavior.

8.6 SRAM Model

In this section, we develop a Verilog model to represent the operation of a static RAM (SRAM). RAM stands for random-access memory, which means that any word in the memory can be accessed in the same amount of time as any other word. Strictly speaking, ROM memories are also random-access, but historically, the term RAM is normally applied only to read-write memories. This model also illustrates the usefulness of the multivalued logic system. Multivalued logic is used to model tristate conditions on the memory data lines.

Figure 8-13 shows the block diagram of a static RAM with n address lines, m data lines, and three control lines. This memory can store 2^n words, each m bits wide. The data lines are bidirectional in order to reduce the required number of pins and the package size of the memory chip. When reading from the RAM, the data lines are outputs; when writing to the RAM, the data lines serve as inputs. The three control lines function as follows:

$\overline{CS}$ When asserted low, chip select selects the memory chip so that memory read and write operations are possible.

$\overline{OE}$ When asserted low, output enable enables the memory output onto an external bus.

$\overline{WE}$ When asserted low, write enable allows data to be written to the RAM.

We say that a signal is asserted when it is in its active state. An active-low signal is asserted when it is low, and an active-high signal is asserted when it is high.

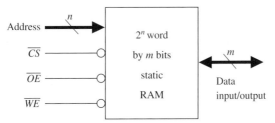

FIGURE 8-13: Block Diagram of Static RAM

The truth table for the RAM (Table 8-2) describes its basic operation. High-Z in the I/O column means that the output buffers have high-Z outputs and the data inputs are not used. In the read mode, the address lines are decoded to select m memory cells, and the data comes out on the I/O lines after the memory access time has elapsed. In the write mode, input data is routed to the latch inputs in the selected memory cells when $\overline{WE}$ is low, but writing to the latches in the memory cells is not completed until either $\overline{WE}$ goes high or the chip is deselected. The truth table does not take memory timing into account.

TABLE 8-2: Truth Table for Static RAM

$\overline{CS}$	$\overline{OE}$	$\overline{WE}$	Mode	I/O pins
H	X	X	not selected	high-Z
L	H	H	output disabled	high-Z
L	L	H	read	data out
L	X	L	write	data in

We now write a simple Verilog model for the memory that does not take timing considerations into account. In Figure 8-14, the RAM memory array is represented by an array of registers (*RAM1*). This memory has 256 words, each of which are 8 bits. The RAM process sets the I/O lines to high-Z if the chip is not selected. If *We_b* = 1, the RAM is in the read mode and *IO* is the data read from the memory array. If *We_b* = 0, the memory is in the write mode, and the data on the I/O lines is stored in *RAM1* on the rising edge of *We_b*. This is a RAM with asynchronous read and synchronous write.

FIGURE 8-14: Simple Memory Model

```verilog
module RAM6116 (Cs_b, We_b, Oe_b, Address, IO);

   input Cs_b;
   input We_b;
   input Oe_b;
   input[7:0] Address;
   inout[7:0] IO;

   reg[7:0] RAM1[0:255];

   assign IO = (Cs_b == 1'b1 | We_b == 1'b0 | Oe_b == 1'b1) ?
           8'bZZZZZZZZ : RAM1[Address] ; // read from RAM

   always @(We_b, Cs_b)
   begin
       @(negedge We_b)  //falling edge of We_b
       if(Cs_b == 1'b0)
       begin
        RAM1[Address] <= IO ;   // write to RAM
       end
   end
endmodule
```

8.7 Model for SRAM Read/Write System

In order to further illustrate the use of multivalued logic, we present an example with a **bidirectional tristate bus**. We will design a memory read-write system that reads the content of 32 memory locations from a RAM, increments each data

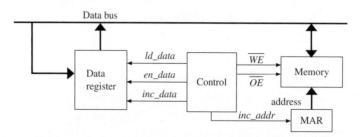

FIGURE 8-15: Block Diagram of RAM Read-Write System

value, and stores it back into the RAM. A block diagram of the system is shown in Figure 8-15. In order to hold the word that we read from memory, we use a **data register**. In order to hold the memory address that we are accessing, we use a memory address register **(MAR)**. The system reads a word from the RAM, loads it into the data register, increments the data register, stores the result back in the RAM, and then increments the memory address register. This process continues until the memory address equals 32.

The data bus is used as a **bidirectional bus**. During the read operation, the memory output appears on the bus, and the data register output to the data bus will be in a tristate condition. During the write operation, the data register output is on the data bus and the memory will use it as input data.

Control signals required to operate the system are as follows:

ld_data	load data register from Data Bus
en_data	enable data register output onto Data Bus
inc_data	increment Data Register
inc_addr	increment MAR
$\overline{WE}$	Write Enable for SRAM
$\overline{OE}$	Output Enable for SRAM

Figure 8-16 shows the SM chart for the system. The SM chart uses four states. In state S_0, the SRAM drives the memory data onto the bus and the memory data is loaded into the Data Register. The control signal $\overline{OE}$ and ld_data are true in this state. The Data Register is incremented in S_1. The en_data control signal is true in state S_2, and hence the Data Register drives the bus. Write enable $\overline{WE}$ is an active-low signal that is asserted low only in S_2; as a result, $\overline{WE}$ is high in the other states. The writing to the RAM is initiated in S_2 and completed on the rising edge of $\overline{WE}$, which occurs during the transition from S_2 to S_3. The memory address is incremented. The process continues until the address is 32. State S_3 checks this and produces a done signal when the address reaches 32.

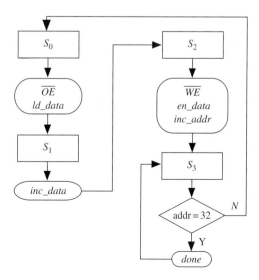

FIGURE 8-16: SM Chart for RAM System

Figure 8-17 shows the Verilog code for the RAM system. The first always statement represents the SM chart, and the second always statement is used to update the registers on the rising edge of the clock. A short delay is added when the address is incremented to make sure the write to memory is completed before the address changes. A concurrent statement is used to simulate the tristate buffer, which enables the Data Register output to go onto the I/O lines.

FIGURE 8-17: Verilog Code for RAM System

```verilog
// SRAM Read-Write System model

module RAM6116_system ();

    reg[1:0] state;
    reg[1:0] next_state;
    reg inc_adrs;
    reg inc_data;
    reg ld_data;
    reg en_data;
    reg Cs_b;
    reg clk;
    reg Oe_b;
    reg done;
    reg We_b;
    reg[7:0] Data;
    reg[7:0] Address;
    wire[7:0] IO;

    initial
    begin
        inc_adrs = 1'b0;
        inc_data = 1'b0;
        ld_data = 1'b0;
        en_data = 1'b0;
        clk = 1'b0;
        Cs_b = 1'b0;
        Oe_b = 1'b0;
        done = 1'b0;
        We_b = 1'b1; //initialize to read mode
        Address = 8'b00000000; //address register
    end

    RAM6116 RAM1 (Cs_b, We_b, Oe_b, Address, IO);

    always @(state or Address)
    begin : control
        ld_data = 1'b0 ;
        inc_data = 1'b0 ;
        inc_adrs = 1'b0 ;
        en_data = 1'b0 ;
        done = 1'b0 ;
```

```verilog
         We_b = 1'b1 ;
         Cs_b = 1'b0 ;
         Oe_b = 1'b1 ;

      case (state)
         0 :
                     begin
                        Oe_b = 1'b0 ;
                        ld_data = 1'b1 ;
                        next_state = 1 ;
                     end
         1 :
                     begin
                        inc_data = 1'b1 ;
                        next_state = 2 ;
                     end
         2 :
                     begin
                        We_b = 1'b0 ;
                        en_data = 1'b1 ;
                        inc_adrs = 1'b1 ;
                        next_state = 3 ;
                     end
         3 :
                     begin
                        if (Address == 8'b00100000)
                        begin
                           done = 1'b1 ;
                           next_state = 3 ;
                        end
                        else
                        begin
                           next_state = 0 ;
                        end
                     end
      endcase
end
always @(posedge clk) //always block to update data register
begin : register_update
   state <= next_state ;
   if (inc_data == 1'b1)
   begin
      Data <= Data + 1 ; //increment data in data register
   end
   if (ld_data == 1'b1)
   begin
      Data <= unsigned(IO) ; //load data register from bus
   end
   if (inc_adrs == 1'b1)
   begin
```

```
            Address <= #1 Address + 1 ;  //delay added to allow completion of memory write
        end
    end

//Concurrent statements
    always #100 clk = ~clk ;
    assign IO = (en_data ==1'b1) ? Data : 8'bZZZZZZZZ ;

endmodule
```

This system can be modified to include all memory locations for testing the correctness of the entire SRAM. Memory systems are often tested by writing checkerboard patterns (alternate 0s and 1s) in all locations. For instance, one can write 01010101 (55 hexadecimal) into all odd addresses and 10101010 (hexadecimal AA) into all even addresses. Then the odd and even locations can be swapped. Developing Verilog code for such a system is left as an exercise problem.

8.8 Rise and Fall Delays of Gates

In Section 8.4, we described built-in primitives with delay specifications. In this section, we show how modules can be created with variable delays. *Parameters* are commonly used to specify constant values for a module in such a way that the parameter values may be specified when the module is instantiated. For example, the rise and fall times for a gate could be specified as parameters, and different numeric values for these parameters could be assigned for each instance of the gate. The example of Figure 8-18 describes a two-input NAND gate whose rise and fall delay times depend on the number of loads on the gate. In the NAND2 module declaration, *Trise*, *Tfall*, and *load* are parameters that specify default values for rise time, fall time, and the number of loads. An internal *nand_value* is computed whenever *a* or *b* changes. If *nand_value* has just changed to a 1, a rising output has occurred and the gate delay time is computed as

Trise + 3 ns * load

where 3 ns is the added delay for each load. Otherwise, a falling output has just occurred, and the gate delay is computed as

Tfall + 2 ns * load

where 2 ns is the added delay for each load.

FIGURE 8-18: Rise/Fall Time Modeling Using Parameter

```
module NAND2 (a, b, c);

    parameter Trise = 3;
    parameter Tfall = 2;
    parameter load  = 1;
```

```
    input a;
    input b;
    output c;

    reg c;

    wire nand_value;

    assign nand_value = ~(a & b) ;

    always @ (nand_value)
    begin
      if (nand_value == 1'b1)
        #(Trise + 3 * load) c = 1'b1;
      else
        #(Tfall + 3 * load) c = 1'b0;
    end
endmodule
module NAND2_test (in1, in2, in3, in4, out1, out2);
    input in1;
    input in2;
    input in3;
    input in4;
    output out1;
    output out2;

    NAND2 #(2, 1, 2) U1 (in1, in2, out1);
    NAND2 U2 (in3, in4, out2);
endmodule
```

The module *NAND2_test* tests the NAND2 component. The parameter declaration in the module specifies default values for *Trise*, *Tfall*, and *load*. When *U1* is instantiated, the parameter map specifies different values for *Trise*, *Tfall*, and *load*. When *U2* is instantiated, no parameter map is included, so the default values are used. Another way to instantiate *U1* is by using **defparam** as follows:

```
    defparam U1.Trise = 2;
    defparam U1.Tfall = 1;
    defparam U1.load = 2;
    NAND2 U1 (in1,in2, out1);
    NAND2 U2 (in3, in4, out2);
```

8.9 Named Association

Up to this point, we have used *positional association* in the port maps and parameter maps that are part of an instantiation statement. For example, assume that the module declaration for a full adder is

```verilog
module FullAdder (Cout, Sum, X, Y, Cin);

output Cout;
output Sum;
input X;
input Y;
input Cin;

.........

endmodule
```

The statement

```verilog
FullAdder FA0 (Co[0], S[0], A[0], B[0], Ci[0]);
```

creates a full adder and connects *A[0]* to the *X* input of the adder, *B[0]* to the *Y* input, Ci[0] to the C_{in} input, Co[0] to the C_{out} output, and *S[0]* to the *Sum* output of the adder. The first signal in the port map is associated with the first signal in the module declaration, the second signal with the second signal, and so on.

As an alternative, we can use *named association*, in which each signal in the port map is explicitly associated with a signal in the port of the module declaration. For example, the statement

```verilog
FullAdder FA0 (.Sum(S[0]), .Cout(Co[0]), .X(A[0]), .Y(B[0]),
               .Cin(Ci[0]));
```

makes the same connections as the previous instantiation statement—that is, *Sum* connects to *S[0]*, C_{out} connects to *Co[0]*, *X* connects to *A[0]*, and so on. When named association is used, the order in which the connections are listed is not important, and any port signals not listed are left unconnected. Use of named association makes code easier to read, and it offers more flexibility in the order in which signals are listed.

When named association is used with a parameter map, any unassociated parameter assumes its default value. For example, if we replace the statement in Figure 8-18 labeled U1 with the following:

```verilog
NAND2 #(.load(3), .Trise(4)) U1 (in1, in2, out1);
```

Tfall would assume its default value of 2 ns.

8.10 Generate Statements

In Chapter 2, we instantiated four full adders and interconnected them to form a 4-bit adder. Specifying the port maps for each instance of the full adder would become very tedious if the adder had 8 or more bits. When an iterative array of identical operations or module instance is required, the **generate** statement provides an easy way of instantiating these components. The example of Figure 8-19 shows how a **generate** statement can be used to instantiate four 1-bit full adders to create a 4-bit adder. A 5-bit vector is used to represent the carries, with C_{in} the same

FIGURE 8-19: Adder4 Using Generate Statement

```verilog
module Adder4 (A, B, Ci, S, Co);
    input[3:0] A; //inputs
    input[3:0] B;
    input Ci;
    output[3:0] S; //outputs
    output Co;

    wire[4:0] C;

    assign C[0] = Ci ;

    genvar i;
    generate
    for (i=0; i<4; i=i+1)
      begin: gen_loop
        FullAdder FA (A[i], B[i], C[i], C[i+1], S[i]);
      end
    endgenerate

    assign Co = C[4] ;

endmodule

module FullAdder (X, Y, Cin, Cout, Sum);
    input X; //inputs
    input Y;
    input Cin;
    output Cout; //outputs
    output Sum;

    assign #10 Sum = X ^ Y ^ Cin ;
    assign #10 Cout = (X & Y) | (X & Cin) | (Y & Cin) ;
endmodule
```

as $C(0)$ and C_{out} the same as $C(4)$. The **for** loop generates four copies of the full adder, each with the appropriate **port map** to specify the interconnections between the adders.

Another example where the **generate** statement would have been very useful is the array multiplier. The Verilog code for the array multiplier (Chapter 4) made repeated use of port map statements in order to instantiate each module instance. They could have been replaced with **generate** statements.

In the preceding example, we used a **generate** statement of the form

```
genvar gen_variable1,…;
generate
for ( for_loop_condition with gen_variables )
  concurrent statement(s)
endgenerate
```

At compile time, a set of concurrent statement(s) is generated for each value of the identifier in the given range. In Figure 8-19, one concurrent statement—a module instance instantiation statement—is used. The statement

```
FullAdder FA (A[i], B[i], C[i], C[i+1], S[i]);
```

inside the **generate** clause creates the effect of the following four statements:

```
FullAdder FA (A[0], B[0], C[0], C[1], S[0]);
FullAdder FA (A[1], B[1], C[1], C[2], S[1]);
FullAdder FA (A[2], B[2], C[2], C[3], S[2]);
FullAdder FA (A[3], B[3], C[3], C[4], S[3]);
```

A **generate** statement itself is defined to be a concurrent statement, so nested generate statements are allowed.

Conditional Generate

A **generate** statement with an **if** clause may be used to conditionally generate a set of concurrent statement(s). This type of **generate** statement has the form

```
generate
if condition
   concurrent statement(s)
endgenerate
```

In this case, the concurrent statements(s) are generated at compile time only if the condition is true. Conditional compilation can also be accomplished using the `` `ifdef `` compiler directive, which is explained in Section 8-12. Conditional compilation is useful for

(i) Selectively including behavioral, structural, or switch-level models as desired.

(ii) Selectively including different timing or structural information.

(iii) Selectively including different stimuli for different runs under different scenarios.

(iv) Selectively adapting the module functionality to similar but different needs from different customers.

Figure 8-20 illustrates the use of conditional compilation using a **generate** statement with an **if** clause. An N-bit left-shift register is created if *Lshift* is true, using the statement

```
generate
  if(Lshift)
    assign shifter = {Q[N - 1:1], Shiftin} ;
  else
    assign shifter = {Shiftin, Q[N:2]} ;
endgenerate
```

If *Lshift* is false, a right-shift register is generated using another conditional **generate** statement. The example also shows how parameters and **generate** statements can be used together. It illustrates the use of parameters to write a Verilog model with parameters so that the size and function can be changed when it is instantiated. It also shows another example of named association.

FIGURE 8-20: Shift Register Using Conditional Compilation

```verilog
module shift_reg (D, Qout, CLK, Ld, Sh, Shiftin);
   parameter N = 4;
   parameter Lshift = 1;

   output[N:1] Qout;
   input[N:1] D;
   input CLK;
   input Ld;
   input Sh;
   input Shiftin;

   reg[N:1] Q;
   wire[N:1] shifter;

   assign Qout = Q ;

   generate
     if(Lshift)
       assign shifter = {Q[N - 1:1], Shiftin} ; //left shift register
     else
       assign shifter = {Shiftin, Q[N:2]} ; //right shift register
   endgenerate

   always @(posedge CLK)
   begin
     if (Ld == 1'b1)
       begin
         Q <= D ;
       end
     else if (Sh == 1'b1)
       begin
         Q <= shifter ;
       end
   end
endmodule
```

8.11 System Functions

In addition to tasks and functions that the user can create, Verilog has tasks and functions at the system level. System tasks and functions in Verilog start with the $ sign. System tasks that we have used so far include **$display, $monitor,** and **$write.** The file I/O functions tasks discussed in Section 8.13 are also system tasks. These system functions are not for synthesis, however; they are intended for convenience during simulation.

Systems tasks mainly include display tasks for outputting text or data during simulation, file I/O tasks, and simulation control tasks such as **$finish** and **$stop.** We describe a few of the system tasks/functions in this section.

8.11.1 Display Tasks

Display tasks are very useful during simulation to check outputs. There are several of them, and there are variations for displaying data in binary, hex, or octal formats (e.g., **$displayb, $displayh, $displayo).** The major tasks are the following:

$display	Immediately outputs text or data with new line.
$write	Immediately outputs text/data without new line.
$strobe	Outputs text or data at the end of the current simulation step.
$monitor	Displays text or data for every event on signal.

8.11.2 File I/O Tasks

File I/O tasks are described in detail in Section 8.13.

8.11.3 Simulation Control Tasks

There are two system tasks to control the simulation: **$stop** and **$finish**. The **$stop** task is a system function to temporarily suspend simulation for interaction. **$finish** is a system task to come out of simulation. The **$finish** system task simply makes the simulator exit and pass control back to the host operating system. Both these tasks take an optional expression argument (0, 1, or 2) that determines what type of diagnostic message is printed.

8.11.4 Simulation Time Functions

It can be beneficial to access simulation time. The following system functions provide access to current simulation time:

$time	Returns an integer that is a 64-bit time, scaled to the timescale unit of the module that invoked it.
$stime	Returns an unsigned integer that is a 32-bit time, scaled to the timescale unit of the module that invoked it. If the actual simulation time does not fit in 32 bits, the low-order 32 bits of the current simulation time are returned.
$realtime	Returns a real number time that, like **$time**, is scaled to the time unit of the module that invoked it.

The following statement can be used to print simulation time:

```
$monitor($time);
```

We can assign it to other variables as follows:

```
time simtime;    // time is one of the variable data types
simtime = $time; // Assign current simulation time to variable
                 // simtime
```

8.11.5 Conversion Functions

There are also system functions to perform conversions between data types.

$signed() and $unsigned()

Two system functions are used to handle type casting on expressions: **$signed()** and **$unsigned()**. These functions evaluate the input expression and return a value with the same size and value of the input expression and the type defined by the function.

```
reg [3:0] regA;
reg signed [7:0] regB;

regA = $unsigned(-4);          // regA = 4'b1100
regB = $signed (8'b11111100);  // regB = -4
```

$realtobits and $bitstoreal

The system functions **$realtobits** and **$bitstoreal** are useful for type conversions. Although Verilog is not a strongly typed language, there are type restrictions on many operators. For instance, the concatenate, replicate, modulus, case equality, reduction, shift, and bit-wise operators cannot work with real-number operands. Real numbers can be converted to bits or vice versa using the afore-mentioned conversion functions. The system functions **$realtobits** and **$bitstoreal** can also be used for passing bit patterns across module ports, when they are represented as real numbers on the other side. For example, consider the following code:

```
module bus_driver (net_bus);
   output net_bus;

   real sig;
   wire [64:1] net_bus = $realtobits(sig);

endmodule
```

The variable sig, which is in real type is converted to bits using the statement

```
wire [64:1] net_bus = $realtobits(sig);
```

and passed to net_bus.

8.11.6 Probablistic Functions

Verilog has system functions to generate probabilistic distributions or random numbers. The **$random** function returns a random signed integer that is 32-bits.

8.12 Compiler Directives

Verilog has several compiler directives that add programming convenience to the development and maintenance of Verilog modules. All Verilog compiler directives are preceded by the (`) character. This character is called *grave accent* (ASCII 0x60). It is different from the character ('), which is the *apostrophe* character

(ASCII 0x27). The scope of a compiler directive extends from the point where it is processed, across all files processed, to the point where another compiler directive supersedes it or the processing completes. There is no semicolon (;) at the end of the line with the compiler directive. This section describes a few compiler directives.

8.12.1 `define

The directive `**define** creates a macro for text substitution. This directive can be used both inside and outside module definitions. After a text macro is defined, it can be used in the source description by using the (`) character, followed by the macro name. This directive can be used to define constants or expressions. For example,

```
`define wordsize 16
```

causes the string `wordsize` to be replaced by 16. Such replacements can increase the readability and modifiability of the code.

In the following example, a macro `max` is defined to represent an expression. In order to invoke the macro, one must write `max. For example, the statement

```
`define max(a,b)((a) > (b) ? (a) : (b))
```

defines the macro. The larger of the two operands is returned by this macro. The statement

```
n = `max(p+q, r+s) ;
```

invokes the macro. This macro expands as

```
n = ((p+q) > (r+s)) ? (p+q) : (r+s) ;
```

Here, the larger of the two expressions out of (p + q) and (r + s) will be evaluated twice.

8.12.2 `include

This compiler directive is used to insert the contents of one source file into another file during compilation (i.e., for file inclusion). The result is as though the contents of the included source file appear in place of the `**include** compiler directive. The `**include** compiler directive can be used to include global or commonly used definitions and tasks without encapsulating repeated code within module boundaries. It helps modular code development and facilitates structured organization of the files in the design. It seriously contributes to the convenience in managing source code of a design.

The compiler directive is followed by the filename to be included. Only white space or a comment may appear on the same line as the `**include** compiler directive. A file included in the source using the `**include** compiler directive may contain other `**include** compiler directives. The number of nesting levels allowed might vary in different Verilog compilers. The following are valid examples of the `**include** compiler directive:

```
`include "lab3/sevenseg.v"
`include "myfile"
`include "myfile"   // including myfile. A comment or only white
                    // space allowed
```

If file `sub.v` contains

```
reg g;
initial
   g = 0;
```

and file `main.v` contains

```
module main;
`include sub.v
always @(posedge clk)
   g <= ~g;
endmodule;
```

it is equivalent to having one file as follows:

```
module main;
reg g;
initial
   g = 0;
always @(posedge clk)
   g <= ~g;
endmodule;
```

8.12.3 `ifdef

Verilog contains several compiler directives for conditional compilation. These directives allow one to optionally include lines of a Verilog HDL source description during compilation. Conditional compilation can also be accomplished with the conditional **generate** statement as in Section 8.10; however, the `**ifdef** compiler directive is convenient due to the existence of similar directives in high-level languages such as C. Many designers are already familiar with similar directives from C.

The `**ifdef** compiler directive is followed by a `text_macro_name` in the code. At compile time, the compiler checks for the definition of the `text_macro_name`. If the `text_macro_name` is defined, then the lines following the `**ifdef** directive are included. Otherwise, the statements in the `**else** clause are included. The end of the construct is marked with the `**endif**. As mentioned in Section 8.10, conditional compilation is useful for

(i) Selectively including behavioral, structural, or switch-level models as desired.
(ii) Selectively including different timing or structural information.
(iii) Selectively including different stimulus for different runs under different scenarios.
(iv) Selectively adapting the module functionality to similar but different needs from different customers.

The example that follows shows a simple usage of an `**ifdef** directive for selecting between a behavioral design and a structural design. If the identifier `behavioral` is defined, a continuous net assignment will be compiled; otherwise, a built-in gate primitive **and** will be instantiated. In **and** a1 (f,a,b), a1 is the instance name of the instantiated gate and **and** is the built-in gate primitive introduced in Section 8.4.

```
module selective_and (f, a, b);
  output f;
  input a, b;

  `ifdef behavioral
     wire f = a & b;
  `else
     and a1 (f,a,b);
  `endif

endmodule
```

8.13 File I/O Functions

The ability to handle files and text is very valuable while testing large Verilog designs. This section introduces file input and output in Verilog. Files are frequently used with test benches to provide a source of test data and to provide storage for test results. The format of the file I/O functions is based on the C standard I/O functions, such as fopen and fclose. A file can be opened for reading or writing using the **$fopen** function as shown in the following:

```
integer file_r, file_w;
file_r = $fopen("filename",r);    // Reading a file
file_w = $fopen("filename",w);    // Writing a file
```

The filename can be either a double-quoted string or a reg having filename information. $fopen can have multiple modes: – 'r' for reading, 'w' for writing, and 'a' for appending. If the file is successfully opened, it returns an integer number indicating the file number. If there is an error for opening the file, NULL value will be returned.

To close an opened file, the $fclose function can be used. If $fclose is used without any arguments, it closes all opened files. If an argument is specified, it will close only a file in which the descriptor is given.

```
$fclose(file_r);    // Closing only one file
$fclose();          // Closing all opened files
```

Verilog supports the following ways of handling a file:

$fopen/$fclose   open/close an existing file for reading or writing.

$feof   tests for end of file. If an end-of-file has been reached while reading a file, a non-zero value is returned; otherwise, a 0 is returned.

```
integer file;
reg eof;

eof = $feof(file);
```

8.13 File I/O Functions

$ferror returns the error status of a file. If an error has occurred while reading a file, $ferror returns a non-zero value; otherwise, it returns a 0.

```
integer file;
reg error;
error = $ferror(file);
```

$fgetc reads a single character from the specified file and returns it. If the end-of-file is reached, $fgetc returns EOF.

```
integer file, char;
char = $fgetc(file);
```

$fputc writes a single character to a file. It returns EOF if there is an error, 0 otherwise.

```
integer stream, r, char;
r = $fputc(stream, char);
```

$fscanf parses formatted text from a file according to the format and writes the results to args.

```
integer file, count, tmp_a;
count = $fscanf(file,"%d", tmp_a);
//     count = $fscanf(file, format, args);
```

$fprintf writes a formatted string to a file.

```
integer file, tmp_a, tmp_b, ret;
file = $fopen("test.log", w);
ret = $fprintf(file, "%d : %x", tmp_a, tmp_b);
```

$fread reads binary data from the file specified by the file descriptor into a register or into a memory.

```
integer rd, file;
reg rd_value;
rd = $fread(file, rd_value);
```

$fwrite writes binary data from a register to the file specified by the file descriptor.

```
integer file;
reg tmp_a, tmp_b;
tmp_a = 0;
tmp_b = 0;
file = $fopen("test.log");
$fwrite(file,"A=%d B=%d",tmp_a,tmp_b);
```

$readmemb
$readmemh To read data from a file and store it in memory, use the functions **$readmemb** and **$readmemh**. The **$readmemb** task reads binary data, and **$readmemh** reads hexadecimal data. Data has to be present in a text file.

```
$readmemb ("file", memory [,start_addr [,finish_addr]]) ;
$readmemh ("file", memory [,start_addr [,finish_addr]]) ;

$readmemb("file.bin", mem);
$readmemh("file.hex", mem);
```

Figure 8-21 gives a Verilog code that reads hexadecimal data from a file using **$readmemh**. Four 32-bit data are stored in a file "data.txt". **$readmemh** will read data and store it to the storage.

FIGURE 8-21: Verilog Code to Read Hexadecimal Data from a File Using $readmemh

```
module example_readmemh;
  reg [31:0] data [0:3];

  initial $readmemh("data.txt", data);

  integer i;

  initial begin
    $display("read hexa_data:");
    for (i=0; i < 4; i=i+1)
    $display("%d:%h",i,data[i]);
  end
endmodule
```

Figure 8-22 shows a Verilog code that read a file with commands, addresses, and data values. The code uses **$fopen** for accessing the file and **$fscanf** for parsing each line. In Figure 8-22, the **disable** statement is used to terminate the block named 'file_read'. The **disable** statement followed by a task name or block name will disable only tasks and named blocks. It cannot disable functions.

FIGURE 8-22: Verilog Code to Read and Parse a File Using $fopen

```
`define NULL 0
`define EOF 32'hffffffff

module file_read;
integer file, ret;
reg [31:0] r_w, addr, data;

initial
  begin : file_read

  file = $fopen("data", r);
  if (file == `NULL)
      disable file_read;
  while (!$feof(file))
      begin
      ret = $fscanf(file, "%s %h %h\n", r_w, addr, data);
      case (r_w)
```

```
            "rd":
                $display("READ mem[%h] => %h", addr, data);
            "wr":
                $display("WRITE mem[%h] <= %h", addr, data);
            default:
                $display("Unknown command '%s'", r_w);
        endcase
        end

    ret = $fclose(file);
    end
endmodule
```

8.14 Timing Checks

Verilog facilitates several timing checks—for example, to check whether setup and hold times of flip-flops are met. The timing check tasks start with a **$**, but they are not system tasks. Timing checks must not appear in procedural code. Timing checks must be within **specify** ..... **endspecify** blocks. Verilog provides several timing checks, but we will discuss only the few following examples:

$setup	This timing check displays a warning **if setup time is not met**. It needs a data event, a reference event, and a limit (the setup time) to be specified as in
	        $setup (data-event, reference_event, limit)
	For example,
	        $setup (D, posedge CLK, 10)

$hold	This timing check displays a warning **if hold time is not met**. It needs a reference event, a data event, and a limit (the hold time) to be specified as in
	        $hold (reference_event, data_event, limit)
	For example,
	        $hold (posedge CLK, D, 2)

$skew	This timing check displays a warning **if skew is above the limit**. It needs a reference event, a data event, and a limit to be specified as in
	        $skew (reference_event, data-event, limit)
	For example,
	        $skew (posedge CLK1, posedge CLK2, 4)

$width	This timing check displays a warning **if pulse width is shorter than limit**. It needs a data event, a reference event, and a limit, but the data event is not explicitly specified. The data event is derived from reference event as a reference event signal with opposite edge. The pulse width has to be greater than or equal to the limit in order to avoid a timing violation. The reference event must be an edge-triggered event. It is specified as in
	        $width (reference event, limit)
	For example,
	        $width (posedge CLK, 20)

Problems

8.1 Write a Verilog function to complete the following module, one that automatically counts the number of shifts. It needs to shift a word to the left until the most significant bit of the input word is equal to 1.

```
module word_shift(in_word, num_shifts);
  input [7:0] in_word;
  output [7:0] num_shifts;
  assign num_shifts = shift_count(in_word);
  //write the function below
  ….
  ….
  ….

endmodule
```

8.2 (a) Write a Verilog function that will create the 2's complement of an *N*-bit vector. Use a call of the form comp2(bit_vec, N), where *bit_vec* is the vector and *N* is the length of the vector. Do the complement on a bit-by-bit basis using a loop. You may declare N as a global parameter in the calling module.

(b) Write a Verilog module that will call the function in part (a).

8.3 The following Verilog code defines a function that computes the factorial of a number recursively. The **automatic** keyword is used to allow the function to be called recursively. Fill in the missing code that follows.

```
module factorial_test;

  function automatic integer factorial;
  input [31:0] num;

  /*
       Insert your code to implement factorial
  */

  endfunction

  integer result;

  initial begin
  /*
       Insert your code to compute the factorial of 9
  */
  $display("factorial = %d", result);
  end
endmodule
```

8.4 (a) *A* and *B* are bit vectors that represent unsigned binary numbers. Write a Verilog function that returns TRUE (1) if $A > B$. The function call should be of the form GT(A, B, N), where *N* is the length of the bit vectors. *Hint*: start comparing the most significant bits of *A* and *B* first and proceed from left to right. As soon as you find a pair of unequal bits, you can determine whether $A > B$. For example, if $A = 1011010$ and $B = 1010110$, you can determine that $A > B$ when you make the fourth comparison. You may declare N as a global parameter in the calling module.

(b) Write a Verilog module that will call the function in part (a).

8.5 What are the major differences between Verilog functions and Verilog tasks?

8.6 Write a Verilog module that could read numbers from a text file line by line and sort them by a user-defined task. The text file is named as "sort.txt", which contains 10 positive integers. The ten inputs should be stored in a global integer array that is declared as

$$\text{integer ARRAY [9:0];}$$

Write a task to sort this global array in ascending order. The sorted array can occupy the same global array. All the temporary space for sorting should be defined inside the task.

8.7 (a) Write a Verilog task that counts the number of 1s in an input bit vector that is up to N bits long ($N \leq 31$). The output should be 5 bits long. The task call should have the following form: (N, A, B) where A is the input and B is the output. You may declare N as a global parameter in the calling module.

(b) Write a Verilog module that will call the task in part (a).

8.8 Write a Verilog module that implements a 4-digit BCD adder with accumulator (see the block diagram that follows). If $LD = 1$, then the contents of *BCDacc* are replaced with *BCDacc* + *BCDin*.

Write a task that adds two BCD digits and a carry and returns a BCD digit and a carry. Call this task concurrently four times in your code.

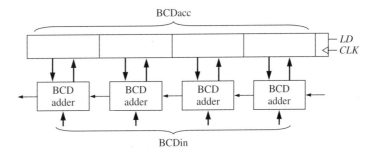

8.9 For the following Verilog code, list the values of B and C at each time a change occurs. Include all deltas and stop your listing when time > 8 ns. Assume that B is changed to 0110 at time 5 ns. Indicate the times at which task P1 is called.

```
module Q1(B,C);
  input[3:0] B;
  output[3:0] C;
  integer i;

  task P1;
  input[3:0] A;
  output reg[3:0] D;
  begin
      for(i = 1; i<= 3;i = i+1)
```

```
                    D[i] <= A[i-1];
            D[0] <= A[3];
        end
    endtask

    always @(B)
    begin
        P1(B,C);
        #1;
        P1(B,C);
    end

    assign C = B;

endmodule
```

8.10 Hamming codes are used for error detection and correction in communication and memory systems. Error detection and correction capability is incorporated in these codes by inserting extra bits into the data word. Addition of one parity bit can detect odd number of bit flips, but no error correction is possible with one parity bit. A (7,4) Hamming code has 4 bits of data but 7 bits in total, including the 3 parity bits. It can detect two errors and correct one error. This code can be constructed as follows: If we denote data bits as $d_4 d_3 d_2 d_1$, the encoded code word would be $d_4 d_3 d_2 p_4 d_1 p_2 p_1$, where p_4, p_2, and p_1 are the added parity bits. These bits must satisfy the following conditions for even parity:

$$p_4 = d_2 \text{ XOR } d_3 \text{ XOR } d_4; \quad\quad\quad\quad\quad\quad (1)$$
$$p_2 = d_1 \text{ XOR } d_3 \text{ XOR } d_4; \quad\quad\quad\quad\quad\quad (2)$$
$$p_1 = d_1 \text{ XOR } d_2 \text{ XOR } d_4; \quad\quad\quad\quad\quad\quad (3)$$

When the 7 bits are received/decoded, an error syndrome $S_3 S_2 S_1$ is calculated as follows:

$$p_4 \text{ XOR } d_2 \text{ XOR } d_3 \text{ XOR } d_4 = S_3; \quad\quad\quad\quad\quad\quad (4)$$
$$p_2 \text{ XOR } d_1 \text{ XOR } d_3 \text{ XOR } d_4 = S_2; \quad\quad\quad\quad\quad\quad (5)$$
$$p_1 \text{ XOR } d_1 \text{ XOR } d_2 \text{ XOR } d_4 = S_1; \quad\quad\quad\quad\quad\quad (6)$$

The syndrome indicates which bit is wrong. For example, if the syndrome is 110, it indicates that bit 6 from right end (i.e., d3) has flipped. If $S_3 S_2 S_1$ is 000, there is no error.

(a) Is there any error in the code word 0110111? If yes, which bit? What was the original data? What must be the corrected code word?

(b) How will these 6 equations get modified for odd parity? Write the 6 equations for odd parity.

(c) Write a Verilog module for error detection without using tasks. The inputs to the module are the 7-bit encoded data word and the type of parity, and the output is the syndrome. The type of parity is encoded as 0 for odd parity and 1 for even parity.

```
module  error_detector(data, PARITY, Syndrome)
{

}
```

(d) Modify the Verilog module in **(c)** to use Verilog tasks. Write required Verilog task(s) that will generate the error syndrome given the code word. The input to the task is data, and output is the syndrome. You have to call necessary task(s) for each type of parity.

```
task   error_p(data, Syndrome)
{

}
```

8.11 Write a Verilog model that uses only built-in primitives to implement the following circuit.

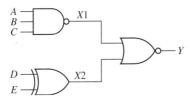

Rise delay of NAND gate is 15 ns.
Rise delay of XOR gate is 14 ns, fall delay of XOR gate is 16 ns.
Rise delay of NOR gate is 12 ns, and fall delay of NOR gate is 14 ns.

8.12 Create a user-defined primitive (UDP) for a J-K flip-flop with asynchronous clear and preset using "?" if needed

8.13 Write a Verilog user-defined primitive (UDP) for generating odd parity for 4-bit input data.

8.14 A Verilog module has inputs A and B and outputs C and D. A and B are initially high. Whenever A goes low, C will go high 5 ns later, and if A changes again, C will change 5 ns later. D will change if B does not change for 3 ns after A changes. *Note:* The timing checks should be done inside: specify .... Endspecify

(a) Write the Verilog module with an always block that determines the outputs C and D.
(b) Write another always block to check that B is stable 2 ns before and 1 ns after A goes high. The always block should also report an error if B goes low for a time interval less than 10 ns.

8.15 Write a Verilog module of an address decoder/address match detector. One input to the address decoder is an 8-bit address, *addr*. The second input is the 6-bit vector check. The address decoder will output $Sel = 1$ if the upper 6 bits of the 8-bit address match the 6-bit check vector. For example, if *addr* = 10001010 and *check* = 1000XX, then $Sel = 1$. Only the 6 leftmost bits of *addr* will be compared; the remaining bits are ignored. An X in the check vector is treated as a don't care.

8.16 Write a Verilog module for one flip-flop in a 74HC374 (octal D-type flip-flop with 3-state outputs. Given the D-type flip-flop setup time = 15 ns, hold time = 5 ns, pulse width time = 15 ns). Assume that all logic values are $x, 0, 1$, or z. Check setup, hold, and pulse-width specs using monitor statements. Unless the output is z, the output should be x if CLK or OC is x, or if an x has been stored in the flip-flop. *Note*: The timing checks should be done inside: specify .... Endspecify

8.17 Write a Verilog function to compare two 8-bit vectors to determine whether they are equal. Report an error if any bit of either vector is not 0, 1, or z. The function call should pass only the vectors. The function should return TRUE (1) if the vectors are equal, else FALSE (0). All bits including zs should match (i.e., z only matches z).

8.18 In the following code, all signals are 1-bit. Draw a logic diagram that corresponds to the code. Assume that a D flip-flop with CE is available.

```
assign F = (EA == 1) ? A: ((EB == 1)? B : Z);
always @(posedge CLK)
begin
 if(Ld == 1)
        A <= B;
 if(Cm == 1)
        A <= ~A;
end
```

8.19 Design a memory-test system to test the first 256 bytes of a static RAM memory. The system consists of simple controller, an 8-bit counter, a comparator, and a memory as shown subsequently. The counter is connected to both the address and data (IO) bus so that 0 will be written to address 0, 1 to address 1, 2 to address 2, ..., and 255 to address 255. Then the data will be read back from address 0, address 1, ..., address 255 and compared with the address. If the data does not match, the controller goes to the fail state as soon as a mismatch is detected; otherwise, it goes to a pass state after all 256 locations have been matched. Assume that $OE\_b = 0$ and $CS\_b = 0$.

(a) Draw an SM chart or a state graph for the controller (five states). Assume that the clock period is long enough so that one word can be read every clock period.

(b) Write Verilog code for the memory-test system.

8.20 Design a memory-test system similar to that of Problem 8.18, except write a checkerboard pattern into memory (01010101 into address 0, 10101010 into address 1, etc.). Draw the block diagram and the SM chart.

8.21 Design a memory tester that verifies the correct operation of a 6116 static RAM (Figure 8-15). The tester should store a checkerboard pattern (alternating 0s and 1s

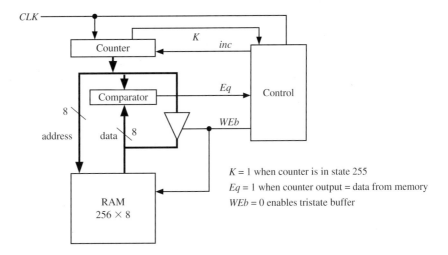

in the even addresses, and alternating 1s and 0s in the odd addresses) in all memory locations and then read it back. The tester should then repeat the test using the reverse pattern.

(a) Draw a block diagram of the memory tester. Show and explain all control signals.
(b) Draw an SM chart or state graph for the control unit. Use a simple RAM model and disregard timing.
(c) Write Verilog code for the tester and use a test bench to verify its operation.

8.22 A clocked T flip-flop has propagation delays from the rising edge of *CLK* to the changes in Q and Q' as follows: if Q (or Q') changes to 1, t_{plh} = 8 ns, and if Q (or Q') changes to 0, t_{phl} = 10 ns. The minimum clock pulse width is t_{ck} = 15 ns, the setup time for the T input is t_{su} = 4 ns, and the hold time is t_h = 2 ns. Write a Verilog model for the flip-flop that includes the propagation delay and that reports if any timing specification is violated. Write the model using parameters with default values. *Note:* The timing checks should be done inside: specify .... Endspecify

8.23 (a) Write a model for a D flip-flop with a direct clear input. Use the following timing parameters: t_{plh}(10 ns), t_{phl}(10 ns), t_{su}(5 ns), t_h(3 ns), and t_{cmin}(20). The minimum allowable clock period is t_{cmin}. Report appropriate errors if timing violations occur. *Note:* The timing checks should be done inside: specify .... Endspecify
(b) Write a test bench to test your model. Include tests for every error condition.

8.24 Write a Verilog model for an *N*-bit comparator using an iterative circuit. In the module, use the parameter *N* to define the length of the input bit vectors *A* and *B*. The comparator outputs should be $EQ = 1$ if $A = B$, and $GT = 1$ if $A > B$. Use a for loop to do the comparison on a bit-by-bit basis, starting with the high-order bits. Even though the comparison is done on a bit-by-bit basis, the final values of *EQ* and *GT* apply to *A* and *B* as a whole.

8.25 Four RAM memories are connected to CPU busses as shown here. Assume that the following RAM component is available.

```
module SRAM(cs-b, we-b, oe-b, address, data);
  input cs-b,we-b,oe-b;
  input[14:0] address;
  inout[7:0] data;
endmodule
```

Write a Verilog code segment that will connect the four RAMs to the busses. Use a **generate** statement and a named association.

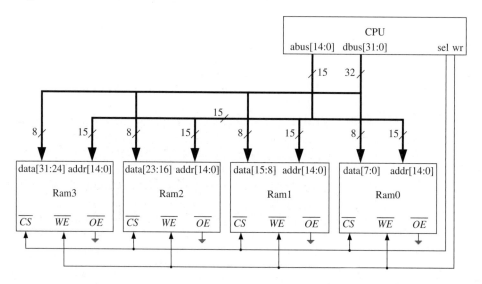

8.26 Write structural Verilog code for a module that is an *N*-bit serial-in, serial-out right-shift register. Inputs to the shift register are bit signals: *SI* (serial input), *Sh* (shift enable), and *CLK*. Your module should have a **generate** statement. Assume that a component for a D flip-flop with clock enable (*CE*) is available.

8.27 Write structural Verilog code for a module that has two inputs: an *N*-bit vector *A*, and a control signal *B* (1 bit). The module has an *N*-bit output vector, *C*. When $B = 1$, $C <= A$. When $B = 0$, C is all 0s. Use parameter to specify the value of *N* (default = 4). To implement the logic, use a **generate** statement that instantiates *N* 2-input AND gates.

8.28 The structural Verilog code that follows is a 2-input NOR gate with the rise/fall time defined as parameters.

```
module NOR2(a, b, c);
  parameter Trise = 3;
  parameter Tfall = 2;
```

```verilog
    parameter load = 1;

    input a, b;
    output reg c;

    wire nor_value;

    assign nor_value = ~(a | b);

    always @(nor_value)
    begin
       if(nor_value == 1'b1)
           #(Trise + 3*load) c = 1'b1;
       else
           #(Tfall + 3*load) c = 1'b0;
    end
endmodule

module NOR2_TEST(in1, in2, in3, in4, out1, out2);
input in1, in2, in3, in4;
output out1, out2;

NOR2 U1 (in1, in2, out1);

/*
                place for your new code
*/

endmodule
```

(a) Instantiate NOR2 (U2) by using the parameter map method (Trise = 5, Tfall = 4, load = 3).
(b) Use the **defparam** to pre-define the timing values and instantiate NOR2 (U3), which has rise time = 4, fall time = 3, and load = 2.
(c) What are the rise time, fall time, and fan-out time of U1?
(d) What are the rise and fall delays of NOR2 (U2)?

8.29 Create a 4 × 4 array multiplier using **generate** statements. Use full adder, half adder, and AND gate components as in Chapter 4.

8.30 B is an integer array declared as integer B [4:0]. Write a Verilog code segment that will read five integers in a line of text from a file named "FILE2" and then write the five integers into array B.

8.31 Write a task that has an integer signal and a file name as parameters. Each line of the file contains a delay value and an integer. The task reads a line from the file, waits for the delay time, assigns the integer value to the signal, and then reads the next line. The task should return when end-of-file is reached.

8.32 Write a task that logs the history of values of a bit vector signal to a text file. Each time the signal changes, write the current time and signal value to the file. Verilog has a built-in funtion called $time that could display the current simulation time.

8.33 Indicate the final decimal value of the regA and regB after executing the following instructions.

```
reg[8:0] regA;
reg[8:0] regB;

regA <= $unsigned -9;
regB <= $signed 9'b100000000;
```

8.34 (a) Complete the following code by defining a macro *Sum*. Use parameters to initialize the numbers to 9 and 11.

```
always @(posedge clk)
begin
    $display("The sum of %d and %d is %d", A, B, Sum(A,B));
    $stop;
    $display("This sentence should be on the screen");
end
```

(b) What will the completed code print on the screen when it executes?

8.35 What is the difference between **$stop** and **$finish**?

8.36 Based on the tri-state bus circuit (Figure 1-56), fill out the table that follows by using 4-valued logic. The values stored in RegA, RegB, and RegC are 8'd5, 8'd10, and 8'd15, respectively. Assume Eni is equal to 0.

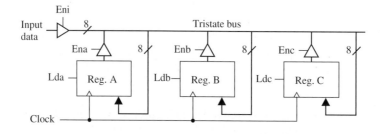

Time	Enc	Ena	Enb	Bus
0	Z	Z	Z	Z
2	1	0	0	
4	0	1	0	
6	0	0	1	
8	1	1	1	
10	1	1	1	

CHAPTER 9

Design of a RISC Microprocessor

A microprocessor is an example of a complex digital system. In this chapter, we will describe a microprocessor from MIPS Technologies, the MIPS R2000, and implement a subset of the MIPS processor's instruction set architecture (ISA). The term instruction set architecture denotes the instructions that are visible to the assembly language programmer—the number of registers, the addressing modes, and the operations (opcodes) available in the particular processor. An introduction to the RISC philosophy is presented first followed by a description of the MIPS ISA. The arithmetic, memory access, and control transfer instructions of the MIPS are presented along with a design to implement a subset of the ISA. A synthesizable Verilog model for the MIPS subset is then presented. Use of a test bench for testing the design is illustrated.

9.1 The RISC Philosophy

Many early microprocessors, such as the Intel 8086 and Motorola 68000, incorporated a variety of powerful instructions and addressing modes. A natural consequence of this was the complexity of the design, especially the control unit complexity. These microprocessors included a microprogrammed control unit because it was difficult to design and debug a hardwired control unit for such complex digital systems. (See Chapter 5 for tradeoffs between microprogramming and hardwiring.)

The value of simplicity became clearer in the late 1970s and early 1980s. The result was the advent of **RISC** or the reduced instruction set computing philosophy. RISC processors are a type of microprocessor that uses a small and simple set of instructions rather than a variety of complex instructions and versatile addressing modes. The first RISC projects came from IBM, Stanford, and UC–Berkeley in the late 1970s and early 1980s. The IBM 801, Stanford MIPS, and Berkeley RISC 1 and 2 were all designed with a similar philosophy, which has become known as RISC. In contrast, earlier processors such as the Intel 8086 and the Motorola 68000/68020 started to be called **CISC** or complex instruction set computing processors after the advent of the RISC philosophy. The first generation of RISC processors included MIPS R2000 from MIPS, SPARC from Sun Microsystems, and RS/6000 from IBM. The IBM RS/6000 has evolved into the POWERPC and POWER architecture.

> **MIPS:** MIPS Technologies is a computer manufacturer that has designed and sold several RISC microprocessors starting with the MIPS R2000 processor in the 1980's. The term MIPS was commonly known to computer designers as a performance metric, the Millions of Instructions Per Second metric. The MIPS in the name of the MIPS Corporation however does not stand for that. The original acronym stood for Microprocessor without Interlocked Pipelined Stages. In a pipelined processor, there must exist a mechanism to enforce dependencies between instructions. So, if one instruction needs the result of the previous one, the second instruction should not proceed until the first instruction's result is ready. Enforcing of this type of dependency is usually done by hardware. The first MIPS processor, however, did not have hardware interlocks. It reflected the early RISC idealism that anything that can be done in software should be done in software. In early MIPS processors, pipeline interlocks were implemented by software by inserting the appropriate number of nop (no operation) instructions between the dependent instructions.

Certain design features have been characteristic of most RISC processors:

- **Uniform instruction length:** All instructions have the same length—32 bits. This is in sharp contrast to previous microprocessors that contained instructions as small as a byte and as large as 16 bytes.
- **Few instruction formats:** The RISC ISAs emphasized having as few instruction formats as possible and encoding the different fields in the instruction as uniformly as possible. This greatly simplifies instruction decoding.
- **Few addressing modes:** Most RISC processors support only one or two memory-addressing modes. Addressing modes offer different ways an instruction can indicate the memory address to be accessed. Examples are direct addressing, immediate addressing, base-plus-offset addressing, based ndexed addressing, indirect addressing, among others. Many RISC processors support only one addressing mode. Typically, this addressing mode specifies addresses with a register and an offset.
- **Large number of registers:** The RISC design philosophy generally incorporates a larger number of registers to prevent the loss of performance by frequently accessing memory. RISC processors are also often called **register-register architectures**. All arithmetic operations operate on register operands. CISC architectures typically contained 8 or 12 registers, whereas most RISC architectures contained 32 registers.
- **Load/store architecture:** RISC architectures are also called load/store architectures. The key idea is the absence of arithmetic instructions that directly operate on memory operands (i.e., arithmetic instructions that take one or more operands from memory). The only instructions that are allowed to access memory are load and store instructions. The load instructions bring the data to registers, and arithmetic operations operate on the data in the registers. These

architectures are also called register-register architectures because input and output operands for computation operations are in registers. A load/store architecture inherently means that it is also a register-register architecture.
- **No implied operands or side effects:** Most earlier ISAs contained implied operands, such as accumulators or implied results (**side effects**), such as flags (condition codes), to indicate such conditions as carry, overflow, negative, and the like. Implied operands and side effects can cause difficulties and challenges in pipelined and parallel implementations. A principle behind RISC architectures is to have minimal implied operands/operations and side effects.

The RISC philosophy has been to adhere to these features and embrace simplicity of design. The terms RISC and CISC are used very often as antonyms, but perhaps it is not clear how "reduced" is the opposite of "complex". It is not even clear that RISC processors have a smaller instruction set than prior CISC processors. Some RISC ISAs have 100+ instructions whereas some CISC processors have only 80 instructions. However, these 80 CISC instructions could assume several addressing modes. A CISC processor, the Motorola 68020 supported up to 20 different addressing modes. Considering all the different forms an instruction could take, most RISC ISAs do contain fewer instructions than CISC ISAs. The key point in the RISC philosophy has been its emphasis on simplicity: having only simple basic operations, simplifying instruction formats, reducing the number of addressing modes, and eliminating complex operations. This computing paradigm could have been called simple instruction set computing (SISC), but SISC sounds like CISC.

CISC architectures are not without advantages. Instruction encoding is denser in CISC than in RISC. The fixed instruction width in RISC leads to using more bits than necessary for some instructions. In CISC ISAs, every instruction is just as wide as it needs to be. Hence, code size is smaller in the CISC case. If instruction memory size has to be kept small, as in embedded environments, CISC ISAs have an advantage.

Most modern microprocessors have RISC ISAs. Some examples are the MIPS R12000, Sun UltraSPARC, IBM PowerPC, HP PA-RISC, among others. The Pentium 4 or the x86 processors in general are examples of modern processors with a CISC ISA. (The term x86 is used to refer to the different processors that have used the ISA that originated with Intel 8086. This list includes Intel 8086, 80286, 80386, 80486; Pentium and AMD K5; K6; Opteron; and others.)

Whether RISC or CISC is better was a topic of intense debate in the 1980s and 1990s. It has now become understood that decoding and processing is easy with a RISC ISA; however, it also has been shown that hardware can translate complex CISC-style instructions into RISC-style instructions and process them. Pentium 4 and other current high-end x86 processors have a CISC ISA; however, they use hardware to convert each CISC instruction to one or more RISC-type instructions or microoperations (called **uops** or **R-ops**) that can be pipelined easily. In spite of all the arguments that have taken place, there is no disagreement about the ease of implementation of RISC ISAs.

The MIPS instruction set architecture is one of the earliest RISC ISAs and one of the simplest ones. It has only one memory-addressing mode. In contrast, another early RISC architecture, the SPARC, has two memory-addressing modes.

The MIPS ISA is described in detail in the book *MIPS RISC Architecture* by Gerry Kane. The MIPS architecture is described in more detail in the book *Computer Organization and Architecture: A Hardware Software Interface,* by David A. Patterson and John L. Hennessy. A very concise description of the MIPS ISA is presented in the following section.

> **The Single Instruction Computer:** It has also been shown in the past that a microprocessor can be designed with a single instruction. This single instruction should be able to access memory operands, do arithmetic operations and do control transfers. A subtract instruction that operates on memory operands, writes results to memory, and branches to an address if the result of the subtraction is negative can be used to write any program. Will such a single instruction microprocessor qualify to be called a RISC? Probably not. Although it is a single instruction computer, it is not a register-register architecture, and it is not an ISA that supports simple operations. We would classify it under a CISC category since every instruction is a complex branch and memory access instruction. More discussion of such a computer and illustration of a program written using the single instruction can be found in [Patterson/Hennessey].

9.2 The MIPS ISA

The MIPS instruction set architecture (ISA) contains a set of simple arithmetic, logical, memory-access, branch, and jump instructions. The architecture emphasizes simplicity and excludes instructions that could possibly take longer than the most common instructions.

There are 32 general-purpose registers in the MIPS architecture. Each register is 32 bits wide. The MIPS registers are often referred to as $0, $1, $2, ..., and $31. The MIPS instructions follow a **three-address format** for ALU instructions, meaning they specify two source addresses and one destination address. For example, an add instruction that adds registers $3 and $4 and writes the result to $5 is written as

```
add $5, $3, $4
```

Each group of instructions is described in the following subsections.

Arithmetic Instructions

The MIPS ISA contains instructions for performing addition, subtraction, multiplication, and division of integers. The various arithmetic instructions are summarized in Table 9-1. Addition and subtraction of signed or unsigned quantities can be accomplished using the *add, addu, sub,* and *subu* instructions. Signed arithmetic instructions detect overflows, whereas unsigned arithmetic instructions do not detect overflows.

For example the instruction

```
sub $5, $3, $4
```

will subtract the value in register $4 from the value in register $3 and write the result to register $5. It is a signed instruction, and overflow will be detected. When an overflow is detected, it is handled as an exception. The address of the instruction that caused the exception is saved, and control is transferred to the operating system, which handles the exception.

TABLE 9-1: Arithmetic Instructions in the MIPS ISA

Instruction	Example	Meaning	Comments
Add	add $s1, $s2, $s3	$s1 = $s2 + $s3	Overflow detected
Subtract	sub $s1, $s2, $s3	$s1 = $s2 − $s3	Overflow detected
Add immediate	addi $s1, $s2, k	$s1 = $s2 + k	k is a 16 bit constant; sign extended and added; 2's complement overflow detected
Add unsigned	addu $s1, $s2, $s3	$s1 = $s2 + $s3	Overflow not detected
Subtract unsigned	subu $s1, $s2, $s3	$s1 = $s2 − $s3	Overflow not detected
Add immediate unsigned	addiu $s1, $s2, k	$s1 = $s2 + k	k is an unsigned 16-bit constant;
Move from co-processor register	mfc0 $s1, $epc	$s1 = $epc	epc is Exception Program Counter
Multiply	mult $s2, $s3	Hi, Lo = $s2 × $s3	64-bit signed product in Hi, Lo
Multiply unsigned	multu $s2, $s3	Hi, Lo = $s2 × $s3	64-bit unsigned product in Hi, Lo
Divide	div $s2, $s3	Lo = $s2 / $s3 Hi = $s2 mod $s3	Lo = quotient, Hi = remainder
Divide unsigned	divu $s2, $s3	Lo = $s2 / $s3 Hi = $s2 mod $s3	Unsigned quotient and remainder
Move from Hi	mfhi $s1	$s1 = Hi	Copy Hi to $s1
Move from Lo	mflo $s1	$s1 = Lo	Copy Lo to $s1

Addition of the contents of a register with an immediate value specified in the instruction can be done using the *addi* and *addiu* instructions. The instruction

```
addi $5, $3, 400
```

will add the value in register $3 to the immediate constant 400 and write the result to register $5. The immediate constant is sign extended before the addition. The action of the *addiu* instruction is similar, except that the *addiu* instruction never causes an overflow exception.

Multiplication of two 32-bit quantities results in a 64-bit result that cannot be contained in one MIPS register. Hence, two special registers called Hi and Lo are used by the MIPS processors to hold the products. Use of implied Hi and Lo registers, is certainly a deviation from the RISC philosophy. Table 9-1 illustrates the multiply and divide instructions in the MIPS ISA along with the use of the Hi and Lo registers. The use of these special registers also necessitate special instructions to transfer data from these registers to the required destination registers. The *mfhi* and *mflo* accomplish this task.

Logical Instructions

The logical instructions in the MIPS ISA are presented in Table 9-2. The MIPS ISA contains logical instructions for performing bit-wise AND and OR of register contents. The *and* and *or* instructions perform these operations for register operands. The *andi* and *ori* instructions can be used when one operand is in a register and the other operand is an immediate constant. The *sll* and *srl* instructions are provided to perform logical left and right shifts of register contents (with zero fill). The number of shifts is encoded as an immediate value in the instruction.

TABLE 9-2: Logical Instructions in the MIPS ISA

Instruction	Example		Meaning	Comments
and	and	$s1, $s2, $s3	$s1 = $s2 AND $s3	logical AND
or	or	$s1, $s2, $s3	$s1 = $s2 OR $s3	logical OR
and immediate	andi	$s1, $s2, k	$s1 = $s2 AND k	k is a 16-bit constant; k is 0-extended first.
or immediate	ori	$s1, $s2, k	$s1 = $s2 OR k	k is a 16-bit constant; k is 0-extended first
shift left logical	sll	$s1, $s2, k	$s1 = $s2 << k	Shift Left by 5-bit constant k
shift right logical	srl	$s1, $s2, k	$s1 = $s2 >> k	Shift right by 5-bit constant k

Memory Access Instructions

The only instructions in the MIPS ISA to access the memory are load and store instructions. A load instruction transfers data from memory to the specified register. A store instruction transfers data from a register to the specified memory address.

The RISC researchers investigated the number of addressing modes that are needed to efficiently code high-level language programs such as those in C. They concluded that one addressing mode with a base register and an offset was sufficient. The only addressing mode that is supported for memory instructions in the MIPS processor is this addressing mode with one base register and an offset. The memory address is computed as the sum of the register contents and the offset specified in the instruction.

Consider the MIPS load instruction

```
lw $5, 100($4)
```

This instruction computes the memory address as the sum of the value in register $4 and the offset 100. So if register $4 contains 4000, the effective address is 4100. The content of memory location 4100 is moved to register $5 in the processor. In the case of sw $6, 100($8), the content of register $6 is written to the memory location pointed to by the sum of the contents of register $8 and 100.

A group of 32 bits is called a word in the MIPS world. MIPS has instructions to load and store words, half-words (16 bits) or bytes (8 bits). These instructions are summarized in Table 9-3.

TABLE 9-3: Memory Access Instructions in the MIPS ISA

Instruction	Example	Meaning	Comments
load word	lw $s1, k($s2)	$s1 = Memory[$s2 + k]	Read a word (32 bits) from memory; memory address = Register content + k ; k is 16 bit offset
store word	sw $s1, k($s2)	Memory[$s2 + k] = $s1	Write a word (32 bits) to memory; memory address = Register content + k; k is 16 bit offset
load half word	lh $s1, k($s2)	$s1 = Memory[$s2 + k]	Read a word (16 bits) from memory; sign-extend and load into register
store half word	sh $s1, k($s2)	Memory[$s2 + k] = $s1	Write a half-word (16 bits) to memory
load byte	lb $s1, k($s2)	$s1 = Memory[$s2 + k]	Read byte from memory; sign extend and load to register
store byte	sb $s1, k($s2)	Memory[$s2 + k] = $s1	Write byte to memory
load byte unsigned	lbu $s1, k($s2)	$s1 = Memory[$s2 + k]	Read byte from memory; byte is 0-extended
load upper immediate	lui $s1, k	$s1 = k * 2^{16}	Loads constant k to upper 16 bits of register

Control Transfer Instructions

Typically program execution proceeds in a sequential fashion, but loops, procedures, functions, and sub-routines change the program control flow. A microprocessor needs branch and jump instructions in order to accomplish transfer of control whenever non-sequential control flow is required. The MIPS ISA includes two conditional branch instructions, *branch on equal (beq)* and *branch on not equal (bne)* as illustrated in Table 9-4.

The MIPS instruction

```
beq $5, $4, 25
```

will compare the contents of $5 and $4 and branch to PC + 4 + 100 if $4 and $5 are equal. The constant offset provided in the branch instruction is specified in terms of the number of instructions from the current PC (program counter). MIPS

TABLE 9-4: Conditional Control–Related Instructions in the MIPS ISA

Instruction	Example	Meaning	Comments
branch on equal	beq $s1, $s2, k	If ($s1 == $s2) go to PC + 4 + k*4	Branch if registers are equal; PC-relative branch; Target = PC+4+Offset*4; k is sign-extended
branch on not equal	bne $s1, $s2, k	If ($s1 != $s2) go to PC + 4 + k*4	Branch if registers are not equal; PC relative branch; Target = PC+4+Offset*4; k is sign-extended
set on less than	slt $s1, $s2, $s3	If ($s2 < $s3) $s1 = 1; else $s1 = 0;	Compare and Set (2's complement)
set on less than immediate	slti $s1, $s2, k	If ($s2 < k) $s1 = 1; else $s1 = 0;	Compare and Set; k is 16-bit constant; sign-extended and compared
set on less than unsigned	sltu $s1, $s2, $s3	If ($s2 < $s3) $s1 = 1; else $s1 = 0;	Compare and Set; natural numbers
set on less than immediate unsigned	sltiu $s1, $s2, k	If ($s2 < k) $s1 = 1; else $s1 = 0;	Compare and set ; natural numbers; K the16-bit constant is sign extended; no overflow

uses byte addressing; hence, the offset in words is multiplied by 4 to get the offset in bytes. The program counter is assumed to point to the next instruction at PC + 4 already; hence, the target address is computed as PC + 4 + 4 * offset. The offset is 16 bits long; however, one bit is used for sign. Branching is thus possible to only +/− 32K.

Having only two conditional branch instructions is in contrast to CISC processors, which provide branch on less than, branch on greater than, branch on higher than, branch on lower than, branch on carry, branch on overflow, branch on negative, and several other such conditional branch instructions. The MIPS philosophy was that only two conditional branch instructions are necessary and that checking of other conditions can be accomplished using separate instructions. In order to facilitate checking of less than and greater than, MIPS ISA provides the set on less than *(slt)* instructions. These are explicit compare instructions that will set an explicit destination register to 1 or 0 depending on the results of the compare. The *slt* instruction is used along with a *bne* or *beq* instruction to create the effect of branch on less than, branch on greater than, and so forth. These instructions are used for implementing **loop** and **if-then-else** statements from high-level languages.

The MIPS ISA also includes three unconditional jump instructions as illustrated in Table 9-5. These instructions are used for implementing function and procedure calls as well as returns.

TABLE 9-5: Unconditional Control Transfer Instructions in the MIPS ISA

Instruction	Example	Meaning	Comments
jump	j addr	Go to addr*4; i.e., PC <= addr*4	Target address = Imm offset * 4; addr is 26-bits
jump register	jr $reg	Go to $reg; i.e., PC <= $reg	$reg contains 32-bit target address
jump and link	jal addr	return address = PC + 4; go to addr*4	Used for procedure call. Return address saved in the link register $31

The **jump (j)** instruction transfers control to the address specified in the instruction. Since the MIPS instruction is 32 bits wide, the number of bits available for encoding the address will be (32 – number of opcode bits). In the MIPS, the opcode consumes 6 bits; therefore, only 26 bits are available for the address in the jump instruction. In order to increase the range of addresses to which control can be transferred, MIPS designers consider the specified address as a word address (instead of a byte address) and multiply the specified address by 4 to obtain the resulting byte address.

The **jump register (jr)** instruction is an **indirect jump**. In contrast, the jump instruction *j* described in the previous paragraph is called a **direct** jump because the target address is directly specified in the instruction itself. In the case of the *jr* instruction, the target address is in the register. This type of branch instruction is very useful for implementing case statements from high-level languages.

The **jump and link (jal)** instruction is specifically designed for procedure calls. It computes the target address from the offset specified in the instruction, but in addition to transferring control to that address, it also saves the return address in link register $31. The return address means the address control should return to after the subroutine or procedure call is completed. The return address is equal to the current PC + 4, since every instruction is four bytes wide and PC + 4 is the address of the instruction following the current instruction (the *jal* instruction).

We have described the major classes of instructions in the MIPS ISA. In order to become familiar with the instructions, let us practice some assembly language programming.

Example

Write a MIPS assembly language program for the following program which adds two arrays x(i) and y(i), each of which has 100 elements.

```
for i=0; i<100; i++     ; repeat 100 times
y(i) = x(i) + y(i)      ; add ith element of the arrays
```

Assume that the x and y arrays start at locations 4000 and 8000 (decimal).

Answer:

```
            andi    $3, $3, 0       ; initialize loop counter $3 to 0
            andi    $2, $2, 0       ; clear register for loop bound
            addi    $2, $2, 400     ; loop bound
$label:     lw      $15, 4000($3)   ; load x(i) to R15
```

```
lw    $14, 8000($3)      ; load y(i) to R14
add   $24, $15, $14      ; x(i) + y(i)
sw    $24, 8000($3)      ; save new y(i)
addi  $3, $3, 4          ; update address register,
                         ;   address= address + 4
bne   $3, $2, $label     ; check if loop counter=loop
                         ;   bound
```

Several microprocessors with the MIPS ISA have been designed since the MIPS R2000 was designed during the 1980s. In those days, the main processor could not integrate the floating-point unit. Hence, the floating-point units were implemented as a math coprocessor, the MIPS R2010. Currently, the floating-point unit is integrated with the main CPU. The MIPS R2000 was followed by the MIPS R3000, R4000, R8000, R10000, R12000. and R14000. They all have the MIPS ISA but have different implementations with different levels of pipelining and different techniques to obtain high performance.

9.3 MIPS Instruction Encoding

Adhering to the RISC philosophy, all instructions in the MIPS processor have the same width, 32 bits. In a move towards simplicity, there are only three different instruction formats for the MIPS instructions. The three formats are called **R-format, I-format,** and **J-format**, as illustrated in Table 9-6.

TABLE 9-6: Instruction Formats in the MIPS ISA

Format	Fields						Comments
	6 bits 31..26	5 bits 25..21	5 bits 20..16	5 bits 15..11	5 bits 10..6	6 bits 5..0	All MIPS instruction are 32 bits.
R-format	opcode	rs	rt	rd	shamt	F_Code (funct)	ALU instructions except immediate, Jump Register (JR)
I-format	opcode	rs	rt	address/immediate			Load, store, Immediate ALU, beq, bne
J-format	opcode	target address					Jump (J)

The **R-format** is primarily for ALU instructions, which require three operands. These ALU instructions have two source operands (input registers) and one destination address (result register) to be specified. The jump register instruction (jr) also uses this format. The instruction consists of six fields, the first of which is the 6-bit **opcode** field. The opcode field is followed by the three register fields *rs, rt,* and *rd*, each of which takes 5 bits. The first two are the source register fields and the third is the destination register field. The next field is called shift amount (**shamt**) field,

9.3 MIPS Instruction Encoding

TABLE 9-7: Instruction Encoding for the MIPS Instructions

Name	Format	Bits 31..26	Bits 25..21	Bits 20-16	Bits 15-11	Bits 10-6	Bits 5..0	Instruction (operation dest, src1, src2)
Add	R	0	2	3	1	0	32	add $1, $2, $s3
Sub	R	0	2	3	1	0	34	sub $1, $2, $3
addi	I	8	2	1	100			addi $1, $2, 100
addu	R	0	2	3	1	0	33	addu $1, $2, $3
subu	R	0	2	3	1	0	35	subu $1, $2, $3
addiu	I	9	2	1	100			addiu $1, $2, 100
mfc0	R	16	0	1	14	0	0	mfc0 $1, $epc
mult	R	0	2	3	0	0	24	mult $2, $3
multu	R	0	2	3	0	0	25	multu $2, $3
div	R	0	2	3	0	0	26	div $2, $3
divu	R	0	2	3	0	0	27	divu $2, $3
mfhi	R	0	0	0	1	0	16	mfhi $1
mflo	R	0	0	0	1	0	18	mflo $1
and	R	0	2	3	1	0	36	and $1, $2, $3
or	R	0	22	3	1	0	37	or $1, $2, $3
andi	I	12	2	1	100			andi $1, $2, 100
ori	I	13	2	1	100			ori $1, $2, 100
sll	R	0	0	2	1	10	0	sll $1, $2, 10
srl	R	0	0	2	1	10	2	srl $1, $2, 10
lw	I	35	2	1	100			lw $1, 100($2)
sw	I	43	2	1	100			sw $1, 100($2)
lui	I	15	0	1	100			lui $1, 100
beq	I	4	1	2	25			beq $1, $2, 100
bne	I	5	1	2	25			bne $1, $2, 100
slt	R	0	2	3	1	0	42	slt $1, $2, $3
slti	I	10	2	1	100			slti $1, $2, 100
sltu	R	0	22	3	1	0	43	sltu $1, $2, $3
sltiu	I	11	2	1	100			sltiu $1, $2, 100
j	J	2	2500					j 10000
jr	R	0	31	0	0	0	8	jr $31
jal	J	3	2500					jal 10000

which is used to specify the amount of shifting to be done in shift instructions. Any number between 0 and 31 can be specified as the shift amount. This field is used only in shift instructions. The last field is an additional opcode field, called the function field **funct** or **F_code**. The first opcode field can encode only 2^6 or 64 instructions. The MIPS processor does have more than 64 instructions considering the different variations of loads (byte load, half-word load, word load, floating point loads, etc.). Hence, more than 6 bits are required to fully specify an instruction. The MIPS designers chose a scheme in which the first 6 bits are 0 for the R-format instructions, and then use an additional field to specifically identify the instruction. The last 6 bits of the instruction are used to further identify the instruction in the case of the R-format ALU instructions.

The **I-format** is for load/store instructions, branch instructions, and ALU instructions that need an immediate constant to be specified in the instruction. These instructions need only two registers to be specified in addition to the immediate constant. The opcode field takes 6 bits, and the two register fields take 5 bits each. The remaining 16 bits are used as an immediate constant to specify an operand for instructions such as *addi* or to specify the offset in a load/store instruction or to specify the branch offset in a conditional branch instruction.

The **J-format** is for jump instructions. The first 6 bits of the instruction word are used for the opcode, and the remaining 26 bits of the instruction are used to specify the branch target. The branch target address is calculated by multiplying the 26-bit target by 4, since the jump offset is specified as a word address rather than byte address. MIPS uses byte addressing for accessing instructions and data.

Table 9-7 illustrates the instruction encoding for the MIPS instructions we have discussed. The opcode, source, and destination are assigned the same field in the instruction format as much as possible. The first 6 bits (bits 31 – 26) are for the opcode in all three formats. The source and destination register fields are in similar positions (bits 25–21, bits 20–16 and bits 15–11) as much as possible. This greatly simplifies decoding.

The encoding is very regular; however, compromises had to be made to accommodate various instructions into the same width. For instance, the destination register appears in different fields in 3-register and 2-register formats. Similarly, in a load instruction, the second register field is a destination register, whereas in a store instruction, it is the source of the data to be stored. In spite of these irregularities, one can say that the encoding is largely regular.

To increase the reader's familiarity with the MIPS instruction encoding, let us practice some machine coding.

Example

Create the machine code equivalent of the following assembly language program.

```
         andi   $3, $3, 0         ; initialize loop counter $3 to 0
         andi   $2, $2, 0         ; clear register for loop bound
         addi   $2, $2, 4000      ; loop bound register
$label:  lw     $15, 4000($3)     ; load x(i) to R15
         lw     $14, 8000($3)     ; load y(i) to R14
         add    $24, $15, $14     ; x(i) + y(i)
```

```
          sw    $24, 8000($3)      ; save new y(i)
          addi  $3, $3, 4          ; update address register,
                                     address= address + 4
          bne   $3, $2, $label
```

Answer: The first instruction

```
          andi  $3, $3, 0
```

can be translated as follows. Table 9-7 shows that the opcode for **andi** is 12. Hence, the first 6 bits for the first instruction will be 001100 as indicated in row 1 (after the header row) of Table 9-8. The source register field is next. It should be 00011, because the source register is $3. The destination register field is next. It should be 00011, because the destination register is $3. The immediate constant is 0, and it leads to sixteen 0s in bits 0 to 15. This explains the contents of row 1. In hex representation, it becomes 3063 0000.

We will also explain the encoding of the last instruction, bne $3, $2, label. The opcode is 5 (i.e., 000101). The next field corresponds to register $3, so it is 00011. The next field is 00010 to indicate the register $2. The byte offset should be −24, but the instruction is supposed to contain the word offset, which is −24 divided by 4 (i.e., −6). In 2's complement representation, it is 1010. Sign extending to fill the sixteen bits, one gets 1111111111111010, which will occupy bits 0 to 15. Machine code corresponding to all the instructions is shown in Table 9-8.

TABLE 9-8: MIPS Machine Code for Example 2; Binary as Well as Hex Representations Are Shown

Instruction	Bits 31–26	Bits 25–21	Bits 20–16	Bits 15–11	Bits 10–6	Bits 5–0	Equivalent Hex
andi $3, $3, 0	001100	00011	00011	00000	00000	000000	3063 0000
andi $2, $2, 0	001100	00010	00010	00000	00000	000000	3042 0000
addi $2, $2, 4000	001000	00010	00010	00001	11110	100000	2042 0FA0
lw $15, 4000($3)	100011	00011	01111	00001	11110	100000	8C6F 0FA0
lw $14, 8000($3)	100011	00011	01110	00011	11101	000000	8C6E 1F40
add $24, $15, $14	000000	01111	01110	11000	00000	100000	01EE C020
sw $24, 8000($3)	101011	00011	11000	00011	11101	000000	B478 1F40
addi $3, $3, 4	001000	00011	00011	00000	00000	000100	2063 0004
bne $3, $2, −6	000101	00011	00010	11111	11111	111010	1462 FFFA

9.4 Implementation of a MIPS Subset

In this section, we describe a simple implementation of a subset of the MIPS ISA, This subset, illustrated in Table 9-9, includes most of the important instructions, including ALU, memory access, and branch instructions. What we present in this section is a naïve implementation of this instruction set. Modern microprocessors implement features such as multiple instruction issue, out-of-order execution, branch prediction, pipelining, and so forth. For the sake of simplicity, what is presented here is a simple in-order, non-pipelined implementation. Some of the exercise problems describe other implementations that will provide better performance.

TABLE 9-9: Subset of MIPS Instructions Implemented in This Chapter

Arithmetic	add subtract add immediate
Logical	and or and immediate or immediate shift left logical shift right logical
Data Transfer	load word store word
Conditional Branch	branch on equal branch on not equal set on less than
Unconditional Branch	jump jump register

Design of the Data Path

In order to design a microprocessor, first we will examine the sequence of operations during execution of instructions. Then we will describe the nature of the hardware required to accomplish the instruction execution. In general, any microprocessor works in the following manner:

1. The processor **fetches** an instruction.
2. It **decodes** the instruction that was fetched. Decoding means identifying what the instruction is.
3. It reads the operands and **executes** the instruction. For a RISC ISA, for arithmetic instructions, the operands are in registers. The registers that contain the input operands are called source registers. For memory access instructions, addresses are computed using registers and memory is accessed. After execution, the processor **writes** the result of the instruction execution into the destination. The destination is a register for all instructions other than the store instruction, which has to write the result into the memory.

Hence, the design must contain a **unit to fetch** the instructions, a **unit to decode** the instructions, an arithmetic and logic unit (**ALU**) to execute the instructions, a **register file** to hold the operands, and the **memory** that stores instructions and data. These components are described in the following subsections.

Instruction Fetch Unit

In general, a microprocessor has a special register called the program counter (PC), which points to the next instruction in the instruction memory (or in the instruction caches). The PC sends this address to the instruction memory, which sends the instruction back. The processor increments the PC to point to the next instruction to be fetched. A block diagram for this unit is shown in Figure 9-1.

FIGURE 9-1: Block Diagram for Instruction Fetch

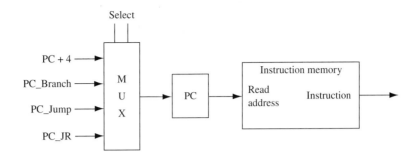

The next PC is one of the following, depending on the current instruction:

(a) **PC + 4:** For instructions other than branch and jump instructions, the next instruction is at address PC + 4, since four bytes are needed for the current instruction.

(b) **PC_Branch:** In the branch (bne and beq) instructions, the next PC is obtained by adding the offset in the instruction to the current PC. In the MIPS ISA, the branch offset is provided as a signed word offset (number of words to jump forward or backward). First the word offset is sign extended, then converted to a byte offset by multiplying by 4, and finally added to the current PC. Thus, the next PC for branch instructions is

$$PC\_Branch = PC + 4 + Offset * 4.$$

(c) **PC_Jump:** In the jump (J) instruction, the jump target address is provided in the instruction itself. In the MIPS ISA, the opcode field takes 6 out of the 32 bits. Hence, the biggest jump address that can be encoded is only 26 bits. In order to compute the 32-bit jump address, first, the 26-bit word address in the instruction is shifted twice to the left, resulting in a 28-bit address, which is a byte address. Then it is concatenated with the four highest bits of the PC, yielding a 32-bit address. Thus, the next PC for jump instructions is

$$PC\_Jump = PC[31..28] \, || \, Address * 4,$$

where || stands for concatenation. Note that this symbol is different from the concatenate symbol in Verilog. This is the symbol typically used in the MIPS ISA references [Kane, Gerry, *MIPS RISC Architecture* Prentice Hall, 1989].

(d) **PC_JR:** In the jump register (JR) instruction, the jump target is obtained from the register specified in the instruction. Thus, the next PC for a JR instruction is

$$PC\_JR = [REG],$$

where [REG] indicates contents of the register.

The appropriate target addresses are computed and fed to the PC. A multiplexer is used to select between the branch target, jump target, jump register target, or PC + 4, depending on the instruction.

There are several choices as to when the target addresses are computed. The default target, PC + 4, can be computed at instruction fetch itself, since it needs no information other than the PC itself. In conditional branch instructions, the branch target (PC_Branch) computation can be done as soon as the instruction is read; however, whether the branch is taken or not will not be known until the registers are read and compared. In the case of the jump instruction, the target (PC_Jump) can be computed as soon as the instruction is fetched, since the information for the target is available in the instruction itself. In a jump register (*jr*) instruction, the branch target (PC_JR) can be computed after the register is read.

Figure 9-1 also shows the instruction memory unit. In the initial design, we will use a separate instruction memory and separate data memory, in alignment with the popular scheme of separate instruction and data caches found in modern processors. We will not be designing a cache memory; however, we will assume the presence of on-chip instruction memory that can be accessed by the processor in one cycle after the address is provided to it.

Instruction Decode Unit

Decoding is fairly simple due to the simplicity of the RISC ISA. One can observe from Table 9-7 that the instruction formats in the MIPS ISA are very regular and uniform. The first 6 bits of the instruction specify the opcode in most cases. But, as described in Section 9.3, for the R-format ALU instructions, the first 6 bits are 0, and the last 6 bits of the instruction, called **F_code**, need to be used to further identify the instruction.

The opcode is used to identify the instruction and the instruction format used by the instruction. The uniformity of the instruction format allows many of the instruction fields to be directly used for register addressing and control-signal generation. The instruction opcode bits are fed to a control unit that generates the various control signals.

Instruction Execution Unit

Once the instruction is identified at the decode stage, the next task is to read the operands and perform the operation. In RISC instruction sets, the operands are in registers. The MIPS architecture contains 32 registers, and these registers are collectively referred to as the **register file**. The register file should have at least two read ports to support reading two operands at the same time, and it should have one write port.

The operation of the register file is as follows. The registers that hold the input operands are called **source registers**, and the register that should receive the result is called the **destination register**. The source register addresses are applied to the register file. The register file will produce the data from the corresponding registers on the output data lines. This data is fed to the arithmetic and logic unit (**ALU**), which executes the instruction. The ALU contains functional units such as adders, shifters, and the like. It may also include more complex units such as multipliers, although our restricted design here does not include multiplication.

In most instructions, the result from the ALU should be written into the destination register. To accomplish this, the ALU result is applied to the input data lines of the register file. The destination register name and the **register write (RegW) command** is applied to the register file. That causes the input data to get written into the destination register.

Figure 9-2 shows a block diagram of the data path that is required to execute the ALU and memory instructions. The data path includes an ALU, which will perform the following operations: add, sub, and, and or. In the case of R-format instructions, both operands for the ALU are read from the register file. In the case of the I-format instructions, the immediate constant in the instruction is sign extended to create the second operand. Since one of the ALU operands comes from either the register file or the sign extender, a multiplexer is required to select the appropriate operand.

FIGURE 9-2: Required Data Path for Computation and Memory Instructions

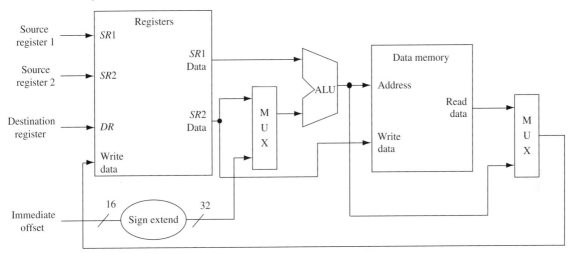

The ALU is also required for non-arithmetic instructions. For memory access instructions, we have to first calculate the address to be accessed. The ALU can be used for calculating the address. For address calculation for load and store instructions, the first operand is obtained from the register specified in the instruction, and the second operand is obtained by sign extending the immediate offset specified in the instruction.

The ALU is required for conditional branch instructions as well. As you know, MIPS has only two branch instructions—branch on equal (*beq*) and branch not equal (*bne*). The comparison for determining whether the registers are equal can be done by the ALU. Both operands for this comparison can be obtained from the register file. The data path must also include a **data memory unit**, because load and store instructions have to access the data memory unit. Modern microprocessors contain on-chip data caches. We will not be designing a cache memory, but we will assume the presence of on-chip data memory that can be accessed by the instructions in one cycle after the address is provided to the memory.

Figure 9-2 also shows use of several multiplexers and how the different bits of the instruction are connected to the register file. As Table 9-6 illustrates, bits 21 to 25 of the instruction contains one of the source register addresses in all ALU instructions. Hence, these bits can be directly connected to the first source register address of the register file. Any instruction with a second register source contains the register address in bits 16 to 20. Hence, these bits can also be directly connected to the source register address of the register file. However, the destination register address appears in different fields in different instructions. In R-format instructions, the destination register address appears in bits 11 to 15. In I-format instructions, however, the destination address is in bits 16 to 20. Hence, a multiplexer is required to choose the appropriate destination register address. Another multiplexer chooses between the immediate operand or the register operand for the ALU. A third multiplexer is used to select whether ALU output or memory data will be written to the destination register.

Overall Data Path

The overall data path is shown in Figure 9-3. It integrates the fetch and execute hardware from Figures 9-1 and 9-2 and adds other required elements for correct operation. In addition, control signals are also shown.

FIGURE 9-3: Overall Data Path

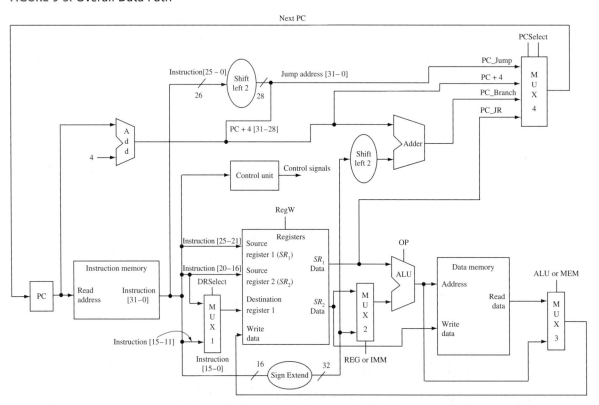

Figure 9-3 also illustrates the details of the computation of the target addresses in the various kinds of instructions. Default next address of PC + 4 is calculated with an adder. Addition of the branch offset to the PC is also done using a separate adder.

Several multiplexers are shown in this data path:

- MUX 1 selects a destination register address from an appropriate register field depending on the instruction format. For R-format instructions, bits 20-16 yield the destination address, and for I-format instructions, bits 15-11 of instruction provide the destination address.

- MUX 2 selects whether the second operand for ALU comes from a register or an immediate constant. For R-format ALU instructions and conditional branch instructions, a register is chosen. For I-format ALU instructions, the immediate constant provides the operand.
- MUX 3 selects between the memory or the ALU output for data to go into the destination register. For load instructions, the memory data is chosen.
- MUX 4 selects between the four possible next PC values depending on the type of instruction.

Instruction Execution Flow

Figure 9-4 illustrates the flow of execution for a possible implementation. The first step is fetch for all instructions. The address in the program counter (PC) is sent to the instruction memory unit. All instructions also need to update the PC to point to the next instruction. While the PC should be updated differently for branch or jump instructions, the vast majority of instructions are in sequence; hence, the PC can be updated to point to the next instruction in sequence. Branch and jump instructions can later modify the PC appropriately.

The second step is decode. Depending on the opcode that is encountered, different actions follow. For R-type instructions, and for some I-type instructions (e.g., *bne* and *beq*), both ALU operands are read from registers. For other I-type instructions, one operand is read from the register file, and the immediate constant in the instruction is sign extended as the other operand. Reading of a register source satisfies the requirements for a jump register *(jr)* instruction,

FIGURE 9-4: Flow Chart for Instruction Processing

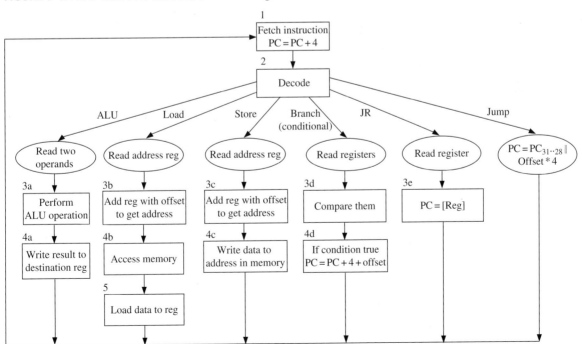

which is an R-type instruction. The ALU operation required for each instruction is identified during the decode step. For instance, the *bne* and *beq* instructions need a subtract operation. The load and store instructions require an add operation. If the jump opcode is encountered, a jump target is calculated. Since the jump instruction does not need any further action, flow of control can go to step 1.

Step 3 is the actual execution of the instructions. Depending on the instruction, different ALU operations are performed during this step. The different actions are shown in boxes labeled 3a, 3b, and so on for the different types of instructions. Each instruction goes through only one of these operations depending on what type of instruction it is. All instructions other than the jump instruction must come to this step. The jump register *(jr)* instruction does not need any arithmetic operation. The content of the register fetched during step 2 must be loaded into the PC. For load and store instructions, the ALU performs an addition to calculate the memory address.

Step 4 varies widely between the instructions. Arithmetic and logic instructions (of R-type and I-type) can write their computation result to the destination register. Branch instructions must examine their condition and decide to take the branch or not. If the branch is to be taken, the branch target address is calculated. For load instructions, a memory read operation is initiated. For memory store instructions, the data from the second source register is steered to the memory, and a memory write operation is initiated. This is the final step for all instructions other than load instructions.

Step 5 is required only for load instructions. The data output from memory is written into the destination register.

One can implement this instruction flow in various ways. In the most naïve implementation, one can have a very slow clock, and the processor will perform all operations required for each instruction during one clock cycle. The disadvantage with this scheme is that all instructions will be as slow as the slowest instruction, because the clock cycle has to be long enough for the slowest instruction. Another option is to do an implementation whereby each instruction takes multiple cycles but just enough cycles to finish all operations for each class of instruction. For instance, Figure 9-4 can be considered as an ASM chart with each box taking one cycle. In this case, a jump instruction can finish in two cycles, while an ALU instruction needs four cycles, and a load instruction takes five cycles. In the following section, we present the Verilog model of such an implementation.

9.5 Verilog Model

The Verilog model for the processor is organized as shown in Figure 9-5. The instruction memory, data memory, and register file are created as components with their architecture and entity descriptions. The main code, the MIPS entity, embeds the control sequencing the instructions through the various stages of its operation.

FIGURE 9-5: Organization of the Verilog Model for the Processor

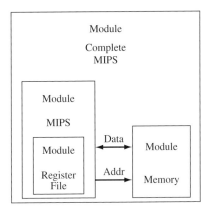

We combine the instruction and data memory units to be a single memory and illustrate the use of the address and data buses. Later we use a test bench to test the functionality of the designed modules. The model we create is synthesizable into hardware. If the models are synthesized into an FPGA, appropriate display modules such as an LCD or 7-segment displays should be included so that we can observe outputs.

Let us model the register and memory components first.

Verilog Model for the Register File

Figure 9-6 shows the Verilog model for the register file. The REG entity is used to represent the 32 MIPS registers. Each register is 32 bits long. The destination register address is *DR*, and the source register addresses are *SR1* and *SR2*. Since there are 32 registers, *DR*, *SR1*, and *SR2* are 5 bits each. The outputs *Reg1* and *Reg2* are the contents of the registers specified by *SR1* and *SR2*. *Reg1* is fed straight to the ALU. *Reg2* can be used as a second ALU input or as the input to data memory in the case of store instructions. The control signal *RegW* is used to control the write operation to the register file. If *RegW* is true, the data on lines *Reg_In* is written into the register pointed to by *DR*.

It is desirable to perform register reads synchronously. It is also desirable to have the register files close to the arithmetic and logical units. Hence in FPGA-based designs, register files should be in the distributed RAM. If the reads are performed synchronously as in the provided code, some synthesis tools may not map the register file into the distributed RAM (it might be mapped into block RAM). Modern synthesis tools allow designers to explicitly choose distributed RAM or block RAM as desired from a menu option. In FPGA-based designs, it is important to check that the major modules are mapped into the regions expected by the designer. One can check synthesis reports from FPGA tools to ascertain this.

Verilog Model for Memory

Figure 9-7 illustrates the Verilog code for the memory unit. The Verilog model is similar to the SRAM model that we created in Chapter 8. Although Figures 9-3 and 9-5 illustrated separate instruction and data memories, for convenience and for

FIGURE 9-6: Verilog Code for Register File

```verilog
module Register(CLK, RegW, DR, SR1, SR2, Reg_In, ReadReg1, ReadReg2);
    input CLK, RegW;
    input [4:0] DR, SR1, SR2;
    input [31:0] Reg_In;
    output reg [31:0] ReadReg1, ReadReg2;

    reg [31:0] REG [0:31];
    integer i;

    initial begin
        ReadReg1 = 0;
        ReadReg2 = 0;
          for (i = 0; i < 32; i = i+1) begin
                REG[i] = 0;
            end
    end
    always @(posedge CLK) begin
        if(RegW == 1'b1)
            REG[DR] <= Reg_In[31:0];
        ReadReg1 <= REG[SR1];
        ReadReg2 <= REG[SR2];
    end
endmodule
```

illustrating use of address and data buses, we have used a unified memory module that stores both instructions and data. The memory consists of 128 locations, each 32 bits wide. We assume that the instructions are the first 64 words in the array, and the other 64 words are allocated for data memory. The signal *ADDR* specifies the location in memory to be read from or stored to. The address bus is actually 32 bits wide, but we use only the 7 lower bits since we implement only a small memory

The address bus will be driven by the processor appropriately for instruction and data access. The address input may come from the program counter for reading the instruction or from the ALU that computes the address to access the data portion of the memory. The chip select (*CS*) and write enable (*WE*) signals allow the processor to control the reads and writes. When *CS* and *WE* are true, the data on *Mem_Bus* gets written to the memory location pointed to by address *ADDR*. The writing is shown to occur synchronously at the negative edge of the clock.

As you know from Chapter 6, the Xilinx Spartan/Virtex FPGAs contain dedicated block RAM. It is desirable to have the block RAM. Modern synthesis tools allow designers to explicitly choose distributed RAM or block RAM as desired.

Loading specific contents into the memory can be accomplished by reading from a file. For instance, in order to load a program from file into our memory module, it can be done using the statement

$readmemh("MIPS_Instructions.txt", RAM);

FIGURE 9-7: Verilog Code for the Unified Instruction/Data Memory

```verilog
module Memory(CS, WE, CLK, ADDR, Mem_Bus);
    input CS, WE, CLK;
    input [31:0] ADDR;
    inout [31:0] Mem_Bus;

    reg [31:0] data_out;
    reg [31:0] RAM [0:127];

    integer i;
    reg[6:0] counter;

    initial begin
        for (i=0; i<128; i=i+1)
            begin
                RAM[i] = 32'd0; //initialize all locations to 0
            end
            $readmemh("MIPS_Instructions.txt", RAM);
            //this optional statement can be inserted to read initial values
            //from a file
    end
    assign Mem_Bus = ((CS == 1'b0) || (WE == 1'b1)) ? 32'bZ : data_out;
    always @(negedge CLK) begin
        if((CS == 1'b1) && (WE == 1'b1))
            RAM[ADDR] <= Mem_Bus[31:0];
        data_out <= RAM[ADDR];
    end
endmodule
```

This statement (currently a comment in the text) can be included into the **initial** block of the code if memory has to be loaded with specific contents. The $readmemh which reads hex values was explained in Chapter 8. In this case, the file MIPS_Instructions.txt is assumed to be a text file with a 32-bit instruction in hex format in each row. In our case, we will load MIPS instructions into the memory. The contents of the file look like

    30000000

    20010006

if the MIPS instructions we want to load into memory are

    andi $0, $0, 0

    addi $1, $0, 6

For simplicity, the address is shown as a word address in the Verilog code for the memory. In the actual MIPS processor, the memory is byte addressable. Therefore, each instruction memory access should obtain the data found in the specified location concatenated with the next three memory locations. For example, if address = 0, the instruction register must be loaded with the contents of MEM[0],

MEM[1], MEM[2], and MEM[3]. The instructions are stored depending on the endian-ness of the machine. Many modern microprocessors support both big-endian and little-endian approaches.

> **Little-Endian and Big-Endian**
> When we store 16-bit or 32-bit data into byte-addressable memory, there are two possible ways to store the data: little-endian and big-endian. In a little-endian system, the least significant byte in the sequence is stored first. In a big-endian system, the most significant byte in the sequence is stored at the lowest storage address (i.e., first). Let us consider how a MIPS instruction will be stored into byte-addressable memory in the two systems. The MIPS instruction **andi $3, $3, 0** will be encoded as **30630000 (hex).** When this instruction is stored at address 2000, depending on whether big-endian or little-endian system is used, the memory will look as follows:
>
> | Address | Big-Endian Representation of 30630000 hex | Little-Endian Representation of 30630000 hex |
> |---|---|---|
> | 2000 | 30 | 00 |
> | 2001 | 63 | 00 |
> | 2002 | 00 | 63 |
> | 2003 | 00 | 30 |

Verilog Code for the Processor CPU

In this section, we present the Verilog code for the central processing unit (CPU) of the microprocessor. The register module that was created in the earlier section is used here. Figure 9-8 shows a Verilog model for the MIPS instructions in Table 9-9. It generally follows the flow in Figure 9-4, implementing the fetch, decode, and execute phases of an instruction. The various statements and variables used in the code are explained in the succeeding pages.

FIGURE 9-8: Verilog Code for the MIPS Subset Implementation

```
`define opcode   instr[31:26]
`define sr1      instr[25:21]
`define sr2      instr[20:16]
`define f_code   instr[5:0]
`define numshift instr[10:6]

module MIPS (CLK, RST, CS, WE, ADDR, Mem_Bus);
    input CLK, RST;
    output reg CS, WE;
```

9.5 Verilog Model

```verilog
    output [31:0] ADDR;
    inout [31:0] Mem_Bus;

    //special instructions (opcode == 000000), values of F code (bits 5-0):
    parameter add  = 6'b100000;
    parameter sub  = 6'b100010;
    parameter xor1 = 6'b100110;
    parameter and1 = 6'b100100;
    parameter or1  = 6'b100101;
    parameter slt  = 6'b101010;
    parameter srl  = 6'b000010;
    parameter sll  = 6'b000000;
    parameter jr   = 6'b001000;
    //non-special instructions, values of opcodes:
    parameter addi = 6'b001000;
    parameter andi = 6'b001100;
    parameter ori  = 6'b001101;
    parameter lw   = 6'b100011;
    parameter sw   = 6'b101011;
    parameter beq  = 6'b000100;
    parameter bne  = 6'b000101;
    parameter j    = 6'b000010;
    //instruction format
    parameter R = 2'd0;
    parameter I = 2'd1;
    parameter J = 2'd2;
    //internal signals
    reg [5:0] op, opsave;
    wire [1:0] format;
    reg [31:0] instr, pc, npc, alu_result;
    wire [31:0] imm_ext, alu_in_A, alu_in_B, reg_in, readreg1, readreg2;
    reg [31:0] alu_result_save;
    reg alu_or_mem, alu_or_mem_save, regw, writing, reg_or_imm,
reg_or_imm_save;
    reg fetchDorI;
    wire [4:0] dr;
    reg [2:0] state, nstate;

//combinational
    assign imm_ext = (instr[15] == 1)? {16'hFFFF, instr[15:0]} : {16'h0000,
            instr[15:0]};//Sign extend immediate field
    assign dr = (format == R)? instr[15:11] : instr[20:16];
//Destination Register MUX (MUX1)
    assign alu_in_A = readreg1;
    assign alu_in_B = (reg_or_imm_save)? imm_ext : readreg2;
//ALU MUX (MUX2)
    assign reg_in = (alu_or_mem_save)? Mem_Bus : alu_result_save;
//Data MUX
    assign format = (`opcode == 6'd0)? R : ((`opcode == 6'd2)? J : I);
    assign Mem_Bus = (writing)? readreg2 : 32'bZZZZZZZZZZZZZZZZZZZZZZZZZZZZZZZZ;
//drive memory bus only during writes
```

```verilog
    assign ADDR = (fetchDorI)? pc : alu_result_save;
//ADDR Mux
    REG Register(CLK, regw, dr, `sr1, `sr2, reg_in, readreg1, readreg2);
    initial begin
        op = and1;
        opsave = and1;
        state = 3'b0;
        nstate = 3'b0;
        alu_or_mem = 0;
        regw = 0;
        fetchDorI = 0;
        writing = 0;
        reg_or_imm = 0;
        reg_or_imm_save = 0;
        alu_or_mem_save = 0;
    end
always @(*)
begin
    fetchDorI = 0; CS = 0; WE = 0; regw = 0; writing = 0; alu_result = 32'd0;
    npc = pc; op = jr; reg_or_imm = 0; alu_or_mem = 0;
    case (state)
        0: begin //fetch
                npc = pc + 32'd1; CS = 1; nstate = 3'd1;
                fetchDorI = 1;
            end
        1: begin //decode
                nstate = 3'd2; reg_or_imm = 0; alu_or_mem = 0;
                if (format == J) begin //jump, and finish
                    npc = {pc[31:26], instr[25:0]};
                    nstate = 3'd0;
                end
                else if (format == R) //register instructions
                    op = `f_code;
                else if (format == I) begin //immediate instructions
                    reg_or_imm = 1;
                    if(`opcode == lw) begin
                       op = add;
                       alu_or_mem = 1;
                    end
                    else if ((`opcode == lw)||(`opcode == sw)||(`opcode == addi))
                           op = add;
                    else if ((`opcode == beq)||(`opcode == bne)) begin
                       op = sub;
                       reg_or_imm = 0;
                    end
                    else if (`opcode == andi) op = and1;
                    else if (`opcode == ori) op = or1;
                end
            end
```

9.5 Verilog Model

```verilog
        2:  begin //execute
                nstate = 3'd3;
                if (opsave == and1) alu_result = alu_in_A & alu_in_B;
                else if (opsave == or1) alu_result = alu_in_A | alu_in_B;
                else if (opsave == add) alu_result = alu_in_A + alu_in_B;
                else if (opsave == sub) alu_result = alu_in_A - alu_in_B;
                else if (opsave == srl) alu_result = alu_in_B >> `numshift;
                else if (opsave == sll) alu_result = alu_in_B << `numshift;
                else if (opsave == slt) alu_result = (alu_in_A < alu_in_B)? 32'd1 :
                        32'd0;
                else if (opsave == xor1) alu_result = alu_in_A ^ alu_in_B;
                if (((alu_in_A == alu_in_B)&&(`opcode == beq)) || ((alu_in_A != alu_
                    in_B)&&(`opcode == bne))) begin
                    npc = pc + imm_ext;
                    nstate = 3'd0;
                end
                else if ((`opcode == bne)||(`opcode == beq)) nstate = 3'd0;
                else if (opsave == jr) begin
                    npc = alu_in_A;
                    nstate = 3'd0;
                end
            end

        3:  begin //prepare to write to mem
                nstate = 3'd0;
                if ((format == R)||(`opcode == addi)||(`opcode == andi)||(`opcode == ori))
                    regw = 1;
                else if (`opcode == sw) begin
                    CS = 1;
                    WE = 1;
                    writing = 1;
                end
                else if (`opcode == lw) begin
                    CS = 1;
                    nstate = 3'd4;
                end
            end

        4:  begin
                nstate = 3'd0;
                CS = 1;
                if (`opcode == lw) regw = 1;
            end
        endcase
end //always

always @(posedge CLK) begin
        if (RST) begin
            state <= 3'd0;
            pc <= 32'd0;
        end
        else begin
```

```
            state <= nstate;
            pc <= npc;
        end
        if (state == 3'd0) instr <= Mem_Bus;
        else if (state == 3'd1) begin
            opsave <= op;
            reg_or_imm_save <= reg_or_imm;
            alu_or_mem_save <= alu_or_mem;
        end
        else if (state == 3'd2) alu_result_save <= alu_result;
    end //always
endmodule
```

For readability of the code, we have also used constant declarations to associate the various opcodes with the corresponding codes from Table 9-7. For instance, the load instruction *lw* has 35 as its opcode, and the store instruction *sw* has 43 as its opcode. Several statements, such as the following, are used in order to denote the various opcodes.

```
parameter lw    = 6'b100011;
parameter sw    = 6'b101011;
```

This could also be done using compiler directive `define as follows:

```
`define lw    6'b100011
`define sw    6'b101011
```

but if `define is used, one should remember to use (`) with each macro substitution. For example,

```
op <= add;
```

in the code should be substituted with

```
op <= `add; // The (`) is required
```

if `define add is used.

The bits of the instruction word are decoded into the relevant fields using continuous assign statements. The most significant 6 bits of the instruction form the *opcode*. The lowest 6 bits of the instruction are denoted with the alias *f_code*. The shift amount in shift instructions is denoted using *numshift*. The two register source fields are aliased to *sr1* and *sr2*. The following statements accomplish this aliasing:

```
`define opcode   instr[31:26]
`define sr1      instr[25:21]
`define sr2      instr[20:16]
`define f_code   instr[5:0]
`define numshift instr[10:6]
```

9.5 Verilog Model

Sign extension of the immediate quantity is accomplished by the following statement:

assign imm_ext = (instr[15] == 1)? {16'hFFFF, instr[15:0]} : {16'h0000, instr[15:0]};//Sign extend immediate field

The following are the signals used in the Verilog model:

MIPS Processor Model Signals

Signal	Description
CLK (input)	Clock
RST (input)	Synchronous reset
CS (output)	Memory chip select. When cs is active and WE is inactive, the memory module outputs the memory contents at the address specified by Addr to mem_bus.
WE (output)	Memory write enable. When WE and cs are active, the memory module stores the contents of mem_bus to the location specified by Addr during the falling edge of the clock.
ADDR (ouput)	Memory address. During state 0 (fetch instruction from memory), ADDR is connected to the PC. Otherwise, it is connected to the ALU result (32 bits).
Mem_Bus (in/out)	Memory bus; carries data to and from the memory module. The MIPS module outputs to the bus during memory writes. The memory module outputs to the bus during memory reads. When not in use, the bus is at 'hi-Z' (32 bits).
Op	ALU operation select; determines the specific operation (e.g, add, and, or) to be performed by ALU. Determined during decode.
Format	Indicates whether the current instruction is of R, I, or J format.
Instr	The current instruction (32 bits)
imm_ext	Sign extended immediate constant from the instruction (32 bits)
Pc	Current program counter (32 bits).
Npc	Next program counter (32 bits).
readreg1	Contents of the first source register (sr1) (32 bits).
readreg2	Contents of the second source register (sr2) (32 bits).
reg_in	Data input to registers. When executing a load instruction, reg_in is connected to the memory bus. Otherwise, it is connected to the ALU result (32 bits).
alu_in_A	First operand for the ALU (32 bits).
alu_in_B	Second operand for the ALU. alu_in_B is connected to imm_ext during immediate mode instructions. Otherwise, it is connected to readreg2 (32 bits).
alu_result	Output of ALU (32 bits).
alu_or_mem	Select signal for the reg_in multiplexer; indicates whether the register input should come from the memory or the ALU.
reg_or_imm	Select signal for the alu_in_B multiplexer; determines whether the second ALU operand is a register output or sign-extended immediate constant.
Regw	Indicates if the destination register should be written to. Some instructions do not write any results to a register (e.g, branch, store).
fetchDorI	Select signal for the Address multiplexer; determines whether ADDR is the location of an instruction to be fetched or the location of data to be read or written.
Writing	Control signal for the MIPS processor output to the memory bus. Except during memory writes, the output is 'hi-Z' so the bus can be used by other modules. *Note*: writing cannot be replaced with WE, because WE is of mode out. writing is used in mode *in* as well.
Dr	Address of destination register (5 bits).
State	Current state.
nState	Next state.

Since we have used separate clock cycles for the fetch operation, the decode operation, the execute operation, and so on, it is necessary to save signals created during each stage for later use. The statements such as

opsave <= op;

reg_or_imm_Save <= reg_or_imm;

alu_or_mem_save <= alu_or_mem;

alu_result_save <= alu_result;

are used in the clocked process for saving (explicit latching) of the relevant signals.

Two always blocks are used in the code: (1) a clocked process for the sequential part of the code and (2) another for the combinational block.

The multiplexer at the input of the program counter is not explicitly coded. The various data transfers are coded behaviorally in the various states. A good synthesizer will be able to generate the multiplexer to accomplish the various data transfers. Similarly, the multiplexer to select the destination register address is also not explicitly coded. If the synthesis tool generates inefficient hardware for this multiplexed data transfer, one can code the multiplexer into the data path and generate control signals for the select signals.

Complete MIPS

The processor module and the memory are integrated to yield the complete MIPS model. Component descriptions are created for the processor and the memory units. These components are instantiated into the Complete_MIPS module. We have also brought out the address and data buses as outputs from the high-level entity. If no outputs are shown in an entity, when the code is synthesized, it results in empty blocks. Depending on the synthesis tool, unused signals (and corresponding nets) may be deleted from the synthesized circuit. Bringing the address and data buses allows the testing of the modules observing these buses.

FIGURE 9-9: Verilog Code Integrating the Processor and Memory Modules

```
module Complete_MIPS(CLK, RST, A_Out, D_Out);
    input CLK;
    input RST;
    output [31:0] A_Out;
    output [31:0] D_Out;

    wire CS, WE;
    wire [31:0] ADDR, Mem_Bus;
    assign A_Out = ADDR;
    assign D_Out = Mem_Bus;

    MIPS CPU(CLK, RST, CS, WE, ADDR, Mem_Bus);
    Memory MEM(CS, WE, CLK, ADDR, Mem_Bus);
endmodule
```

9.5 Verilog Model

We synthesized the foregoing model. The Xilinx ISE tools targeted for a Spartan 3 FPGA yield 1108 4-input LUTs, 660 slices, 111 flip-flops, and one block RAM. The register file takes 194 4-input LUTs. Since one LUT can give 16 bits of storage, thirty-two 32-bit registers would need the storage from 64 LUTs. Since the register file has two read ports, it would need 128 LUTs. Additional LUTs are required for the address decoder and the control signals. In order to implement the design on a prototyping board, interface to the input and display modules should be added.

Testing the Processor Model

The overall MIPS Verilog model is tested using a test bench. The test bench must verify the proper operation of each implemented instruction. The test bench consists of a MIPS program with test instructions and Verilog code to load the program into memory and verify the program's output. We use a constant array of instructions that we want to write into the memory and a constant array of expected outputs to which we will compare the result of the processor execution.

The model was tested for the following instructions. The expected action of each instruction is also shown as follows:

```
x"30000000",   -- andi $0, $0, 0    => $0 = 0
x"20010006",   -- addi $1, $0, 6    => $1 = 6
x"34020012",   -- ori $2, $0, 18    => $2 = 18
x"00221820",   -- add $3, $1, $2    => $3 = $1 + $2 = 24
x"00412022",   -- sub $4, $2, $1    => $4 = $2 - $1 = 12
x"00222824",   -- and $5, $1, $2    => $5 = $1 and $2 = 2
x"00223025",   -- or $6, $1, $2     => $6 = $1 or $2 = 22
x"0022382A",   -- slt $7, $1, $2    => $7 = 1 because $1<$2
x"00024100",   -- sll $8, $2, 4     => $8 = 18 * 16 = 288
x"00014842",   -- srl $9, $1, 1     => $9 = 6/2 = 3
x"10220001",   -- beq $1, $2, 1     => Will not branch. $10 wrong if fails.
x"8C0A0004",   -- lw $10, 4($0)     => $10 = 5th instr = x"00412022" = 4268066
x"14620001",   -- bne $1, $2, 1     => Branch to PC+1+1 (not PC+4+4 because
                                        word-addressed memory instead of byte-
                                        addressed)
x"30210000",   -- andi $1, $1, 0    => $1 = 0 (skipped if bne worked correctly)
x"08000010",   -- j 16              => PC = $16.  $2 wrong if j malfunctioned
x"30420000",   -- andi $2, $2, 0    => $2 = 0 (skipped if j worked correctly)
x"00400008",   -- jr $2             => PC = $2 = 18. $3 wrong if jr malfunction
x"30630000",   -- andi $3, $3, 0    => $3 = 0 (skipped if jr worked correctly)
x"AC010040",   -- sw $1, 64($0)     => Mem(64) = $1 = 6
x"AC020041",   -- sw $2, 65($0)     => Mem(65) = $2 = 18
x"AC030042",   -- sw $3, 66($0)     => Mem(66) = $3 = 24
x"AC040043",   -- sw $4, 67($0)     => Mem(67) = $4 = 12
x"AC050044",   -- sw $5, 68($0)     => Mem(68) = $5 = 2
x"AC060045",   -- sw $6, 69($0)     => Mem(69) = $6 = 22
x"AC070046",   -- sw $7, 70($0)     => Mem(70) = $7 = 1
x"AC080047"    -- sw $8, 71($0)     => Mem(71) = $8 = 288
x"AC090048",   -- sw $9, 72($0)     => Mem(72) = $9 = 3
x"AC0A0049",   -- sw $10, 73($0)    => Mem(73) = $10 = x00412022
```

File I/O is used to load these instructions into the memory module. The file "MIPS_Instructions.txt" contains the following hex data:

```
30000000
20010006
34020012
00221820
00412022
00222824
00223025
0022382A
00024100
00014842
10220001
8C0A0004
14620001
30210000
08000010
30420000
00400008
30630000
AC010040
AC020041
AC030042
AC040043
AC050044
AC060045
AC070046
AC080047
AC090048
AC0A0049
```

The memory module in Figure 9-7 must include the following line in order to appropriately initialize the code memory:

$readmemh("MIPS_Instructions.txt", RAM);

Note that now the memory is connected to the processor and test bench, and that means both our test bench and the processor will try to control the two signals at the same time. One way to resolve this is to put MUXes at the input ports of the memory. There are a few MUXes for that purpose: *Address_Mux* (for choosing the address), *CS_Mux* for choosing the *CS* signal, and *WE_Mux* (for choosing the WE signal). The select signal for the MUXes is *init*. When the *init* signal is 1, the three MUXes select the address and the *CS* and *WE* signals from the test bench. Otherwise, these signals from the processor are chosen. We also assert the reset of our CPU throughout the initialization process to make sure the CPU does not run until the test bench finishes writing the instructions into the memory. When *init* is 0, the CPU and memory are connected for normal operation.

As the MIPS program executes, each test instruction stores its result in a different register. The last 10 instructions perform a series of *sw* operations that store registers 1-10 to memory. Each of these instructions places the contents of a different register onto the bus as it executes. During the memory write stage, the testbench will compare the value of these registers (by looking at the bus value) with the expected output. No explicit checks for branch instructions are in the test sequence; however, if a branch does not execute as expected, an error will be detected because the result register values for the instructions after the branch instructions will be incorrect.

In the actual MIPS processor, register $0 is always 0. We did not implement that in the register file. Hence we clear register $0 using an instruction. The first instruction in the test sequence does that. In normal MIPS processor code, you will not find instructions with register $0 as the destination. Essentially, writes to register $0 are ignored.

FIGURE 9-10: Test Bench for the Processor Model

```verilog
module mips_testbench;
    reg CLK;
    wire CS, WE;

    parameter N = 10;
    reg[31:0] expected[N:1];

    wire[31:0] Address, Address_Mux, Mem_Bus_Wire;
    reg[31:0] AddressTB;
    wire WE_Mux, CS_Mux;
    reg init, rst, WE_TB, CS_TB;

    integer i;

    MIPS CPU(CLK, rst, CS, WE, Address, Mem_Bus_Wire);
    Memory MEM(CS_Mux, WE_Mux, CLK, Address_Mux, Mem_Bus_Wire);

    assign Address_Mux = (init)? AddressTB : Address;
    assign WE_Mux = (init)? WE_TB : WE;
    assign CS_Mux = (init)? CS_TB : CS;

    always
        #10 CLK = ~CLK;

    initial begin
        expected[1] = 32'h00000006; // $1 content=6 decimal
        expected[2] = 32'h00000012; // $2 content=18 decimal
        expected[3] = 32'h00000018; // $3 content=24 decimal
        expected[4] = 32'h0000000C; // $4 content=12 decimal
        expected[5] = 32'h00000002; // $5 content=2
        expected[6] = 32'h00000016; // $6 content=22 decimal
```

```verilog
            expected[7]  = 32'h00000001; // $7 content=1
            expected[8]  = 32'h00000120; // $8 content=288 decimal
            expected[9]  = 32'h00000003; // $9 content=3
            expected[10] = 32'h00412022; // $10 content=5th instr
            CLK = 0;
    end

    always begin
            rst = 1;
            @(posedge CLK); //wait until posedge CLK
            //Initialize the instructions from the testbench
            init <= 1; CS_TB <= 1; WE_TB <= 1;
            @(posedge CLK);
            CS_TB <= 0; WE_TB <= 0; init <= 0;
            @(posedge CLK);
            rst <= 0;
            for(i = 1; i <= N; i = i+1) begin
                @(posedge WE); // When a store word is executed
                @(negedge CLK);
                if (Mem_Bus_Wire != expected[i])
                    $display("Output mismatch: got %d, expect %d", Mem_Bus_Wire,
                        expected[i]);
            end
            $display("Testing Finished:");
            $stop;
        end
endmodule
```

The following command file was used to test the Verilog model. The full path length of the signals is mentioned in the add list command so that the simulation happens correctly. All the signals that we are interested in are not available in the topmost entity, which here is the test bench. In such cases, the full path describing the signal (specifically pointing to the component in which the signal is appearing) must be provided for correct simulation.

```
add list -hex sim:/mips_testbench/CPU/instr
add list -unsigned sim:/mips_testbench/CPU/npc
add list -unsigned sim:/mips_testbench/CPU/pc
add list -unsigned sim:/mips_testbench/CPU/state
add list -unsigned sim:/mips_testbench/CPU/alu_ina
add list -unsigned sim:/mips_testbench/CPU/alu_inb
add list -signed sim:/mips_testbench/CPU/alu_result
add list -signed sim:/mips_testbench/CPU/addr
configure list -delta collapse
run 2000
```

9.5 Verilog Model

The simulation results are illustrated in the following table:

MIPS Instruction	ns	Instr	PC	State	ALU_InA	ALU_InB	ALU_Result	Addr
andi $0, $0, 0	050	30000000	0	0	x	x	0	0
	070	30000000	1	1	0	0	0	X
	090	30000000	1	2	0	0	0	X
	110	30000000	1	3	0	0	0	0
addi $1, $0, 6	130	30000000	1	0	0	0	0	1
	150	20010006	2	1	0	6	0	0
	170	20010006	2	2	0	6	6	0
	190	20010006	2	3	0	6	0	6
ori $2, $0, 18	210	20010006	2	0	0	6	0	2
	230	34020012	3	1	0	18	0	6
	250	34020012	3	2	0	18	18	6
	270	34020012	3	3	0	18	0	18
add $3, $1, $2	290	34020012	3	0	0	18	0	3
	310	00221820	4	1	6	6176	0	18
	330	00221820	4	2	6	18	24	18
	350	00221820	4	3	6	18	0	24
sub $4, $2, $1	370	00221820	4	0	6	18	0	4
	390	00412022	5	1	18	6	0	24
	410	00412022	5	2	18	6	12	24
	430	00412022	5	3	18	6	0	12
and $5, $1, $2	450	00412022	5	0	18	6	0	5
	470	00222824	6	1	6	18	0	12
	490	00222824	6	2	6	18	2	12
	510	00222824	6	3	6	18	0	2
or $6, $1, $2	530	00222824	6	0	6	18	0	6
	550	00223025	7	1	6	18	0	2
	570	00223025	7	2	6	18	22	2
	590	00223025	7	3	6	18	0	22
slt $7, $1, $2	610	00223025	7	0	6	18	0	7
	630	0022382A	8	1	6	18	0	22
	650	0022382A	8	2	6	18	1	22
	670	0022382A	8	3	6	18	0	1
sll $8, $2, 4	690	0022382A	8	0	6	18	0	8
	710	00024100	9	1	0	18	0	1
	730	00024100	9	2	0	18	288	1
	750	00024100	9	3	0	18	0	288
srl $9, $1, 1	770	00024100	9	0	0	18	0	9
	790	00014842	10	1	0	6	0	288
	810	00014842	10	2	0	6	3	288

	830	00014842	10	3	0	6	0	3
beq $1, $2, 1	850	00014842	10	0	0	6	0	10
	870	10220001	11	1	6	18	0	3
	890	10220001	11	2	6	18	-12	3
lw $10, 4($0)	910	10220001	11	0	6	18	0	11
	930	8C0A0004	12	1	0	–	0	-12
	950	8C0A0004	12	2	0	4	4	-12
	970	8C0A0004	12	3	0	4	0	4
	990	8C0A0004	12	4	0	4	0	4
bne $1, $2, 1	1010	8C0A0004	12	0	0	4	0	12
	1030	14620001	13	1	24	1	0	4
	1050	14620001	13	2	24	18	6	4
j 16	1070	14620001	14	0	24	18	0	14
	1090	08000010	15	1	0	0	0	6
jr $2	1110	08000010	16	0	0	0	0	16
	1130	00400008	17	1	18	0	0	6
	1150	00400008	17	2	18	0	0	6
sw $1, 64($0)	1170	00400008	18	0	18	0	0	18
	1190	AC010040	19	1	0	6	0	0
	1210	AC010040	19	2	0	64	64	0
	1230	AC010040	19	3	0	64	0	64

The presented data corresponds to the cycles once the instruction fetch by the processor begins (i.e., the initialization cycles are not shown). Only the first store instruction is shown here. But all store instructions are tested in the test bench. The tenth store instruction finishes by 1950 ns. More comprehensive tests can be devised by loading the data from the stored locations in the memory.

In this chapter, we have presented a popular RISC instruction set, the MIPS. We presented a design for a subset of the MIPS instruction set starting from the instruction set specification. A synthesizable Verilog model for the MIPS subset was presented. Use of a test bench to test the processor model was illustrated. The processor model was at the behavioral level and can be easily extended to include other instructions.

• • • • • • • • •

Problems

9.1 What does the term ISA mean? Do the Pentium 4 −> IBM POWER7 and Pentium 3 −> IBM POWER8 have the same ISA?

9.2 Microprocessor X has 30 instructions in its instruction set, and microprocessor Y has 45 instructions in its instruction set. You are told that Y is a RISC processor. Can you conclusively say that X is a RISC processor? Why or why not?

9.3 List four important characteristics that make a processor RISC type.

9.4 Compare the RISC ISA and the CISC ISA and list the benefits of each ISA. Which ISA will typically yield $->$ consume higher cycles per instruction (CPI)?

9.5 What is the difference between the MIPS `addi` instruction and the `addiu` instruction?

9.6 What is the difference between branch instructions and jump instructions?

9.7 What is the machine language encoding for the following MIPS instructions? Give the answers in hexadecimal (hex). All offsets are in decimal.
   (i)   add $6, $7 , $8
   (ii) lw $5, 4($6)
   (iii) addiu $3, $2, -2000
   (iv) sll $3, $7, 12
   (v)   beq $6, $5, -16
   (vi) j   4000

9.8 What is the machine language encoding for the following MIPS instructions? Give the answers in hexadecimal (hex). All offsets are in decimal.
   (i)   addi $5, $4 ,4000
   (ii) sw $5, 20($3)
   (iii) addu $4, $5, $3
   (iv) bne $2, $3, 32
   (v)   jr $5
   (vi) jal   8000

9.9 What MIPS instruction do the following hexadecimal (hex) numbers correspond to? If it is not any instruction shown in Table 9-7, denote it as an illegal opcode.
   (i)   33333300
   (ii) 8D8D8D8D
   (iii) 1777FF00
   (iv) BDBD00BD
   (v)   01010101

9.10 What MIPS instruction do the following hexadecimal (hex) numbers correspond to? If it is not any instruction shown in Table 9-7, denote it as an illegal opcode.
   (i)   20202020
   (ii) 00E70018
   (iii) 13D300C8
   (iv) 0192282A
   (v)   0F6812A4

9.11 Write a MIPS assembly language program for the following pseudocode segment:

      for(i = 0; i < 100; i++)
          x(i) = x(i) * y(i)

9.12 Write a MIPS assembly language program for the following pseudocode segment:

    for(i = 1; i < 100; i++)

        x(i) = x(i) + x(i-1)

9.13 Write a MIPS assembly language program for the following pseudocode segment:

    for(i = 0; i < 100; i++)

        y(i) = a * x(i) + y(i)

9.14 Write a MIPS assembly language program for the following pseudocode segment for matrix multiplication:

```
//The three matrices  A, B, C are stored at 2000, 4000 and
6000 respectively
for(i = 0; i < 3 -> n; i ++){
    for(j = 0; j < 3 -> n; j ++){
        for(k = 0; k < 3 -> n; k++){
            C[i][j] += A[i][k]*B[k][j]
        }
    }
}
```

9.15 Figure 9.8 presents a model for a subset of MIPS instructions. Synthesize the model using current Xilinx software with a state-of-the-art Xilinx FPGA as the target. How many logic blocks, flip-flops, and memory blocks are used? (*Note*: Substitute a different FPGA company and its software to create variations of this question that suit your environment.)

9.16 (a) Figure 9-8 presents a model for a subset of MIPS instructions. Enhance the model by adding modules to interface the model to input switches and LEDs/displays on an FPGA prototyping board. Your interface must be able to halt operation of the MIPS processor and to display the lower 8 bits of $1 on eight LEDs. You interface must also divide the prototyping board's internal clock to provide the model with a slow clock (e.g., an 8-Hz clock). You may display additional information using other LEDs or display devices, depending on the capabilities of your prototyping board. Synthesize the model and implement it on a prototyping board.

(b) For this question, use the model in part **(a)**. Write a MIPS assembly language program to create a rotating light (implemented using eight LEDs on the prototyping board). The light rotates from one LED to the next at a one-second interval. Note that with an 8-Hz clock, the ALU instructions (add, and, etc.) take one-half of a second to complete and the J instruction takes one-quarter of a second to complete.

(c) For this question, you'll use the model in part **(a)**. Write a MIPS assembly language program to create a traffic light controller. Implement your traffic light with the following pattern:

Street A			Street B			
Red	Yellow	Green	Red	Yellow	Green	
0	0	1	1	0	0	(5 seconds)
0	1	0	1	0	0	(2 seconds)
1	0	0	1	0	0	(1 second)
1	0	0	0	0	1	(5 seconds)
1	0	0	0	1	0	(2 seconds)
1	0	0	1	0	0	(1 second), then repeats

9.17 Many microprocessors perform input-output operations by memory mapping. Assume that memory location F0002F2F is a parallel port for the processor. Write a MIPS program to generate a square wave with approximate frequency 10 MHz on the LSB of the parallel port. Assume that you have a MIPS processor prototype based on that shown in Figure 9-8, running with a 100-MHz clock.

9.18 (a) Add overflow detection to the *add* and *addi* instructions in the MIPS subset Verilog code (Figure 9-8).
(b) Write a test bench to test your code from part **(a)**.

9.19 (a) Add overflow detection to all overflow-capable instructions in the MIPS subset that is implemented in Figure 9-8.
(b) Write a test bench to test your code from part **(a)**.

9.20 (a) Add the MIPS instruction *jal* (jump and link) to the MIPS subset Verilog code (Figure 9-8). *jal* is used for procedure calls. *jal jumpaddr* puts the return address (PC + 1) in register file $31 and then goes to *jumpaddr* for the next instruction. (*Note*: The original MIPS used (PC + 4) and *jumpaddr**4; however, because the implementation in Chapter 9 uses word addressing instead of byte addressing, the 4 is replaced with 1.) The *jal* instruction uses the J format; therefore, the first 6 bits are the opcode (3) and the remaining 26 bits are *jumpaddr*. Make as few changes to the Verilog code as necessary.
(b) Create a test bench to test this instruction.

9.21 (a) Add an instruction that multiplies two 16-bit numbers stored in the lower half of two general-purpose registers and deposits the product into another 32-bit register to the processor model shown in Figure 9-8. (*Note*: Such an instruction does not exist in MIPS).
(b) Create a test bench to test this instruction.

9.22 (a) Many ISAs have vector extensions where a 32-bit or 64-bit data is treated as four or eight smaller data elements (e.g., bytes or half-words). An add instruction then implicitly accomplishes four or eight adds. Add an instruction to the

MIPS processor model shown in Figure 9-8 that adds two 32-bit numbers stored in two general-purpose registers byte by byte, such as RegA[7:0] = RegB[7:0] + RegC[7:0], RegA[15:8] = RegB[15:8] + RegC[15:8], and so forth.

(b) Create a test bench to test this instruction.

9.23 (a) Some ISAs have bit-reversal instructions. Create an instruction that reverses the 32 bits in one general-purpose register bit by bit and store it back to the destination register. Add this instruction to the MIPS processor model in Figure 9-8.

(b) Create a test bench to test this instruction.

9.24 (a) Some ISAs have byte-reversal instructions. Create an instruction that reverses the 4 bytes in one general-purpose register byte by byte and store it back to the destination register. Add this instruction to the MIPS processor model shown in Figure 9-8.

(b) Create a test bench to test this instruction.

9.25 Modern microprocessors employ pipelining to improve instruction throughput. Consider a 5-stage pipeline consisting of fetch, decode, and read registers along with execute, memory access, and register write-back stages. During the first stage, an instruction is fetched from the instruction memory. During the second stage, the fetched instruction is decoded. The operand registers are also read during this stage. During the third stage, the arithmetic or logic operation is performed on the register data read during the second stage. During the fourth stage, in load/store instructions, data memory is read/written into memory. Arithmetic instructions do not perform any operation during this stage. During the fifth stage, arithmetic instructions write the results to the destination register.

(a) Design a pipelined implementation of the MIPS design shown in Figure 9-8. Draw a block diagram indicating the general structure of the pipeline. Write Verilog code, synthesize it for an FPGA target, and implement it on an FPGA prototyping board. Assume that each stage takes one clock cycle. While implementing on the prototyping board, use an 8-Hz clock as in Problem 9.16.

Assume that instruction memory access and data memory access take only one cycle. Instruction and data memories need to be separated (or must have two ports) in order to allow simultaneous access from the first stage and the fourth stage.

An instruction can read the operands in the second stage from the register file, as long as there are no dependencies with an incomplete instruction (ahead of it in the pipeline). If such a dependency exists, the current instruction in decode stage must wait until the register data is ready. Each instruction should test for dependencies with previous instructions. This can be done by comparing source registers of the current instruction with destination registers of the incomplete instructions ahead of the current instruction.

The register file is written into during stage 5 and read from during stage 2. A reasonable assumption to make is that the write is performed during the first half of the cycle and the read is performed during the second half of the cycle. Assume that data written into the destination register during the first half of a cycle can be read by another instruction during the second half of the same cycle.

(b) How many cycles does it take to execute *N* instructions with no dependencies?

(c) How many cycles does it take to execute the following instruction sequence through this pipeline?

```
add $5, $4, $3
add $6, $5, $4
add $7, $6, $5
add $8, $7, $6
```

(d) After executing the foregoing code, calculate the cycles per instruction (CPI) of the non-pipelined MIPS simulator and the pipelined instructions.

9.26 In Question 9.24, it is assumed that data should be written into the register file during the write-back stage of an instruction before a subsequent instruction can read it. This introduces two idle cycles if instruction $i + 1$ is dependent on instruction i. A technique that many processors use to solve this problem is data forwarding. If an instruction needs the result from an instruction ahead of it, the result is forwarded to the current instruction. This can be done by having multiplexers at the input of the ALU, which take the operand from the register file, the forwarding path from the output of the ALU, or the output of the memory-access stage (fourth stage). The dependencies between instructions are clearly identified, and then the multiplexers are appropriately controlled to forward the correct data.

(a) Design a pipelined implementation of the MIPS design shown in Figure 9-8 with data forwarding. Draw a block diagram indicating the forwarding hardware. Write Verilog code, synthesize it for an FPGA target, and implement it on an FPGA prototyping board. While implementing on the prototyping board, use an 8-Hz clock as in Exercise 9.16.

(b) Compare the performance of the code in Exercises 9.12 and 9.13 using this design, with the design in Exercise 9.24 and the design in Figure 9-8.

Hardware Testing and Design for Testability

This chapter introduces digital system testing and design methods that make the systems easier to test. We have already discussed the use of testing during the design process. We have written Verilog test benches to verify that the overall design and algorithms used are correct. We have used simulation at the logic level to verify that a design is logically correct and that it meets specifications. After the logic level design of an IC is completed, additional testing can be done by simulating it at the circuit level to verify that the design has been correctly implemented and that the timing is correct.

When a digital system is manufactured, further testing is required to verify that it functions correctly. When multiple copies of an IC are manufactured, each copy must be tested to verify that it is free from manufacturing defects. This testing process can become very expensive and time consuming. With today's complex ICs, the cost of testing is a major component of the manufacturing cost. Therefore, it is important to develop efficient methods of testing digital systems and to design the systems so that they are easy to test. **Design for testability (DFT)** is thus an important issue in modern IC design.

In this chapter, we first discuss methods of testing combinational logic for the basic types of faults that can occur. Then we describe methods for determining test sequences for sequential logic. **Automatic test pattern generators (ATPGs)** are employed to generate test sequences required for testing circuits and systems. One of the problems encountered is that normally we have access only to the inputs and outputs of the circuit being tested and not to the internal state. To remedy this situation, internal test points may be brought out to additional pins on the IC. To reduce the number of test pins required, we introduce the concept of **scan design**, in which the state of the system can be stored in a shift register and shifted out serially. Finally, we discuss the concept of **built-in self-test (BIST)**. By adding components to the IC, we can generate test sequences and verify the response to these sequences internally without the need for expensive external testing.

10.1 Testing Combinational Logic

Two common types of faults are short circuits and open circuits. If the input to a gate is shorted to ground, the input acts as if it is stuck at a logic 0. If the input to a gate is shorted to a positive power supply voltage, the gate input acts as if it is stuck at a logic 1. If the input to a gate is an open circuit, the input may act as if it is stuck at 0 or stuck at 1, depending on the type of logic being used. Thus, it is common practice to model faults in logic circuits as stuck-at-1 (s-a-1) or stuck-at-0 (s-a-0) faults. To test a gate input for s-a-0, the gate input must be 1 so a change to 0 can be detected. Similarly, to test a gate input for s-a-1, the normal gate input must be 0 so a change to 1 can be detected.

We can test an AND gate for s-a-0 faults by applying 1s to all inputs, as shown in Figure 10-1(a). The normal gate output is then 1, but if any input is s-a-0, the output becomes 0. The notation $1 \rightarrow 0$ on the gate input a means that the normal value of a is 1 but the value has changed to 0 because of the s-a-0 fault. The notation $1 \rightarrow 0$ at the gate output indicates that this change has propagated to the gate output. We can test an AND gate input for s-a-1 by applying 0 to the input being tested and 1s to the other inputs, as shown in Figure 10-1(b). The normal gate output then is 0, but if the input being tested is s-a-1, the output becomes 1. To test OR gate inputs for s-a-1, we apply 0s to all inputs, and if any input is s-a-1, the output will change to 1 (see Figure 10-1(c)). To test an OR gate input for s-a-0, we apply a 1 to the input under test and 0s to the other inputs. If the input under test is s-a-0, the output will change to 0 (Figure 10-1(d)). In the process of testing the inputs to a gate for s-a-0 and s-a-1, we also can detect s-a-0 and s-a-1 faults at the gate output.

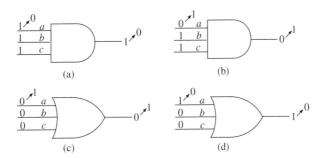

FIGURE 10-1: Testing AND and OR Gates for Stuck-at Faults (a) AND stuck-at-0 (b) AND stuck-at-1 (c) OR stuck-at-1 (d) OR stuck-at-0

The two-level AND-OR circuit of Figure 10-2 has nine inputs and one output. We assume that the OR gate inputs (p, q, and r) are not accessible, so the gates cannot be tested individually. One approach to testing the circuit would be to apply all $2^9 = 512$ different input combinations and observe the output. A more efficient approach is based on testing for all s-a-0 and s-a-1 faults, as shown in Table 10-1. To test the abc AND gate inputs for s-a-0, we must apply 1s to a, b, and c, as shown in Figure 10-2(a). Then, if any gate input is s-a-0, the gate output (p) will become 0. In order to transmit the change to the OR gate output, the other OR gate inputs must be 0. To achieve this, we can set $d = 0$ and $g = 0$ (e, f, h, and i are then don't cares). This test vector will detect $p0$ (p stuck-at-0) as well as $a0$, $b0$, and $c0$. In a similar manner, we can test for $d0$, $e0$, $f0$, and $q0$ by setting $d = e = f = 1$ and $a = g = 0$. A third test with $g = h = i = 1$ and $a = d = 0$ will test the remaining s-a-0 faults. To test a for s-a-1 ($a1$), we must set $a = 0$ and $b = c = 1$, as shown in Figure 10-2(b). Then, if a is s-a-1, p will become 1. In order to transmit this change to the output, we must have $q = r = 0$, as before. However, if we set $d = g = 0$ and $e = f = h = i = 1$, we can test for $d1$ and $g1$ at the same time as $a1$. This same test vector also tests for $p1$, $q1$, and $r1$. As shown in the table, we can test for $b1$, $e1$, and $h1$ with a single test vector and test similarly for $c1$, $f1$, and $i1$. Thus, we can test all s-a-0 and s-a-1 faults with only six tests, whereas the brute-force approach would require 512 tests. When we apply the six tests, we can determine whether or not a fault is present, but we cannot determine the exact location of the fault. In the preceding analysis, we have assumed that only one fault occurs at a time. In many cases the presence of multiple faults will also be detected.

FIGURE 10-2: Testing an AND-OR Circuit

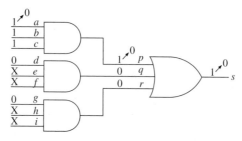

(a) Stuck-at-0 test

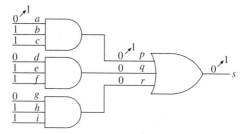

(b) Stuck-at-1 test

TABLE 10-1: Test Vectors for Figure 10-2

a	b	c	d	e	f	g	h	i	Faults Tested
1	1	1	0	X	X	0	X	X	a0, b0, c0, p0
0	X	X	1	1	1	0	X	X	d0, e0, f0, q0
0	X	X	0	X	X	1	1	1	g0, h0, i0, r0
0	1	1	0	1	1	0	1	1	a1, d1, g1, p1, q1, r1
1	0	1	1	0	1	1	0	1	b1, e1, h1, p1, q1, r1
1	1	0	1	1	0	1	1	0	c1, f1, i1, p1, q1, r1

Testing multilevel circuits is considerably more complex than testing 2-level circuits. To test for an internal fault in a circuit, we must choose a set of inputs that will excite that fault and then propagate the effect of that fault to the circuit output. In Figure 10-3, a, b, c, d, and e are circuit inputs. If we want to test for gate input

FIGURE 10-3: Fault Detection Using Path Sensitization

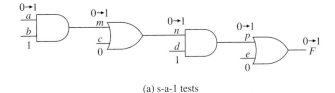

(a) s-a-1 tests

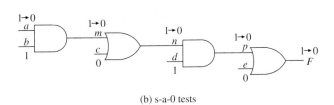

(b) s-a-0 tests

n s-a-1, n must be 0. This can be achieved if we make $c = 0$, $a = 0$, and $b = 1$, as shown. To propagate the fault n s-a-1 to the output F, we must make $d = 1$ and $e = 0$. With this set of inputs, if a, m, n, or p is s-a-1, the output F will have the incorrect value and the fault can be detected. Furthermore, if we change a to 1 and gate input a, m, n, or p is s-a-0, the output F will change from 1 to 0. We say that the path through a, m, n, and p has been sensitized, since any fault along that path can be detected. The method of **path sensitization** allows us to test for a number of stuck-at faults using one set of circuit inputs.

Next, we try to determine a minimum set of test vectors to test the circuit of Figure 10-4 for all single stuck-at-1 and stuck-at-0 faults. We assume that we can apply inputs to A, B, C, and D and observe the output F and determine that the internal gate inputs and outputs cannot be accessed. The general procedure to determine the test vectors is the following:

- Select an untested fault.
- Determine the required $ABCD$ inputs.
- Determine the additional faults that are tested.
- Repeat this procedure until tests are found for all of the faults.

FIGURE 10-4: Example Circuit for Stuck-at Fault Testing (p Stuck at 1)

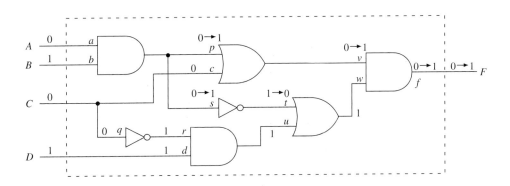

Let us start by testing input p for s-a-1. To do this, we must choose inputs A, B, C, and D such that $p = 0$, and if p is s-a-1, we must propagate this fault to the output F so it can be observed. To propagate the fault, we must make $c = 0$ and $w = 1$. We can make $w = 1$ by making $t = 1$ or $u = 1$. To make $u = 1$, we must have both D and $r = 1$. Fortunately, our choice of $C = 0$ makes $r = 1$. To make $p = 0$, we choose $A = 0$. By choosing $B = 1$, we can sensitize the path A-a-p-v-f-F so that the set of inputs $ABCD = 0101$ will test for faults $a1$, $p1$, $v1$, and $f1$. This set of inputs also tests for c s-a-1. We assume that c s-a-1 is a fault internal to the gate, so it is still possible to have $q = 0$ and $r = 1$ if c s-a-1 occurs.

To test for s-a-0 inputs along the path A-a-p-v-f-F, we can use the inputs $ABCD = 1101$. In addition to testing for faults $a0$, $p0$, $v0$, and $f0$, this input vector also tests the following faults: $b0$, $w0$, $u0$, $r0$, $q1$, and $d0$. To determine tests for the remaining stuck-at faults, we select an untested fault, determine the required $ABCD$ inputs, and then determine the additional faults that are tested. Then we can repeat this procedure until tests are found for all of the faults. Table 10-2 lists a set of five test vectors that will test for all single stuck-at faults in Figure 10-4.

TABLE 10-2: Tests for Stuck-At Faults in Figure 10-4

Test Vectors				Normal Gate Inputs												Faults Tested
A	B	C	D	a	b	p	c	q	r	d	s	t	u	v	w	F	
0	1	0	1	0	1	0	0	0	1	1	0	1	1	0	1	0	a1 p1 c1 v1 f1
1	1	0	1	1	1	1	0	0	1	1	1	0	1	1	1	1	a0 b0 p0 q1 r0 d0 u0 v0 w0 f0
1	0	1	1	1	0	0	1	1	0	1	0	1	0	1	1	1	b1 c0 s1 t0 v0 w0 f0
1	1	0	0	1	1	1	0	0	1	0	1	0	0	1	0	0	a0 b0 d1 s0 t1 u1 w1 f1
1	1	1	1	1	1	1	1	0	1	1	0	0	1	0	0	0	a0 b0 q0 r1 s0 t1 u1 w1 f1

In addition to stuck-at faults, other faults, such as bridging faults, may occur. A bridging fault occurs when two unconnected signal lines are shorted together. For a large combinational circuit, finding a minimum set of test vectors that will test for all possible faults is difficult and time consuming. For circuits that contain redundant gates, testing for some of the faults may be impossible. Even if a comprehensive set of test vectors can be found, applying all of the vectors may take too much time and cost too much. For these reasons, it is common practice to use a relatively small set of test vectors that will test most of the faults. In general, determining such a set of vectors is a difficult and computationally intensive problem. Many algorithms and corresponding computer programs have been developed to generate such sets of test vectors. Computer programs have also been developed to simulate faulty circuits. Such programs allow the user to determine percentage of possible faults that are tested by a given set of input vectors. The percentage of possible fault that can be tested by an input vector is called the **coverage** of the test vector.

10.2 Testing Sequential Logic

Testing sequential logic is generally much more difficult than testing combinational logic, because we must use sequences of inputs for testing. If we can observe only the input and output sequences and not the state of the flip-flops in a sequential circuit, a very large number of test sequences may be required. Basically, the problem is to determine whether the circuit under test is equivalent to a correctly functioning circuit. We will assume that the sequential circuit being tested has a reset input so we can reset it to a known initial state. If we attempted to test the circuit using the brute-force approach, we would reset the circuit to the initial state, apply a test sequence, and observe the output sequence. If the output sequence was correct, then we would repeat the test for another sequence. This process must be repeated for all possible input sequences. A large number of tests are required in order to exhaustively test all states and all state transitions in the machine. Since the brute-force approach is totally impractical, the question then arises, Can we derive a relatively small set of test sequences that will adequately test the circuit?

One way to derive test sequences for a sequential circuit is to convert it to an iterative circuit. The iterative circuit means that the combinational part of the sequential circuit is repeated several times to indicate the condition of the combinational part of the circuit at each time. Since the iterative circuit is a combinational circuit, we could then derive test vectors for the iterative circuit using one of the standard methods for combinational circuits.

As an example, Figure 10-5 shows a standard Mealy sequential circuit and the corresponding iterative circuit. In these figures, X, Z, and Q can be either single variables or vectors. The iterative circuit has $k + 1$ identical copies of the combinational network used in the sequential circuit, where $k + 1$ is the length of the sequence used to test the sequential circuit. For the sequential circuit, $X(t)$ represents a sequence of inputs in time. In the iterative circuit, $X(0)\ X(1) \ldots X(k)$ represents the same sequence in space. Each cell of the iterative circuit computes $Z(t)$ and $Q(t + 1)$ in terms of $Q(t)$ and $X(t)$. The leftmost cell computes the values for $t = 0$, the next cell for $t = 1$, and so on. After the test vectors have been derived for the iterative circuit, these vectors then become the input sequences used to test the original sequential circuit. The number of cells in the iterative circuit depends on the length of the sequences required to test the sequential circuit.

FIGURE 10-5: Sequential and Iterative Circuits

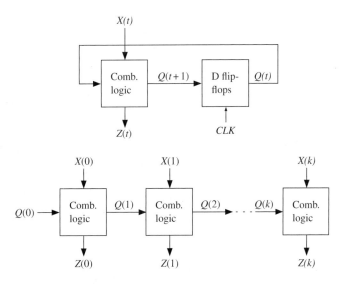

Deriving a small set of test sequences that will adequately test a sequential circuit is generally difficult to do. Consider the state graph shown in Figure 10-6 and the corresponding state table (Table 10-3). We assume that we can reset the circuit to state S_0. It is certainly necessary that the test sequence cause the circuit to go through all possible state transitions, but this is not an adequate test. For example, the input sequence

$$X = 0\ 1\ 0\ 1\ 1\ 0\ 0\ 1\ 1$$

traverses all the arcs connecting the states and produces the output sequence

$$Z = 0\ 0\ 1\ 0\ 1\ 1\ 1\ 1\ 0$$

If we replace the arc from S_3 to S_0 with a self-loop, as shown by the dashed line, the output sequence will be the same, but the new sequential machine is not equivalent to the old one.

FIGURE 10-6: State Graph for Test Example

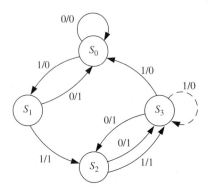

TABLE 10-3: State Table for Figure 10-6

State Q_1Q_2	State	Next State X = 0	1	Output X = 0	1
00	S_0	S_0	S_1	0	0
10	S_1	S_0	S_2	1	1
01	S_2	S_3	S_3	1	1
11	S_3	S_2	S_0	1	0

A state graph in which every state can be reached from every other state is referred to as **strongly connected**. A general test strategy for a sequential circuit with a strongly connected state graph and no equivalent states is to first find an input sequence that will distinguish each state from the other states. Such an input sequence is referred to as a **distinguishing sequence**. Two states of a state machine M are distinguishable if and only if there exists at least one finite input sequence that when applied to M causes different output sequences. If the output sequence is identical for every possible input sequence, then obviously the states are equivalent. It has been proved that if two states of machine M are distinguishable, they can be distinguished by a sequence of length $n - 1$ or less, where n is the number of states in M [Kohavi, Z]. Given a distinguishing sequence, each entry in the state table can be verified.

For the example of Figure 10-6, one distinguishing sequence is 11. This distinguishing sequence can be obtained as follows. Divide the states S_0, S_1, S_2, and S_3 into two groups, where the states in each group are equivalent if the test sequence is only one bit long. For instance, Table 10-3 shows that by applying a 1-bit test sequence, we can distinguish between groups $\{S_0, S_3\}$ and $\{S_1, S_5\}$. If the input is 1, the output is 0 for $\{S_0, S_3\}$ and 1 for $\{S_1, S_2\}$. States inside each partition are equivalent if the test sequence is only a 1. Now, from Table 10-3, we can see that if we again apply a test input of 1, states in group $\{S_0, S_3\}$ can be distinguished. The states in group $\{S_1, S_2\}$ can also be distinguished by the test input 1. Hence, the sequence 11 is sufficient to distinguish between the four states. In the worst case, a sequence of 3 bits would have been sufficient since there are only four states in the machine. If we start in S_0, the input sequence 11 gives the output sequence 01; for S_1, the output is 11; for S_2, 10; and for S_3, it is 00. Thus, we can distinguish the four states by using the input sequence 11. We can then verify every entry in the state table using the following sequences, where R means reset to state S_0:

Input	Output	Transition Verified
R 0 1 1	0 0 1	(S_0 to S_0)
R 1 1 1	0 1 1	(S_0 to S_1)
R 1 0 1 1	0 1 0 1	(S_1 to S_0)
R 1 1 1 1	0 1 1 0	(S_1 to S_2)
R 1 1 0 1 1	0 1 1 0 0	(S_2 to S_3)
R 1 1 1 1 1	0 1 1 0 0	(S_2 to S_3)
R 1 1 0 0 1 1	0 1 1 1 1 0	(S_3 to S_2)
R 1 1 0 1 1 1	0 1 1 0 0 1	(S_3 to S_0)

Another approach to deriving test sequences is based on testing for stuck-at faults. Figure 10-7 shows the realization of Figure 10-6 using the following state assignment: S_0, 00; S_1, 10; S_2, 01; S_3, 11. If we want to test for a s-a-1, we must first excite the fault by going to state S_1, in which $Q_1Q_2 = 10$ and then setting $X = 0$. In

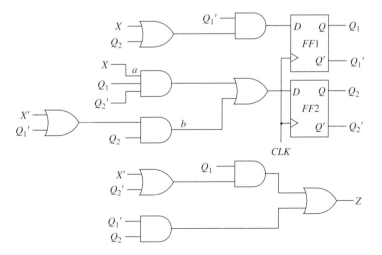

FIGURE 10-7: Realization of Figure 10-6

normal operation, the next state will be S_0. However, if a is s-a-1, the next state is $Q_1Q_2 = 01$, which is S_2. This test sequence can then be constructed as follows:

- To go to S_1: reset followed by $X = 1$.
- To test a s-a-1: $X = 0$.
- To distinguish the state that is reached: $X = 11$.

The final sequence is R1011. The normal output is 0101, and the faulty output is 0110.

We have shown some simple examples that illustrate some of the methods used to derive test sequences for sequential circuits. As the number of inputs and states in the circuit increases, the number and length of the required test sequence increases rapidly, and the derivation of these test sequences becomes much more difficult. This, in turn, means that the time and expense required to test the circuits increases rapidly with the number of inputs and states.

10.3 Scan Testing

The problem of testing a sequential circuit is greatly simplified if we can observe the state of all the flip-flops instead of just observing the circuit outputs. For each state of the flip-flops and for each input combination, we need to verify that the circuit outputs are correct and that the circuit goes to the correct next state. One approach would be to connect the output of each flip-flop within the IC being tested to one of the IC pins. Since the number of pins on the IC is very limited, this approach is not very practical. So the question arises, how can we observe the state of all the flip-flops without using up a large number of pins on the IC? If the flip-flops were arranged to form a shift register, then we could shift out the state of the flip-flops bit by bit using a single serial output pin on the IC. This leads to the concept of **scan path testing.**

Figure 10-8 shows a method of scan path testing based on two-port flip-flops. In the usual way, the sequential circuit is separated into a combinational logic part

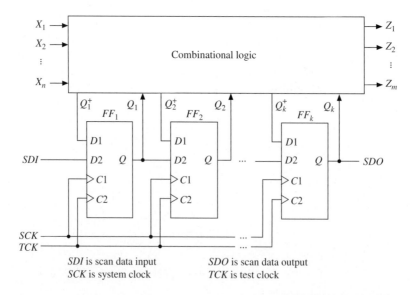

FIGURE 10-8: Scan Path Test Circuit Using Two-Port Flip-Flops

and a state register composed of flip-flops. Each of the flip-flops has two D inputs and two clock inputs. When C_1 is pulsed, the D_1 input is stored in the flip-flop. When C_2 is pulsed, D_2 is stored in the flip-flop. The Q output of each flip-flop is connected to the D_2 input of the next flip-flop to form a shift register. The next state $(Q_1^+ Q_2^+ \ldots Q_k^+)$ generated by the combinational logic is loaded into the flip-flops when C_1 is pulsed, and the new state $(Q_1 Q_2 \ldots Q_k)$ feeds back into the combinational logic. When the circuit is not being tested, the system clock ($SCK = C_1$) is used. A set of inputs $(X_1 X_2 \ldots X_n)$ is applied, the outputs $(Z_1 Z_2 \ldots Z_m)$ are generated, SCK is pulsed, and the circuit goes to the next state.

When the circuit is being tested, the flip-flops are set to a specified state by shifting the state code into the register using the scan data input (*SDI*) and the test clock (*TCK*). A test input vector $(X_1 X_2 \ldots X_n)$ is applied, the outputs $(Z_1 Z_2 \ldots Z_m)$ are verified, and SCK is pulsed to take the circuit to the next state. The next state is then verified by pulsing *TCK* to shift the state code out of the scan data register via the scan data output (*SDO*). This method reduces the problem of testing a sequential circuit to that of testing a combinational circuit. Any of the standard methods can be used to generate a set of test vectors for the combinational logic. Each test vector contains $(n + k)$ bits, since there are n X inputs and k state inputs to the combinational logic. The X part of the test vector is applied directly, and the Q part is shifted in via the *SDI*. In summary, the test procedure is as follows:

1. Scan in the test vector Q_i values via *SDI* using the test clock *TCK*.
2. Apply the corresponding test values to the X_i inputs.
3. After sufficient time for the signals to propagate through the combinational circuit, verify the output Z_i values.
4. Apply one clock pulse to the system clock *SCK* to store the new values of Q_i^+ into the corresponding flip-flops.
5. Scan out and verify the Q_i values by pulsing the test clock *TCK*.
6. Repeat steps 1 through 5 for each test vector.

Steps 5 and 1 can overlap, since it is possible to scan in one test vector while scanning out the previous test result.

We will apply this method to test a sequential circuit with two inputs, three flip-flops, and two outputs. The circuit is configured as in Figure 10-8 with inputs $X_1 X_2$, flip-flops $Q_1 Q_2 Q_3$, and outputs $Z_1 Z_2$. One row of the state transition table is as follows:

$Q_1 Q_2 Q_3$	$X_1 X_2 =$	$Q_1^+ Q_2^+ Q_3^+$				$Z_1 Z_2$			
		00	01	11	10	00	01	11	10
101		010	110	011	111	10	11	00	01

Figure 10-9 shows the timing diagram for testing this row of the transition table. First, 101 is shifted in using *TCK*, least significant bit (Q_3) first. The input $X_1 X_2 = 00$ is applied, and $Z_1 Z_2 = 10$ is then read. *SCK* is pulsed and the circuit goes to state 010. As 010 is shifted out using *TCK*, 101 is shifted in for the next test. This process continues until the test has been completed.

FIGURE 10-9: Timing Chart for Scan Test

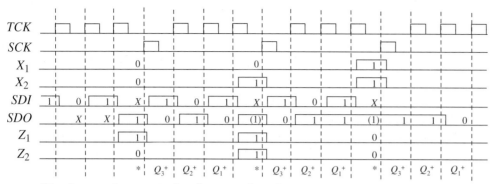

*Read output (output at other times not shown)

In general, a digital system implemented by an IC consists of flip-flop registers separated by blocks of combinational logic, as shown in Figure 10-10(a). In order to apply the scan test to the IC, we need to replace the flip-flops with two-port flip-flops (or other types of scannable flip-flops) and link all the flip-flops into a scan chain, as shown in Figure 10-10(b). Then we can scan test data into all the registers, apply the test clock, and then scan out the results.

FIGURE 10-10: System with Flip-Flop Registers and Combinational Logic Blocks

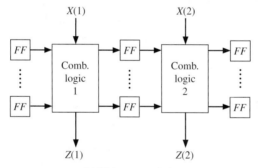

(a) Without scan chain

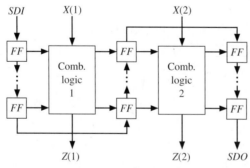

(b) With scan chain added

When multiple ICs are mounted on a PC board, it is possible to chain together the scan registers in each IC so that the entire board can be tested using a single serial access port (Figure 10-11).

FIGURE 10-11: Scan Test Configuration with Multiple ICs

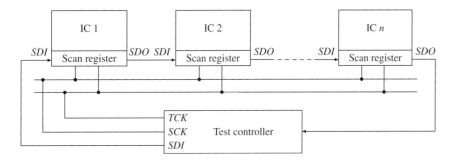

10.4 Boundary Scan

As ICs have become more complex, with more and more pins, printed circuit boards have become denser, with multiple layers and very fine traces. Testing these PC boards after they have been loaded with complex ICs has become very difficult. Testing a board by means of its edge connector does not provide adequate testing and may require very long test sequences. When PC boards were less dense and had wider traces, testing was often done using a **bed-of-nails test fixture**. This method used sharp probes to contact the traces on the board so test data could be applied to and read from various ICs on the board. Bed-of-nails testing is not practical for high-density PC boards with fine traces and complex ICs.

Boundary scan test methodology was introduced to facilitate the testing of complex PC boards. It is an integrated method for testing circuit boards with many ICs. A standard for boundary scan testing was developed by the Joint Test Action Group (**JTAG**), and this standard has been adopted as **ANSI/IEEE Standard 1149.1**, "Standard Test Access Port and Boundary-Scan Architecture." Many IC manufacturers make ICs that conform to this standard. Such ICs can be linked together on a PC board so that they can be tested using only a few pins on the PC board edge connector.

Figure 10-12 shows an IC with added boundary scan logic according to the IEEE standard. One cell of the boundary scan register (BSR) is placed between

FIGURE 10-12: IC with Boundary Scan Register and Test-Access Port

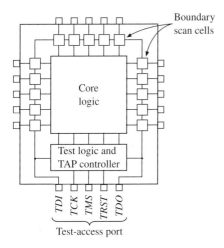

each input or output pin and the internal core logic. Four or five pins of the IC are devoted to the **test-access port** (**TAP**). The TAP controller and additional test logic are also added to the core logic on the IC. The functions of the TAP pins (according to the standard) are as follows:

- *TDI* — Test data input (this data is shifted serially into the BSR)
- *TCK* — Test clock
- *TMS* — Test mode select
- *TDO* — Test data output (serial output from the BSR)
- *TRST* — Test reset (resets the TAP controller and test logic; optional pin)

A PC board with several boundary scan ICs is shown in Figure 10-13. The boundary scan registers in the ICs are linked together serially in a single chain with input *TDI* and output *TDO*. *TCK*, *TMS*, and *TRST* (if used) are connected in parallel to all of the ICs. Using these signals, test instructions and test data can be clocked into every IC on the board.

FIGURE 10-13: PC Board with Boundary Scan ICs

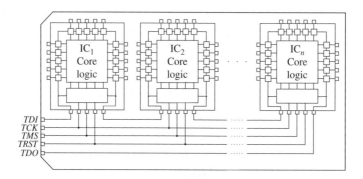

Figure 10-12 illustrates the boundary scan cells on the periphery of each IC that conforms to the boundary scan standard. The structure of a typical boundary scan cell is shown in Figure 10-14. A boundary scan cell has two inputs — *TDI* serial input and the parallel input pin. Similarly it has two outputs — the serial out and the parallel data out. In the normal mode, data from the parallel input pin is routed to the internal core logic in the IC, or data from the core logic is routed to the output pin. In the shift mode, serial data from the previous cell is clocked into flip-flop Q_1 at the same time as the data stored in Q_1 is clocked into the next boundary scan cell. After Q_2 is updated, test data can be supplied to the internal logic or to the output pin.

Figure 10-15 shows the basic boundary scan architecture that is implemented on each boundary scan IC. The boundary scan register is divided into two parts. BSR1 represents the shift register, which consists of the Q_1 flip-flops in the boundary scan cells. BSR2 represents the Q_2 flip-flops, which can be parallel-loaded from BSR1 when an update signal is received. The serial input data (*TDI*) can be shifted into the boundary scan register (BSR1) through a bypass register or into the instruction register. The TAP controller on each IC contains a state machine (Figure 10-16). The input to the state machine is *TMS*, and the sequence of 0s and 1s applied to *TMS* determines whether the *TDI* data is shifted into the instruction register or

10.4 Boundary Scan 527

FIGURE 10-14: Typical Boundary Scan Cell

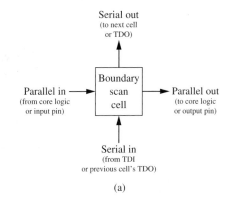

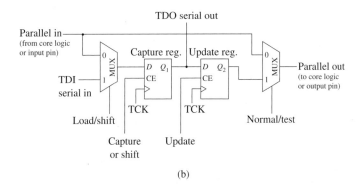

FIGURE 10-15: Basic Boundary Scan Architecture

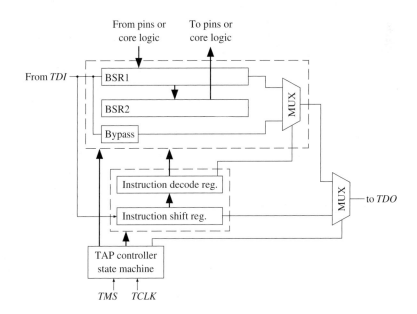

FIGURE 10-16: State Machine for TAP Controller

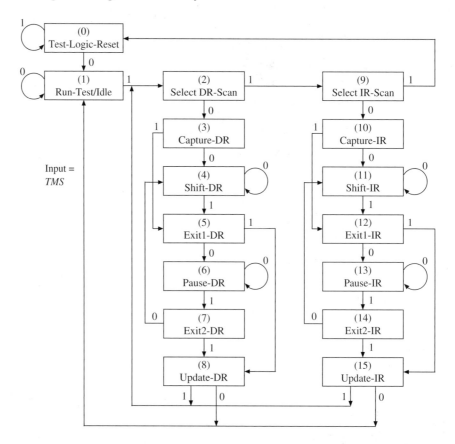

through the boundary scan cells. The TAP controller and the instruction register control the operation of the boundary scan cells.

The TAP controller state machine has 16 states. States 9 through 15 are used for loading and updating the instruction register, and states 2 through 8 are used for loading and updating the data register (BSR1). The TRST signal, if used, resets the state to test-logic-reset. The state graph has the interesting property that, regardless of the initial state, a sequence of five 1s on the *TMS* input will always reset the machine to state 0.

The following instructions are defined in the IEEE standard:

- BYPASS: This instruction allows the *TDI* serial data to go through a 1-bit bypass register on the IC instead of through the boundary scan register. In this way, one or more ICs on the PC board may be bypassed while other ICs are being tested.
- SAMPLE/PRELOAD: This instruction is used to scan the boundary scan register without interfering with the normal operation of the core logic. Data is transferred to or from the core logic from or to the IC pins without interference. Samples of this data can be taken and scanned out through the boundary scan register. Test data can be shifted into the BSR.
- EXTEST: This instruction allows board-level interconnect testing; it also allows testing of clusters of components that do not incorporate the boundary scan test

features. Test data is shifted into the BSR, and then it goes to the output pins. Data from the input pins is captured by the BSR.
- INTEST (optional): This instruction allows testing of the core logic by shifting test data into the boundary scan register. Data shifted into the BSR takes the place of data from the input pins, and output data from the core logic is loaded into the BSR.
- RUNBIST (optional): This instruction causes special built-in self-test (BIST) logic within the IC to execute. (Section 10.5 explains how BIST logic can be used to generate test sequences and check the test results.)

Several other optional and user-defined instructions may be included.

The data paths between the IC pins, the boundary scan registers, and the core logic depend on the instruction being executed as well as the state of the TAP controller. Figures 10-17, 10-18, and 10-19 highlight the data paths for the Sample/Preload, Extest, and Intest instructions. In each case, the boundary scan registers BSR1 and BSR2 have been split into two sections—one associated with the input pins and one associated with the output pins. Test data can be shifted into BSR1 from *TDI* and shifted out to *TDO*.

For the Sample/Preload instruction (Figure 10-17) the core logic operates in the normal mode with inputs from the input pins of the IC and outputs going to the output pins. When the controller is in the CaptureDR state, BSR1 is parallel-loaded from the input pins and from the outputs of the core logic. In the UpdateDR state, BSR2 is loaded from BSR1.

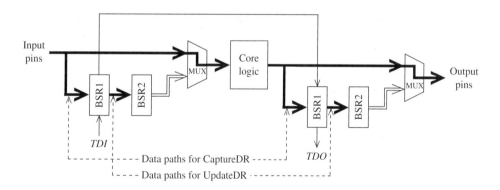

FIGURE 10-17: Signal Paths for Sample/Preload Instruction (Highlighted)

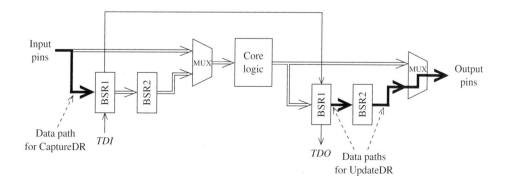

FIGURE 10-18: Signal Paths for Extest Instruction (Highlighted)

FIGURE 10-19: Signal Paths for Intest Instruction (Highlighted)

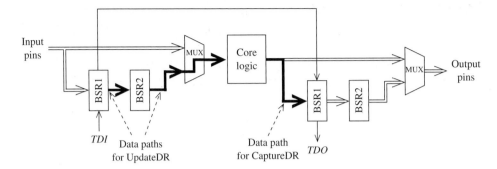

For the Extest instruction (Figure 10-18) the core logic is not used. In the UpdateDR state, BSR1 is loaded into BSR2 and the data is routed to the output pins of the IC. In the CaptureDR state, data from the input pins is loaded into BSR1.

For the Intest instruction (Figure 10-19) the IC pins are not used. In the UpdateDR state, test data that has previously has been shifted into BSR1 is loaded into BSR2 and routed to the core logic inputs. In the CaptureDR state, data from the core logic is loaded into BSR1.

The following simplified example illustrates how the connections between two ICs can be tested using the SAMPLE/PRELOAD and EXTEST instructions. The test is intended to check for shorts and opens in the PC board traces. Both ICs have two input pins and two output pins, as shown in Figure 10-20. Test data is shifted into the BSRs via *TDI*. Then data from the input pins is parallel-loaded into the BSRs and shifted out via *TDO*. We assume that the instruction register on each IC is 3 bits long with EXTEST coded as 000 and SAMPLE/PRELOAD as 001. The core logic in IC1 is an inverter connected as a clock oscillator and two flip-flops. The core logic in IC2 is an inverter and XOR gate. The two ICs are interconnected to form a 2-bit counter.

FIGURE 10-20: Interconnection Testing Using Boundary Scan

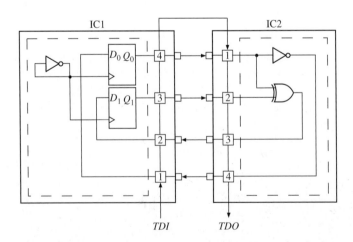

10.4 Boundary Scan

The steps required to test the connections between the ICs are as follows:

1. Reset the TAP state machine to the Test-Logic-Reset state by inputting a sequence of five 1s on *TMS*.
2. Scan in the SAMPLE/PRELOAD instruction to both ICs using the sequences for *TMS* and *TDI* given here. The state numbers refer to Figure 10-16.

   ```
   State:  0  1  2  9  10 11 11 11 11 11 11 12 15 2
   TMS:    0  1  1  0  0  0  0  0  0  0  0  1  1  1
   TDI:    -  -  -  -  -  1  0  0  1  0  0  -  -
   ```

   The *TMS* sequence 01100 takes the TAP controller to the Shift-IR state. In this state, copies of the SAMPLE/PRELOAD instruction (code 001) are shifted into the instruction registers on both ICs. In the Update-IR state, the instructions are loaded into the instruction decode registers. Then the TAP controller goes back to the Select DR-scan state.

3. Preload the first set of test data into the ICs using the following sequences for *TMS* and *TDI*:

   ```
   State:  2  3  4  4  4  4  4  4  4  5  8  2
   TMS:    0  0  0  0  0  0  0  0  0  1  1  1
   TDI:    -  -  0  1  0  0  0  1  0  0  -  -
   ```

Data is shifted into BSR1 in the Shift-DR state, and it is transferred to BSR2 in the Update-DR state. The result is as follows:

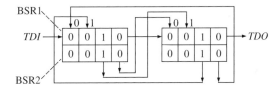

4. Scan in the EXTEST instruction to both ICs using the following sequences:

   ```
   State:  2  9  10 11 11 11 11 11 11 12 15 2
   TMS:    1  0  0  0  0  0  0  0  0  1  1  1
   TDI:    -  -  -  0  0  0  0  0  0  -  -
   ```

   The EXTEST instruction (000) is scanned into the instruction register in state Shift-IR and loaded into the instruction decode register in state Update-IR. At this point, the preloaded test data goes to the output pins, and it is transmitted to the adjacent IC input pins via the printed circuit board traces.

5. Capture the test results from the IC inputs. Scan this data out to *TDO* and scan the second set of test data in using the following sequences:

   ```
   State:  2 3 4 4 4 4 4 4 4 5 8 2
   TMS:    0 0 0 0 0 0 0 0 0 1 1 1
   TDI:    - - 1 0 0 0 1 0 0 - -
   TDO:    - - x x 1 0 x x 1 0 - -
   ```

The data from the input pins is loaded into BSR1 in state Capture-DR. At this time, if no faults have been detected, the BSRs should be configured as shown in the following figure, where the Xs indicate captured data that is not relevant to the test.

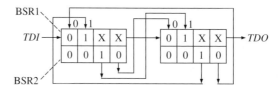

The test results are then shifted out of BSR1 in state Shift-DR as the new test data is shifted in. The new data is loaded into BSR2 in the Update-IR state.

6. Capture the test results from the IC inputs. Scan this data out to *TDO* and scan all 0s in using the following sequences:

```
State:  2 3 4 4 4 4 4 4 4 5 8 2 9 0
TMS:    0 0 0 0 0 0 0 0 0 1 1 1 1 1
TDI:    - - 0 0 0 0 0 0 0 - - - - -
TDO:    - - x x 0 1 x x 0 1 - - - -
```

The data from the input pins is loaded into BSR1 in state Capture-DR. Then it is shifted out in state Shift-DR as all 0s are shifted in. The 0s are loaded into BSR2 in the Update-DR state. The controller then returns to the Test-Logic-Reset state, and normal operation of the ICs can then occur. The interconnection test passes if the observed *TDO* sequences match the ones given in the preceding list.

Verilog code for the basic boundary scan architecture of Figure 10-15 is given in Figure 10-21. Only the three mandatory instructions (EXTEST, SAMPLE/PRELOAD, and BYPASS) are implemented using a 3-bit instruction register. These instructions are coded as 000, 001, and 111, respectively. The number of cells in the *BSR* is a generic parameter. A second generic parameter, *CellType*, is a bit_vector that specifies whether each cell is an input cell or output cell. The case statement implements the TAP controller state machine. The instruction code is scanned in and loaded into *IDR* in states Capture-IR, Shift-IR, and Update-IR. The instructions are executed in states Capture-DR, Shift-DR, and Update-DR. The actions taken in these states depend on the instruction being executed. The register updates and state changes all occur on the rising edge of *TCK*. The Verilog code implements most of the functions required by the IEEE boundary scan standard, but it does not fully comply with the standard.

Verilog code that implements the interconnection test example of Figure 10-20 is given in Figure 10-22. The *TMS* and *TDI* test patterns are the concatenation of the test patterns used in steps 2 through 6. A copy of the basic boundary scan architecture is instantiated for IC1 and for IC2. The external connections and internal logic for each IC are then specified. The internal clock frequency was arbitrarily chosen to be different from the test clock frequency. The test process runs the internal logic, then runs the scan test, and then runs the internal logic again. The test results verify that the IC logic runs correctly and that the scan test produces the expected results.

FIGURE 10-21: Verilog Code for Basic Boundary Scan Architecture

```verilog
// Verilog for Boundary Scan Architecture of Figure 10-15
module BS_arch (TCK, TMS, TDI, TDO, BSRin, BSRout, CellType);
   parameter[6:0] NCELLS  = 2;                  // ranege 2 to 120
   input TCK;
   input TMS;
   input TDI;
   output TDO;

   input[1:NCELLS] BSRin;
   inout[1:NCELLS] BSRout;
   wire[1:NCELLS] BSRout;
   wire[1:NCELLS] BSRout_xhdl0;
   input[1:NCELLS] CellType;

   reg[1:3] IR;                                 //instruction registers
   reg[1:3] IDR;
   reg[1:NCELLS] BSR1;                          //boundary scan cells
   reg[1:NCELLS] BSR2;
   reg BYPASS;                                  //bypass bit

   //TAP Controller State
   parameter[3:0] TestLogicReset = 0;
   parameter[3:0] RunTest_Idle = 1;
   parameter[3:0] SelectDRScan = 2;
   parameter[3:0] CaptureDR = 3;
   parameter[3:0] ShiftDR = 4;
   parameter[3:0] Exit1DR = 5;
   parameter[3:0] PauseDR = 6;
   parameter[3:0] Exit2DR = 7;
   parameter[3:0] UpdateDR = 8;
   parameter[3:0] SelectIRScan = 9;
   parameter[3:0] CaptureIR = 10;
   parameter[3:0] ShiftIR = 11;
   parameter[3:0] Exit1IR = 12;
   parameter[3:0] PauseIR = 13;
   parameter[3:0] Exit2IR = 14;
   parameter[3:0] UpdateIR = 15;
   reg[3:0] St;

   assign BSRout = BSRout_xhdl0;
   always @(TCK)
   begin
      if (TCK == 1'b1)
      begin
         case (St)
            TestLogicReset :
                     begin
                        if (TMS == 1'b0)
                           St <= RunTest_Idle ;
                        else
                           St <= TestLogicReset ;
                     end
```

```verilog
                  RunTest_Idle :
                        begin
                           if (TMS == 1'b0)
                               St <= RunTest_Idle ;
                           else
                               St <= SelectDRScan ;
                        end
                  SelectDRScan :
                        begin
                           if (TMS == 1'b0)
                               St <= CaptureDR ;
                           else
                               St <= SelectIRScan ;
                        end
                  CaptureDR :
                        begin
                           if (IDR == 3'b111)
                               BYPASS <= 1'b0 ; //EXTEST (input cells capture pin data)
                           else if (IDR == 3'b000)
                               BSR1 <= (~CellType & BSRin) | (CellType & BSR1) ;
                           else if (IDR == 3'b001) //SAMPLE/PRELOAD
                               BSR1 <= BSRin ; all cells capture cell input data
                           if (TMS == 1'b0)
                               St <= ShiftDR ;
                           else
                               St <= Exit1DR ;
                        end
                  ShiftDR :
                        begin
                           if (IDR == 3'b111)
                               BYPASS <= TDI ; //shift data through bypass reg
                           else
                               BSR1 <= {TDI, BSR1[1:NCELLS - 1]} ; //shift data into BSR
                           if (TMS == 1'b0)
                               St <= ShiftDR ;
                           else
                               St <= Exit1DR ;
                        end
                  Exit1DR :
                        begin
                           if (TMS == 1'b0)
                               St <= PauseDR ;
                           else
                               St <= UpdateDR ;
                        end
                  PauseDR :
                        begin
                           if (TMS == 1'b0)
                              St <= PauseDR ;
                           else
                               St <= Exit2DR ;
                        end
```

10.4 Boundary Scan

```verilog
            Exit2DR :
                begin
                    if (TMS == 1'b0)
                        St <= ShiftDR ;
                    else
                        St <= UpdateDR ;
                end
            UpdateDR :
                begin
                  if (IDR == 3'b000) //EXTEST (update output reg. for output cells)
                        BSR2 <= (CellType & BSR1) | (~CellType & BSR2) ;
                    else if (IDR == 3'b001) //SAMPLE/PRELOAD
                        BSR2 <= BSR1 ; //update output reg. in all cells
                    if (TMS == 1'b0)
                        St <= RunTest_Idle ;
                    else
                        St <= SelectDRScan ;
                end
            SelectIRScan :
                begin
                    if (TMS == 1'b0)
                        St <= CaptureIR ;
                    else
                        St <= TestLogicReset ;
                end
            CaptureIR :
                begin
                    IR <= 3'b001 ; //load 2 LSBs of IR with 01 as required by the
                                    standard
                    if (TMS == 1'b0)
                        St <= ShiftIR ;
                    else
                        St <= Exit1IR ;
                end
            ShiftIR :
                begin
                    IR <= {TDI, IR[1:2]} ; //shift in instruction code
                    if (TMS == 1'b0)
                        St <= ShiftIR ;
                    else
                        St <= Exit1IR ;
                end
            Exit1IR :
                begin
                    if (TMS == 1'b0)
                        St <= PauseIR ;
                    else
                        St <= UpdateIR ;
                end
            PauseIR :
                begin
                    if (TMS == 1'b0)
                        St <= PauseIR ;
```

```verilog
                                else
                                    St <= Exit2IR ;
                            end
                    Exit2IR :
                            begin
                                if (TMS == 1'b0)
                                    St <= ShiftIR ;
                                else
                                    St <= UpdateIR ;
                            end
                    UpdateIR :
                            begin
                                IDR <= IR ; //update instruction decode register
                                if (TMS == 1'b0)
                                    St <= RunTest_Idle ;
                                else
                                    St <= SelectDRScan ;
                            end
            endcase
        end
    end
    assign TDO = (St == ShiftDR & IDR == 3'b111) ? BYPASS : (St == ShiftDR) ?
                BSR1[NCELLS] : (St == ShiftIR) ? IR[3] : TDO;
    assign BSRout = (St == TestLogicReset | ~(IDR == 3'b000)) ? BSRin : BSR2 ; //define
                    cell outputs
endmodule
```

FIGURE 10-22: Verilog Code for Interconnection Test Example

```verilog
// Boundary Scan Tester
module system ();
    reg TCK;
    reg TMS;
    reg TDI;
    wire TDO;
    wire TDO1;
    reg Q0;
    reg Q1;
    reg CLK1;
    wire[1:4] BSR1in;
    wire[1:4] BSR1out;
    wire[1:4] BSR2in;
    wire[1:4] BSR2out;
    integer count;
    parameter[0:62] TMSpattern =
            63'b011000000011100000000111100000001110000000011100000000111111;
```

```verilog
        parameter[0:62] TDIpattern =
                63'b000001001000000010001000000000000000100010000000000000000000;

    initial
    begin
        count = 0;
        TCK = 0;
        CLK1 = 0;
    end
    BS_arch BS1 (TCK, TMS, TDI, TDO1, BSR1in, BSR1out, 4'b0011);
    BS_arch BS2 (TCK, TMS, TDO1, TDO, BSR2in, BSR2out, 4'b0011);
    //each BSR has two input cells and two output cells
    assign BSR1in[1] = BSR2out[4] ;                    //IC1 external connections
    assign BSR1in[2] = BSR2out[3] ;
    assign BSR1in[3] = Q1 ;                            //IC1 internal logic
    assign BSR1in[4] = Q0 ;

    always #7 CLK1 <= ~CLK1;                           //internal clock

    always @(posedge CLK1)
    begin
        Q0 <= BSR1out[1] ;                             //D flip-flops
        Q1 <= BSR1out[2] ;
    end
    assign BSR2in[1] = BSR1out[4] ;                    //IC2 external connections
    assign BSR2in[2] = BSR1out[3] ;
    assign BSR2in[3] = BSR2out[1] ^ BSR2out[2] ;       //IC2 internal logic
    assign BSR2in[4] = ~BSR2out[1] ;
    always #5 TCK <= ~TCK;                             //test clock

    always
    begin
        TMS <= 1'b1 ;
        #70;                                           //run internal logic
        @(posedge TCK);
        begin : xhdl_12
            integer i;
            for(i = 0; i <= 62; i = i + 1)             //run scan test
            begin
                TMS <= TMSpattern[i] ;
                TDI <= TDIpattern[i] ;
                #0;
                count <= i + 1 ;                       //count triggers listing output
                @(posedge TCK);
            end
        end
        #70;                                           //run internal logic
        forever #100000;                               //wait for manual termination
    end
endmodule
```

10.5 Built-In Self-Test

As digital systems become more and more complex, they become much harder and more expensive to test. One solution to this problem is to add logic to the IC so that it can test itself. This is referred to as built-in self-test (BIST). Figure 10-23 illustrates the general method for using BIST. An on-chip test generator applies test patterns to the circuit under test. The resulting output is observed by the response monitor, which produces an error signal if an incorrect output pattern is detected.

FIGURE 10-23: Generic BIST Scheme

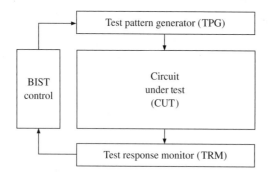

BIST is often used for testing memory. The regular structure of a memory chip makes it easy to generate test patterns. Figure 10-24 shows a block diagram of a self-test circuit for a RAM. The BIST controller enables the write-data generator and address counter so that data is written to each location in the RAM. Then the address counter and read-data generator are enabled, and the data read from each RAM location is compared with the output of the read-data generator to verify that it is correct. Memory is often tested by writing **checkerboard patterns** (alternating 0s and 1s) in all memory locations and reading them back. For instance, one could first write alternating 0s and 1s in all even addresses and alternating 1s and 0s in all odd addresses. After reading these back, the odd and even address patterns can be swapped to complete the test. In another test, the **March test**, each cell is read and the complemented value

FIGURE 10-24: Self-Test Circuit for RAM

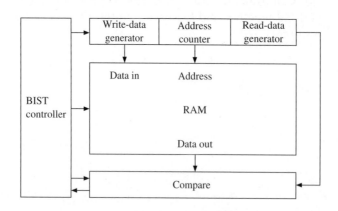

is written. This process is continued until the entire memory array has been traversed. The, the process is repeated in the reverse order of addresses.

The test circuit can be simplified by using a signature register. The signature register compresses the output data into a short string of bits called a **signature,** and this signature is compared with the signature for a correctly functioning component. A **multiple-input signature register (MISR)** combines and compresses several output streams into a single signature. Figure 10-25 shows a simplified version of the RAM self-test circuit. The read-data generator and comparator have been eliminated and replaced with a MISR. One type of MISR simply forms a check sum by adding up all the data bytes stored in the RAM. When testing a ROM, Figure 10-25 can be simplified further, since no write-data generator is needed.

FIGURE 10-25: Self-Test Circuit for RAM with Signature Register

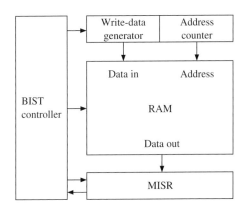

Linear feedback shift registers (LFSRs) are often used to generate test patterns and to compress test outputs into signatures. An LFSR is a shift register whose serial input bit is a linear function of some bits of the current shift register content. The bit positions that affect the serial input are called **taps**. The general form of a LFSR is a shift register with two or more flip-flop outputs XOR'ed together and fed back into the first flip-flop. The term **linear** comes from the fact that exclusive OR is equivalent to modulo-2 addition, and addition is a linear operation. Figure 10-26 shows an example of a LFSR. The outputs from the first and fourth flip-flops are XOR'ed together and fed back into the D input of the first flip-flop; the taps are positions 1 and 4.

FIGURE 10-26: 4-bit Linear Feedback Shift Register (LFSR)

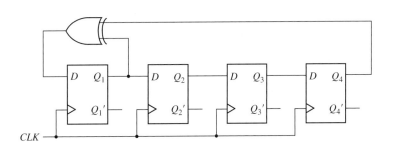

By proper choice of the outputs that are fed back through the XOR gate, it is possible to generate $2^n - 1$ different bit patterns using an n-bit shift register. All possible patterns can be generated except for all 0s. The patterns generated by the LFSR of Figure 10-26 are

1000, 1100, 1110, 1111, 0111, 1011, 0101, 1010, 1101, 0110, 0011,

1001, 0100, 0010, 0001, 1000, . . .

These patterns have no obvious order, and they have certain randomness properties. Such a LFSR is often referred to as a **pseudo-random pattern generator**, or **PRPG**. PRPGs are obviously very useful for BIST since they can generate a large number of test patterns with a small amount of logic circuitry. Table 10-4 gives one feedback combination that will generate all $2^n - 1$ bit patterns for some LFSRs with lengths in the range $n = 4$ to 32.

TABLE 10-4: Feedback for Maximum-Length LFSR Sequence

n	Feedback
4,6,7	$Q_1 \oplus Q_n$
5	$Q_2 \oplus Q_5$
8	$Q_2 \oplus Q_3 \oplus Q_4 \oplus Q_8$
12	$Q_1 \oplus Q_4 \oplus Q_6 \oplus Q_{12}$
14,16	$Q_3 \oplus Q_4 \oplus Q_5 \oplus Q_n$
24	$Q_1 \oplus Q_2 \oplus Q_7 \oplus Q_{24}$
32	$Q_1 \oplus Q_2 \oplus Q_{22} \oplus Q_{32}$

If the all-0s test pattern is required, an n-bit LFSR can be modified by adding an AND gate with $n - 1$ inputs, as shown in Figure 10-27 for $n = 4$. When in state 0001, the next state is 0000; when in state 0000, the next state is 1000; otherwise, the sequence is the same as for Figure 10-26.

FIGURE 10-27: Modified LFSR with 0000 State

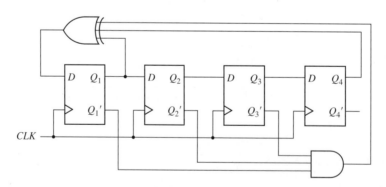

An MISR can be constructed by modifying an LFSR by adding XOR gates, as shown in Figure 10-28. The test data $(Z_1 Z_2 Z_3 Z_4)$ is XORed into the register with each clock, and the final result represents a signature that can be compared with the signature for a known correctly functioning component. This type of signature analysis

FIGURE 10-28: Multiple-Input Signature Register (MISR)

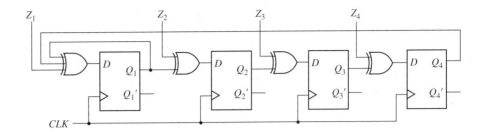

will catch many (but not all) possible errors. An n-bit signature register maps all possible input streams into one of the 2^n possible signatures. One of these is the correct signature, and the others indicate that errors have occurred. The probability that an incorrect input sequence will map to the correct signature is of the order of $1/2^n$.

For the MISR of Figure 10-28, assume that the correct input sequence is 1010, 0001, 1110, 1111, 0100, 1011, 1001, 1000, 0101, 0110, 0011, 1101, 0111, 0010, 1100. This sequence maps to the signature 0010 assuming the initial contents of the MISR to be 0000. Any input sequence that differs in one bit will map to a different signature. For example, if 0001 in the sequence is changed to 1001, the resulting sequence maps to 0000. Most sequences with two errors will be detected, but if we change 0001 to 1001 and 0010 to 0110 in the original sequence, the result maps to 0010, which is the correct signature, so the errors would not be detected.

Several types of architectures have been proposed for BIST. Two popular examples are the STUMPS architecture and the BILBO architecture, which are discussed in the following paragraphs.

STUMPS stands for **S**elf-**T**esting **U**sing **M**ISR and **P**arallel **SRSG**. **SRSG** in turn stands for **S**hift **R**egister **S**equence **G**enerator. STUMPS is a BIST architecture that uses scan chains. An overview of the STUMPS architecture is shown in Figure 10-29. A pseudo-random pattern generator feeds test stimulus to the scan chains, and after a capture cycle, the test response analyzer receives the test responses. The test procedure in STUMPS is the following:

(i)   Scan in patterns from the test pattern generator (LFSR) into all scan chains.
(ii) Switch to normal function mode and clock once with system clock.
(iii) Shift out scan chain into test response analyzer (MISR) where test signature is generated.

FIGURE 10-29: The STUMPS Architecture

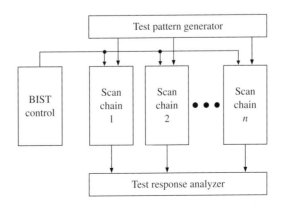

If the scan chain contains 100 scan cells, steps 1 and 3 will take 100 clocks. All scan chains should first be filled by the pseudo-random generator; hence, long scan chains necessitate long testing times. Since one test is done per scan, the STUMPS architecture is called a **test-per-scan** scheme. To reduce the testing time, a large number of parallel scan chains can be used, which reduces the time for filling the scan chains with the test, since all scan chains can be loaded in parallel.

The STUMPS architecture was originally developed for self-testing of multi-chip modules [Bardell 1982]. The scan chain on each logic chip (module) is loaded in parallel from the pseudo-random pattern source. The number of clock cycles required is equal to the number of flip-flops in the longest scan chain. If there are m scan cells in the longest scan chain, it will take *2m + 1* cycles to perform 1 test (*m* cycles for scan-in, 1 for capture, and *m* cycles for scan out). The shorter scan chains will overflow into the MISR, but that will not affect the final correct signature.

To reduce test-times, steps 1 and 3 can be overlapped. When the scan chain is unloaded into the MISR after one test, simultaneously the next pseudo-random pattern set from the SRSG can be loaded into the scan chain—that is, when test response from test *I* is being shifted out, the test pattern for test *I + 1* can be shifted in. Assuming overlap between scan-out of a test and scan-in of the following test, each test vector will take *m + 1* cycles, and it will take *n(m + 1) + m* cycles to apply *n* test vectors, including the m cycles taken for the last scan-out.

As opposed to the test-per-scan scheme just discussed, a *test-per-clock* scheme can be used for faster testing. One such scheme is called the **built-in logic block observer (BILBO)** technique.

In BILBO schemes, the scan register is modified so that parts of the scan register can serve as a state register, pattern generator, signature register, or shift register. When used as a shift register, the test data can be scanned in and out in the usual way. During testing, part of the scan register can be used as a pattern generator (PRPG) and part as a signature register (MISR) to test one of the combinational blocks. The roles can then be changed to test another combinational block. When the testing is finished, the scan register is placed in the state-register mode for normal operation. After the BILBO registers are initialized, since there is no loading of test patterns as in the case of scan chains, a test can be applied in each clock cycle. Hence, this is categorized as a **test-per-clock** BIST scheme. BILBO involves shorter test lengths, but more test hardware.

Figure 10-30 shows the placement of BILBO registers for testing a circuit with two combinational blocks. Combinational circuit 1 is tested when the first BILBO is used as a PRPG, and combinational circuit 2 is tested when the BILBO is used as an MISR. The roles of the registers are reversed to test combinational circuit 2. In the normal operating mode, both BILBOs serve as registers for the associated combinational logic. To scan data in and out, both BILBOs operate in the shift register mode.

Figure 10-31 shows the structure of one version of a 4-bit BILBO register. The control inputs B_1 and B_2 determine the operating mode. Si and So are the serial input and output for the shift register mode. The Zs are inputs from the combinational logic. The equations for this BILBO register are

$$D_1 = Z_1 B_1 \oplus (Si\ B_2' + FB\ B_2)(B_1' + B_2)$$
$$D_i = Z_i B_1 \oplus Q_{i-1}(B_1' + B_2) \quad (i > 1)$$

FIGURE 10-30: BIST Using BILBO Registers

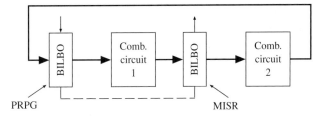

(a) Testing combinational circuit 1

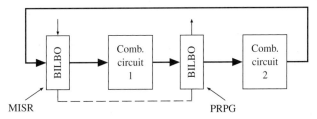

(b) Testing combinational circuit 2

FIGURE 10-31: 4-bit BILBO Register

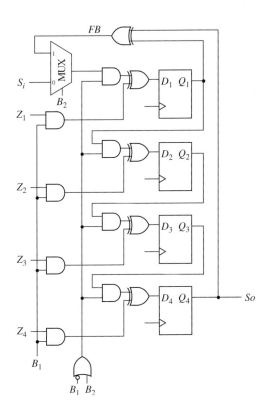

When $B_1 = B_2 = 0$, these equations reduce to

$$D_1 = S_i \text{ and } D_i = Q_{i-1} \ (i > 1)$$

which corresponds to the shift register mode. When $B_1 = 0$ and $B_2 = 1$, the equations reduce to

$$D_1 = FB, \qquad D_i = Q_{i-1}$$

which corresponds to the PRPG mode, and the BILBO register is equivalent to Figure 10-26. When $B_1 = 1$ and $B_2 = 0$, the equations reduce to

$$D_1 = Z_1, \qquad D_i = Z_i$$

which corresponds to the normal operating mode. When $B_1 = B_2 = 1$, the equations reduce to

$$D_1 = Z_1 \oplus FB, \qquad D_i = Z_i \oplus Q_{i-1}$$

which corresponds to the MISR mode, and the BILBO register is equivalent to Figure 10-28. In summary, the BILBO operating modes are as follows:

B1B2	Operating Mode
00	shift register
01	PRPG
10	normal
11	MISR

Figure 10-32 shows the Verilog description of an n-bit BILBO register. NBITS, which equals the number of bits, is a generic parameter in the range 4 through 8. The register is functionally equivalent to that shown in Figure 10-31, except that we have added a clock enable (*CE*). The feedback (*FB*) for the LFSR depends on the number of bits.

FIGURE 10-32: Verilog Code for BILBO Register of Figure 10-31

```verilog
module BILBO (Clk, CE, B1, B2, Si, So, Z, Q);
    parameter[2:0] NBITS = 4;
    input Clk;
    input CE;
    input B1;
    input B2;
    input Si;
    output So;

    input[1:NBITS] Z;
    inout[1:NBITS] Q;

    reg[1:NBITS] Q_tmp;

    wire FB;
    wire[1:0] mode;

    initial
      begin
        Q_tmp = 0;
      end
```

```verilog
    assign mode = {B1,B2};
    assign Q = Q_tmp;

    assign FB = Q[1] ^ Q[NBITS] ;

    always @(Clk)
    begin : xhdl_4
    if (Clk == 1'b1 & CE == 1'b1)
    begin
         case (mode)
              2'b00 :                               //Shift register mode
                    begin
                       Q_tmp <= {Si, Q[1:NBITS - 1]} ;
                    end
              2'b01 :                               //Pseudo Random Pattern Generator mode
                    begin
                       Q_tmp <= {FB, Q[1:NBITS - 1]} ;
                    end
              2'b10 :                               //Normal Operation mode
                    begin
                       Q_tmp <= Z ;
                    end
              2'b11 :                               //Multiple Input Signature Register mode
                    begin
                       Q_tmp <= Z[1:NBITS] ^ ({FB, Q[1:NBITS - 1]}) ;
                    end
         endcase
      end
   end
   assign So = Q[NBITS] ;
endmodule
```

The system shown in Figure 10-33 illustrates the use of BILBO registers. In this system, registers A and B can be loaded from the D bus using the LdA and LdB signals. Then the registers are added, and the sum and carry are stored in register C. When B_1 & $B_2 = 10$, the registers are in the normal mode ($Test = 0$), and loading of the registers is controlled by LdA, LdB, and LdC. To test the adder, we first set B_1 & $B_2 = 00$ to place the registers in the shift register mode and scan in initial values for A, B, and C. Then we set B_1 & $B_2 = 01$, which places registers A and B in PRPG mode and register C in MISR mode. After 15 clocks, the test is complete. Then we can set B_1 & $B_2 = 00$ and scan out the signature. The Verilog code for the system is given in Figure 10-34, and a test bench is given in Figure 10-35.

The system uses three BILBO registers and the 4-bit adder of Figure 8-20. The test bench scans in a test vector to initialize the BILBO registers and then it runs the test with registers A and B used as PRPGs and register C as an MISR. The resulting signature is shifted out and compared with the correct signature.

FIGURE 10-33: System with BILBO Registers and Tester

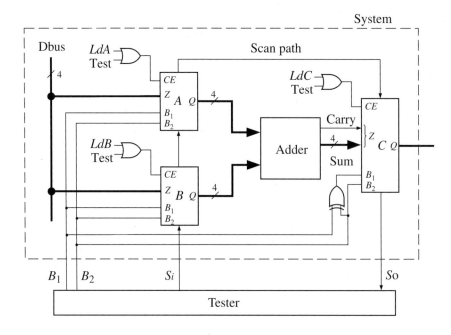

FIGURE 10-34: Verilog Code for System with BILBO Registers and Tester

```verilog
module BILBO_System (Clk, LdA, LdB, LdC, B1, B2, Si, So, DBus, Output);
    input Clk;
    input LdA;
    input LdB;
    input LdC;
    input B1;
    input B2;
    input Si;
    output So;

    input[3:0] DBus;
    inout[4:0] Output;
    wire[4:0] Output;
    wire[4:0] Output_xhd10;

    wire[3:0] Aout;
    wire[3:0] Bout;
    wire[4:0] Cin;
    wire ACE;
    wire BCE;
    wire CCE;
    wire CB1;
    wire Test;
    wire S1;
    wire S2;
```

```
    assign Output = Output_xhdl0;
    assign Test = ~B1 | B2 ;
    assign ACE = Test | LdA ;
    assign BCE = Test | LdB ;
    assign CCE = Test | LdC ;
    assign CB1 = B1 ^ B2 ;
    assign Cin = {Carry,Sum};

    BILBO #(4) RegA(Clk, ACE, B1, B2, S1, S2, DBus, Aout);

    BILBO #(4) RegB(Clk, BCE, B1, B2, Si, S1, DBus, Bout);

    BILBO #(5) RegC(Clk, CCE, CB1, B2, S2, So, Cin, Output);

    Adder4 Adder (Aout, Bout, 1'b0, Sum, Carry);
endmodule
```

FIGURE 10-35: Test Bench for BILBO System

```
// System with BILBO test bench
module BILBO_test ();
    reg Clk;
    wire LdA;
    wire LdB;
    wire LdC;
    reg B1;
    reg B2;
    reg Si;
    wire So;
    wire[3:0] DBus;
    wire[4:0] Output;
    reg[4:0] Sig;
    parameter[0:12] test_vector = 13'b1000110000000;
    parameter[4:0] test_result = 5'b01011;
    integer i;
    initial
    begin
        B1 = 1'b0;
        B2 = 1'b0;
        Si = 1'b0;
        Clk = 1'b0;
        Sig = 0;
    end

    always #25 Clk = ~Clk ;
    BILBO_System Sys (Clk, LdA, LdB, LdC, B1, B2, Si, So, DBus, Output);

    always
    begin
```

```verilog
            B1 <= 1'b0 ;                                    //Shift in test vector
            B2 <= 1'b0 ;
            begin : xhdl_2
                for(i = 0; i <= 12; i = i + 1)
                begin
                    Si <= test_vector[i] ;
                    @(posedge Clk);
                end
            end
            B1 <= 1'b0 ;                                    //Use PRPG and MISR
            B2 <= 1'b1 ;
            begin : xhdl_3
                integer i;
                for(i = 1; i <= 15; i = i + 1)
                begin
                    @(posedge Clk);
                end
            end
            B1 <= 1'b0 ;                                    //Shift signature out
            B2 <= 1'b0 ;
            begin : xhdl_4
                integer i;
                for(i = 0; i <= 4; i = i + 1)
                begin
                    $display("I is %d, So is %d", I, So);
                    $display($time);
                    @(posedge Clk);
                    Sig <= {So, Sig[4:1]} ;
                end
            end
            if (Sig == test_result)                         //Compare signature
            begin
                $display("System passed test. (NO ERROR)");
            end
            else
            begin
                $display("System did not pass test! (ERROR)");
            end
            forever #100000;                                //wait for manual termination
    end
endmodule
```

In this chapter, we introduced the subject of testing hardware, including combinational circuits, sequential circuits, complex ICs, and PC boards. Use of scan techniques for testing and built-in self-test has become a necessity as digital systems have become more complex. It is very important that design for testability be considered early in the design process so that the final hardware can be tested efficiently and economically.

Problems

10.1 (a) Determine the necessary inputs to the following circuit to test for u stuck-at-0.
(b) For this set of inputs, determine which other stuck-at faults can be tested.
(c) Repeat (a) and (b) for r stuck-at-1.

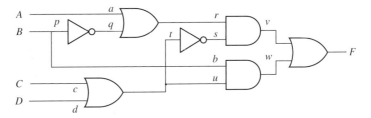

10.2 For the following circuit,
(a) Determine the values of A, B, C, and D necessary to test for e s-a-1. Specify the other faults tested by this input vector.
(b) Repeat (a) for g s-a-0.

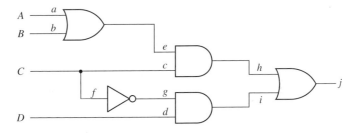

10.3 Find a minimum set of tests that will test all single stuck-at-0 and stuck-at-1 faults in the following circuit. For each test, specify which faults are tested for s-a-0 and for s-a-1.

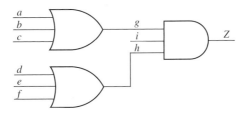

10.4 Give a minimum set of test vectors that will test for all stuck-at faults in the following circuit. List the faults tested by each test vector.

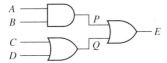

10.5 For the following circuit, specify a minimum set of test vectors for a, b, c, d, and e that will test for all stuck-at faults. Specify the faults tested by each vector.

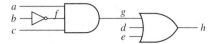

10.6 For the following circuit, find a minimum number of test vectors that will test all s-a-0 and s-a-1 faults at the AND and OR gate inputs. For each test vector, specify the values of A, B, C, and D, and the stuck-at faults that are tested.

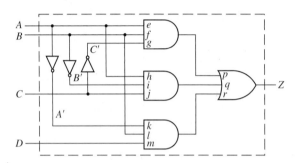

10.7 Find a test sequence to test for b s-a-0 in the sequential circuit of Figure 10-7.

10.8 A sequential circuit has the following state graph:

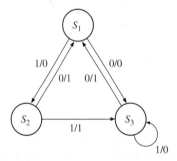

The three states can be distinguished using the input sequence 11 and observing the output. The circuit has a reset input, R, that resets the circuit to state S_1. Give a set of test sequences that will test every state transition and give the transition tested by each sequence. (When you test a state transition, you must verify that the output and the next state are correct by observing the output sequence.)

10.9 State graphs for two sequential machines are given here. The first graph represents a correctly functioning machine, and the second represents the same machine with a malfunction. Assuming that the two machines can be reset to their starting states (S_0 and T_0), determine the shortest input sequence that will distinguish the two machines.

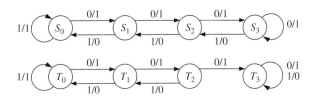

10.10 When testing a sequential circuit, what are the major advantages of using scan-path testing as compared with applying input sequences and observing output sequences?

10.11 A scan path test circuit of the type shown in Figure 10-8 has three flip-flops, two inputs, and two outputs. One row of the state table of the sequential circuit to be tested is as follows:

$Q_1 Q_2 Q_3$	$X_1 X_2 =$	\multicolumn{4}{c}{$Q_1^+ Q_2^+ Q_3^+$}	\multicolumn{4}{c}{$Z_1 Z_2$}
		00	01	11	10	00	01	11	10
011		010	110	011	111	10	11	00	01

For this row of the table, complete a timing chart similar to that shown in Figure 10-9 to show how the circuit can be tested to verify the next states and outputs for inputs 00, 01, and 10. Show the expected Z_1 and Z_2 outputs only at the time when they should be read.

10.12 **(a)** Redraw the code converter circuit of Figure 1-26 in the form of Figure 10-8 using dual-port flip-flops.
(b) Determine a test sequence that will verify the first two rows of the transition table of Figure 1-24(b). Draw a timing diagram similar to that shown in Figure 10-9 for your test sequence.

10.13 **(a)** Write Verilog code for a dual-port flip-flop.
(b) Write Verilog code for your solution to Problem 10.12(a).
(c) Write a test bench that applies the test sequence from Problem 10.12(b) and compare the resulting waveforms with your solution to Problem 10.12(b).

10.14 Instead of using dual-port flip-flops of the type shown in Figure 10-8, scan testing can be accomplished using standard D flip-lops with a MUX on each D input to select D_1 or D_2. Redraw the circuit of Figure 1-22 to establish a scan chain using D flip-flops and MUXes. A test signal (T) should control the MUXes.

10.15 Referring to Figure 10-16, determine the sequence of *TMS* and *TDI* inputs required to load the instruction register with 011 and the boundary scan register BSR2 with 1101. Start in state 0 and end in state 1. Give the sequence of states along with the *TMS* and *TDI* inputs.

10.16 The INTEST instruction (code 010) allows testing of the core logic by shifting test data into the boundary scan register (BSR1) and then updating BSR2 with this test data. For input cells this data takes the place of data from the input pins. Output data from the core logic is captured in BSR1 and then shifted out. For this problem, assume that the BSR has three cells.

(a) Referring to Figure 10-16, give the sequence for *TMS* and *TDI* that will load the instruction register with 010 and BSR2 with 011. In addition, give the state sequence, starting in state 0.

(b) In the code of Figure 10-21, what changes or additions must be made in the last BSRout assignment statement, in the CaptureDR state, and in the UpdateDR state to implement the INTEST instruction?

10.17 Based on the Verilog code of Figure 10-21, design a 2-cell boundary scan register. The first cell should be an input cell, and the second cell an output cell. Do not design the TAP controller; just assume that the necessary control signals such as shift-DR, capture-DR, and update-DR are available. Do not design the instruction register or instruction decoding logic; just assume that the following signals are available: EXT (EXTEST instruction is being executed), SPR (sample/preload instruction is being executed), and BYP (bypass instruction is being executed). Use two flip-flops for BSR1, two flip-flops for BSR2, and one BYPASS flip-flop. In addition to the control signals mentioned previously, the inputs are Pin1 (from a pin), Core2 (from the core logic), *TDI*, and *TCK*; outputs are Core1 (to core logic), Pin2 (to a pin), and *TDO*. Use *TCK* as the clock input for all of the flip-flops. Draw a block diagram showing the flip-flops, MUXes, and so forth. Then give the logic equations or connections for each flip-flop D input, each *CE* (clock enable), and each MUX control input.

10.18 Simulate the boundary scan tester of Figure 10-22 and verify that the results are as expected. Change the code to represent the case where the lower input to IC1 is shorted to ground; simulate again and interpret the results.

10.19 Write Verilog code for the boundary scan cell of Figure 10-14(b). Rewrite the Verilog code of Figure 10-21 to use this boundary scan cell as a component in place of some of the behavioral code for the BSR. Use a generate statement to instantiate NCELLS copies of this component. Test your new code using the boundary scan tester example of Figure 10-22.

10.20 (a) Draw a circuit diagram for an LFSR with $n = 5$ that generates a maximum-length sequence.
(b) Add logic so that 00000 is included in the state sequence.
(c) Determine the actual state sequence.

10.21 (a) Draw a circuit diagram for an LFSR with $n = 6$ that generates a maximum-length sequence.
(b) Add logic so that 000000 is included in the sequence.
(c) Determine the 10 elements of the sequence starting in 101010.

10.22 (a) Write Verilog code for an 8-bit MISR that is similar to that shown in Figure 10-28.
(b) Design a self-test circuit, similar to that shown in Figure 10-25, for a 6116 static RAM (see Figure 8-15). The write-data generator should store data in the following sequence: 00000000, 10000000, 11000000,...,11111111, 01111111, 00111111,...,00000000.
(c) Write Verilog code to test your design. Simulate the system for at least one example with no errors, one error, two errors, and three errors.

10.23 In the system of Figure 10-33, A, B, and C are BILBO registers. The B_1 and B_2 inputs to each of the registers determine its BILBO operating mode as follows:

$B_1B_2 = 00$, shift register; $B_1B_2 = 01$, PRPG (pattern generator);
$B_1B_2 = 10$, normal system mode; $B_1B_2 = 11$, MISR (signature register).

The shifting into A, B, and C is always LSB first. When in the test mode, the Dbus is not used. Specify the sequence of the tester outputs (B_1, B_2, and S_i) needed to perform the following operations:

1. Load A with 1011 and B with 1110; clear C.
2. Test the system by using A and B as pattern generators and C as a signature register for four clock times.
3. Shift the C register output into the tester.
4. Return to the normal system mode.

$$B_1 \, B_2 \, S_i = 0\,0\,0,$$

10.24 Given the BILBO register that follows, specify B_1 and B_0 for each of the following modes:

normal mode
shift register mode
PRPG (LSFR) mode
MISR mode

When in the PRPG mode, what sequence of states would be generated for Q_1, Q_2, and Q_3, assuming that the initial state is 001?

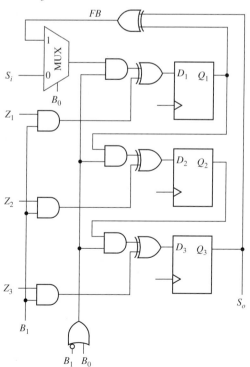

Verilog Language Summary

Disclaimer: This Verilog summary is not complete and contains some special cases. Only Verilog statements used in this text are listed. For a complete description of Verilog syntax, refer to *Verilog Language Reference Manual (LRM)*, IEEE Standard for Verilog Hardware Description Language, IEEE Std 1364 2005

Notes:

- Verilog is case sensitive.
- Signal names and other identifiers may contain letters, numbers, the underscore (_) character, and dollar ($) character.
- An identifier must start with a letter or underscore (_).
- An identifier must not start with a number or dollar ($) symbol.
- The dollar ($) character is reserved as the first character for system tasks.
- Declaration and assignment statements must be terminated with a semicolon.
- Verilog is not a strongly typed language. So mixing of some data types is allowed with many operators.

Legend

bold   reserved word
[ ]   optional items
{ }   repeated zero or more times
| or

1. Module

```
module module-name (module interface list);
[list-of-interface-ports]
[port-declarations]
[net and register declarations]
[constants]
[functional-specification-of-module]
endmodule
Functional specification of module can contain:
assign statements, initial blocks, always blocks, function
... endfunction, task   ...   endtask, specify ... endspecify,
instantiations, primitives, modules
```

2. Concurrent Or Parallel Statements

```
assign [#delay] signal_name = expression;
assign signal_name = condition ? expression_T : expression_F;

initial
begin
     sequential-statements
end

always @(sensitivity-list); // Sensitivity list can be @(A,B)
                            //   @(A or B)    @(*)
begin
     sequential-statements
end
```

3. Sequential Statements (Procedural Statements)

(a) Blocking Statements
```
always @(sensitivity list)
begin
  blocking statements using the = operator
end
```

(b) Non-blocking Statements
```
always @(sensitivity list)
begin
  non-blocking statements using the <= operator
end
```

(c)
```
if (condition)
    sequential statements
    //0 or more else if clauses may be included
{else if (condition)
    sequential statements}
[else  sequential statements]
```

(d) wait (Boolean-expression)

(e)
```
case expression
    choice1 : sequential statements1
    choice2 : sequential statements2
    . . .
    [default : sequential statements]
endcase;

case_statement ::=
case (expression)
case_item {case_item} endcase
| casez (expression)
case_item {case_item} endcase
| casex (expression)
case_item {case_item} endcase
case_item ::=
```

```
           expression { , expression } : statement_or_null
                      | default [ : ] statement_or_null
```

In a case expression comparison, the comparison only succeeds when each bit matches exactly with respect to the values 0, 1, x, and z.
casez treats high-impedance values (z) as do-not-cares, and casex treats both high-impedance and unknown (x) values as do-not-cares.

(f) **LOOPS:**
```
    loop_statement ::=
    forever statement
    | repeat (expression) statement
    | while (expression) statement
    | for (variable_assignment ; expression ; variable_assignment)
          statement

          for (initial_statement; expression; incremental_statement)
          begin
             sequential statement(s);
          end

          while condition
          begin
             sequential statements;
          end

          initial
          begin
             clk = 1'b0;
             forever #10 clk = ~clk;
          end

          repeat(n)
          begin
             statements to be repeated
          end
```

(g) ```
    genvar gen_variable1,...;
    generate
    for (for_loop_condition with gen_variables)
       concurrent statement(s)
    endgenerate
    ```

(h) ```
    generate
    if condition
       concurrent statement(s)
    endgenerate
    ```

(i) ```
    function <range or type> function-name
    input [declarations]
       <declarations>   //reg, parameter, integer, etc.
    begin
       sequential statements
    end
    endfunction
    ```
    ```

The general form of a function call is: **function_name(input-argument-list)**—*function call is not a statement by itself, it is part of a statement*

Note: *A function call is used within (or in place of) an expression.*

Key Features

One or more Inputs, but only one return value (no Outputs or Inouts)
Returns a value by assigning the value to the function name
Can call other functions, but cannot call tasks
Executes in zero time
Cannot embed delays, wait statements or any time-controlled statements
Invoked in concurrent assignments or procedural blocks
Can be recursive
Cannot contain non-blocking assignments

(j)   **task** task_name
     input [declarations]
     output [declarations]

     <declarations>   //reg, parameter, integer, etc.
     begin
       sequential statements
     end
     endtask

The formal-parameter-list specifies the inputs and outputs to the task and their types. A task call is a sequential or concurrent statement of the form:

`task_name(actual-parameter-list);`

Key Features

Any number of Inputs, Outputs or Inouts
Outputs need not use task name
Can call other functions or tasks
May contain time-controlled statements
Invoked only in procedural blocks
Can be recursive

4. Predefined Data Types

Unlike VHDL, all data types are predefined by the Verilog language and not by the user. Some of the popular predefined types are

nets	connections between hardware elements (declared with keywords such as **wire**)
variables	data storage elements that can retain values (declared with keywords such as **reg**)
integer	an integer is a variable data type (declared with the keyword **integer**)
real	real number constants and real variable data types for floating-point number (declared with the keyword **real**)

time   a special variable data type to store time information (declared with the keyword **time**)

vectors   wire or reg data types can be declared as vectors (multiple bits)

5. **Verilog Operators By Decreasing Precedence**

  1. Unary sign and reduction operators:
     - `+ −`   Unary sign operators
     - `&`   Reduction AND (unary operator to AND bits in a vector and reduce to one bit)
     - `~&`   Reduction NAND
     - `|`   Reduction OR (Unary operator to OR bits in a vector and reduce to one bit)
     - `~|`   Reduction NOR
     - `^`   Reduction XOR
     - `~^ or ^~`   Reduction XNOR
     - `!`   Logical negation
     - `~`   Bit-wise negation

  2. Arithmetic: Exponent
     - `**`   Power (exponent)

  3. Arithmetic: Multiplying, Modulus operators:
     - `*`   Multiply
     - `/`   Divide
     - `%`   Modulus

  4. Arithmetic: Addition:
     - `+`   Add
     - `−`   Subtract

  5. Shift operators:
     - `<<`   Logical left shift
     - `>>`   Logical right shift
     - `<<<`   Arithmetic left shift
     - `>>>`   Arithmetic right shift

  6. Relational operators:
     - `>`   Greater than
     - `<`   Less than
     - `>=`   Greater than or equal
     - `<=`   Less than or equal

  7. Logical and bitwise operators: Equality and Inequality
     - `==`   Logical equality
     - `!=`   Logical inequality
     - `===`   Case equality
     - `!==`   Case inequality

8. Bitwise operators:
       &   Bit-wise AND (binary operator)
9. Logical and bitwise operators:
       ^   Bit-wise XOR (binary operator)
       ^~ or ~^ Bit-wise EQUIVALENCE (binary operator)
       | Bit-wise OR (binary operator)
10. Logical AND:
        &&   Logical AND
11. Logical OR:
        || Logical OR
12. Conditional
        ? :   Conditional
13. Concatentation and Replication
        {}   Concatenation
        {{}}   Replication

- It is legal to use the concatenate operator on the left side of the assignment, e.g., {Cout, Sum} = A + B;
- For the *logical equality* and *logical inequality* operators (== and !=), if, due to unknown or high-impedance bits in the operands, the relation is ambiguous, then the result shall be a 1-bit unknown value (x).
- For the *case equality* and *case inequality* operators (=== and !==), bits that are x or z shall be included in the comparison and shall match for the result to be considered equal. The result of these operators shall always be a known value, either 1 or 0.

6. Declaration Examples

```
input    A,B,C;
output   D;
wire     [7:0]   A;
reg  C;
reg  [7:0]    eight_bit_register;
reg signed    [7:0] A = 8'hA5;
parameter     constant_name = constant_value;
localparam    constant_name = constant_value;

`define       constant_name   constant_value;
parameter [0:15] OT = {1'b1, 1'b0, 1'b0, 1'b1, 1'b0, 1'b1,
1'b1, 1'b0, 1'b0, 1'b1, 1'b1, 1'b0, 1'b1, 1'b0, 1'b0, 1'b1};

INSTANTIATION EXAMPLE: c74163 ct1(LdN, ClrN, P, T1, Clk, Din1,
Carry1, Qout1);

NAMED ASSOCIATION: FullAdder FA0 (.Sum(S[0]), .Cout(Co[0]),
.X(A[0]), .Y(B[0]), .Cin(Ci[0]))
```

7. Built In Primitives

Built-in Primitive Type	Primitives
n-input gates	and, nand, nor, or, xnor, xor
n-output gates	buf, not
Three-state gates	bufif0, bufif1, notif0, notif1
Pull gates	pulldown, pullup
MOS switches	cmos, nmos, pmos, rcmos, rnmos, rpmos
Bidirectional Switches	rtran, rtranif0, rtranif1, tran, transif0, tranif1

8. Examples Of Delays

```
wire D;
assign #5 D = A && B;                    // explicit continuous
                                         //   assignment

wire #5 D = A && B;                      // implicit continuous
                                         //   assignment

wire #5 D;                               // net declaration
assign D = A && B;

always @ (X)
begin
   Z1 <= #10 (X);                        // transport delay
end
assign #10 Z2 = X;                       // inertial delay
and #(10) a1 (out, in1, in2);            // primitive with only one
                                         //   delay (so rise delay=10)
and #(10,12) a2 (out, in1, in2);         // primitive with rise
                                         //   delay=10 and fall
                                         //   delay=12
bufif0 #(10,12,11) b3 (out, in, ctrl);   // rise, fall, and turn-off
                                         //   delays
#10 A = B + C; is equivalent to #10      // inter assignment delay
                    A = B + C;           // also called delayed
                                         //   evaluation
A = #10 B + C; is equivalent to
                   temp = B + C;         // intra assignment delay
                   #10 A = temp;         // also called delayed
                                         //   assignment
```

9. System Tasks

$display	Immediately outputs text/data with new line
$write	Immediately outputs text/data without new line
$strobe	Outputs text/data at the end of the current simulation step
$monitor	Displays text/data for every event on signal
$stop	temporarily suspend simulation for interaction
$finish	system task to come out of simulation
$time	Returns an integer that is a 64-bit time, scaled to the timescale unit of the module that invoked it.
$stime	Returns an unsigned integer that is a 32-bit time, scaled to the timescale unit of the module that invoked it. If the actual simulation time does not fit in 32 bits, the low order 32 bits of the current simulation time are returned.

$realtime	Returns a real number time that, like **$time**, is scaled to the time unit of the module that invoked it.
$signed(expr)	Evaluate expression and return value in signed form
$unsigned(expr)	Evaluate expression and return value in unsigned form
$realtobits	Type conversion from real to bits
$bitstoreal	Type conversion from bits to real
$random	Probabilistic function that returns a random signed integer which is 32-bits.

10. Compiler Directive Examples

`define	`define wordsize 16
`include	`include "lab3/sevenseg.v"
`ifdef	`ifdef …. `else …. `endif

11. File I/O Functions

$fopen, $fclose, $feof, $ferror, $fgetc, $fputc, $fscanf, $fprintf, $fread, $fwrite, $readmemb, $readmemh eg: $readmemh("file.hex", mem);

12. Timing Checks

Timing checks must be in specify…..endspecify block

$setup	$setup (data_event, reference_event, limit)
$hold	$hold (reference_event, data_event, limit)
$skew	$skew (reference_event, data_event, limit)
$width	$width (reference_event, limit)

13. Miscellaneous

Resolution Function for 4-valued logic:

Resolution Function	x	0	1	z
x	x	x	x	x
0	x	0	x	0
1	x	x	1	1
z	x	0	1	z

AND and OR functions for 4-valued logic:

AND	x	0	1	z
x	x	0	x	x
0	0	0	0	0
1	x	0	1	x
z	x	0	x	x

OR	x	0	1	z
x	x	x	1	x
0	x	0	1	x
1	1	1	1	1
z	x	x	1	x

List of Keywords (Verilog IEEE Standard 2005)

always
and
assign
automatic
begin
buf
bufif0
bufif1
case
casex
casez
cell
cmos
config
deassign
default
defparam
design
disable
edge
else
end
endcase
endconfig
endfunction
endgenerate
endmodule
endprimitive
endspecify
endtable
endtask
event

for
force
forever
fork
function
generate
genvar
highz0
highz1
if
ifnone
incdir
include
initial
inout
input
instance
integer
join
large
liblist
library
localparam
macromodule
medium
module
nand
negedge
nmos
nor
noshowcancelled
not

notif0
notif1
or
output
parameter
pmos
posedge
primitive
pull0
pull1
pulldown
pullup
pulsestyle_onevent
pulsestyle_ondetect
rcmos
real
realtime
reg
release
repeat
rnmos
rpmos
rtran
rtranif0
rtranif1
scalared
showcancelled
signed
small
specify
specparam
strong0

strong1	tri0	wait
supply0	tri1	wand
supply1	triand	weak0
table	trior	weak1
task	trireg	while
time	unsigned	wire
tran	use	wor
tranif0	uwire	xnor
tranif1	vectored	xor
tri		

References

References 13, 20, 21, 27, 28, 33, 35 and 36 are general references on digital logic and digital system design. References 2, 3, 4, 10, 11, 14, 17, 22, 31, 32, 34, 37, and 38 provide information on PLDs, FPGAs, and CPLDs. References 9, 12, 16, 30, and 35 contain basic treatment of Verilog constructs and primitives. Reference 29 contains VHDL equivalents for majority of the code examples in this book. References 1, 5, 6, 15, 24, and 25 are related to hardware testing and design for testability. For more details on logic minimization, one can refer to reference 7. The MIPS ISA and MIPS processor architectures are described in references 8, 18, 19, and 26. Reference 26 provides an excellent introduction to various computer organization topics, which will enhance the understanding of chapter 9.

1. Abromovici, M., Breuer, M., and Friedman, F. *Digital Systems Testing and Testable Design.* Indianapolis, IN: Wiley-IEEE Press, 1994.
2. Actel Corporation, Actel Technical Documental, www.actel.com/techdocs/.
3. Altera Corporation, Altera Literature, www.altera.com/literature/lit-index.html.
4. Atmel Corporation, www.atmel.com.
5. Bardell, P. H. and McAnney, W. H. "Self-Testing of Logic Modules," *Proceedings of the International Test Conference.* 1982, pp. 200–204.
6. Bleeker, H., van den Eijnden, P., and de Jong, F. *Boundary Scan Test—A Practical Approach.* Boston, MA: Kluwer Academic Publishers, 1993.
7. Brayton, R. K., et al. *Logic Minimization for Algorithms for VLSI Synthesis.* Boston, MA: Kluwer Academic Publishers, 1984.
8. Britton, R. *MIPS Assembly Language Programming.* Upper Saddle River, NJ: Pearson Prentice Hall, 2003.
9. Brown, S. and Vranesic, Z. *Fundamentals of Digital Logic with Verilog Design.* New York, NY: McGraw Hill, 2003.
10. Brown, S. D., Francis, R. J., Rose, J. and Vranesic, Z. G. *Field-Programmable Gate Arrays.* Boston, MA: Kluwer Academic Publishers, 1992.
11. Chan, P. and Mourad, S. *Digital Design using Field Programmable Gate Arrays.* Englewood Cliffs, NJ: Prentice Hall, 1994.

12. Ciletti, M. D., *Advanced Digital Design with the Verilog HDL, 2/E.* Upper Saddle River, NJ: Pearson Prentice Hall, 2011.
13. Comer, D. J. *Digital Logic and State Machine Design, 3/E.* New York: Oxford University Press, 1995.
14. *Cypress Semiconductor Programmable Logic Documentation,* www.cypress.com.
15. Larsson, E., *Introduction to Advanced System-on-Chip Test Design and Optimization.* Springer, 2005.
16. *Verilog Language Reference Manual (LRM),* IEEE Standard for Verilog Hardware Description Language, IEEE Std 1364™ 2005, April 2006.
17. Jenkins, J. H. *Designing with FPGAs and CPLDs.* Englewood Cliffs, N. J.: Prentice Hall, 1994.
18. Kane, G. and Heinrich, J. *MIPS RISC Architecture.* Englewood Cliffs, NJ: Prentice Hall, 1991.
19. Kane, G. *MIPS RISC Architecture.* Englewood Cliffs, NJ: Prentice Hall, 1989.
20. Katz, R. H. *Contemporary Logic Design, 2/E.* Upper Saddle River, NJ: Prentice Hall, 2004.
21. Kohavi, Z. *Switching and Finite Automata Theory.* Boston, MA: McGraw Hill, 1979.
22. Lattice Semiconductors, www.latticesemi.com.
23. LOGICAID Software, Available with Roth, C. H. and John, L. K. *Digital Systems Design Using VHDL, 2/E.* Mason, OH: Cengage Learning, 2008.
24. McCluskey, E. J. *Logic Design Principles with Emphasis on Testable Semicustom Circuits.* Englewood Cliffs, N.J.: Prentice Hall, 1986.
25. Parker, K. P. *The Boundary Scan Handbook, 3/E.* New York, NY: Springer, 2003.
26. Patterson, D. A. and Hennessey, J. L. *Computer Organization and Design: The Hardware Software Interface, 3/E.* San Francisco, CA: Morgan Kaufmann, an imprint of Elsevier, 2005.
27. Prosser, F. P. and Winkel, D. E. *The Art of Digital Design: An Introduction to Top-Down Design, 2/E.* Englewood Cliffs, NJ: Prentice Hall, 1987.
28. Roth, C.H. *Fundamentals of Logic Design, 7/E.* Mason, OH: Cengage Learning, 2014.
29. Roth, C. H., and John, L. K., *Digital Systems Design Using VHDL, 2/E,* Cengage Learning, 2008
30. Rosen, K. H. *Elementary Number Theory and its Applications, 6/E.* Boston, MA: Pearson, Addison Wesley, 2011.
31. Rucinski, A. and Hludik, F. *Introduction to FPGA-Based Microsystem Design,* Dallas, TX: Texas Instruments, 1993.
32. Salcic, C., and Smailagic, A. *Digital Systems Design and Prototyping Using Field Programmable Logic,* 2/E. Boston, MA: Kluwer Academic Publishers, 2000.
33. Smith, M. J. S. *Application-Specific Integrated Circuits,* Boston, MA: Addison Wesley, 1997.
34. Tallyn, K. "Reprogrammable Missile: How an FPGA Adds Flexibility to the Navy's TomaHawk," *Military and Aerospace Electronics,* April 1990, Article reprinted in *The Programmable Gate Array Data Book,* Xilinx, 1992.

35. Vahid, F. *Digital Design with RTL Design, VHDL, and Verilog*, Hoboken, NJ; John Wiley & Sons, 2011.
36. Wakerly, J. F. *Digital Design Principles and Practices 4/E*. Upper Saddle River, NJ: Pearson Prentice Hall, 2006.
37. Xilinx Corporation. *Xilinx Documentation and Literature.* www.xilinx.com/support/library.htm.
38. Xilinx, Inc. *The Programmable Logic Data Book,* 1996. http://www.xilinx.com.

Index

0-hazard, 13
1-hazard, 13
Abbreviations, common, 64
Adder4 using generate statement, 453
Add function, 433
Adding multiple bits, 436
Adding redundant terms, 5–6
Addition, floating-point arithmetic, 417–425
    flowchart, 419
    overview, floating point adder, 420
    Verilog code, 421–424
Addition, two BCD numbers, 213
Add and Shift Multiplier, 232
Altera Stratix IV logic module, 356
Always blocks using event control statements, 82–83
AND and OR functions for 4-valued logic Verilog language, 561
AND-OR circuit testing, 516
AND-OR circuit test vectors, 516
Antifuse programming technology, 188
Applications, FPGAs, 201–203
    as final product in medium-speed systems, 202
    glue logic, 203
    hardware accelerators/coprocessors, 203
    rapid prototyping, 201
    reconfigurable circuits and systems, 202–203
Architecture, FPGAs, 184
Architecture, technology, logic block types of commercial FPGAs, 193
Area, power, and delay optimizations, 385
Area Time (AT) Bound, 385
Area Time^2 (AT^2) Bound, 385
Arithmetic instructions, MIPS ISA, 476–478
Array multiplier, 238–241

4-bit multiplier partial products, 238
4 × 4 array multiplier, block diagram, 239
4 × 4 array multiplier, Verilog code, 240–241
Verilog coding, 239–240
Array of AND gates, 68
Arrays, 124–126
    look-up table method using arrays and parameters, 125–126
    matrices, 125
    parity code generator contents, 126
    parity code generator using the LUT method, 126
ASM Charts, see SM Charts
Assignments, Verilog, 73–74
Assign statements, 65
ATPG, see Automatic test pattern generator
Automatic test pattern generator (ATPG), 514

Basic gates, 2
BCD adder, 212–214
    addition, two BCD numbers, 213
    Verilog code for, 213
BCD to excess-3 code converter, 20–24
BCD to 7-segment display decoder, 211–212
    behavioral Verilog code, 212
    block diagram, 211
    segment display, 211
Behavioral and structural Verilog, 112–123
    A block diagram, 113
    code converter using a single process, behavioral model for, 118–119
    code converter, waveforms for, 123
    D flip-flop, 121
    excess-3 code converter, behavioral model, 115–117

Behavioral and structural Verilog (*Continued*)
  excess-3 code converter, simulator output for, 118
  implementation of $F = AB + BC$
  levels of abstraction, AND device, 112
  modeling sequential machine, 114–115
  sequential machine model using equations, 120
  sequential machine, state table and block diagram
  sequential machine, structural model, 122
  3-input NAND gate, 121
Behavioral Verilog for multiplier controller, 294–295
Big-endian, 496
BILBO, 543
Binary dividers, 264–277
  divider control circuit state diagram, 267
  divider control circuit state table, 267
  parallel binary divider block diagram, 265
  signed divider, 267–277
  signed divider block diagram, 268
  signed divider control circuit state graph, 270
  signed divider simulation test results, 276
  signed divider test bench, 273–276
  32-bit signed divider Verilog model, 270–273
  unsigned divider, 264–267
Binary fixed-point fractions, 242–243
Binary multiplier, 293
  block diagram, 233
  chart, 294
  controller implementation, 307
  control, state graph, 234
  with serial state assignment for single address, 321
BIST, *see* Built-in self-test, 538-548
Black box view of the 2-gate module, 69
Blocking and non-blocking assignments, 75–77
Block RAM, 359
Boolean algebra and algebraic simplification, 3–6
  adding redundant terms, 5–6
  combining terms, 5
  eliminating literals, 5
  eliminating terms, 5
  laws and theorems of Boolean algebra, 4
Booth's Algorithm, 282
Boundary scan, 525–538
  boundary scan architecture, 527
  boundary scan architecture, Verilog code, 533–536
  boundary scan cell, 527
  IC with boundary scan register and test-access port, 525

  interconnection testing using boundary scan, 530
  interconnection test, Verilog code, 536–537
  PC boards with boundary scan ICs, 526
  signal path, extest instruction, 529
  signal path, intest instruction, 530
  signal path, sample/preload instruction, 529
  TAP controller state machine, 528
Built-in gate and switch primitives, 441
Built-in primitives, 439–441
  built-in gate and switch primitives, 441
  Verilog array of tristate buffers, 440
  Verilog gate module, 440
  Verilog tristate buffer module, 441
Built-in primitives Verilog language, 560
Built-in self-test (BIST), 538–548
  with BILBO registers, 543
  BILBO registers and tester system, 546
  BILBO registers and tester system Verilog code, 546–547
  BILBO register Verilog code, 544–545
  BILBO system test bench, 547–548
  built-in logic block observer (BILBO) technique, 542
  circuit for RAM, 538
  circuit for RAM with signature register, 539
  4-bit BILBO register, 543
  4-bit linear feedback shift register (LFSR), 539
  maximum-length LSFR sequence feedback, 540
  modified LSFR with 0000 state, 540
  multiple-input signature register (MISR), 541
  scheme, generic, 538
  STUMPS architecture, 541
  test-per-clock BIST scheme, 542
  test-per-scan scheme, 542

CAD design flow, 376
Carry chains in FPGAs, 352–353
Carry look-ahead adders, 214–216
Cascade chains in FPGAs, 353–354
Cascaded 2-to-1 MUXes, conditional assignment, 102
Case statement, Verilog, 103
Circuit for Mealy sequence detector, 20
Circuit for RAM, 538
Circuit for RAM with signature register, 539
Circuits to avoid, 45
Circuit with three flip-flops, 38
Characteristic equation, 15
Clocked D flip-flop with rising-edge trigger, 15

Index

Clocked J-K flip-flop, 15
Clocked T flip-flop, 16
Clock skew, 41, 195
Clock-to-Q delay, 30, 32
Code converter, 20
Code converter using a single process,
    behavioral model for, 118–119
Code converter, waveforms for, 123
Code, example of Verilog, 99
Code, Verilog, that will not
    synthesize, 101
Coding schemes for serial data
    transmission, 26
Combinational circuits, Verilog description
    of, 64–68
  array of AND gates, 68
  assign statements, 65
  concurrent statements, 65
  continuous assignments, 65
  gates with common input and different
    delays, 67
  identifiers, 66
  inverter with feedback, 66
  simple gate circuit, 65
  system tasks, 66–67
Combinational logic, 1–3
  basic gates, 2
  full adder, 1
  truth table, 2
Combinational logic with blocking
    assignments, always block, 77
Combinational logic testing, 515–518
  AND-OR circuit testing, 516
  AND-OR circuit test vectors, 516
  fault detection using path sensitization,
    517
  stuck-at-fault testing circuit, 517
  stuck-at-fault tests, 518
Combining terms, 5
Command file and simulation results
    for $+5/8$ by $-3/8$, 249
Comparison, ripple-carry and carry
    look-ahead adders, 220
Compilation, simulation, synthesis, Verilog
    code, 87–93
  event queue, 88–89
  multiple processes (initial or always
    block), 90
  non-determinism, 90
  process simulation, 90–91
  signal drivers, simulation, 91
Compiler directives, 457–460
  `define, 458
  `ifdef, 459–460
  `include, 458–459
Compiler directives Verilog language, 561

Complex programmable logic devices
    (CPLDs), 176–180
  CPLD function block and macrocell, 178
  CPLD implementation of Mealy
    machine, 179
  CPLD implementation of parallel adder
    with accumulator, 179–180
  major CPLDs, approximate capacity, 177
  n-bit parallel adder with
    accumulator, 179
  Xilinx CoolRunner, 177–179
Computation of squares and its call,
    434–435
Computer-aided design (CAD), 59–61
  design entry, 59
  design flow in modern digital system, 59
  schematic capture, 59–60
  spectrum of design technologies, 61
Concurrent or parallel statements Verilog
    language, 555
Concurrent statements, 65
Conditional generate, 454
Conditional operator, 101
Conditional output elimination, 317
Configuration, bits number, 371
Constants, 123–124
Content addressable memories (CAM), 201
Continuous assignments, 65
Control circuit state graphs, 225–226
Controller, keypad scanner, 258
Control store, 315
Control transfer instructions, MIPS ISA,
    479–481
Conversion functions, 457
Conversion of AND-OR circuit to NAND
    gates, 12
Conversion of NOR-NOR to AND-OR, 168
Conversion to NOR gates, 11
Correct control signal gating, rising-edge
    devices, 47
Cost of programmability, 369–371
  configuration bits number, 371
  logic blocks with programmable SRAM
    cells, 370
  programmable points in FPGA I/O
    block
Counter model, 108
Counter operation, 108
CPLDs, 177–180
  capacity, approximate, 177
  function block and macrocell, 178
  implementation of Mealy machine, 179
  implementation of parallel adder with
    accumulator, 179–180
Critical path, 385
Cyclic shift register, 104

Data path, MIPS subset implementation, 486–490
    design, 486
    overall, 490
    required, computation and memory instruction, 489
Data transfer using tristate bus, 48
Data types and operators, Verilog, 93–98
    net data, 93
    operators+, 94–98
    variable data, 93
Debouncer, keypad scanner, 257
Debouncing and synchronizing circuit, keypad scanner, 257
Debouncing mechanical switches, 230
Declaration examples Verilog language, 559
Declaration, Verilog, 70
Decoder, keypad scanner, 257–258
Decoder truth table, keypad scanner, 257
Dedicated arithmetic units, FPGAs, 199
Dedicated memory, FPGAs, 199, 357–368
    dedicated memory, inferred from Verilog code, 361
    Distributed Memory, 359
    embedded RAMs, FPGAs, 358
    4 × 4 multiplier look-up table based, 362–368
    LUT-based memory, inferred from Verilog code, 361
    LUT-based RAM, 360
    memory creation from LUTs, 359
    RAM in example FPGAs, 358
    Verilog models for inferring memory in FPGAs, 360
Dedicated memory, inferred from Verilog code, 361
Dedicated multipliers in FPGAs, 200, 368–369
Dedicated specialized components in FPGAs, 199–201
    arithmetic units, 199
    content addressable memories (CAM), 201
    digital signal processing (DSP) blocks, 200–201
    embedded microprocessors, 200
    embedded processors, 200–201
    memory, 199
    multipliers, 200
    variable-width RAM aspect ratios, 199
`define, 458
Delays in Verilog, 84–87
    inertial and transport delays, 85
    net delays, 85–87
Delays Verilog language, 560
Denormalized numbers, 404–405

Derivation of SM charts, 293–305
    behavioral Verilog for multiplier controller, 294–295
    binary multiplier, 293
    binary multiplier chart, 294
    dice game, 296–305
    dice game block diagram, 296
    dice game controller behavioral model, 299–301
    dice game controller state graph, 299
    dice game flowchart, 297
    dice game SM chart, 298
    dice game test bench, 302
    dice game tester, 305
    dice game test module, 303–305
Derivations of J-K input equations, 24
Design entry, 59
Design examples, 210–287
    array multiplier, 238–241
    BCD adder, 212–214
    BCD to 7-segment display decoder, 211–212
    binary dividers, 264–277
    control circuit state graphs, 225–226
    keypad scanner, 255–264
    scoreboard and controller, 226–230
    separation, data path and controller, 210
    shift-and-add multiplier, 232–238
    signed integer/fraction multiplier, 241–255
    synchronization and debouncing, 230–232
    32-bit adders, 214–220
    traffic light controller, 220–225
Design flow for PLDs, 176
Design flow in modern digital system, 59
Designing with NAND and NOR gates, 11–12
    conversion of AND-OR circuit to NAND gates, 12
    conversion to NOR gates, 11
    NAND and NOR gates, 11
Design of Moore sequential circuit, 25–27
    coding schemes for serial data transmission, 26
    Moore circuit for NRZ-to-Manchester conversion, 27
    Moore sequence detector, 25–26
    state graph of Moore sequence detector, 25
    state table for Moore sequence detector, 26
    timing for Moore circuit, 27
    transition table for Moore sequence detector, 26
Designs, field programmable gate arrays (FPGAs), 341–398

carry chains in FPGAs, 352–353
cascade chains in FPGAs, 353–354
cost of programmability, 369–371
dedicated memory, FPGAs, 357–368
dedicated multipliers in FPGAs, 368–369
design translation, 375–385
implementing functions in FPGAs, 341–346
implementing functions using Shannon's decomposition, 347–351
logic blocks, commercial FPGAs, 355–357
mapping, placement, and routing, 385–389
maximum gates versus usable gates, 373–375
one-hot state assignment, 371–373
Design translation, 375–385
area, power, and delay optimizations, 385
Area Time (AT), 385
Area Time^2 (AT^2), 385
CAD design flow, 376
critical path, 385
Energy Delay (ED), 385
Energy Delay^2 (ED^2), 385
netlist, 375
synthesis, 375
synthesis and corresponding hardware, 383–384
synthesis of arithmetic components, 383
synthesis of case statement, 377–378
synthesis of if statements, 381–383
unintentional latch creation, 378–380
D flip-flop, 121
D flip-flop, user-defined primitives, 444
D flip-flop with asynchronous clear, 80
Dice controller, 322
comparison of different implementations, 326
single-address microcode, 324
two-address microcode implementation, 323
Dice game, 296–305, 309–314, 322–327
block diagram, 296
complete, 314
controller behavioral model, 299–301
controller state graph, 299
counter, 313
data flow model, 312–313
equivalent SM blocks, 290
equivalent SM charts, combinational circuit, 290
flowchart, 297
linked SM charts, 328
maps, 311
microprogram with single-address microcoding, 326

MUX for single-address microcoding, 326
MUX for two-address microcoding, 324
realization of controller, 309
SM block example, 289
SM block principal components, 289
SM chart, 298
state transition table, 310
test bench, 302
tester, 305
test module, 303–305
two-address microprogram, 324
Digital signal processing (DSP) blocks, FPGAs, 200–201
Direct interconnects, 194, 195
Display tasks, 456
Distributed Memory, 359
Divider control circuit state diagram, 267
Divider control circuit state table, 267
Division, floating-point arithmetic, 425–426

8-bit counter using 4-bit counter modules, 109
8-to-3 priority encoder, 164
8-word × 4-bit ROM, 162
Eliminating literals, 5
Eliminating terms, 5
Elimination of 1-hazard, 14
Embedded microprocessors, FPGAs, 200
Embedded processors, FPGAs, 200–201
Embedded RAMs, FPGAs, 358
Energy Delay (ED), 385
Energy Delay^2 (ED^2), 385
EPROM connections for SRAM FPGA initialization, 204
EPROM/EEPROM programming technology, 187–188
Equivalent gate count, 373
Equivalent states and reduction of state tables, 28–30
sequential circuits, 28
state table reduction, 29
Event queue, 88–89
Excess-3 code converter, behavioral model, 115–117
Excess-3 code converter, simulator output for, 118
Excitation table for J-K flip-flop, 23
Exponent adder, multiplication, floating-point arithmetic, 407–409

Faster multiplier, block diagram, 246
Fault detection using path sensitization, 517
Field-programmable gate arrays (FPGAs), 180–205
applications, 201–203
commercial FPGAs, 181

Field-programmable gate arrays (*Continued*)
dedicated specialized components in FPGAs, 199–201
design flow for FPGAs, 203–205
I/O blocks in FPGAs, 197–199
interconnects, 192–197
logic block architectures, 189–192
organization of FPGAs, 182–185
programming technologies, 185–189
File I/O functions, 460–463
hexadecimal data from file using $readmemh, 462
Verilog code to read and parse file using $fopen, 462–463
File I/O functions Verilog language, 561
File I/O tasks, 456
Flip-flop registers and combinational logic blocks system, 524
Flip-flops and latches, 15–17
clocked D flip-flop with rising-edge trigger, 15
clocked J-K flip-flop, 15
clocked T flip-flop, 16
implementation of D-latch, 17
S-R latch, 16
transparent D-latch, 16
Floating-point arithmetic, 399–430
addition, 417–425
division, 425–426
multiplication, 406–417
representation of floating-point numbers, 399–405
subtraction, 425
Forever loop (infinite loop), 127
For loop, 127–128
Four-bit adder, structural description, 71
4-bit adder test bench, 130–132
4-bit BILBO register, 543
4-bit CLA block diagram, 216
4-bit CLA Verilog description, 217–218
Four-bit full adder, 70–71
4-bit linear feedback shift register (LFSR), 539
4-bit multiplier partial products, 238
4-to-1 multiplexer, 102
4-to-1 multiplexer, programmable logic block, 344
4-to-1 MUX paths, 343
4-valued logic system, 437
Four-variable Karnaugh maps, 7
4 × 4 array multiplier, block diagram, 239
4 × 4 array multiplier, Verilog code, 240–241
4 × 4 binary multiplier, behavioral model, 235–236

4 × 4 multiplier look-up table based, 362–368
Fraction multiplier, floating-point arithmetic, 409
Full adder, 1
Full adder module, 70
Function implementation, 4-variable function generators, 348
Functions, Verilog, 431–435
add function, 433
parity generation using a function, 432–433
computation of squares and its call, 434–435

GAL, *see* Generic Array Logic Generic Array Logic, 160, 174
Gates with common input and different delays, 67
General model of Mealy sequential machine, 17
General purpose interconnect, 192–193
Generate statements, 452–455
adder4 using generate statement, 453
conditional generate, 454
shift register using conditional compilation, 455
Generic Array Logic, 160, 174
Genvar, Verilog statement, 453
Glitches in sequential circuits, 39–40
Global lines, 194–195
Glue logic, FPGAs, 203
Greedy algorithms, 387
Guard bit, 405

Hardware accelerators/coprocessors, FPGAs, 203
Hardware arrangement, 316
Hardware description languages, 62–64
background, 62–63
common abbreviations, 64
learning a language, 63–64
Hardware testing and design for testability (DFT), 514–553
automatic test pattern generators (TTPGs), 514
boundary scan, 525–538
built-in self—test, 538–548
combinational logic testing, 515–518
scan testing, 522–525
sequential logic testing, 518–522
Hardware, Verilog code, 100
Hardwiring, 314
Hazards in combinational circuits, 13–14
elimination of 1-hazard, 14
simple circuits containing hazards, 13
Hexadecimal data from file using $readmemh, 462

Hierarchical architectures, FPGAs, 184
Highlighting paths for function F_1, 190
Hold time in flip-flop path, 34
Hold-time violation, 32

IC with boundary scan register and test-access port, 525
Identifiers, 66
IEEE 754 floating-points formats, 401–405
    double precision format, 403–404
    single precision format, 402–403
    special cases, 404
`ifdef, 459–460
If-else or case statement in always block, 103
If statement, Verilog, 80
Implementation of dice game, 309–314
    complete, 314
    counter, 313
    data flow model, 312–313
    maps, 311
    realization of controller, 309
    state transition table, 310
Implementation of D-latch, 17
Implementation of $F = AB + BC$
Implementation of full adder using PAL, 173
Implementing functions in FPGAs, 342–346
    building blocks, 342
    4-to-1 multiplexer, programmable logic block, 344
    4-to-1 MUX paths, 343
    programmable logic, three look-up tables, 344
    ring counter, 345
    shift register implementation, LUT-based FPGA, 346
Implementing functions using Shannon's decomposition, 347–351
    function implementation, 4-variable function generators, 348
    realization of functions by decomposition, 347
    7-variable function, 4-input LUTs and 2-to-1 MUXes, 350
    7-variable function, four Xilinx Spartan slices, 351
    Xilinx Spartan slice, 350
Implication table, 24, 28–30
Implication chart, 28–30
`include, 458–459
Incorrect clock gating, rising-edge devices, 47
Inertial and transport delays, 85
Infinity, floating-point arithmetic, 405

Initial statements, 74–75
Inout mode, use of, 72–73
Instruction, MIPS subset implementation
    decode unit, 488
    execution flow, MIPS subset implementation, 491–492
    execution unit, MIPS subset implementation, 488–490
    fetch block diagram, 487
    fetch unit, 486–488
Instruction Set Architecture, 473–476
    ISA, *see* Instruction Set Architecture
Integrating processor and memory modules, Verilog code, 502
Interconnection testing using boundary scan, 530
Interconnection test, Verilog code, 536–537
Interfacing of signals, 129
Inverter with feedback, 66
Iterative improvement, 387

J-K flip-flop model, 82

Karnaugh maps (K-maps), 7–10
    four-variable Karnaugh maps, 7
    multiple-output, 170
    partial truth table, 6-variable function, 9
    selection of prime implicants, 8
    simplification using map-entered variables, 9–10
    for state graph and table for code converter, 23
Keypad scanner, 255–264
    block diagram, 256
    controller, 258
    debouncer, 257
    debouncing and synchronizing circuit, 257
    decoder, 257–258
    decoder truth table, 257
    scanner, 257
    scanner modules, 256
    state graph, 258, 259
    test bench, 262
    test bench interface, 262
    test bench, Verilog, 263–264
    three column, four row keypad scanner, 256
    Verilog code, 259–262
K-maps for next states and output of sequence detector, 19

Laws and theorems of Boolean algebra, 4
Layout, FPGAs, 182
Left shift register with synchronous clear and load, 107, 110

Levels of abstraction, AND device, 112
Linear Feedback Shift Register (LFSR), 539–540
LFSR, see Linear Feedback Shift Register
Little-endian, 496
Logical instructions, MIPS ISA, 478
Logic blocks, commercial FPGAs, 355–357
    Altera Stratix IV logic module, 356
    Microsemi Fusion VersaTile, 356–357
    Xilinx Kintex slice, 355
Logic blocks with programmable SRAM cells, 370
Logic design fundamentals, 1–57
    Boolean algebra and algebraic simplification, 3–6
    combinational logic, 1–3
    designing with NAND and NOR gates, 11–12
    design of Moore sequential circuit, 25–27
    equivalent states and reduction of state tables, 28–30
    flip-flops and latches, 15–17
    hazards in combinational circuits, 13–14
    Karnaugh maps (K-maps), 7–10
    Mealy sequential circuit design, 17–25
Logic design fundamentals (*Continued*)
    sequential circuit timing, 30–47
    tristate logic and busses, 47–48
Long lines, 195
Look-up table (LUT), 163, 181, 189, 190
    LUT, see Look-up table
    JTAG standard, 525
Look-up-table–based programmable logic blocks, 189–191
Look-up table method using arrays and parameters, 125–126
Loops, 127–129
    forever loop (infinite loop), 127
    for loop, 127–128
    while loop, 128–129
LUT-based memory, inferred from Verilog code, 361
LUT-based RAM, 360

Main control unit, multiplication, floating-point arithmetic, 409–411
Mapping, 386
Mapping, placement, and routing, 385–389
    greedy algorithms, 387
    iterative improvement, 387
    mapping, 386
    placement, 386
    routed FPGA, 389
    routing, 386
    simulated annealing, 387
    simulated annealing versus iterative improvement algorithms, 388
    standard cell approach, 386, 387
Mask Programmable Gate Array (MPGA), 61
MPGA, see Mask Programmable Gate Array
March test, 538
    BYPASS, boundary scan instruction, 528, 532, 533
Matrices, 125
Matrix-based architectures, FPGAs, 183
Maximum gates versus usable gates, 373–375
    equivalent gate count, 373
    Programmable Electronics Performance Company (PREP), 374–375
Maximum-length LSFR sequence feedback, 540
Mealy sequence detector, 17–20
Mealy sequential circuit design, 17–25
    BCD to excess-3 code converter, 20–24
    block diagram of sequence detector, 17
    circuit for sequence detector, 20
    code converter, 20
    derivations of J-K input equations, 24
    excitation table for J-K flip-flop, 23
    general model of sequential machine, 17
    Karnaugh maps for state graph and table for code converter, 23
    K-maps for next states and output of sequence detector, 19
    realization of code converter, 23
    sequence detector, 17–20
    sequence detector block diagram, 17
    state assignment for BCD to excess-3 Mealy code converter, 22
    state graph and table for Mealy code converter, 21
    state graph for sequence detector, 18
    state table for Mealy sequence detector, 19
    transition table for Mealy sequence detector, 19
Mealy sequential circuit with ROM, realization of, 165
Mealy state graph for sequence detector, 18
Medium-speed systems, FPGAs, 202
Memory access instructions, MIPS ISA, 478–479
Memory creation from LUTS, 359
Memory, Verilog model, 493–496
Microinstruction, 316
Microprogram memory, 315
Microprogramming, 314–327
    binary multiplier controller implementation, 307
    binary multiplier with serial state assignment for single address, 321

comparison, different implementations of multiplier control, 322
conditional output elimination, 317
control store, 315
hardware arrangement, 316
hardwiring, 314
microinstruction, 316
microprogram memory, 315
Moore outputs, one qualifier per state, 323
multiple controller state transition table, 308
multiplexer for microprogramming multiplier, 322
multiplier, after state minimization, 319
multiplier, no conditional outputs, 318
multiplier operational flow, 315
multiplier, two-address microprogram, 319
MUX for microprogramming multiplier, 319
qualifiers, one per state, 317
sequencing, 315
serial state assignment and added X-state, 325
single address microcode, 320
single address microprogram for multiplier, 322
single qualifier, single-address micromodel, 320–322
SM chart transformations, 317
Microsemi Fusion VersaTile, 356–357
MIPS, 474
MIPS instruction encoding, 482–485
    format, 482
    machine code, 485
    with MIPS instructions, 483
MIPS ISA, 476–482
    arithmetic instructions, 476–478
    control transfer instructions, 479–481
    logical instructions, 478
    memory access instructions, 478–479
MIPS processor model signals, Verilog model, 501
MIPS subset implementation, 485–492
    data path design, 486
    decode unit instruction, 488
    execution flow, 491–492
    execution unit instruction, 488–490
    fetch block diagram instruction, 487
    fetch unit instruction, 486–488
    overall data path, 490
    required data path, computation and memory instruction, 489
Modeling flip-flops using always block, 78–82
    D flip-flop with asynchronous clear, 80
    J-K flip-flop model, 82
    nested ifs and else ifs flow chart, 80

simple D flip-flop, 79
transparent latch, Verilog code for, 79
Modeling registers and counters using Verilog always statements, 104–112
    counter model, 108
    counter operation, 108
    cyclic shift register, 104
    8-bit counter using 4-bit counter modules, 109
    left shift register with synchronous clear and load, 107, 110
    register with synchronous clear and load, 106
    simple synchronous counter, 107
    two 74163 counters cascaded, 8-bit counter, 109
    unwanted latches, avoiding, 110–111
Modeling sequential machine, 114–115
Models for multiplexers, 101–103
    cascaded 2-to-1 MUXes, conditional assignment, 102
    conditional operator, 101
    4-to-1 multiplexer, 102
    if-else or case statement in always block, 103
    2-to-1 multiplexer, 101
ModelSim Verilog simulator, 71–72
Modified LSFR with 0000 state, 540
Modules, Verilog, 68–73
    black box view of the 2-gate module, 69
    description, 68–69
    four-bit adder, structural description, 71
    four-bit full adder, 70–71
    full adder module, 70
    inout mode, use of, 72–73, 73
    ModelSim Verilog simulator, 71–72
    output as input signal, use of, 73
    program structure, 69
    two gates, 68
Module Verilog language, 554
Moore circuit for NRZ-to-Manchester conversion, 27
Moore outputs, one qualifier per state, 323
Moore sequence detector, 25–26
Moore sequential circuit design, 25–27
    circuit for NRZ-to-Manchester conversion, 27
    coding schemes for serial data transmission, 26
    sequence detector, 25–26
    state graph of sequence detector, 25
    state table for sequence detector, 26
    timing for circuit, 27
    transition table for sequence detector, 26
Multiple controller state transition table, 308

Multiple ICs scan test configuration, 525
Multiple-input signature register (MISR), 541
Multiple processes (initial or always block), 90
Multiplexer-based logic blocks in FPGAs, 191–192
Multiplexer for microprogramming multiplier, 322
Multiplexer implementing function F_1, 192
Multiplication, floating-point arithmetic, 406–417
   components, 409
   exponent adder, 407–409
   fraction multiplier, 409
   main control unit, 409–411
   multiplier control state graph, 411
   SM chart, 410
   test data and simulation results, 416–417
   2's complement fractions/exponents, 408
   Verilog code, 412–416
Multiplier control state graph, 411
Multiplier control with counter, 237
   after state minimization, 319
   no conditional outputs, 318
   operational flow, 315
   operation using control, 237
   two-address microprogram, 319
Multivalued logic and signal resolution, 436–439
   4-valued logic system, 437
   tristate buffers Verilog code, 437–438
   tristate buffers with active-high output enable, 437
MUX for microprogramming multiplier, 319

Named association, 451–452
NAND and NOR gates, 11
n-bit parallel adder with accumulator, 179
Negative skew, 43
Nested ifs and else ifs flow chart, 80
Net data, 93
Net delays, 85–87
Netlist, 375
nMOS NOR gate, 168
Non-determinism, 90
Not a number (NaN), 405

One-hot state assignment, 371–373
Operators+, 94–98
Operators by decreasing precedence, 558–559
Organization of FPGAs, 182–185
   hierarchical architectures, 184
   layout of FPGA, 182
   matrix-based architectures, 183
   row-based architecture, 184
   sea-of-gates architecture, 185
   typical architecture, 184
Organization, Verilog model, 493
Output as input signal, use of, 73
Overall data path, MIPS subset implementation, 490

PAL segment, 172
Parallel binary divider block diagram, 265
Parity code generator contents, 126
Parity code generator using the LUT method, 126
Parity generation using a function, 432–433
Partial truth table, 6-variable function, 9
PC boards with boundary scan ICs, 526
Placement, 386
PLA realization of equations 3-4, 170
PLA table, 171
PLA table for equations 3-1, 169
PLA with three inputs, five product terms, and four outputs, logic level, 167
PLA with three inputs, five product terms, and four outputs, transistor level, 168
PLD output macrocell, 175
Positional association, 71
   EXTEST, boundary scan instruction, 528, 529
   INTEST, boundary scan instruction, 528, 530
   IEEE 1149 standard, 525
Predefined data types Verilog language, 557–558
Probabilistic functions, 457
Procedural assignments, 74–78
   blocking and non-blocking assignments, 75–77
   combinational logic with blocking assignments, always block, 77
   initial statements, 74–75
   sensitivity list, 77–78
   wire and reg, 78
Processor CPU, Verilog code, 496
Process simulation, 90–91
Programmable array logic (PALs), 171–174
   implementation of full adder using PAL, 173
   PAL segment, 172
   segment of sequential PAL, 173
Programmable Electronics Performance Company (PREP), 374–375
Programmable interconnects, 192–197
   architecture, technology, logic block types of commercial FPGAs, 193
   clock skew, 195
   direct interconnects, 194, 195

general purpose interconnect, 192–193
global lines, 194–195
routing matrix, general-purpose interconnection, 194
row-based FPGA interconnects, 195–197
symmetric array FPGA interconnects, 192–194
Programmable I/O blocks in FPGAs, 197–199
Programmable logic array structure, 166–171
conversion of NOR-NOR to AND-OR, 168
Karnaugh maps, multiple-output, 170
nMOS NOR gate, 168
PLA realization of equations 3–4, 170
PLA table, 171
PLA table for equations 3–1, 169
PLA with three inputs, five product terms, and four outputs, logic level, 167
PLA with three inputs, five product terms, and four outputs, transistor level, 168
reduced PLA table, 170
Programmable logic block architectures, 189–192
highlighting paths for function F_1, 190
look-up-table–based programmable logic blocks, 189–191
multiplexer-based logic blocks in FPGAs, 191–192
multiplexer implementing function F_1, 192
Programmable logic devices, 158–209
comparison, 161
complex programmable logic devices (CPLDs), 176–180
elements, 160
field-programmable gate arrays (FPGAs), 180–205
history, 159–160
major programmable logic devices, 159
simple programmable logic devices (SPLDs), 161–176
Programmable logic devices (PLDs)/generic array logic (GALs), 174–176
block diagram, 22V10, 174
design flow for PLDs, 176
PLD output macrocell, 175
Programmable logic, three look-up tables, 344
Programmable points in FPGA I/O block
Programming technologies, 185–189
antifuse programming technology, 188
characteristics, FPGA technologies, 188
comparison, programming technologies, 188–189
EPROM/EEPROM programming technology, 187–188
routing with static RAM programming, 186
6-transistor SRAM cell, 186
SRAM programming technology, 185
Program structure, 69
Propagation delays, setup, and hold times, 30–31

Qualifiers, one per state, 317

RAM in example FPGAs, 358
RAM read-write system block diagram, 446
RAM system SM chart, 447
RAM system Verilog code, 448–450
Rapid prototyping, FPGAs, 201
Read-only memory (ROM), 162–166
Read-only memory with n inputs and m outputs, 162
Realization of functions by decomposition, 347
Realization of Mealy code converter, 23
Realization of SM charts, 306–309
binary multiplier controller implementation, 307
example for implementation, 306
multiple controller state transition table, 308
Reconfigurable circuits and systems, FPGAs, 202–203
Reduced PLA table, 170
Reduction operator, Verilog, 94
Register chains in FPGA's, 353–354
Register file, Verilog code, 494
Register file, Verilog model, 493
Register with synchronous clear and load, 106
Representation of floating-point numbers, 399–405
denormalized numbers, 404–405
IEEE double precision format, 403–404
IEEE 754 floating-points formats, 401–405
IEEE single precision format, 402–403
IEEE special cases, 404
infinity, 405
not a number (NaN), 405
rounding, 405
2's complement form, 399–401
zero, 404
Required data path, computation and memory instruction, MIPS subset implementation, 489
Resolution function, 4-valued logic Verilog language, 561
Ring counter, 345
Ripple-carry adder, 214

RISC microprocessor design, 473–513
    MIPS, 474
    MIPS instruction encoding, 482–485
    MIPS ISA, 476–482
    MIPS subset implementation, 485–492
    philosophy, 473–476
    single instruction computer, 476
    Verilog model, 492–508
Rise and fall delays of gates, 450–451
ROM implementation of a 2-bit full adder, 164
ROM truth table, 166
Rounding, floating-point arithmetic, 405
Routed FPGA, 389
Routing matrix, general-purpose interconnection, 194
Routing with static RAM programming, 186
Row-based architecture, FPGAs, 184
Row-based FPGA interconnects, 195–197

Safe regions for input changes, 37
Scanner modules, keypad scanner, 256
Scan testing, 522–525
    flip-flop registers and combinational logic blocks system, 524
    multiple ICs scan test configuration, 525
    scan path test circuit, 522
Schematic capture, 59–60
Scoreboard and controller, 226–230
    controller, 227
    data path, 226–227
    Verilog code, 228–230
    Verilog model, 227–228
Sea-of-gates architecture, FPGAs, 185
Segment of sequential PAL, 173
Selection of prime implicants, 8
Sensitivity list, 77–78
Separation, data path and controller, 210
Sequencing, 315
Sequential circuits, 28
Sequential circuit state table, 165
Sequential circuit timing, 30–47
    circuits to avoid, 45
    circuit with three flip-flops, 38
    clock skew, 41
    clock-to-Q delay, 30, 32
    correct control signal gating, rising-edge devices, 47
    glitches in sequential circuits, 39–40
    hold time in flip-flop path, 34
    hold-time violation, 32
    incorrect clock gating, rising-edge devices, 47
    negative skew, 43
    propagation delays, setup, and hold times, 30–31
    safe regions for input changes, 37
    setup and hold times for changes in X, 36
    setup and hold times for D-flip-flop, 31
    setup time margin, 33
    setup time violation, 32
    simple frequency divider, 36
    skew and timing violations, 42
    slack, 32
    static timing analysis (STA), 31–32
    synchronous design, 40–47
    synchronous digital system, 41
    techniques used to synchronize control signals, 45
    timing chart for system with falling-edge devices, 45
    timing chart, system with rising-edge devices, 46
    timing conditions for proper operation, 31–39
    timing diagram, code converter, 40
    timing diagram, realization of code converter, 40
    timing paths, 31
    timing rules, circuits with skew, 41–44
    timing rules, flip-flop to flip-flop paths, 32–35
    timing rules, input to flip-flop paths, 35–39
Sequential logic testing, 518–522
    iterative circuits, 519
    realization, 521
    state graph, 520
    state table, 520
Sequential machine
    model using equations, 120
    state table and block diagram, 114
    structural model, 122
Sequential statements (procedural statements) Verilog language, 555–557
Serially linked state machines, 327
Serial state assignment and added X-state, 325
Setup and hold times for changes in X, 36
Setup and hold times for D-flip-flop, 31
Setup time margin, 33
Setup time violation, 32
7-variable function, 4-input LUTs and 2-to-1 MUXes, 350
7-variable function, four Xilinx Spartan slices, 351
Shift-and-add multiplier, 232–238
    binary multiplier, block diagram, 233
    binary multiplier control, state graph, 234

4 × 4 binary multiplier, behavioral model, 235–236
    multiplier control with counter, 237
    operation, multiplier using control, 237
Shift register implementation, LUT-based FPGA, 346
Shift register using conditional compilation, 455
Signal drivers, simulation, 91
Signal path, extest instruction, 529
Signal path, intest instruction, 530
Signal path, sample/preload instruction, 529
Signed divider, 266–277
    block diagram, 268
    control circuit state graph, 270
    simulation test results, 276
    test bench, 273–276
Signed integer/fraction multiplier, 241–255
    binary fixed-point fractions, 242–243
    command file and simulation results for +5/8 by −3/8, 249
    faster multiplier, block diagram, 246
    interface, multiplier and test bench, 250
    signed multiplier command file and simulation, 252–253
    signed multiplier test bench, 250–251
    2's complement multiplier, behavioral model, 247–248
    2's complement multiplier, block diagram, 245
    2's complement multiplier, state graph, 246
    2's complement multiplier with control signals model, 253–255
Signed multiplier command file and simulation, 252–253
Signed multiplier test bench, 250–251
Simple circuits containing hazards, 13
Simple D flip-flop, 79
Simple frequency divider, 36
Simple gate circuit, 65
Simple memory model, 446
Simple programmable logic devices (SPLDs), 161–176
    block diagram and truth table of a 2-bit adder, 164
    8-to-3 priority encoder, 164
    8-word × 4-bit ROM, 162
    Mealy sequential circuit with ROM, realization of, 165
    programmable array logic (PALs), 171–174
    programmable logic array structure, 166–171
    programmable logic devices (PLDs)/ generic array logic (GALs), 174–176
    read-only memory (ROM), 162–166
    read-only memory with n inputs and m outputs, 162
    ROM implementation of a 2-bit full adder, 164
    ROM truth table, 166
    sequential circuit state table, 165
Simple synchronous counter, 107
Simple synthesis, 98–101
    block diagram for Verilog code, 99
    hardware, Verilog code, 100
    simulation and synthesis, different outputs result, 98
    Verilog code example, 99
    Verilog code that will not synthesize, 101
Simplification using map-entered variables, 9–10
Simulated annealing, 387
Simulated annealing versus iterative improvement algorithms, 388
Simulation and synthesis, different outputs result, 98
Simulation control tasks, 456
Simulation results, Verilog model, 507–508
Simulation time functions, 456
Single address microcode, 320
Single address microprogram for multiplier, 322
Single instruction computer, 476
Single pulser, 231–232
Single pulser and synchronizer circuit, 232
Single pulse, state diagram, 231
Single qualifier, single-address micromodel, 320–322
16-bit CLA block diagram, 217
6-transistor SRAM cell, 186
Skew and timing violations, 42
Slack, 32
Slate machine (SM) charts, 288–293
    block example, 289
    block principal components, 289
    equivalent blocks, 290
    equivalent charts, combinational circuit, 290
SM charts and microprogramming, 288–340
    derivation of charts, 293–305
    implementation, dice game, 309–314
    linked state machines, 327–329
    microprogramming, 314–327
    realization of charts, 306–309
    slate machine charts, 288–293
Spectrum of design technologies, 61
SRAM model, 444–446
    simple memory model, 446
    static RAM block diagram, 445
    static RAM truth table, 445

SRAM programming technology, 185
SRAM read/write system model, 446–450
    RAM read-write system block diagram, 446
    RAM system SM chart, 447
    RAM system Verilog code, 448–450
S-R latch, 16
State assignment for BCD to excess-3 Mealy code converter, 22
Static hazard, 13
Static 0-hazard, 13
Static 1-hazard, 13
Static RAM block diagram, 445
Static RAM truth table, 445
Static timing analysis (STA), 31–32
Sticky bit, 405
Stuck-at-fault testing circuit, 517
Stuck-at-fault tests, 518
STUMPS architecture, 541
Subtraction, floating-point arithmetic, 425
Symmetric array, FPGA interconnects, 192–194
Synchronization and debouncing, 230–232
    debouncing mechanical switches, 230
    single pulser, 231–232
    single pulser and synchronizer circuit, 232
    single pulse, state diagram, 231
Synchronous design, 40–47
Synchronous digital system, 41
Synthesis, 375
    of arithmetic components, 383
    of case statement, 377–378
    corresponding hardware, 383–384
    of if statements, 381–383
System functions, 455–457
    conversion functions, 457
    display tasks, 456
    file I/O tasks, 456
    probabilistic functions, 457
    simulation control tasks, 456
    simulation time functions, 456
System tasks, 66–67
System tasks Verilog language, 455, 560

TAP controller state machine, 528
Tasks, 435–436
    adding multiple bits, 436
    form, 435
Techniques used to synchronize control signals, 45
Technology characteristics, FPGAs, 188
Test bench, Verilog model, 505–506
Testing, 129–132
    interfacing of signals, 129
    test bench for 4-bit adder, 130–132
Testing, Verilog model, 503–505
Test-per-clock BIST scheme, 542
Test-per-scan scheme, 542
32-bit adders, 214–220
    behavioral model, 219
    carry look-ahead adders, 214–216
    comparison, ripple-carry and carry look-ahead adders, 220
    4-bit CLA block diagram, 216
    4-bit CLA Verilog description, 217–218
    ripple-carry adder, 214
    16-bit CLA block diagram, 217
32-bit signed divider Verilog model, 270–273
Three column, four row keypad scanner, 256
3-input NAND gate, 121
Timing
    chart for system with falling-edge devices, 45
    chart, system with rising-edge devices, 46
    checks, 463
    conditions for proper operation, 31–39
    diagram, code converter, 40
    diagram, realization of code converter, 40
    for Moore circuit, 27
    paths, 31
    rules, circuits with skew, 41–44
    rules, flip-flop to flip-flop paths, 32–35
    rules, input to flip-flop paths, 35–39
Timing checks Verilog language, 463, 561
Traffic light controller, 220–225
    block diagram, 221
    state graph, 221
    test results, 224
    Verilog code, 222–224
Transition table for Mealy sequence detector, 19
Transition table for Moore sequence detector, 26
Transparent D-latch, 16
Transparent latch, Verilog code for, 79
Tristate buffers, 48
Tristate buffers Verilog code, 437–438
Tristate buffers with active-high output enable, 437
Tristate logic and busses, 47–48
    buffers, 48
    data transfer using tristate bus, 48
Truth table, 2
22V10 block diagram, 174
2-bit adder block diagram truth table, 164
Two gates, 68
2's complement form, floating-point numbers, 399–401

Index

2's complement fractions/exponents, 408
2's complement multiplier
    behavioral model, 247–248
    block diagram, 245
    with control signals model, 253–255
    state graph, 246
Two 74163 counters cascaded, 8-bit counter, 109
2-to-1 multiplexer, 101
2-to-1 multiplexer, user-defined primitives, 442–443
2-to-1 multiplexer using ?, user-defined primitives, 443

Unary operator, Verilog, 94
Unified instruction/data memory, Verilog code, 495
Unintentional latch creation, 378–380
Unsigned divider, 264–267
Unwanted latches, avoiding, 110–111
User-defined primitives, 442–444
    for a D flip-flop, 444
    2-to-1 multiplexer, 442–443
    2-to-1 multiplexer using ?, 443

Variable data, 93
Variable-width RAM aspect ratios, FPGAs, 199
Verilog code block diagram, 99
Verilog code to read and parse file using $fopen, 462–463
Verilog coding, array multiplier, 239–240
Verilog gate module, 440
Verilog introduction, 58–157
    always blocks using event control statements, 82–83
    arrays, 124–126
    assignments, 73–74
    behavioral and structural Verilog, 112–123
    combinational circuits, Verilog description of, 64–68
    compilation, simulation, synthesis, Verilog code, 87–93
    computer-aided design (CAD), 59–61
    constants, 123–124
    data types and operators, 93–98
    delays in Verilog, 84–87
    hardware description languages, 62–64
    loops, 127–129
    modeling flip-flops using always block, 78–82
    modeling registers and counters using Verilog always statements, 104–112
    models for multiplexers, 101–103
    modules, 68–73
    procedural assignments, 74–78
    simple synthesis, 98–101
    strategies, 132–135
    testing, 129–132
    uses, 132–133
Verilog key words, 562–563
Verilog language summary, 554–561
    AND and OR functions for 4-valued logic, 561
    built-in primitives, 560
    compiler directives, 561
    concurrent or parallel statements, 555
    declaration examples, 559
    delays, 560
    file I/O functions, 561
    module, 554
    operators by decreasing precedence, 558–559
    predefined data types, 557–558
    resolution function, 4-valued logic, 561
    sequential statements (procedural statements), 555–557
    system tasks, 560
    timing checks, 561
Verilog model, RISC microprocessor design, 492–508
    big-endian, 496
    integrating processor and memory modules, code, 502
    little-endian, 496
    memory, 493–496
    MIPS, complete, 502
    MIPS processor model signals, 501
    MIPS subset implementation, code, 496–500
    organization, 493
    processor CPU, code, 496
    register file, 493
    register file, code, 494
    simulation results, 507–508
    test bench, 505–506
    testing, 503–505
    unified instruction/data memory, code, 495
Verilog models for inferring memory in FPGAs, 360
Verilog topics, 431–472
    built-in primitives, 439–441
    compiler directives, 457–460
    file I/O functions, 460–463
    functions, 431–435
    generate statements, 452–455
    multivalued logic and signal resolution, 436–439
    named association, 451–452
    rise and fall delays of gates, 450–451

SRAM model, 444–446
SRAM read/write system model, 446–450
system functions, 455–457
tasks, 435–436
timing checks, 463
user-defined primitives, 442–444
Verilog tristate buffer module, 441

While loop, 128–129
Wire and reg, 78

Xilinx CoolRunner, 177–179
Xilinx Kintex slice, 355
Xilinx Spartan slice, 350

Zero, floating-point arithmetic, 404